Advances in Experimental Medicine and Biology

Volume 1467

Advances in Experimental Medicine and Biology provides a platform for scientific contributions in the main disciplines of the biomedicine and the life sciences. This series publishes thematic volumes on contemporary research in the areas of microbiology, immunology, neurosciences, biochemistry, biomedical engineering, genetics, physiology, and cancer research. Covering emerging topics and techniques in basic and clinical science, it brings together clinicians and researchers from various fields.

Advances in Experimental Medicine and Biology has been publishing exceptional works in the field for over 40 years, and is indexed in SCOPUS, Medline (PubMed), EMBASE, Reaxys, EMBiology, the Chemical Abstracts Service (CAS), and Pathway Studio.

2023 CiteScore: 5.9

Stephen H. Tsang • Tarun Sharma • Vlad Diaconita

Editors

Atlas of Inherited Retinal Diseases

Second Edition

Springer

Editors
Stephen H. Tsang
Jonas Children's Vision Care, Bernard & Shirlee Brown Glaucoma Laboratory, Columbia Stem Cell Initiative-Departments of Ophthalmology Biomedical Engineering, Pathology & Cell Biology, Institute of Human Nutrition, Vagelos College of Physicians and Surgeons Columbia University
New York, NY, USA

Vlad Diaconita
Department of Ophthalmology
Edward S. Harkness Eye Institute
NewYork-Presbyterian Hospital Columbia University
New York, NY, USA

Tarun Sharma
Department of Ophthalmology
Edward S. Harkness Eye Institute
NewYork-Presbyterian Hospital Columbia University
New York, NY, USA

ISSN 0065-2598 ISSN 2214-8019 (electronic)
Advances in Experimental Medicine and Biology
ISBN 978-3-031-72229-5 ISBN 978-3-031-72230-1 (eBook)
https://doi.org/10.1007/978-3-031-72230-1

This Springer imprint is published by the registered company Springer Nature Switzerland AG
The registered company address is: Gewerbestrasse 11, 6330 Cham, Switzerland

Preface to the First Edition

This Atlas was conceived to fulfill the unmet needs of medical students, residents, fellows, PhD students, postdocs, and practicing retinal specialists who manage patients with inherited retinal diseases (IRD), whose aim is to keep pace with evolving genetic testing practice patterns, their influence on counseling, and emerging gene therapy trials.

The diagnosis of IRDs poses a great challenge to physicians because of their underlying genetic and phenotypic heterogeneity. As a group of genetic eye disorders, IRDs are a common cause of blindness in children and adults. They affect about 1 in 4000 people, nearly half of whom have no known family history of similar disorders.

Recent advances in molecular genetics have revealed that most IRDs have an underlying genetic basis. So far, over 250 different causative genes have been identified, and a causative mutation can be identified in up to 60–80% of patients with IRDs. In the future, physicians will need to understand not only the clinical or phenotypic characteristics of the patient's disease, but also their correlation with underlying genetic mutations (the genotype-phenotype correlation).

There has also been a surge in phase I and phase II gene therapy trials aimed at managing several of the IRDs, like X-linked retinitis pigmentosa (RP), choroideremia, retinoschisis, achromatopsia, Best disease, Stargardt disease, Usher syndrome, and *MERTK*-associated autosomal RP. In December 2017, the FDA approved for the first time the use of gene therapy to treat patients with Leber congenital amaurosis, caused by a mutation in the *RPE65* gene.

This Atlas is comprised of eight sections. The first section gives an overview of basic principles of various imaging modalities that are being used to manage patients with IRDs. The second, third, and fourth sections describe IRDs based on their inheritance pattern. The fifth section is dedicated to those IRDs that have associated systemic manifestations, including syndromic retinitis pigmentosa. The sixth section deals with phakomatoses with causative genetic bases. The seventh section illustrates various phenocopies of IRDs (particularly RP) that have no underlying genetic mutations. The last section provides a practical approach to IRDs and a basic understanding of genetic testing and its interpretation. This Atlas provides readers with images of various IRDs and concise text describing clinical features and associated genetic mutations.

The content of this Atlas will be readily understood by clinicians and students studying IRDs.

New York, NY, USA — Stephen H. Tsang
New York, NY, USA — Tarun Sharma

Contents

Part I Basic Knowledge

1 Retinal Histology and Anatomical Landmarks 3
Andrea Andrade, Stephen H. Tsang, Tarun Sharma, and Vlad Diaconita
Retinal Layers 4
Bruch's Membrane 4
Choroid 4
Anatomical Landmarks of the Macula 5
Further Reading 5

2 Fluorescein Angiography 7
Andrea Andrade, Stephen H. Tsang, Tarun Sharma, and Vlad Diaconita
Basic Principles and Techniques 8
Dye-Transit Phases 8
Abnormal Fluorescence 10
Side Effects and Contraindications 11
Indocyanine Green Angiography 11
Further Reading 11

3 Optical Coherence Tomography 13
Stephen H. Tsang and Tarun Sharma
Historical Perspective 14
Principle 14
Normal OCT Scans 14
Further Reading 15

4 Fundus Autofluorescence 17
Andrea Andrade, Jorge Pincay, Promie Faruque, Stephen H. Tsang, Tarun Sharma, and Vlad Diaconita
Basic Principles and Techniques 18
Interpretation and Clinical Application 18
Quantitative Autofluorescence 19
Further Reading 20

5 Electroretinography 21
Julie Shabto, Stephen Tsang, Tarun Sharma, and Vlad Diaconita
Full-Field (Ganzfeld) Electroretinography (ffERG) 22
Electronegative ffERG 23
Multifocal Electroretinography (mfERG) 24
Pattern Electroretinography (pERG) 25
Photopic On-Off ERG 25
Further Reading 26

6 Electrooculography 27
Julie Shabto, Stephen H. Tsang, Tarun Sharma, and Vlad Diaconita
Further Reading 28

7 Glossary of Relevant Genetic and Molecular/Cell Biology 29
Duncan Yi-Te Wang, Megan Soucy, Stephen H. Tsang, Tarun Sharma, and Vlad Diaconita
Genetics 30
Genetic Disorders 31
Molecular Biology 32
Inheritance 33
Diagnosis and Treatment 33

Part II X-Linked Forms

8 X-Linked Retinitis Pigmentosa 37
Stephen H. Tsang and Tarun Sharma
General Features 38
Female Carriers 39
Molecular Genetics 40
RPGR Gene 40
RP2 Gene 41
Gene Therapy Trials 41
Did You Know? 41
Further Reading 41

9 X-Linked Choroideremia 43
Ashley Zhou, Stephen H. Tsang, Tarun Sharma, and Vlad Diaconita
General Features 44
Imaging and Tests 45
Female Carriers 48
Molecular Genetics 49
Gene Therapy Trials 49
Further Reading 49

10 X-Linked Juvenile Retinoschisis 51
Leticia Pinheiro, Stephen H. Tsang, and Tarun Sharma
General Features 52
Imaging and Tests 53
Molecular Genetics 55
Treatments 55
Oral Acetazolamide 55
Vitreoretinal Surgery 55
Gene Therapy 56
Further Reading 56

11 X-Linked Ocular Albinism 57
Benjamin Kuang-Chien Chiang, Stephen H. Tsang, and Tarun Sharma
General Features 58
Female Carriers 58
Oculocutaneous Albinism 60
Molecular Genetics 60
Ocular Albinism 60
Oculocutaneous Albinism 60
Further Reading 60

12 Progressive Cone Dystrophy and Cone-Rod Dystrophy 61
Benjamin Kuang-Chien Chiang, Stephen H. Tsang, Tarun Sharma, and Vlad Diaconita
General Features 62
Molecular Genetics 65
Specific Genotype and Phenotype Correlation 65
ABCA4 65
ADAM9 (CORD9) 66
RPGR1P1 (CORD13, Leber Congenital Amaurosis [LCA6]) 66
KCNV2 (Potassium Voltage-Gated Channel Modifier Subfamily V Member 2) 66
SEMA4A (Semaphoring 4A) (CORD10 or RP35) 66
GUCY2D (CORD6 or LCA1 or RCD2) 66
GUCA1A (Guanylate Cyclase Activator 1A) (COD4 or CORD14) 66
PRPH2 (Peripherin 2 or RDS) 66
CRX (CORD2 or LCA7 or CRD) 66
RIMS1 (Regulating Synaptic Membrane Exocytosis 1, CORD7) 67
PITPNM3 (CORD5) 67
RPGR (COD1 or CORCX1 or RP3 or RP15 or XLRP3) 67
DRAM2 (Damage-Regulated Autophagy Modulator 2, CORD 21) 67
Further Reading 67

13 Congenital Stationary Night Blindness 69
Justin Cheong, Stephen H. Tsang, and Tarun Sharma
Congenital Stationary Night Blindness with Normal Fundi 70
Types 70
Molecular Genetics 71
Congenital Stationary Night Blindness with Abnormal Fundi 71
Fundus Albipunctatus 71
Oguchi Disease 71
Further Reading 72

14 Blue Cone Monochromatism 73
Kristina J. Hartung, Stephen H. Tsang, Tarun Sharma, and Vlad Diaconita
General Features 74
Imaging and Tests 74
Molecular Genetics 75
Further Reading 75

Part III Autosomal Dominant Forms

15 Autosomal Dominant Retinitis Pigmentosa 79
Duncan Yi-Te Wang, Stephen H. Tsang, Tarun Sharma, and Vlad Diaconita
General Features 80
Molecular Genetics 81
RHO Gene (RP4, Opsin-2) 81
PRPF3 Gene (RP18, SNRNP90) 82
PRPF8 Gene (RP13, SNRNP220) 82
RP1 Gene (ORP1) 82
Further Reading 85

16 Best Vitelliform Macular Dystrophy 87
Laura A. Jenny, Stephen H. Tsang, and Tarun Sharma
General Features 88
Imaging and Tests 90
Subtypes 90

Adult-Onset Foveomacular Vitelliform Dystrophy (AFMD) 90
Autosomal Recessive Bestrophinopathy (ARB) 90
Autosomal Dominant Vitreoretinochoroidopathy (ADVIRC) 90
Molecular Genetics 91
BEST1 Gene (*VMD2* Gene or *RP50* or *BMD*) 91
Further Reading 91

17 Pattern Dystrophy 93
Benjamin Kuang-Chien Chiang, Stephen H. Tsang, and Tarun Sharma
General Features 94
Butterfly-Type Pattern Dystrophy 94
Adult-Onset Foveomacular Vitelliform Dystrophy 97
Sjogren's Reticular-Type Pattern Dystrophy 97
Fundus Pulverulentus 97
Molecular Genetics 97
PRPH2 Gene (*Peripherin/RDS* or *rd2* or *Peripherin-2* or *RDS*) 97
Further Reading 98

18 Doyne Honeycomb Retinal Dystrophy (Malattia Leventinese, Autosomal Dominant Drusen) 99
Kristina J. Hartung, Stephen H. Tsang, Tarun Sharma, and Vlad Diaconita
General Features 100
Molecular Genetics 105
EFEMP1 Gene (*DHRD* or *DRAD* or *FBLN3* or *FIBL-3*) 105
Further Reading 105

19 Occult Macular Dystrophy 107
Kristina J. Hartung, Stephen H. Tsang, Tarun Sharma, and Vlad Diaconita
General Features 108
Molecular Genetics 109
Further Reading 109

20 Sorsby Pseudoinflammatory Fundus Dystrophy 111
Kristina J. Hartung, Stephen H. Tsang, Tarun Sharma, and Vlad Diaconita
General Features 112
Molecular Genetics 114
TIMP3 Gene (K222 or SFD) 114
Further Reading 114

21 North Carolina Macular Dystrophy 115
Amber Diaz, Kristina J. Hartung, and Kent Small
General Features 116
Molecular Genetics 117
References 117

22 Pigmented Paravenous Chorioretinal Atrophy (PPCRA) 119
Justin Cheong, Stephen H. Tsang, and Tarun Sharma
General Features 120
Molecular Genetics 121
Further Reading 121

23 Late-Onset Retinal Degeneration 123
Stephen H. Tsang and Tarun Sharma
General Features 124
Molecular Genetics 124
C1QTNF5 Mutation (*CTRP5*) 124
Further Reading 124

Part IV Autosomal Recessive Forms

24 Achromatopsia ... 127
Kristina J. Hartung, Stephen H. Tsang, Tarun Sharma, and Vlad Diaconita
General Features ... 128
Stages ... 128
Molecular Genetics ... 132
Clinical Trials ... 133
Further Reading ... 133

25 Retinitis Pigmentosa (Non-syndromic) ... 135
Ahmet Hondur, Stephen Tsang, Tarun Sharma, and Vlad Diaconita
General Features ... 136
Genes and Mutations ... 138
PDE6 (Phosphodiesterase) ... 138
RP25 or *EYS* (Eyes Shut Homolog) ... 138
TULP1 (Tubby-Like Protein 1) Gene or *RP14* or *TUBL1* or *LCA15* ... 138
CERKL (Ceramide Kinase-Like) Gene or *RP26* ... 140
RPE65 Gene ... 140
Other Genes Associated with AR-RP ... 140
Further Reading ... 141

26 Leber Congenital Amaurosis ... 143
Stephen H. Tsang and Tarun Sharma
General Features ... 144
Molecular Genetics ... 144
Specific Genotype and Phenotype Correlations ... 144
LCA1 ... 144
LCA2 ... 144
LCA3 ... 144
LCA4 ... 146
LCA5 ... 146
LCA6 ... 146
LCA7 ... 146
LCA8 ... 146
LCA9 ... 147
LCA10 ... 147
LCA11 ... 147
LCA12 ... 147
LCA14 ... 148
LCA15 ... 148
LCA16 ... 149
RP38 ... 149
LCA57 ... 149
SLSN5 ... 149
ABCA-4 ... 149
Further Reading ... 149

27 Stargardt Disease ... 151
Leticia Pinheiro, Stephen H. Tsang, and Tarun Sharma
General Features ... 152
Clinical Stages and Groupings ... 152
Stages, Based on Fundus Appearance ... 152
Groupings, Based on Electroretinography (ERG) ... 155

Types, Based on FAF 155
Classes, Based on Optical Coherence Tomography (OCT) 155
Molecular Genetics 158
ABCA4 Gene (RIM Protein or ABCR) 158
Other Mutations 158
Genotype-Phenotype Correlation 158
Further Reading 161

28 Enhanced S-Cone Syndrome (Goldmann-Favre Syndrome) 163
Laura A. Jenny, Stephen H. Tsang, and Tarun Sharma
General Features 164
Molecular Genetics 166
Further Reading 167

29 Best Vitelliform Macular Dystrophy 169
Justin Cheong, Stephen H. Tsang, Tarun Sharma, and Vlad Diaconita
General Features 170
Further Reading 170

Part V Systemic Disorders

30 Mitochondrial Disorder: Kearns-Sayre Syndrome 173
Benjamin Kuang-Chien Chiang, Stephen H. Tsang, Alicia R. P. Aycinena, and Tarun Sharma
General Features 174
Molecular Genetics 175
Further Reading 175

31 Mitochondrial Disorder: Maternally Inherited Diabetes and Deafness 177
Fabiha Ahmed, Stephen H. Tsang, Alicia R. P. Aycinena, and Tarun Sharma
General Features 178
Molecular Genetics 180
Further Reading 180

32 Ciliopathy: Usher Syndrome 181
Benjamin Kuang-Chien Chiang, Stephen H. Tsang, Alicia R. P. Aycinena, and Tarun Sharma
General Features 182
Molecular Genetics 183
USH1 183
USH2 183
USH3 183
Therapy Undergoing Clinical Trials 184
Further Reading 184

33 Ciliopathy: Bardet-Biedl Syndrome 185
Ahmet Hondur, Stephen Tsang, Alicia R. P. Aycinena, Tarun Sharma, and Vlad Diaconita
General Features 186
Molecular Genetics 187
Management 188
Further Reading 188

34 Ciliopathy: Senior-Løken Syndrome 189
Stephen H. Tsang, Alicia R. P. Aycinena, and Tarun Sharma
General Features 190
Molecular Genetics 191
Further Reading 191

35 Ciliopathy: Alström Syndrome . . . 193
Stephen H. Tsang, Alicia R. P. Aycinena, and Tarun Sharma
General Features . . . 194
Molecular Genetics . . . 194
Further Reading . . . 194

36 Ciliopathy: Sjögren-Larsson Syndrome . . . 195
Stephen H. Tsang, Alicia R. P. Aycinena, and Tarun Sharma
General Features . . . 196
Molecular Genetics . . . 196
Further Reading . . . 196

37 Inborn Errors of Metabolism: Gyrate Atrophy . . . 197
Kristina J. Hartung, Stephen H. Tsang, Tarun Sharma, and Vlad Diaconita
General Features . . . 198
Molecular Genetics . . . 200
Further Reading . . . 200

38 Inborn Errors of Metabolism: Pseudoxanthoma elasticum . . . 201
Justin Cheong, Stephen H. Tsang, and Tarun Sharma
General Features . . . 202
Molecular Genetics . . . 203
Further Reading . . . 203

39 Inborn Errors of Metabolism: Refsum Disease . . . 205
Sitara Hirji, Stephen Tsang, Tarun Sharma, and Vlad Diaconita
General Features . . . 206
Molecular Genetics . . . 206
Further Reading . . . 206

40 Inborn Errors of Metabolism: Bietti Crystalline Dystrophy . . . 207
Laura A. Jenny, Stephen H. Tsang, and Tarun Sharma
General Features . . . 208
Molecular Genetics . . . 209
Further Reading . . . 210

41 Extracellular Matrix: Alport Syndrome . . . 211
Jennifer Adeghate, Stephen Tsang, and Tarun Sharma
General Characteristics . . . 212
Molecular Genetics . . . 212
Ocular Features . . . 212
Further Reading . . . 213

42 Stickler Syndrome . . . 215
Jennifer Hu, Stephen Tsang, Tarun Sharma, and Vlad Diaconita
General Features . . . 216
Molecular Features of Stickler Syndrome . . . 216
Management of Stickler Syndrome . . . 216
Further Reading . . . 218

43 *OTX2* Syndromes . . . 219
Chloe Li, Stephen Tsang, Tarun Sharma, and Vlad Diaconita
General Features . . . 220
Molecular Genetics . . . 221
Specific Genotype and Phenotype Correlations . . . 221
Further Reading . . . 221

Part VI Phakomatoses

44 Von Hippel-Lindau Disease 225
Luanna Chan, Stephen H. Tsang, Tarun Sharma, and Vlad Diaconita
General Features of Phakomatoses 226
General Features of Von Hippel-Lindau Disease 226
Molecular Genetics of Von Hippel-Lindau Disease 227
Further Reading 227

45 Tuberous Sclerosis 229
Vlad Diaconita, Amy Shteyman, Anna Jolia, Stephen H. Tsang, Tarun Sharma, and Brian Marr
General Features 230
Molecular Genetics 233
Further Reading 233

46 Neurofibromatosis 235
Jennifer Adeghate, Stephen Tsang, and Tarun Sharma
General Features 236
Neurofibromatosis Type 1 236
Neurofibromatosis Type 2 236
Molecular Genetics 237
Further Reading 237

Part VII Phenocopies

47 Rubella Retinopathy 241
Jin Kyun Oh, Stephen H. Tsang, and Tarun Sharma
General Features 242
Further Reading 243

48 Syphilis 245
Kristina J. Hartung, Stephen H. Tsang, Tarun Sharma, and Vlad Diaconita
General Features 246
Further Reading 247

49 Autoimmune Retinopathy 249
Fabiha Ahmed, Stephen H. Tsang, and Tarun Sharma
General Features: Autoimmune Retinopathies 250
Paraneoplastic Retinopathy 250
Non-paraneoplastic Autoimmune Retinopathy 250
Further Reading 252

50 Drug-Induced Retinal Toxicity 253
Charu Vyas, Anna Jolia, Stephen H. Tsang, Tarun Sharma, and Vlad Diaconita
General Features 254
Damage to RPE and Photoreceptor Complex 254
Damage to the Vascular Bed or Microvasculopathy/Occlusion 260
Damage to Ganglion Cell Layer or Optic Nerve 260
Drugs Causing Cystoid Macular Edema 260
Drugs Causing Crystalline Retinopathy 260
Drugs Causing Uveitis 262
Further Reading 262

51 Noninfectious Uveitis Syndromes . . . 263
Kristina J. Hartung, Omar Moussa, Anna Jolia, Naomi Rivka Goldberg, Royce Chen, Vlad Diaconita, and Jose Pulido
General Features . . . 264
Further Reading . . . 271

52 Diffuse Unilateral Subacute Neuroretinitis (DUSN) . . . 273
Flavius Beca, Maximilian Kong, Anna Jolia, Vlad Diaconita, and Jose Pulido
General Features . . . 274
Further Reading . . . 275

53 Central Serous Chorioretinopathy . . . 277
Jin Kyun Oh, Vlad Diaconita, and Tongalp H. Tezel
General Features . . . 278
Further Reading . . . 280

54 Pathologic Myopia . . . 281
Sophie Z. Gu and Stanley Chang
Further Readings . . . 283

Part VIII Managing IRDs in Clinics

55 A Practical Approach to Retinal Dystrophies . . . 287
Ahmet Hondur, Irena Tsui, Brian J. Song, Chyuan-Sheng Lin, Stephen Tsang, Tarun Sharma, and Vlad Diaconita
Examination and Testing . . . 289
Stationary Retinal Dystrophies . . . 290
  Stationary Rod Dystrophies . . . 290
  Stationary Cone Dysfunction Syndromes . . . 291
Progressive Retinal Dysfunction . . . 291
Central Dystrophies . . . 293
  Stargardt Disease . . . 293
  Other Central Dystrophies . . . 294
Treatments . . . 299
  Dietary Treatment and Vitamins . . . 299
  Other Treatments . . . 300
  Treatment of Complications . . . 300
  Gene Therapy . . . 300
  Other Emerging Therapies . . . 301
Summary . . . 302
References . . . 302

56 Genetic Testing for Inherited Retinal Dystrophies: Basic Understanding . . . 305
Megan E. Soucy
What Are the Structure and Functions of the Genome? . . . 306
What Is Genomic or Genetic Variation? . . . 306
What Is Genetic Code? . . . 307
What Is a Mutation? . . . 307
What Is a Polymorphism? . . . 307
What Is Genetic Testing? . . . 307
Who Are Genetics Professionals? . . . 308
Which Retinal Disease Would Benefit from Genetic Testing? . . . 308

What Is the Role of the Ophthalmologist in Genetic Testing? 308
What Are the Benefits of Genetic Testing? 309
What Are the Risks of Genetic Testing? 310
Does Everyone with the Same Diagnosis Have the Same Genetic Mutation? 310
What Information Is Given by Genetic Testing? 310
What Are the Types of Genetic Testing? 311
Does Every Variant Result in Disease? 312
References 312

57 Genetic Testing for Inherited Retinal Dystrophies: Genetic Reports 313
Megan E. Soucy
How Are Genomic Variants Classified in a Genetic Report? 314
How Do I Read and Understand the Genetic Testing Report? 314
How Are Variants Reported or Described? 315
Are Further Investigations Needed After the Genetic Testing? 316

Part IX Novel Therapies for Inherited Retinal Disease

58 The Genetic Basis of IRDS and the Role of Genetic Testing 319
Gareth D. Mercer, Brian G. Ballios, and Peter J. Kertes
References 321

59 "Seeing Stars": The Gene Therapy Revolution 323
Gareth D. Mercer, Brian G. Ballios, and Peter J. Kertes
Vectors for DNA Delivery 326
Case Vignette: Gene Replacement Therapy for *RPE65*-Associated Retinal Dystrophy 328
Voretigene Neparvovec-rzyl 331
Other Gene Therapies for IRDs 334
References 334

60 Mutation Specific Treatments for Inherited Retinal Diseases 337
Gareth D. Mercer, Brian G. Ballios, and Peter J. Kertes
Gene Editing 338
Post-transcriptional Gene Silencing (Antisense Oligonucleotides) 339
References 340

61 Global Treatment Approaches Part 1: Neuroprotection and Stem Cell Therapy 343
Gareth D. Mercer, Brian G. Ballios, and Peter J. Kertes
Neuroprotection 344
Stem Cell and Precursor Cell Therapies 347
References 350

62 Global Treatment Approaches Part 2: Optogenetics and Retinal Implants 353
Gareth D. Mercer, Brian G. Ballios, and Peter J. Kertes
Optogenetics 354
Bioelectronic Retinal Implants 355
References 356

Index 357

Part I
Basic Knowledge

CHAPTER 1

Retinal Histology and Anatomical Landmarks

Andrea Andrade, Stephen H. Tsang, Tarun Sharma, and Vlad Diaconita

Retinal Layers 4
Bruch's Membrane 4
Choroid 4
Anatomical Landmarks of the Macula 5
Further Reading 5

A. Andrade
Universidade Federal de Pernambuco, Recife, Brazil

S. H. Tsang (✉) · T. Sharma · V. Diaconita
Department of Ophthalmology, Columbia University, New York, NY, USA
e-mail: sht2@cumc.columbia.edu

S. H. Tsang et al. (eds.), *Atlas of Inherited Retinal Diseases*, Advances in Experimental Medicine and Biology 1467, https://doi.org/10.1007/978-3-031-72230-1_1

The wall of the eye consists of three layers: the sclera (the outermost layer), the choroid (the middle layer), and the retina (the innermost layer). In between choroid and retina, Bruch's membrane can be found. Each of these layers has specific characteristics and can be subdivided into different layers. In this chapter, retina, Bruch's membrane, and choroid will be detailed.

Retinal Layers

The retina is the light-sensitive innermost layer of the eye. It is a structure of two laminae, an outer, composed by the Retinal Pigment Epithelium (RPE), and an inner, neurosensory retina. The RPE is a monolayer of pigmented cells that contributes to retinal function. The neurosensory retina can be divided histologically into nine layers (Fig. 1.1). From the vitreous to the scleral side, these layers are organized as follows:

1. **Internal Limiting Membrane (ILM)** is formed by the end feet of the Müller cells and astrocytes.
2. **Nerve Fiber Layer (NFL)** consists of the axons of ganglion cells, retinal vessels, and glial cells.
3. **Ganglion Cell Layer (GCL)** contains the nucleus of ganglion cells and glial cells.
4. **Inner Plexiform Layer (IPL)** has the synapses of bipolar, amacrine, and ganglion cells.
5. **Inner Nuclear Layer (INL)** comprises the nuclei from bipolar, horizontal, amacrine, and Müller cells.
6. **Outer Plexiform Layer (OPL)** has the synapses of photoreceptor cells, bipolar cells, and horizontal cells.
7. **Outer Nuclear Layer (ONL)** contains the nuclei of photoreceptor cells.
8. **Outer Limiting Membrane (OLM)** is formed by the junctional complexes between Müller cells, and between Müller and photoreceptor cells.
9. **Photoreceptor Layer (PL)** has tightly stacked rods and cones.

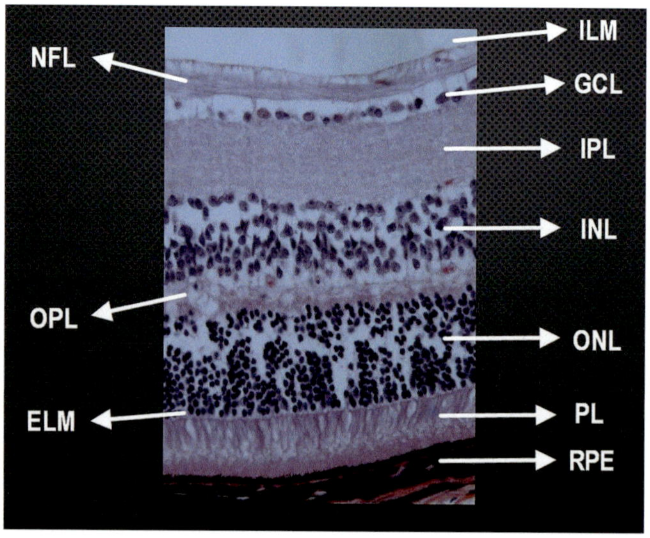

Fig. 1.1 Histological section of a mouse retinae. Hematoxylin/eosin, magnification 20×. (Courtesy Joel Alan Imventarza and Peter M.J. Quinn. Affiliation: Edward S. Harkness Eye Institute, Department of Ophthalmology, Columbia University Irving Medical Center, New York, NY, USA). *ILM* Internal Limiting Membrane, *NFL* Nerve Fiber Layer, *GCL* Ganglion Cell Layer, *IPL* Inner Plexiform Layer, *INL* Inner Nuclear Layer, *OPL* Outer Plexiform Layer, *ONL* Outer Nuclear Layer, *ELM* External Limiting Membrane, *PR* Photoreceptor Layer, *RPE*: Retinal Pigment Epithelium

The neural cells of the retina include photoreceptors (rods and cones), bipolar cells, interneurons (horizontal and amacrine cells), and ganglion cells. These cells are involved in the light transmission through retina. The human retina contains about 120 million rods and 6 million cones, with the highest density of rods located about 12° from fixation. Additionally, there are approximately 4–6 million retinal pigment epithelium (RPE) cells, with a ratio of 45 photoreceptors to 1 RPE cell. Each photoreceptor cell has about 1000 discs in the outer segments, with about 100 shed per day.

Bruch's Membrane

Bruch's membrane is a thin, protective layer that separates the choroid and the retina. The BM is made up of five layers:

1. Basement membrane of the choriocapillaris
2. Outer collagenous layer
3. Central elastic layer
4. Inner collagenous layer
5. Basement membrane of the RPE

Choroid

The choroid is the middle layer of the eye wall. It consists of three layers of blood vessels:

1. **Choriocapillaris:** innermost
2. **Sattler layer:** middle and mid-caliber
3. **Haller layer:** outer and large-caliber

The choroid is about 0.2 mm thick at the macula and becomes progressively thinner anteriorly, to about 0.1 mm.

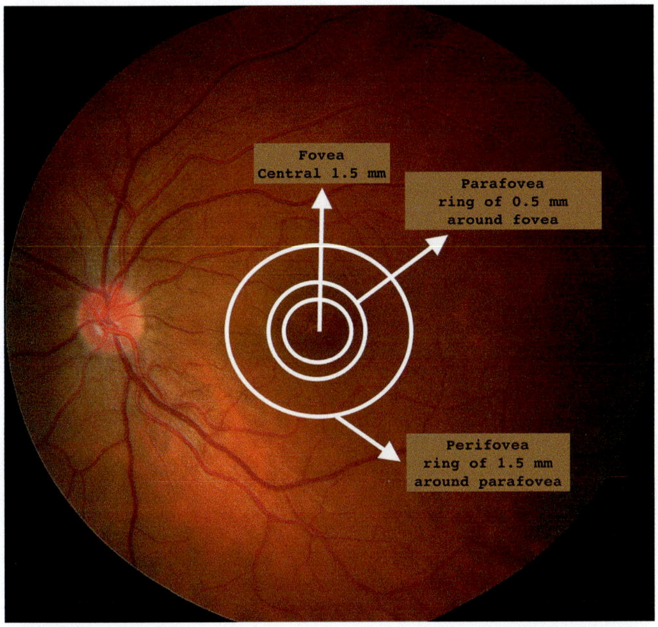

Fig. 1.2 Anatomical landmarks of the macula. Fovea corresponding to a diameter of 1.5 mm, the surrounding Parafovea forming a 0.5 mm ring around the fovea, and perifovea encompassing the parafovea within a 1.5 mm ring. (Courtesy José Ronaldo Lima de Carvalho Júnior. Affiliation: Hospital das Clínicas—Empresa Brasileira de Serviços Hospitalares (Ebserh), Federal University of Pernambuco, Recife, Pernambuco, Brazil)

Anatomical Landmarks of the Macula

Figure 1.2 illustrates the anatomical landmarks of the macula:

- **Macula or central retina or posterior pole or macula lutea (from Latin, *macula* [spot] and *lutea* [yellow])**: The area between temporal arcades, about 5.5 mm in diameter from the temporal to optic disc; it has two or more layers of ganglion cells. The yellow color is due to lutein and zeaxanthin (xanthophyll carotenoids).
- **Fovea**: The central 1.5 mm (about 1 disc diameter or central 5°) of the macula is called the *fovea*. The central part of the fovea is called *foveola* (about 0.35 mm, 350 μm), while the small center of the foveola is called the *umbo* (about 0.15–0.20 mm, 150–200 μm). The foveola has no rods and the fovea contains rods, cones, and processes of Muller cells. The foveal avascular zone (FAZ) is an area at the center of the macula (about 0.5–0.6 mm, 500–600 μm) that is devoid of any blood vessels.
- **Parafovea**: A ring surrounding the fovea, of 0.5 mm in width. The retina is thickest here; the OPL is called the *Henle layer*. The Henle layer is obliquely placed here, accounting for radial pattern of the macular star in hypertensive retinopathy. Beyond the macula, the OPL is perpendicular, not oblique.
- **Perifovea**: A ring surrounding the parafovea, 1.5 mm in width.

Further Reading

Hogan MJ, Alvarado JA, Weddell JE. Histology of the human eye. Philadelphia: Saunders; 1971. Chapters 5, 8, 9, 11.

Cantor LB, Rapuano CJ, McCannel CA. Retina and Vitreous, Basic and Clinical Science Course. American Academy of Ophthalmology, 2019-2020. Basic Anatomy: 29-47.

Purves D, Augustine GJ, Fitzpatrick D, et al., editors. Neuroscience. 2nd edition. Sunderland (MA): Sinauer Associates; 2001. The Retina.

Tschulakow AV, Oltrup T, Bende T, Schmelzle S, Schraermeyer U. The anatomy of the foveola reinvestigated. PeerJ. 2018 Mar 12;6:e4482. https://doi.org/10.7717/peerj.4482. PMID: 29576957; PMCID: PMC5853608.

CHAPTER 2

Fluorescein Angiography

Andrea Andrade, Stephen H. Tsang, Tarun Sharma, and Vlad Diaconita

Basic Principles and Techniques 8
Dye-Transit Phases 8
Abnormal Fluorescence 10
Side Effects and Contraindications 11
Indocyanine Green Angiography 11
Further Reading 11

A. Andrade
Universidade Federal de Pernambuco, Recife, Brazil

S. H. Tsang (✉) · T. Sharma · V. Diaconita
Department of Ophthalmology, Columbia University, New York, NY, USA
e-mail: sht2@cumc.columbia.edu

S. H. Tsang et al. (eds.), *Atlas of Inherited Retinal Diseases*, Advances in Experimental Medicine and Biology 1467, https://doi.org/10.1007/978-3-031-72230-1_2

Basic Principles and Techniques

Fluorescein angiography (FA) or Intravenous Fluorescein Angiography (IVFA) is an imaging modality used to study the circulation of the retina and choroid. It is based on a physical principle where molecules are excited by light within a certain wavelength range, and the emitted wavelength is longer than the excitation light. When this emitted energy occurs in the visible spectrum is called luminescence. The type of excited molecule determines the imaging technique used. In FA, an exogenous molecule is excited, while in autofluorescence, endogenous molecules are excited and detected.

To perform FA, about 5 mL of 10% sterile aqueous solution of fluorescein dye (or 2–3 mL of a 25% solution) is injected in the antecubital vein. After injection, 80% of the fluorescein binds to proteins and is unavailable for fluorescence, while the remaining 20% stays unbound and available for fluorescence. The eye has two blood-retina barriers (BRB): the inner BRB, formed by the tight junctions of endothelial cells of retinal blood vessels, and the outer BRB, formed by the retinal pigment epithelial (RPE) cells. Both of these barriers prevent the fluorescein molecule from passing through them in normal conditions. However, the choriocapillaris vessels have fenestrated walls that allow the dye to leak freely.

The fluorescein molecules excited by a blue light (peak excitation 465–490 nm) emit green-yellow light (peak emission 520–530 nm). The blue reflected light and the green-yellow fluorescent light are passed through a barrier filter that blocks the blue reflected light and allows only the green-yellow fluorescent light to pass through. This allows for the evaluation of only the true fluorescent light. If nonfluorescent light passes through the filter system, pseudofluorescence occurs.

Image acquisition can be made using different equipment and systems. The widespread adoption of digital imaging for FA has been propelled by significant advancements in digital cameras and software, despite its initial association with film-based techniques. Film-based images posed inherent difficulties, necessitating specialized training and laborious processing and duplication procedures. Moreover, the management, storage, transmission, and sharing of film-based images proved to be unpractical for clinical practices. However, the continual progress in digital imaging technology has effectively addressed these challenges. Presently, digital cameras offer enhanced image resolution and the ability to facilitate stereoscopic viewing, thereby rivaling the primary advantages of film-based imaging. Consequently, all modern fundus cameras now incorporate digital technology. With advancements, the field of view has also evolved, leading to the development of various imaging platforms that provide a widefield view of the retina. The first commercially available fundus camera had a field of view of 20°. However, modern systems can now capture images with an approximately 200-degree field of view. The imaging of the peripheral retina has become crucial in the diagnosis, classification, and management of various retinal diseases, including diabetic retinopathy, retinal vascular diseases, and retinal detachment, among others.

It is important to note that Fundus Autofluorescence (FAF) imaging should be obtained prior to initiating FA imaging acquisition. This is because FAF imaging provides important information about the distribution and intensity of endogenous fluorophores in the retina and can help identify areas of abnormal autofluorescence that may indicate underlying pathology. By performing FAF imaging prior to FA, a more comprehensive evaluation of structures can be obtained.

Dye-Transit Phases

When fluorescein is injected into a peripheral vein, it reaches the eye via the ophthalmic artery. A timer should also be used to determine time to fill of each phase of the circulation. The choroid is the first part of the eye to fluoresce, typically beginning in 10–12 s, and typically fills about 1–3 s before retinal circulation. The fluorescence initially has a patchy and irregular pattern, which is referred to as choroidal flush. After the initial 10 s, the choroidal fluorescence becomes bright and regular. However, the pigment in the retinal pigmented epithelium (RPE) can interfere with the fluorescence from the choroid. This is called masking effect. The next phase is the arterial phase, which shortly follows the choroidal flush, and is typically seen in the first 10–13 s. After reaching the central retinal artery, fluorescein flows from arteries to veins: retinal arteries, precapillary arterioles, capillaries, postcapillary venules, and retinal veins. The arteriovenous phase is usually seen between 10 and 20 s, followed by the recirculation phase in which all arteries and veins are visible. In the late phase, which occurs about 10 min after injection, most of the vessels no longer contain dye. However, large choroidal vessels appear as a silhouette against the relatively brighter choroid and the stained sclera (Fig. 2.1). Fluorescein is eliminated by liver and kidney within 24 h.

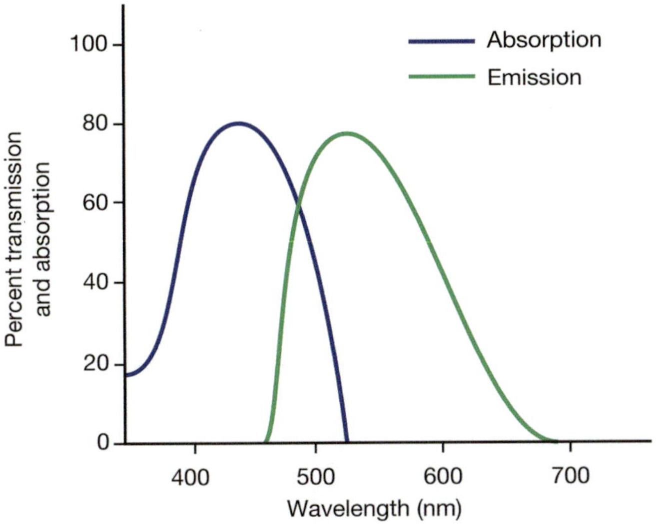

Fig. 2.1 Fluorescein angiogram, late phase (10 min and 24 s)

The retinal circulation goes through various dye-transit phases:

1. **Arterial phase**, which follows the patchy choroidal phase (Fig. 2.2a)
2. **Arteriovenous phase**, with complete filling of arteries and capillaries and laminar flow in the veins (Fig. 2.2b); peak fluorescence is seen later in both artery and vein (Fig. 2.2c)
3. **Recirculation phase**, when fluorescence gradually declines, about 30 s after injection (Fig. 2.2d)

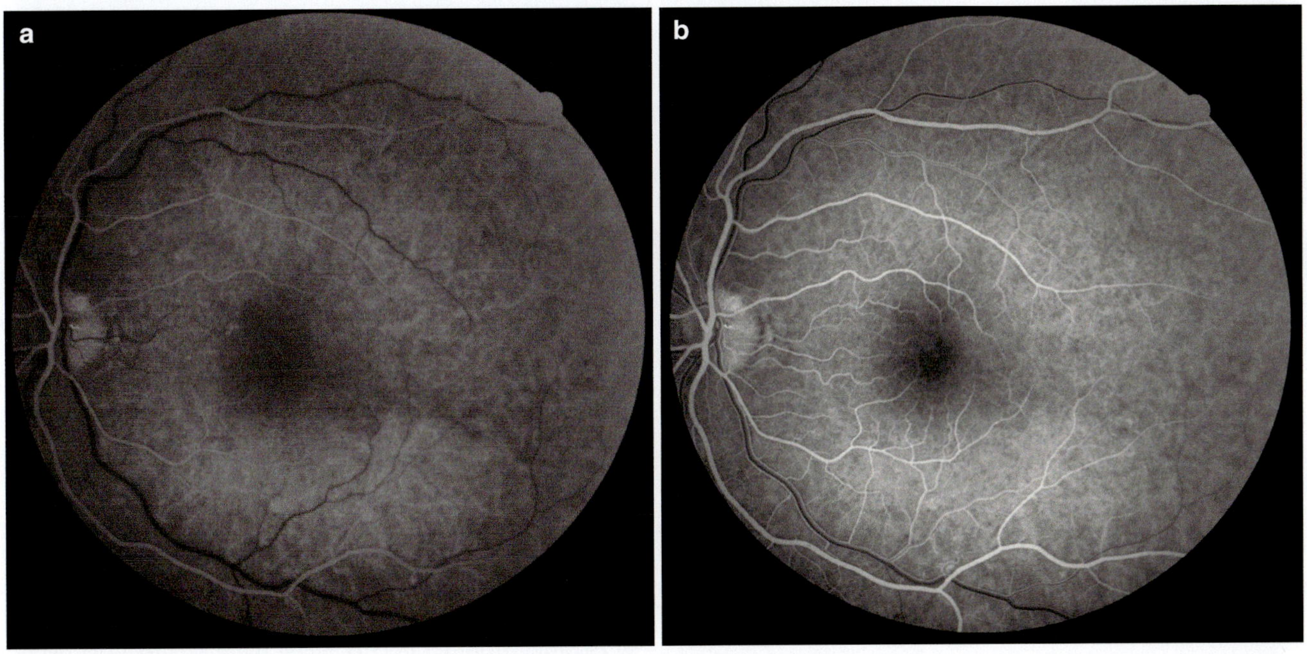

Fig. 2.2 Sequence of dye-transit phases. (**a**) Patchy choroidal and early arterial phase. (**b**) Laminar flow in early venous phase. (**c**) Peak venous phase. (**d**) Recirculation phase. (Courtesy José Ronaldo Lima de Carvalho Júnior. Affiliation: Hospital das Clínicas—Empresa Brasileira de Serviços Hospitalares (Ebserh), Federal University of Pernambuco, Recife, Pernambuco, Brazil)

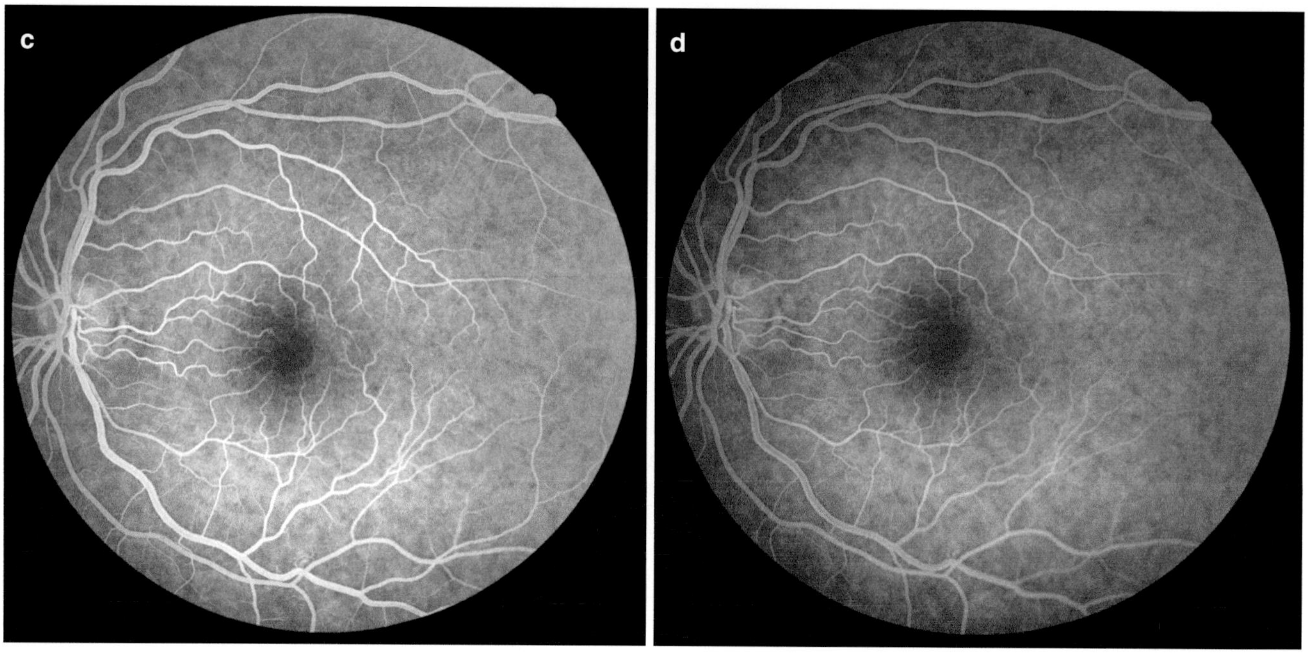

Fig. 2.2 (continued)

Abnormal Fluorescence

Abnormal fluorescence in the eye can be classified into two main categories: hypofluorescence and hyperfluorescence. Hypofluorescence occurs when the fluorescence is reduced or absent and can be classified into two main patterns: vascular filling defects or blocked fluorescence.

1. Vascular filling defects: Hypofluorescence can occur due to the lack of dye perfusion to an area of the retina or choroid, which is referred to as a vascular filling defect. This can happen due to various reasons such as occlusion or stenosis of blood vessels or decreased blood flow in the area. The resulting hypofluorescence can be helpful in identifying areas of ischemia or infarction.
2. Blocked fluorescence: Another mechanism for hypofluorescence is when the fluorescence is blocked by overlying structures such as hemorrhages, exudates, or pigmented lesions. The material can be present in retina or choroid. In these cases, the hypofluorescence is due to the masking effect of the overlying structures rather than a lack of dye perfusion.

Hyperfluorescence occurs when the fluorescence is abnormally excessive and can be divided into five patterns: leakage, staining, pooling, window defect, and autofluorescence.

1. Leakage: This occurs when dye leaks out across blood retina barrier into the surrounding tissue, with an increase in the area and intensity of the fluorescence during the exam. Leakage can be focal or diffuse and can be seen in conditions such as diabetic macular edema, choroidal neovascularization, and uveitis.
2. Staining: Staining refers to diffuse and uniform hyperfluorescence of an area due to abnormal accumulation of dye in the tissue. This can be seen in conditions such as fibrotic scar tissue and drusen.
3. Pooling: Pooling occurs when there is a focal collection of dye in a specific fluid-filled space. This can be seen in conditions such as pigment epithelial detachment (PED) in patients with central serous chorioretinopathy.
4. Window defect: This occurs when there is absence or attenuation of pigment in the RPE, allowing increased transmission of normal choroidal and choriocapillaris fluorescence. This can be seen in conditions such as choroidal folds and macular drusen.
5. Autofluorescence: Autofluorescence refers to intrinsic fluorescence of the tissue, without the use of exogenous dyes. Areas of hyperautofluorescence can be seen in con-

Table 2.1 Classification and clinical examples of abnormal fluorescence

Abnormal fluorescence	Classification	Diseases
Hypofluorescence	Vascular filling defects	Vascular obstruction
	Blocked fluorescence	Hemorrhages, exudates, pigmented lesions
Hyperfluorescence	Leakage	Diabetic macular edema, choroidal neovascularization, uveitis
	Staining	Fibrotic scar tissue, drusen
	Pooling	Pigmenta epithelial detachment (PED)
	Window defect	Choroidal folds, drusen
	Autofluorescence	Astrocytic hamartoma, optic disc drusen

ditions such as astrocytic hamartoma and optic disc drusen (Table 2.1).

Side Effects and Contraindications

After the injection of fluorescein, most patients experience a temporary yellowing of the skin and conjunctiva, which typically lasts for 6–12 h. The most common adverse effects include nausea and vomiting, occurring in approximately 5% of injections, and the development of hives, also in approximately 5% of injections. Nausea is self-limiting and resolves without treatment. For hives, except very mild cases, diphenhydramine is usually prescribed.

Although serious adverse effects such as hypotension, shock, laryngeal spasm, or even death have been reported, they are rare occurrences. Patients with prior urticarial reactions are at a higher risk of experiencing a similar reaction after subsequent injections. However, premedication with antihistamines, corticosteroids, or both can lower this risk. Extravasation of the dye into the skin during injection can be painful and requires the application of ice-cold compresses to the affected area for 5–10 min. It is advisable to closely monitor the patient over hours or days until the edema, pain, and redness subside.

Although teratogenic effects have not been identified, most ophthalmologists avoid using FA in pregnant women during the first trimester unless it is absolutely necessary. In lactating women, fluorescein can be transmitted to breast milk. There are no known restrictions on the use of fluorescein injection in patients with a history of heart disease, cardiac arrhythmia, or cardiac pacemakers.

Indocyanine Green Angiography

Indocyanine green angiography (ICGA) is an imaging technique that utilizes the intravenous injection of indocyanine green (ICG), a water-soluble, tricarbocyanine dye that is almost completely protein-bound (98%) after injection. ICGA enables visualization of the choroidal vasculature due to the limited diffusion through the small fenestrations of the choriocapillaris caused by its protein-bound nature, besides its low permeability. Moreover, ICG dye absorbs and emits light in the near-infrared range, which can penetrate deeper into the tissue compared to the visible light used in fluorescein angiography. ICGA can be particularly useful in imaging conditions such as central serous chorioretinopathy, choroidal neovascularization, and polypoidal choroidal vasculopathy, where the choroidal circulation is affected. ICGA is generally considered safe, and serious side effects are rare. However, some patients may experience mild discomfort or a warm sensation during injection of the dye. In rare cases, allergic reactions can occur, including hives, rash, difficulty breathing, and swelling of the face or throat. ICG is also contraindicated in patients with a known hypersensitivity to the dye or iodine, as well as in patients with severe liver dysfunction or impairment. Pregnant women should also avoid ICG angiography unless absolutely necessary, as the effects of ICG on the developing fetus are unknown. Nursing mothers should discuss the risks and benefits with their healthcare provider before undergoing ICG angiography.

Further Reading

Berkow JW, Flower RW, Orth DH, Kelley JS. Fluorescein and indocyanine green angiography. 2nd ed. San Francisco: American Academy of Ophthalmology; 1997.

Chopdar A, Aung T. Multimodal retinal imaging. JP Medical Ltd; 2014 Aug 22.

Weiss, S.J., Papakostas, T.D. (2021). Fluorescein Angiography. In: Albert, D., Miller, J., Azar, D., Young, L.H. (eds) Albert and Jakobiec's Principles and Practice of Ophthalmology. Springer, Cham. https://doi.org/10.1007/978-3-319-90495-5_138-2

Paques M, Simonutti M, Roux MJ, Picaud S, Levavasseur E, Bellman C, Sahel JA. High resolution fundus imaging by confocal scanning laser ophthalmoscopy in the mouse. Vision research. 2006 Apr 1;46(8-9):1336-45.

Schmitz-Valckenberg S, Pfau M, Fleckenstein M, Staurenghi G, Sparrow JR, Bindewald-Wittich A, Spaide RF, Wolf S, Sadda SR, Holz FG. Fundus autofluorescence imaging. Progress in retinal and eye research. 2021 Mar 1;81:100893.

Witmer MT, Parlitsis G, Patel S, Kiss S. Comparison of ultra-widefield fluorescein angiography with the Heidelberg Spectralis® noncontact ultra-widefield module versus the Optos® Optomap®. Clinical ophthalmology (Auckland, NZ). 2013;7:389.

CHAPTER 3

Optical Coherence Tomography

Stephen H. Tsang and Tarun Sharma

Historical Perspective 14
Principle 14
Normal OCT Scans 14
Further Reading 15

S. H. Tsang (✉) · T. Sharma
Department of Ophthalmology, Columbia University,
New York, NY, USA
e-mail: sht2@cumc.columbia.edu

S. H. Tsang et al. (eds.), *Atlas of Inherited Retinal Diseases*, Advances in Experimental Medicine and Biology 1467, https://doi.org/10.1007/978-3-031-72230-1_3

Historical Perspective

Optical coherence tomography (OCT) has enhanced our understanding and management of retinal diseases, ever since the time-domain OCT was introduced in the early 2000s. The introduction of spectral-domain OCT in the mid-2000s and the later introduction of swept-source OCT provided faster scanning strategies and high-resolution images of the retina and choroid.

Principle

OCT uses the principle of low-coherence interferometry. A low-coherence light beam is directed onto the retina and choroid, and the back-reflected light is combined with a second or reference beam, which was split off from the original light beam. The interference patterns thus produced construct an axial A-scan, and multiple A-scans from adjacent points reconstruct a cross-sectional image of the target tissue, known as a B-scan.

Normal OCT Scans

Figure 3.1 shows the spectral-domain cross-sectional view (tomogram) of the retina and choroid, on a gray scale. Those tissues that are highly reflective (retinal pigment epithelium [RPE] or nerve fiber layer) appear brighter, and tissues with low reflectivity (vitreous or subretinal fluid) appear darker. Areas with intermediate reflectivity (retinal layers or edema) appear as shades of gray. The long wavelength of swept-source OCT (Fig. 3.2) allows a clear image of the deep ocular structures such as choroid.

In inherited retinal dystrophy, the ellipsoid zone (outer segment/inner segment junction) is an important landmark. If this zone is disrupted, discontinuous, or disorganized, it is an early indication of cell death. Progressive loss of photoreceptor cells is reflected by thinning of the outer segment, and later, of the outer nuclear layer.

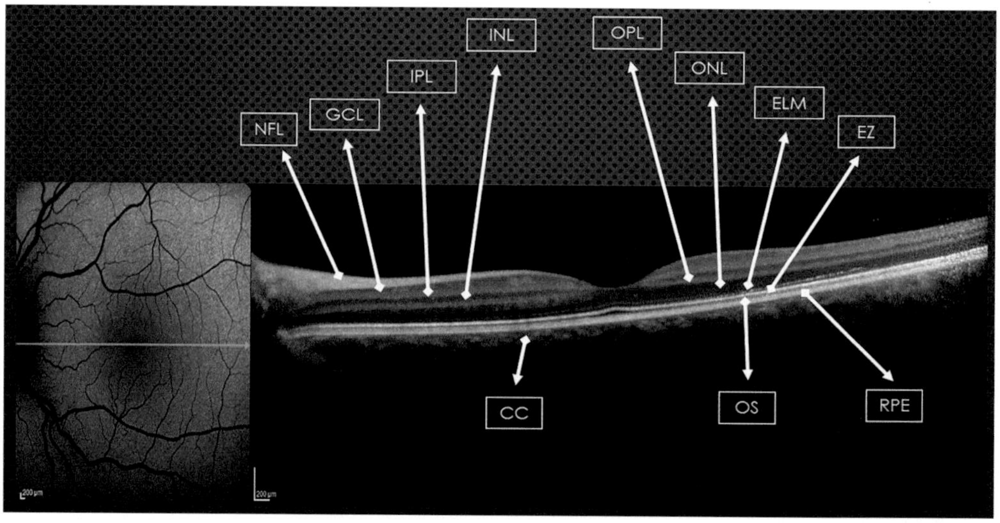

Fig. 3.1 Spectral-domain optical coherence tomography (SD-OCT) image of retinal layers. *CC* choriocapillaris, *EZ* ellipsoid zone, *GCL* ganglion cell layer, *INL* inner nuclear layer, *IPL* inner plexiform layer, *NFL* nerve fiber layer, *OLM* outer limiting membrane, *ONL* outer nuclear layer, *OPL* outer plexiform layer, *OS* outer segment, *RPE* retinal pigment epithelium

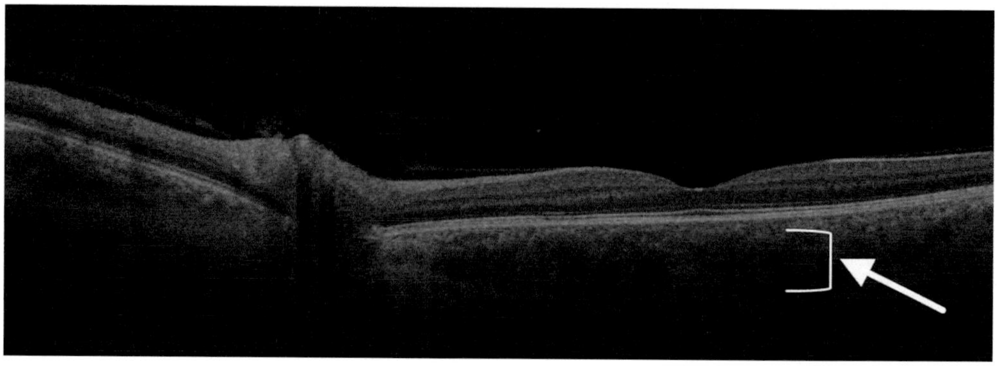

Fig. 3.2 Swept-source OCT showing enhanced visualization of choroidal blood vessels (*arrow*)

Further Reading

Drexler W, Morgner U, Ghanta RK, Kärtner FX, Schuman JS, Fujimoto JG. Ultrahigh-resolution ophthalmic optical coherence tomography. Nat Med. 2001;7:502–7. Erratum in: Nat Med 2001;7:636.

Khurana RN, Bhisitkul RB, Foster BS. Retinal optical coherence tomography. Focal points: clinical modules for ophthalmologists. Module 5. San Francisco: American Academy of Ophthalmology; 2014.

Lumbroso B, Rispoli M. Guide to interpreting spectral domain optical coherence tomography. 2nd ed. New Delhi, India: Jaypee Brothers Medical; 2011.

Fundus Autofluorescence

Andrea Andrade, Jorge Pincay, Promie Faruque, Stephen H. Tsang, Tarun Sharma, and Vlad Diaconita

Basic Principles and Techniques 18
Interpretation and Clinical Application 18
Quantitative Autofluorescence 19
Further Reading 20

A. Andrade
Universidade Federal de Pernambuco, Recife, Brazil

J. Pincay · S. H. Tsang (✉) · T. Sharma · V. Diaconita
Department of Ophthalmology, Columbia University, New York, NY, USA
e-mail: sht2@cumc.columbia.edu

P. Faruque
Columbia University, New York, NY, USA

S. H. Tsang et al. (eds.), *Atlas of Inherited Retinal Diseases*, Advances in Experimental Medicine and Biology 1467, https://doi.org/10.1007/978-3-031-72230-1_4

Basic Principles and Techniques

Fundus autofluorescence (FAF) imaging is a noninvasive and rapid technique to visualize fluorophores in the fundus. The physical principle is based on the excitation of certain molecules by a specific wavelength of light and emission through a longer wavelength.

When excited with short-wavelength (SW) light, retina emits an intrinsic autofluorescence. Short-wavelength fundus autofluorescence (SW-AF) can be excited in vivo in the spectral range between 400 and 590 nm, with a peak excitation at 490–510 nm. The fluorescence is emitted in a range between 520 and 800 nm, with a peak at 600 nm. In healthy eyes, the main source of accumulated fluorophores originates from the family of lipofuscin, located in the cytoplasm of retinal pigment epithelial (RPE) cells, which are produced as a by-product during the digestion of photoreceptor outer segments. Bisretinoid compounds accumulation is toxic to the retina and can impair normal cell function. Gradual accumulation can occur during aging. Excessive accumulation can represent a pathogenic mechanism in different retinal diseases entities.

Imaging techniques can be accomplished using a variety of equipment and software. Fundus cameras, scanning laser ophthalmoscopes (SLOs), and wide-field imaging are some examples of fundus autofluorescence imaging techniques.

The technique for obtaining a fundus camera image has evolved over time, aiming to enhance both its quality and its applicability in the clinical setting. Nowadays, the system function by using green spectrum (495–570 nm) excitation and capturing emissions in the yellow-orange spectrum (570–620 nm). The fundus camera determines the field of view for the image, which typically extends up to 50°. It is possible to obtain wide-field images by using analysis software, employing a seven-field panorama survey.

The use of confocal scanning laser ophthalmoscopy (SLO) for image acquisition effectively overcomes the challenges posed by the low intensity of the autofluorescence signal and the interference caused by the crystalline lens. SLO technique is based on continuous scanning of the fundus at low light intensities in a raster pattern. The intensity of the reflected light is detected by a photomultiplier, after passing through a confocal pinhole or ellipsoid mirror system. With confocal SLO FAF imaging, excitation wavelength is 488 nm and emission filter is used between 500 and 700 nm, to detect autofluorescence signal. The most widely used confocal SLO system for FAF imaging to date is Heidelberg retina angiograph (HRA)/Spectralis HRA (Heidelberg Engineering GmbH, Germany). The standard image field is of 30° × 30°. It is possible to use additional lenses allowing images of a 55° field or larger images using the composite mode.

On the other hand, low confocal SLO, such as Optos, utilizes an ellipsoidal mirror to generate a virtual point of focus positioned behind the pupillary plane. It can acquire ultra-widefield images with approximately 200 internal degrees of view and has an excitation wavelength of 532 nm and emission is over 600 nm. As excitation wavelength is in the green spectrum, there is significant reduction of the FAF signal in the macular area due to less absorption by macular pigment, comparing to the blue wavelength range.

Near-infrared autofluorescence (NIR-FAF) images are obtained with excitation and emission wavelength beyond the red end of the visible spectrum (excitation wavelength at 795 nm and emission with barrier filter over 800 nm). It has been suggested that the signal is largely derived from RPE melanin and less from choroidal melanocytes. Multi-modal imaging, which incorporates NIR-FAF, has revealed that the decrease in melanin signal from the RPE corresponds more accurately with the loss of the ellipsoid zone compared to the signal obtained from short-wavelength FAF.

Interpretation and Clinical Application

Normal macula (Figs. 4.1 and 4.2) shows decreased AF at the center of the macula due to blockage by luteal pigments such as lutein and zeaxanthin. The rest of the macula shows

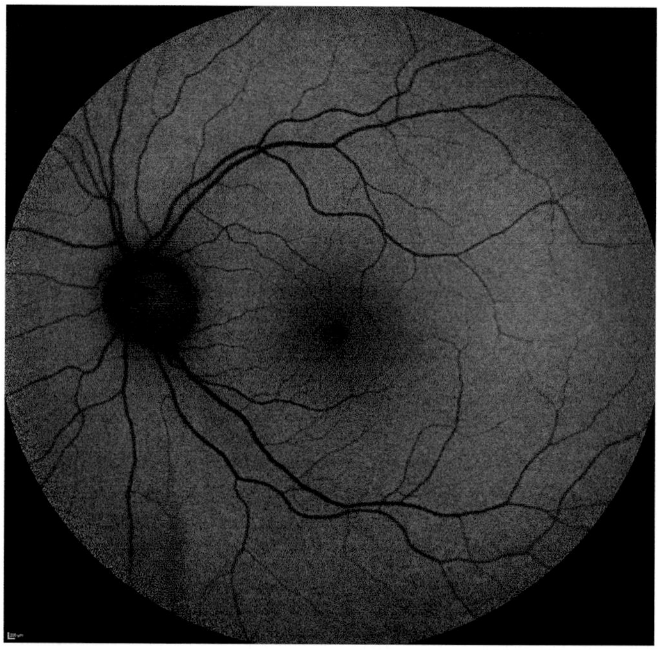

Fig. 4.1 Normal fundus short-wavelenght autofluorescence

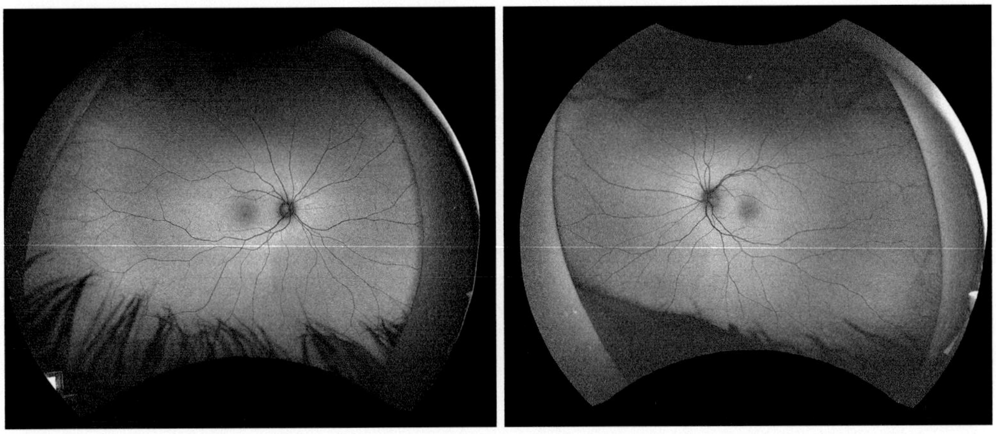

Fig. 4.2 Normal fundus OPTOS autofluorescence

a diffuse, homogenous signal. Additionally, hemoglobin present within retinal blood vessels naturally blocks both the excitation light and the resulting autofluorescence wavelengths emitted by RPE fluorophores. The absence of autofluorescent material in the optic nerve head results in a very low signal and a dark appearance in FAF.

The presence of any fluorophore or filter along the light path can potentially disrupt the detected signal, both in normal and pathological conditions. The lens, in particular, poses a significant confounding factor for the signal, especially in cases of progressive nuclear cataract and when using short-wavelength light (such as blue light) for excitation. This not only obstructs the excitation light from reaching the posterior pole but also generates its own fluorescence light, which further scatters along the light path, thereby interfering with the detection of the FAF signal from the ocular fundus. The development of new technologies for imaging acquisition overcomes this challenge, as can be seen with the use of confocal scanning laser ophthalmoscopy.

FAF can represent distinct patterns, according to retinal diseases. For example, in Stargardt disease, a deficiency in ABCA4 leads to accumulation of lipofuscin, resulting in higher intensities of short-wavelength FAF. Best macular dystrophy and adult vitelliform dystrophy are also associated with an intense locally increased FAF signal. In choroideremia, the progression leads to photoreceptor degeneration with some remaining island with preserved RPE. In this situation, FAF represents a useful outcome measure by quantifying the residual autofluorescence. In summary, FAF imaging plays a crucial role in the clinical assessment, management and research of retinal diseases, offering valuable insights into disease progression, treatment response, and prognosis.

Quantitative Autofluorescence

Although the presence of abnormal AF signals can be readily observed, detecting subtle deviations from normal or tracking progressive changes during subsequent visits can be challenging through qualitative assessment alone. To address the need for reproducible and comparable measurements of FAF signals, there has been a growing demand for standardization and quantification. This has led to the development of quantitative autofluorescence, a methodology that aims to provide standardized and quantifiable measurements of FAF signals.

Quantitative autofluorescence (qAF) combines measurements of fundus autofluorescence (AF) intensity with spatial information. In order to generate qAF values, a specific formula is utilized, considering factors such as laser power, the laser off-set, detector sensitivity, refractive errors, and anterior media structures. To obtain accurate measurements, qAF relies on several factors, including good image quality, the presence of a skilled ophthalmic photographer, the installation of an internal fluorescent reference, and the availability of appropriate software. These factors can pose challenges when implementing this technique in clinical routines.

Different diseases have been analyzed by qAF, such as retinitis pigmentosa, Stargardt disease, pattern dystrophy, bestrophin-1 associated disease, age related macular degeneration, and acute zonal occult outer retinopathy. The presentation of different retinal entities in qAF can increase the understanding of disease mechanisms and progression, being also a valuable tool to guide clinical diagnosis and genetic testing.

Further Reading

Oishi M, Oishi A, Ogino K, Makiyama Y, Gotoh N, Kurimoto M, Yoshimura N. Wide-field fundus autofluorescence abnormalities and visual function in patients with cone and cone-rod dystrophies. Invest Ophthalmol Vis Sci. 2014;55:3572–7.

Schmitz-Valckenberg S, Holz FG, Bird AC, Spaide RF. Fundus autofluorescence imaging: review and perspectives. Retina. 2008;28:385–409.

Paques M, Simonutti M, Roux MJ, Picaud S, Levavasseur E, Bellman C, Sahel JA. High resolution fundus imaging by confocal scanning laser ophthalmoscopy in the mouse. Vision research. 2006 Apr 1;46(8-9):1336-45.

Schmitz-Valckenberg S, Pfau M, Fleckenstein M, Staurenghi G, Sparrow JR, Bindewald-Wittich A, Spaide RF, Wolf S, Sadda SR, Holz FG. Fundus autofluorescence imaging. Progress in retinal and eye research. 2021 Mar 1;81:100893.

Greenberg, J. P., Duncker, T., Woods, R. L., Smith, R. T., Sparrow, J. R., & Delori, F. C. (2013). Quantitative fundus autofluorescence in healthy eyes. *Investigative ophthalmology & visual science*, *54*(8), 5684-5693.

Sparrow JR, Duncker T, Schuerch K, Paavo M, de Carvalho Jr JR. Lessons learned from quantitative fundus autofluorescence. Progress in retinal and eye research. 2020 Jan 1;74:100774.

Pole, C., & Ameri, H. (2021). Fundus Autofluorescence and Clinical Applications. *Journal of ophthalmic & vision research*, *16*(3), 432–461. https://doi.org/10.18502/jovr.v16i3.9439

CHAPTER 5

Electroretinography

Julie Shabto, Stephen Tsang, Tarun Sharma, and Vlad Diaconita

Full-Field (Ganzfeld) Electroretinography (ffERG) 22
Electronegative ffERG 23
Multifocal Electroretinography (mfERG) 24
Pattern Electroretinography (pERG) 25
Photopic On-Off ERG 25
Further Reading 26

J. Shabto · S. Tsang (✉) · T. Sharma · V. Diaconita
Department of Ophthalmology, Columbia University,
New York, NY, USA
e-mail: jms2338@cumc.columbia.edu; sht2@cumc.columbia.edu

S. H. Tsang et al. (eds.), *Atlas of Inherited Retinal Diseases*, Advances in Experimental Medicine and Biology 1467, https://doi.org/10.1007/978-3-031-72230-1_5

The electroretinogram (ERG) is used to measure electrical activity of various cell types in several layers of the retina, including rods, cones, inner retinal cells, and ganglion, in response to a light stimulus. ERG therefore provides objective information about visual system function and can provide diagnostic information for inherited and acquired retinal disorders. It can also provide objective measurements of disease progression, including development of retinal toxicity due to medications, and treatment response.

- The *active electrode* is in contact with the bulbar conjunctiva, cornea, or skin below the lower eyelid
- The *reference electrode* is placed at another site, e.g., outer canthus, ear, forehead, or temple

Three main types of ERG:

1. Full-field (Ganzfeld) ERG (ffERG)
2. Multifocal ERG
3. Pattern ERG

The ffERG is generated predominantly by the peripheral retina while the multifocal ERG and pattern ERG are generated from the macula. The International Society for Clinical Electrophysiology of Vision (ISCEV) establishes protocols for the different forms of ERGs.

Full-Field (Ganzfeld) Electroretinography (ffERG)

A Ganzfeld bowl is used to illuminate the entire retina in a uniform fashion and the response from each eye is recorded (Fig. 5.1). The testing protocol put forth by the International Society for Clinical Electrophysiology of Vision (ISCEV) and most recently updated in 2022 involves five different stimuli and associated responses.

- The nomenclature is based on:
  - Adaptive state of the eye: light-adapted (LA) or dark-adapted (DA)
  - Flash strength in candela-seconds per meter squared ($cd.s.m^{-2}$)
- Responses:
  - a-wave: cornea-negative potential; generated by outer retina, rods, and cones
  - b-wave: cornea-positive potential, generated by inner retina, ON- and OFF-bipolar cells
  - Oscillatory potentials (Ops): The ascending portion of the b-wave can have multiple wavelets called oscillatory potentials, generated by amacrine cells
  - The implicit time of the a-wave or b-wave is measured from the time of the stimulus to the trough or peak of the a- or b-wave, respectively.

Scotopic (rod pathway): Measurements targeting rod function are made from the dark-adapted eye.

1. **Dark-adapted 0.01 (rod-specific)**:
   (a) Stimulus: diffuse, weak, full-field flash of light
   (b) Response: from ON-bipolar cells of the rod system, so it has only a b-wave; no detectable a-wave
2. **Dark-adapted 3.0 (maximum scotopic)**:
   (a) Stimulus: standard flash of light (stronger than dark-adapted 0.01 flash)
   (b) Response: from both rods and cones; contains both a- and b-waves. The two a-wave peaks occur at about 15 and 21 ms. See below for information about electronegative ffERG. As only the first 8 ms represents photoreceptor hyperpolarization, ISCEV recommends an additional DA 10.0 testing.

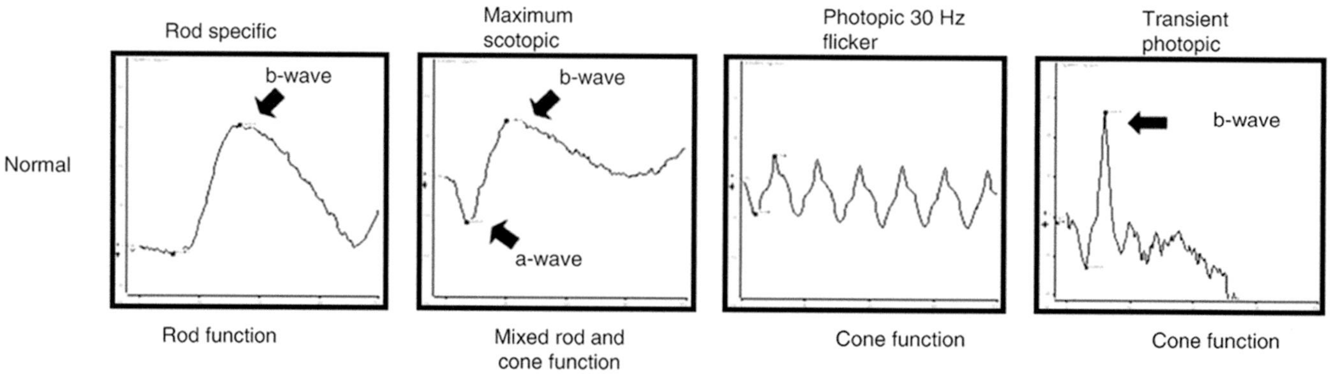

Fig. 5.1 Normal electroretinogram (ERG) tracings. The a-wave and b-wave are noted where applicable

3. **Dark-adapted 10.0 (mixed rod and cone)**:
   (a) Stimulus: a stronger flash of light relative to the Dark-adapted 3.0.
   (b) Response: with this stronger flash, the a-wave amplitude is larger and peak time is shorter compared to the dark-adapted 3.0 response, providing a better measure of rod photoreceptor function. The oscillatory potentials are larger as well. This stimulus-response helps to differentiate abnormality between photoreceptor and inner retina.
   - If there is reduction of both a-wave and b-wave amplitude, dysfunction can be localized to the rod photoreceptors.
   - If there is reduction of b-wave amplitude with sparing of the a-wave, dysfunction can be localized to the inner retinal level.

   Photopic (cone pathway): Measurements targeting cone function are made from the light-adapted eye.
4. **Light-adapted 3.0 single flash (cone-specific)**:
   (a) Stimulus: strong flash against a light background
   (b) Response: reduced b/a-wave ratio suggests post-phototransduction dysfunction
5. **Light-adapted 3.0 30-Hz flicker (maximum cone)**:
   (a) Stimulus: rapid flicker
   (b) Response: this is a more sensitive measure of cone function because rod photoreceptors cannot follow rapid flicker; both time and amplitude of response are important. A delayed time response suggests generalized cone dysfunction, whereas reduced amplitude with normal peak time is indicative of restricted cone dysfunction.

Electronegative ffERG

- Certain conditions produce an irregular response with Dark-adapted 3.0 stimulus in which the a-wave has normal amplitude while the b-wave is markedly reduced. The OPs can also be reduced, delayed, or absent.
- This "electronegative" response can provide diagnostic information for conditions with inner retinal rod dysfunction, like photoreceptor dystrophy, congenital stationary night blindness, retinal ischemia, juvenile X-linked retinoschisis, melanoma-associated retinopathy, and paraneoplastic autoimmune retinopathy (PAIR) (Fig. 5.2).

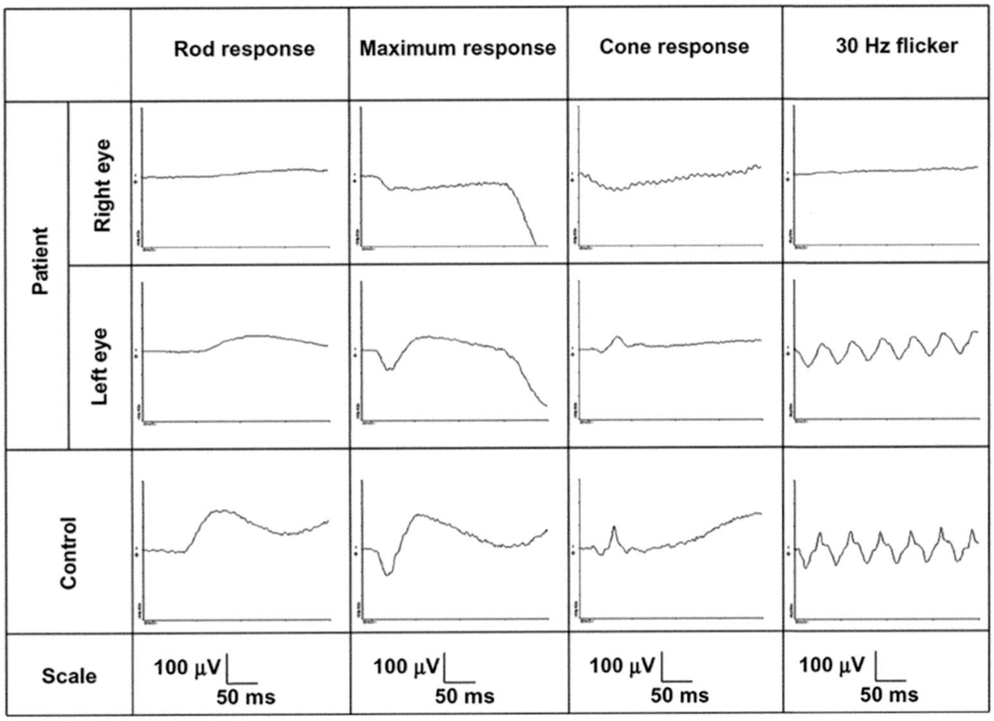

Fig. 5.2 Full-field ERG with electronegative response in right eye and generalized dysfunction of left eye of patient with unilateral central retinal artery occlusion (CRAO). Note the right eye's rod response and maximum response both have a markedly reduced b-wave amplitude

Multifocal Electroretinography (mfERG)

The multifocal ERG (Fig. 5.3) helps in evaluating macular function (cone system) that might be missed with ffERG. For example, mfERG is useful in the early detection of hydroxychloroquine toxicity. With parafoveal photoreceptor damage, these patients have reduced amplitude of the second ring averaged waveform response relative to the central ring.

- Stimulus: multiple hexagons (61 or 103), which are smaller in the center than in the periphery
  - The hexagons are displayed on a television monitor, and each hexagon flashes on and off with a pseudorandom binary sequence (an M-sequence).
- Response: Responses are recorded with light-adapted conditions to assess the cone pathway. The tested area is about 20–30° on either side of the fovea.
  - Each response from the hexagon is calculated via a cross-correlation technique, generating multiple cone system ERG waveforms, thereby providing spatial information about the central cone system and aiding in the detection of localized abnormalities within the macula.
  - The waveform response is similar to the waveform seen with light-adapted ffERG, with an initial negative deflection (N1) followed by a positive deflection (P1) and another negative deflection (N2).

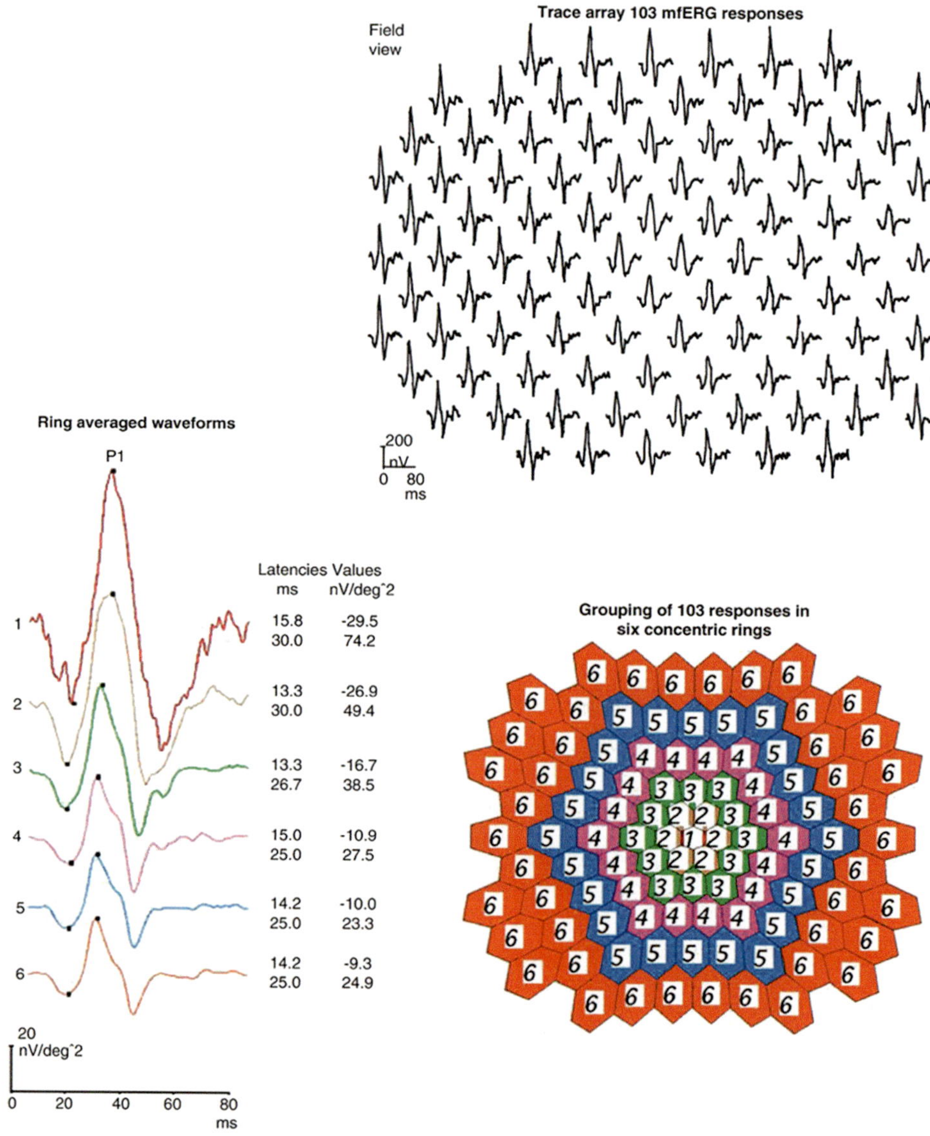

Fig. 5.3 Multifocal ERG tracing of a normal patient. The trace array of 103 mfERG responses shows the response elicited from each hexagon. The ring averaged waveforms represent the average mfERG response within concentric rings defined by the differently colored and numbered hexagon stimuli

Pattern Electroretinography (pERG)

The central retinal function can also be assessed by the pattern ERG (pERG), elicited by using high-contrast checkerboard reversal. The pERG is a useful test for evaluating retinal ganglion cell function in conditions like glaucoma or ischemic optic neuropathy as well as conditions that cause retinal ganglion cell apoptosis such as Autosomal Dominant Optic Atrophy (ADOA) and Leber hereditary optic neuropathy.

- Stimulus: an example of a pERG stimulus is shown (Fig. 5.4). Note the symmetry around the center, the sharp contrast between the squares, and the even luminance levels when averaged over any section of the pattern. In a reversal test, the pattern would reverse such that the top-left square would be white, and then reverse back to black; the retina would be stimulated at each reversal, or twice per cycle. Typically, the "transient pERG" stimulus reverses at a rate of 4 reversals per second.
- Response: the waveform response is comprised of two key components (Fig. 5.5).
  - P50 component: a positive deflection near 50 ms, which is thought to arise from macular photoreceptors, although a significant part arises in the retinal ganglion cells.

    Reduced amplitude or delayed response of the P50 component suggests macular cone dysfunction
  - N95 component: a negative deflection near 95 ms that arises in the retinal ganglion cells.

    Reduction of amplitude in the N95 component with preservation of the P50 component suggests dysfunction of retinal ganglion cells.

Photopic On-Off ERG

The On-Off ERG can be used to evaluate the relative involvement of the cone-driven on- and off-system in retinal dystrophies and disorders that cause post-phototransduction or post-receptoral dysfunction, such as X-linked retinoschisis, Batten disease, Duchenne muscular dystrophy, and quinine toxicity.

- Stimulus: a light with duration of 150–200 ms in the presence of a rod-suppressing background

Fig. 5.4 Sample of an appropriate contrast reversing pattern ERG stimulus

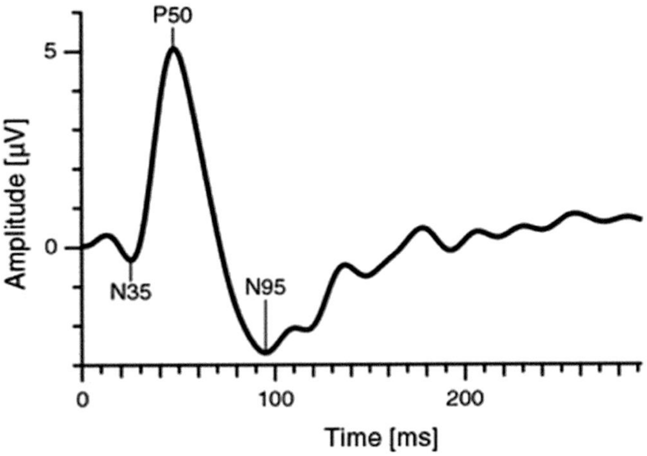

Fig. 5.5 Pattern ERG waveform response in a visually healthy person

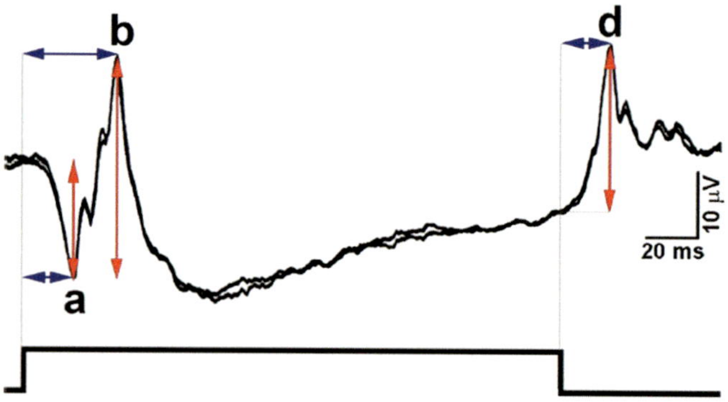

Fig. 5.6 Photopic On-Off ERG with three main component waves. Stimulus duration is indicated at the bottom of the figure

- Response:
  - on-response: elicited after stimulus onset; has a negative a-wave and positive b-wave
  - off-response: evoked by stimulus offset; comprised of a positive d-wave (Fig. 5.6)

Further Reading

Holder GE. Pattern electroretinography (PERG) and an integrated approach to visual pathway diagnosis. Prog Retin Eye Res. 2001;20:531–61.

Jiang X, Mahroo OA. Negative electroretinograms: genetic and acquired causes, diagnostic approaches and physiological insights. Eye. 2021;35:2419-2437.

Sustar M, Holder GE, Kremers J, Barnes CS, Lei B, Khan NW, Robson AG. ISCEV extended protocol for the photopic on-off ERG. Doc Ophthalmol. 2018;136:199-206.

Robson AG, Frishman LJ, Grigg J, Hamilton R, Jeffrey BG, Kondo M, Li S, McCulloch DL, International Society for Clinical Electrophysiology of Vision. ISCEV Standard for full-field clinical electroretinography (2022 update). Doc Ophthalmol. 2022;144:165–177.

Robson AG, Nilsson J, Li S, Jalali S, Fulton AB, Tormene AP, Holder GE, Brodie SE. ISCEV guide to visual electrodiagnostic procedures. Doc Ophthalmol. 2018;136:1-26.

Vincent A, Robson AG, Holder GE. Pathognomic (diagnostic) ERGs. A review and update. Retina. 2013;33:5–12.

CHAPTER 6

Electrooculography

Julie Shabto, Stephen H. Tsang, Tarun Sharma, and Vlad Diaconita

Further Reading 28

J. Shabto · S. H. Tsang (✉) · T. Sharma · V. Diaconita
Department of Ophthalmology, Columbia University,
New York, NY, USA
e-mail: jms2338@cumc.columbia.edu; sht2@cumc.columbia.edu

S. H. Tsang et al. (eds.), *Atlas of Inherited Retinal Diseases*, Advances in Experimental Medicine and Biology 1467, https://doi.org/10.1007/978-3-031-72230-1_6

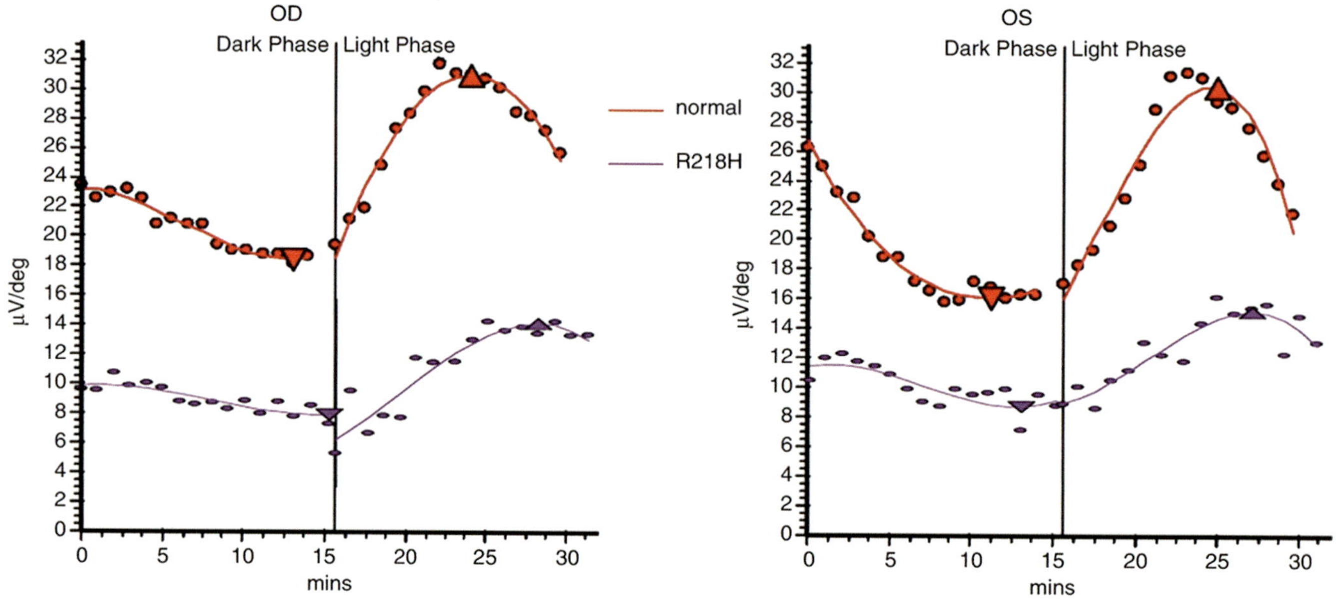

Fig. 6.1 Lack of EOG light rise is seen in a patient with R218H mutation (Best Disease)

The electrooculogram (EOG) assesses the function and health of the retinal pigment epithelium (RPE), and also the interaction between the RPE and photoreceptors. This is done by noting the changes in the corneo-retinal standing potential (the difference between the retina and electropositive cornea) during the dark-adapted and light-adapted states.

During the test, the patient sits in a Ganzfeld bowl and makes fixed 30° lateral eye movements during a 15-min period of dark adaptation, followed by a 15-min period of light adaptation. Electrodes are placed at the medial and lateral canthi. The amplitude of the signal recorded between the electrodes (the standing potential) reaches a minimum during dark adaptation (dark trough) and a maximum during light adaptation (light peak). In a normal individual, the development of this light peak requires normally functioning photoreceptors in contact with a normally functioning RPE and is due to progressive depolarization of the RPE basal membrane. The mechanism of this depolarization is rather poorly understood; it may be caused by a bestrophin protein for opening up of chloride channels. The light peak-to-dark trough ratio (Arden ratio) measures RPE/photoreceptor function with a normal EOG Arden ratio greater than 170%.

This EOG is supplementary to the ERG and can aid in the diagnosis of certain conditions with normal ERG. For example, in Best vitelliform macular dystrophy (Fig. 6.1), the Arden ratio is less than 170%, representing an absent light rise, but the full-field electroretinogram (ERG) is normal. In adult vitelliform macular dystrophy, the EOG is normal or mildly subnormal. In autosomal recessive bestrophinopathy (ARB), both the EOG and ERG are abnormal, but the carriers (unlike in Best) do not show EOG abnormality. The EOG is also abnormal in any condition that has widespread photoreceptor degeneration, such as retinitis pigmentosa. A marked reduction in light rise in the absence of a comparable degree of rod ERG abnormality is also seen in subjects with acute zonal occult outer retinopathy (AZOOR), and this is helpful in distinguishing AZOOR from other causes of white dot syndrome.

Further Reading

Arden GB, Constable PA. The electro-oculogram. Prog Retin Eye Res. 2006;25:207–48.

Burgess R, Millar ID, Leroy BP, Urquhart JE, Fearon IM, De Baere E, et al. Biallelic mutation of BEST1 causes a distinct retinopathy in humans. Am J Hum Genet. 2008;82:19–31.

Constable PA, Bach M, Frishman LJ, Jeffrey BG, Robson AG. ISCEV Standard for clinical electro-oculography (2017 update). Doc Ophthal. 2017;134:1-9.

Heckenlively JR, Arden GB, editors. Principles and practice of clinical electrophysiology of vision. 2nd ed. Cambridge, MA: MIT Press; 2006.

Robson AG, Nilsson J, Li S, Jalali S, Fulton AB, Tormene AP, Holder GE, Brodie SE. ISCEV guide to visual electrodiagnostic procedures. Doc Ophthalmol. 2018;136:1-26.

Glossary of Relevant Genetic and Molecular/Cell Biology

Duncan Yi-Te Wang, Megan Soucy, Stephen H. Tsang, Tarun Sharma, and Vlad Diaconita

Genetics 30
Genetic Disorders 31
Molecular Biology 32
Inheritance 33
Diagnosis and Treatment 33

D. Y.-T. Wang
School of Medicine, College of Medicine, National Taiwan University, Taipei, Taiwan

M. Soucy · S. H. Tsang (✉) · T. Sharma · V. Diaconita
Department of Ophthalmology, Columbia University, New York, NY, USA
e-mail: sht2@cumc.columbia.edu

S. H. Tsang et al. (eds.), *Atlas of Inherited Retinal Diseases*, Advances in Experimental Medicine and Biology 1467, https://doi.org/10.1007/978-3-031-72230-1_7

Genetics

Genes: The basic physical and functional unit of heredity; made up of DNA, they act as instructions to make proteins. There are about 20,000–25,000 genes in humans, with size varying from a few hundred DNA bases to more than 2 million bases. Most genes are the same in all people; less than 1% differ. Each person has two copies of each gene, one inherited from each parent. Each chromosome contains many genes.

Chromosome: Thread-like structures in the nucleus of each cell, in which the DNA molecules and a protein called *histone* are packaged. Each chromosome has a constriction point, the *centromere*, which divides the chromosome into a short arm ("p arm") and a long arm ("q arm").

In humans, each cell contains 23 pairs of chromosomes, for a total of 46. Twenty-two of these pairs, called *autosomes*, look the same in both males and females. In the 23rd pair, the sex chromosome, females have two copies of the X chromosome and males have one X and one Y chromosome. The chromosomes are numbered by size. The picture of the human chromosome lined up in pairs is called a *karyotype*.

Genetics: The study of genes, genetic variation, and heredity in living organisms.

Genotype: The genetic constitution of an individual.

Phenotype: The set of observable or manifest characteristics of an individual due to an interaction of its genotype with the environment.

Genome: The complete set of genes or genetic material in a cell.

Cell: Basic building block of living organisms; the human body has trillions of cells. Cells contain genetic (hereditary) material and can make copies of themselves.

Cytoplasm: A jelly-like fluid (cytosol) within each cell; it contains many organelles (organized or specialized structures, like endoplasmic reticulum, Golgi apparatus, or mitochondria).

Endoplasmic reticulum: Organelle that transports molecules created by the cell to their destination either inside or outside the cell.

Golgi apparatus: Organelle that packages molecules processed by the endoplasmic reticulum and transports them out of the cell.

Nucleus: The command center of a cell, which contains DNA (nuclear DNA); it sends direction to the cell to grow, divide, mature, or die.

Mitochondria: Organelle that produces energy through a process called oxidative phosphorylation, creating adenosine triphosphate (ATP) from oxygen and simple sugars. They have their own genetic material (mitochondrial DNA) that is separate from the DNA in the nucleus and make copies of themselves. Each cell contains hundreds to thousands of mitochondria.

Ribosomes: Organelle in the cytoplasm that makes protein.

GDB Human Genome Database: The official central repository for genomic mapping data resulting from the Human Genome Project.

DNA or deoxyribonucleic acid: The hereditary material in humans and all other organisms. Nearly every cell has the same DNA, located in the nucleus or mitochondria. The information in the DNA is stored as a code made up of four chemical bases: adenine (A), guanine (G), cytosine (C), and thymine (T). These bases pair up with each other, A with T and C with G, to form units called *base pairs*.

Nucleotide: Each base pair is attached to a sugar molecule and a phosphate molecule; together, a base, sugar, and phosphate are called a *nucleotide*. Nucleotides are the building block of DNA.

DNA structure: Nucleotides are arranged in two long strands that form a spiral (double helix). The structure of the double helix is like a ladder, with the base pairs forming the ladder's rungs and the sugar and phosphate molecules forming the vertical sidepieces of the ladder.

Human genome: Each cell contains approximately 3 billion base pairs, which reside in the 23 pairs of chromosome within the nucleus. Each chromosome contains hundreds to thousands of genes, which carry the instructions for making proteins. More than 99% of these bases are the same in all people.

DNA sequence: The order or sequence of the bases determines the information available for building and maintaining an organism, similar to the way in which letters of the alphabet appear in a certain order to form words and sentences. An important property of DNA is that it can make copies of itself (replicate); this is critical when cells divide, because each new cell needs to have an exact copy of the DNA present in the old cell.

Mitochondrial DNA: Besides producing energy, mitochondria also help in regulating apoptosis, the self-destruction of cells, and they are essential to produce substances like cholesterol and heme (the component of hemoglobin that carries oxygen in the blood). Mitochondrial DNA has 37 genes: 13 provide instructions for making enzymes involved in oxidative phosphorylation, and the others provide instructions for making molecules called transfer RNAs (tRNAs) and ribosomal RNAs (rRNAs), both of which help assemble protein building blocks (amino acids) into functioning proteins.

Genetic Disorders

Aneuploidy: A gain or loss of a chromosome from the normal 46 is called aneuploidy.

Trisomy: Having three copies of a particular chromosome in cells instead of the normal two copies. People with Down syndrome typically have three copies of chromosome 21 in each cell, for a total of 47 chromosomes per cell.

Monosomy: Having one copy of a particular chromosome in cells instead of the normal two copies. Women with Turner syndrome usually have one copy of the X chromosome in every cell, for a total of 45 chromosomes per cell.

Gene mutation: A permanent change or alteration in the DNA sequence that makes up a gene. Mutations range in size from a single base pair to a large segment of chromosome that includes multiple genes.

Hereditary or germline mutations: Mutations inherited from a parent (via egg or sperm cells, called *germ cells*) and present in virtually every cell in the body throughout a person's life.

Acquired or somatic mutation: Mutation that occurs at some time during a person's life and is present only in certain cells, not in every cell. These changes can be caused by environmental factors such as ultraviolet radiation or may result from a mistake made as DNA copies itself during a cell division. These mutations cannot be passed to the next generation.

***De novo* (new) mutation**: Mutation can be either germline or somatic. The germline mutation may occur in an egg, sperm cell, or fertilized egg; this may explain why a genetic disorder is present in a child, but the parents remain unaffected (no family history), whereas the somatic mutation does not happen in egg or sperm cells or the fertilized egg but occurs a bit later in an embryo. All cells that arise from a cell with the mutated gene will have the mutation, but other cells will not; this situation is called *mosaicism*. Depending on the mutation and how many cells are affected, mosaicism may or may not cause health problems.

Genetic disorder: A condition caused by mutations in one or more genes. Each cell depends on thousands of proteins doing their jobs in the right place at the right time, and gene mutations can prevent one or more of these proteins from working properly, causing it either to malfunction or to be missing. Not all mutations affect health and development. Some alter a gene's DNA sequence but do not change the function of the protein made by the gene.

DNA repair: An important process by which the body protects itself from disease. As DNA can be damaged or mutated in many ways, each cell has a number of pathways/enzymes which recognize and repair mistakes in DNA.

Beneficial mutations: A very small percentage of all mutations actually have a positive effect; a new version of proteins may help an individual to adapt better to changes in the environment.

Variants of unknown significance (VOUS or VUS): Because a person's genetic code can have a large number of mutations with no effect on health, diagnosing a genetic condition can be difficult. Sometimes, genes thought to be related to a particular genetic condition have mutations that are known as *variants of unknown significance* if it has not been determined whether the changes are involved in the development of the condition.

Missense mutation: A change in one DNA base pair (point mutation, replacement of a single nucleotide) that results in the substitution of one amino acid for another in the protein made by a gene.

Nonsense mutation: A change in one DNA base pair by which the altered DNA sequence prematurely signals the cell to stop building a protein. The result is a shortened protein that may not function properly.

Insertion: The addition of one nucleotide, changing the amino acid sequence that follows; hence, the protein may not function properly.

Deletion: Removal of one base pair or a few base pairs within a gene (small deletion) or removal of an entire gene or several neighboring genes (large deletion), altering the resulting protein.

Duplication: Abnormal copying of a piece of DNA one or more times.

Frameshift mutation: The addition or loss of DNA bases that changes a gene's reading frame (group of three bases that code for one amino acid). Insertions, deletions, and duplications can all be frameshift mutations.

Repeat expansion: Nucleotide repeats are short DNA sequences that are repeated a number of times in a row. A trinucleotide repeat is made up of sequences of three base pairs, and a tetranucleotide repeat is made up of sequences of four base pairs. The resultant protein is malfunctioning.

Complex of multifactorial disorders: Common medical problems such as heart disease, diabetes, and obesity, which are associated with multiple gene defects as well as lifestyle and environmental factors, rather than having a single genetic cause.

Gene names: The HUGO Gene Nomenclature Committee (HGNC) designates an official name and symbol (an abbreviation of the name) for each known human gene.

Molecular Biology

Proteins: Large, complex molecules that do most of the work in cells and are needed for the structure, function, and regulation of the body's tissues and organs. Proteins are made up of hundreds or thousands of smaller units called *amino acids*, the building block of proteins, which are attached to each other in long chains. There are 20 different amino acids that can be combined to make a protein.

Gene expression: The complex and tightly controlled journey from gene to protein within each cell; it involves two steps, transcription and translation.

Transcription: The transfer of the information stored in a gene's DNA to another molecule (RNA) in the cell nucleus. This RNA carries the information or message from the DNA out of the nucleus into the cytoplasm, called messenger RNA (mRNA).

Translation: In the cytoplasm, the mRNA interacts with a specialized complex, a ribosome, which reads the sequence of mRNA bases. Each sequence of three bases, called a codon, usually codes for one particular amino acid. A type of RNA called transfer RNA (tRNA) assembles the protein, one amino acid at a time; the protein assembly continues until the ribosome encounters a stop codon (a sequence of three bases that does not code for an amino acid).

Central Dogma: Explanation of the flow of information from DNA to RNA to proteins; one of the fundamental principles of molecular biology.

Gene regulation: Each cell expresses or turns ON only a fraction of genes; the rest are repressed or turned OFF. Gene regulation is an important part of normal development (for example, making a brain cell different from a muscle cell).

Transcription factor: Protein activated by signals from the environment or from other cells. These proteins bind to regulatory regions of a gene and increase or decrease the level of transcription, thereby determining the amount of protein product that is made by a gene at any given time.

Epigenome: A multitude of the chemical compounds that surround the genome; these are not part of DNA sequence, but are attached to DNA. These compounds can tell the genome what to do and can affect gene activity. Epigenetic changes can help determine whether genes are turned ON or OFF and can influence the production of proteins in certain cells. Some epigenetic changes can be passed on from generation to generation. Environmental factors such as diet and exposure to pollutants can also impact the epigenome.

Epigenomic modification or methylation: When methyl groups are added to a particular gene, that gene is turned OFF or silenced, and no protein is produced from that gene. Conditions like cancers, metabolic disorders, and degenerative disorders have all been found to be related to epigenetic errors.

Human Epigenome Project: A multinational project with the aim of identifying, cataloging, and interpreting genome-wide DNA methylation patterns of all human genes in all major tissues.

Cytogenetic location of a gene: Describing the location of a particular gene on a chromosome. The cytogenetic location is based on a distinctive pattern of bands created when chromosomes are stained with certain chemicals. The combination of numbers and letters provide a gene's address on a chromosome. For example, in 15p12, the first number (15) describes the chromosome number. The next letter describes the arm of the chromosome, short (p) or long (q), and the next number describes the location on the p or q arm, designated by two digits, representing a region and a band. These are sometimes followed by a decimal point and one or two additional digits (representing sub-bands within a light or dark area). Therefore, 15p12 represents position 12 on the short arm of chromosome 15; 15p11 is closer to the centromere than 15p12.

Molecular location of a gene: Molecular address that pinpoints the location of that gene in terms of base pairs, and also the size of the gene. The Human Genome Project determined the sequence of base pairs for each human chromosome. This sequence information allows the researcher to provide a more specific address than the cytogenetic location for many genes.

Inheritance

Hereditary disease: Disease that results from a particular genetic composition and is passed from one generation to another.

Genetic disease: Disease caused by a genetic defect, either acquired or inherited; not passed onto subsequent generations, and therefore, not hereditary.

Familial disease: Disease that occurs in more than one member of a family. It may be hereditary, but not necessarily; it could be due, for example, to a common exposure to infectious agents, as in adenoviral conjunctivitis.

Mendelian disorder (single-gene disorder): A trait or medical disorder that follows patterns of inheritance suggesting that it is determined by a gene at a single locus (autosomal dominant, autosomal recessive, or X-linked recessive inheritance).

Autosomal dominant inheritance: One mutated copy of the gene in each cell is sufficient for a person to be affected. The trait appears in multiple generations (vertical transmission).

Autosomal recessive inheritance: Both copies of the gene in each cell must have mutations for the person to be affected. Typically, the trait appears only in the person's siblings, not the parents or offspring, and the trait is not seen in every generation of an affected family.

X-linked recessive inheritance: Inheritance of traits caused by mutations in genes on the X chromosome. In males (who have only one X chromosome), one altered copy of the gene in each cell is sufficient to cause the condition. An affected father cannot pass X-linked traits to his sons (no male-to-male transmission), but all of his daughters are carriers.

Codominant inheritance: Two different versions (alleles) of a gene are expressed at the same place on a chromosome, and each version makes a slightly different protein. Both alleles influence the genetic trait or determine the characteristics of the genetic disease.

Mitochondrial or maternal inheritance: Because only egg cells contribute mitochondria to the developing embryo, only females can pass on mitochondrial mutations to their children. The trait can appear in every generation of a family and can affect both males and females.

Reduced or incomplete penetrance: Only some people with a particular genetic change or mutation exhibit signs or symptoms of the genetic disorder.

Variable expressivity: A range of signs or symptoms occurring in different people with the same genetic condition.

Clinical heterogeneity: The production of different phenotypes by different mutations at the same locus (location of a gene on a chromosome); for example, mutation of peripherin/RDS can give rise to macular dystrophy or retinitis pigmentosa (same locus, different diseases).

Locus heterogeneity: The production of a similar phenotype by mutations at different loci; for example, X-linked retinitis pigmentosa can result from *RP2* at Xp11 or *RP3* at Xp21 (different locus, same disease).

Allelic heterogeneity: The production of the same phenotypic expression by different mutations within a single gene locus (same locus, same disease).

Anticipation: The tendency for the signs and symptoms of some genetic conditions to become more severe and appear at an earlier age as the disorder is passed from one generation to the next.

Digenic inheritance: Simultaneous inheritance of two nonallelic mutant genes; for example, retinitis pigmentosa caused by mutations of both the *ROM1* and peripherin/*RDS* genes.

Diagnosis and Treatment

Genetic testing: A medical test that identifies changes in chromosomes, genes, or protein.

Molecular genetic test or gene tests: Study of single genes or short lengths of DNA.

Chromosomal genetic tests: Analysis of whole chromosomes or long lengths of DNA.

Biochemical genetic tests: Study of the amount or activity level of proteins.

Positive genetic test: Laboratory finding of a change in a particular gene, chromosome, or protein of interest. Depending on the purpose of the test, the result confirms a diagnosis, indicates that a person is a carrier of a particular genetic mutation, identifies an increased risk of developing a disease (e.g., cancer) in the future, or suggests a need for further testing. A positive test may require similar genetic testing in the family, but it cannot predict the course or severity of a condition.

Negative genetic test: Failure of the laboratory to find a change in the gene, chromosome, or protein under consideration. This result can indicate that a person is not affected by a particular disease, is not a carrier of a specific genetic mutation, or does not have an increased risk of developing a certain disease. It is possible, however, that the test missed a disease-causing genetic alteration, as many genetic tests cannot detect all genetic changes, so further testing may be required to confirm a negative result.

Uninformative or indeterminate test result: Because everyone has common, natural variations in their DNA (polymorphisms) that do not affect health, a genetic test may find a change in DNA that has not been associated with a disorder in other people, and it can be difficult to tell whether the change is a natural polymorphism or a disease-causing mutation. This kind of uninformative test result cannot confirm or rule out a specific diagnosis.

Open reading frame (ORF): Any part of the genome that could be translated into a protein sequence due to an absence of stop codons (e.g., exon).

Single nucleotide polymorphisms (SNPs): The most common type of genetic variation among people. Each SNP represents a difference in a single DNA building block, the nucleotide. SNPs occur normally throughout a person's DNA—on average, once in every 300 nucleotides, so there are roughly 10 million SNPs in the human genome. Most have no effect on health or development, but if SNPs occur within a gene or in a regulatory region near a gene, they may play a more direct role in disease by affecting the gene's function.

Genome-wide association study (GWAS): Method of searching the genome for small variations, SNPs; these changes occur more frequently in people with a particular disease than in people without the disease. Each study can look at thousands of SNPs at the same time.

Whole exome sequencing and whole genome sequencing: Determining the order or sequence of DNA building blocks (nucleotides) in an individual is called *DNA sequencing*. Whole exome sequencing tests just the exons region (the exome), the protein-coding region of DNA that gives instructions to make proteins. (Exons are thought to make up just 1% of a person's genome.)

Whole genome sequencing tests both the exome and introns (non-coding region), as some DNA variations outside the exome can affect the gene activity and protein production and may lead to a genetic disorder.

Gene therapy: The treatment or prevention of diseases by inserting a gene into a patient's cell instead of using drugs or surgery. This therapy can include replacing a mutated gene with a healthy copy of the gene, inactivating/knocking out a mutated gene, or introducing a new gene into the body to help fight a disease. Because a gene that is inserted directly into a cell usually does not function, a carrier called a *vector* is genetically engineered to deliver the gene; certain viruses are often used as vectors as they can deliver the new gene by infecting the cell. Retroviruses integrate their genetic material into a chromosome in the human cell; adenoviruses introduce their DNA into the nucleus of the cell, but the DNA is not integrated into a chromosome. The vector can be injected into the vitreous cavity or the subretinal space.

Precision medicine: An emerging approach for disease treatment and prevention that takes into account individual variability in genes, environment, and lifestyle for each person. This approach will allow more accurate prediction of which treatment and prevention strategies for a particular disease will work in the individual person.

Spatially resolved techniques: The state-of-the-art approaches of deciphering tissue context with the preservation of spatial information. Traditional techniques, such as RNA-seq, can collect massive data from samples, but the geographic information of cells is lost. With well-designed targeting probes, these approaches can unveil the spatial heterogeneity of tissue context, such as spatial gradients of gene expression levels and distribution patterns of different cell types, and thus provide new insights into complex datasets.

Part II
X-Linked Forms

CHAPTER 8

X-Linked Retinitis Pigmentosa

Stephen H. Tsang and Tarun Sharma

General Features 38
Female Carriers 39
Molecular Genetics 40
RPGR Gene 40
RP2 Gene 41
Gene Therapy Trials 41
Did You Know? 41
Further Reading 41

S. H. Tsang (✉) · T. Sharma
Department of Ophthalmology, Columbia University,
New York, NY, USA
e-mail: sht2@cumc.columbia.edu

S. H. Tsang et al. (eds.), *Atlas of Inherited Retinal Diseases*, Advances in Experimental Medicine and Biology 1467, https://doi.org/10.1007/978-3-031-72230-1_8

General Features

X-linked retinitis pigmentosa (XLRP) is considered to be one of the most severe forms of retinitis pigmentosa (RP). It accounts for about 6–20% of all RP cases, including about 10% in the United States and 25% in England.

It primarily affects males. Patients usually present with night blindness in the first decade of life, with reduction in the visual fields in the second decade, and severe visual loss (less than about 20/200) by the fourth decade.

Fundi show all the characteristic features of RP: bone spicule–like pigment deposit in the retina (Fig. 8.1), attenuation of retinal arterioles, waxy pallor of the optic disc, posterior subcapsular lens opacities, and grossly reduced or absent amplitude on electroretinogram (ERG).

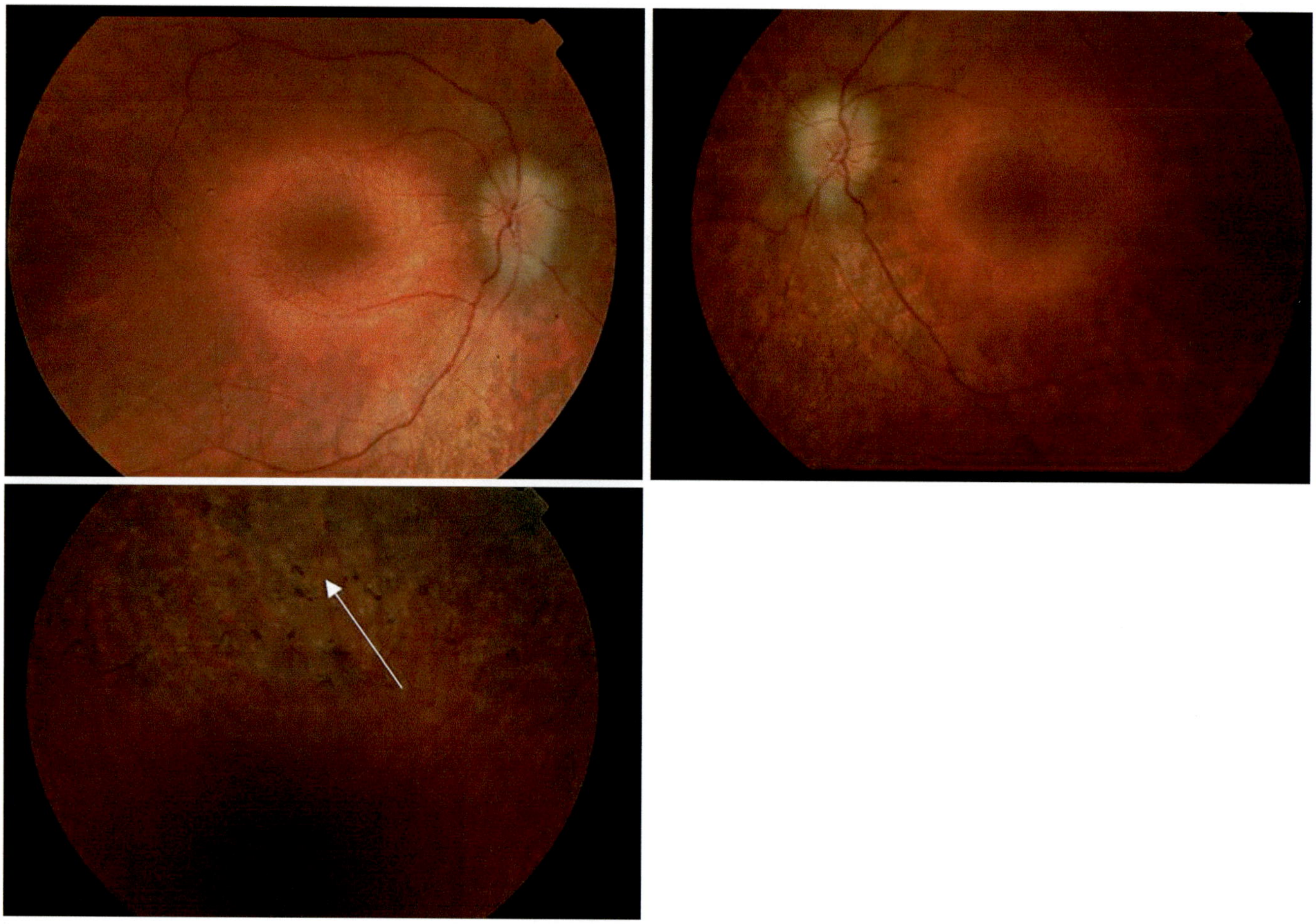

Fig. 8.1 X-linked retinitis pigmentosa X-linked retinitis pigmentosa (XLRP). Color fundus photographs show pigmentary changes (*arrow*) in the periphery, with a relatively healthy macula, in a 10-year-old boy

Female Carriers

The female carrier shows only peripheral retinal pigment epithelium (RPE) changes, including pigmentary deposits (Fig. 8.2). ERG may show a reduced amplitude to white light or delayed cone-wave implicit times (Fig. 8.3). The carrier may also show a characteristic tapetal reflex, a golden, metallic luster in the perimacular area (Fig. 8.4). About 50% of female carriers may have these abnormalities: tapetal reflex, pigmentary alteration, and ERG changes.

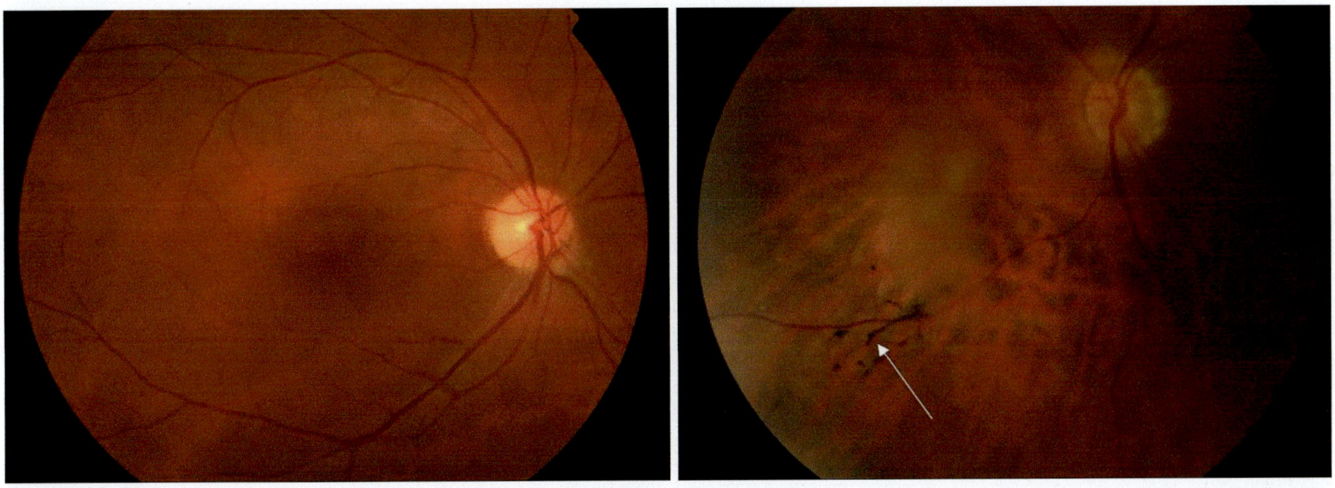

Fig. 8.2 Pigmentary deposits in a female carrier of X-linked retinitis pigmentosa. The mother of the patient in Fig. 8.1 was asymptomatic, but the peripheral fundus of her left eye shows pigmentary changes in the lower nasal quadrant

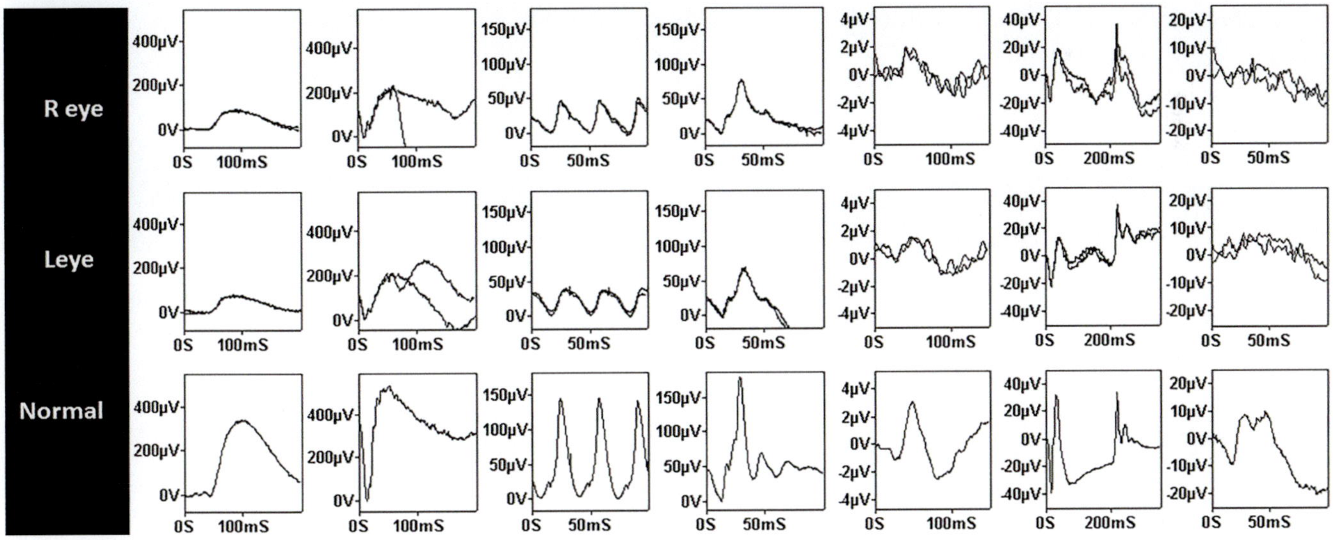

Fig. 8.3 Electroretinogram (ERG) in a female obligate carrier of X-linked retinitis pigmentosa. This ERG from the same asymptomatic mother as in Fig. 8.2 shows below-normal amplitude

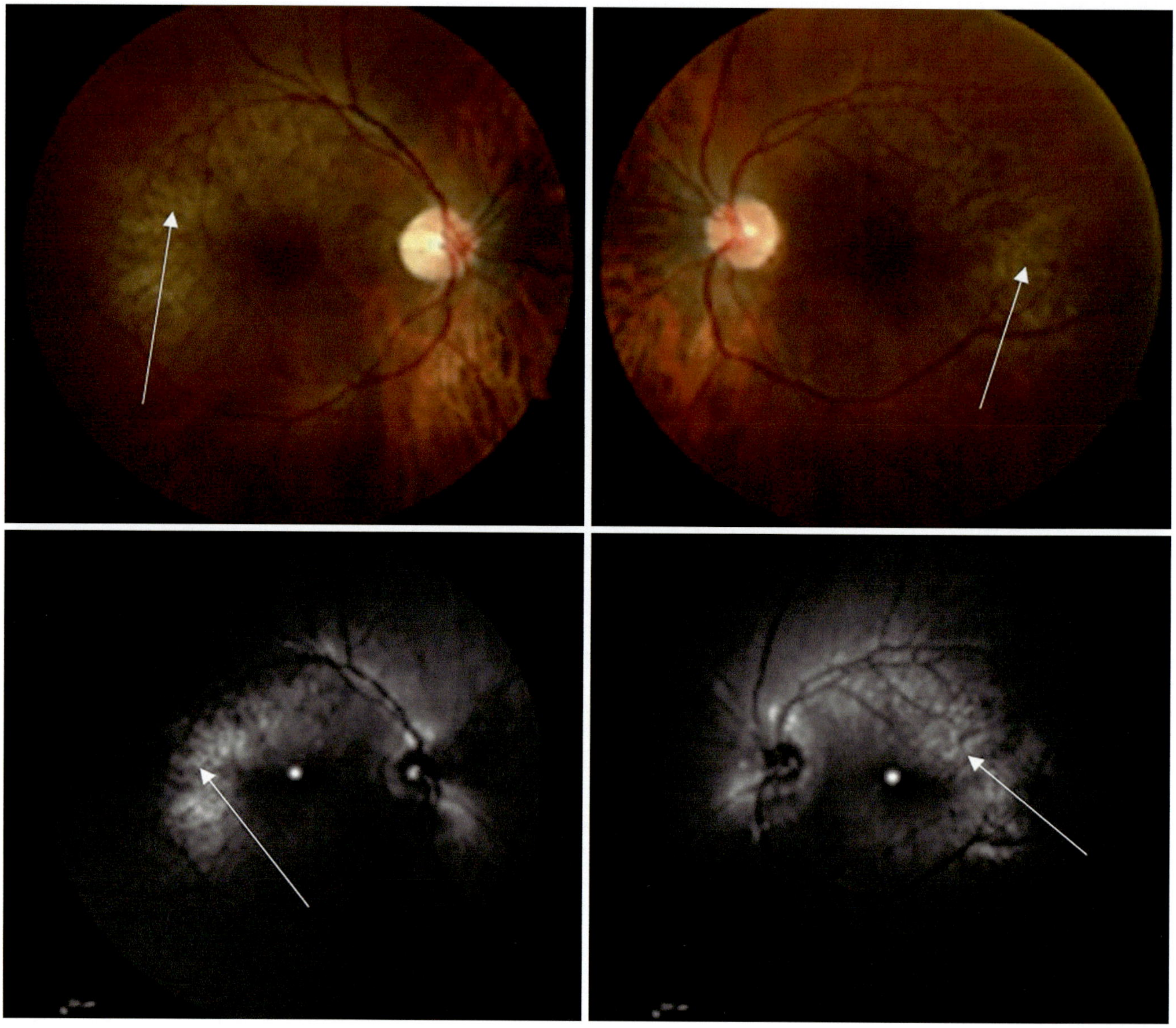

Fig. 8.4 Tapetal reflex in a female carrier of X-linked retinitis pigmentosa. The tapetal reflex can be appreciated in the perimacular region (upper temporal and temporal to fovea, *arrows*) as elongated lines formed by tiny, point-like reflexes

Molecular Genetics

The most frequent single cause of X-linked RP is a mutation in the Retinitis Pigmentosa GTPase Regulator gene (*RPGR*), accounting for about 15–20% of cases in Caucasians. The other gene involved in the mutation is *RP2*. Of about 300 mutations in *RPGR* gene, nearly 95% are associated with XLRP. *RPGR* mutations affect about 15,000 people in the United States and tens of thousands more around the world. Of the reported *RPGR* mutations, 35–55% occur in a glutamic acid–rich domain within exon open reading frame 15 (*ORF15*).

RPGR Gene

RPGR gene produces different versions of RPGR protein, one of which contains an ORF15 exon. This is expressed predominantly in the photoreceptors and plays an important role by regulating the function of cilia. Mutation results in an abnormally short, malfunctioning protein.

The cytogenetic location of the gene is Xp11.4 (short arm of the X chromosome at position 11.4).

RP2 Gene

The RP2 protein may be involved in transporting proteins within photoreceptor cells. More than 70 mutations have been identified in subjects with XLRP; the mutations result in a short version of the RP2 protein. A few mutations change single building blocks (amino acids) in the RP2 protein. The cytogenetic location of the gene is Xp11.3 (short arm of the X chromosome at position 11.3).

Gene Therapy Trials

The *ClinicalTrials.gov* website lists three ongoing or planned trials of gene therapy for XLRP (as of January 2, 2018):

- A Clinical Trial of Retinal Gene Therapy for X-linked Retinitis Pigmentosa (Recruiting, ClinicalTrials.gov Identifier: NCT03116113)
- Gene Therapy for X-linked Retinitis Pigmentosa (XLRP) Retinitis Pigmentosa GTPase Regulator (RPGR) (Recruiting, ClinicalTrials.gov Identifier: NCT03252847)
- Safety and Efficacy of rAAV2tYF-GRK1-RPGR in Subjects With X-linked Retinitis Pigmentosa Caused by *RPGR-ORF15* Mutations (Not yet recruiting, ClinicalTrials.gov Identifier: NCT03316560)

Did You Know?

- Although the *RPGR* mutation causes XLRP most of the time, a few mutations in the *ORF15* exon result in cone-rod dystrophy, cone dystrophy, and atrophic macular degeneration.
- Some affected individuals with the *RPGR* mutation have chronic respiratory and sinus infections, recurrent otitis media, and hearing loss. These effects could be due to disruption of the function of cilia.

Further Reading

Comander J, Weigel-DiFranco C, Sandberg MA, Berson EL. Visual function in carriers of X-linked retinitis pigmentosa. Ophthalmology. 2015;122:1899–906.

Deng WT, Dyka FM, Dinculescu A, Li J, Zhu P, Chiodo VA, *et al*. Stability and safety of an AAV vector for treating *RPGR-ORF15* X-linked retinitis pigmentosa. Hum Gene Ther. 2015;26:593–602.

Lyraki R, Megaw R, Hurd T. Disease mechanisms of X-linked retinitis pigmentosa due to *RP2* and *RPGR* mutations. Biochem Soc Trans. 2016;44:1235–44.

CHAPTER 9

X-Linked Choroideremia

Ashley Zhou, Stephen H. Tsang, Tarun Sharma, and Vlad Diaconita

General Features 44
Imaging and Tests 45
Female Carriers 48
Molecular Genetics 49
Gene Therapy Trials 49
Further Reading 49

A. Zhou
Vagelos College of Physicians and Surgeons,
New York, NY, USA
e-mail: az2682@cumc.columbia.edu

S. H. Tsang (✉) · T. Sharma · V. Diaconita
Department of Ophthalmology, Columbia University,
New York, NY, USA
e-mail: sht2@cumc.columbia.edu

S. H. Tsang et al. (eds.), *Atlas of Inherited Retinal Diseases*, Advances in Experimental Medicine and Biology 1467, https://doi.org/10.1007/978-3-031-72230-1_9

General Features

Choroideremia (CHM) is the most common X-linked hereditary choroidal dystrophy, characterized by progressive degeneration of the choriocapillaris, retinal pigment epithelium (RPE), and retina. Its prevalence is about 1 in 50,000–100,000 people.

Affected males present with impairment of night vision, usually in the first or second decade of life. Gradual peripheral field loss occurs over the next three to five decades. The disease starts with patchy areas of depigmentation ("salt and pepper" mottling) in the mid periphery; these spread centripetally. In the involved scalloped areas, there is diffuse atrophy of the choriocapillaris and RPE (Figs. 9.1 and 9.2). Underlying larger choroidal and overlying retinal blood vessels are preserved. Unlike in retinitis pigmentosa, patients with choroideremia do not show intraretinal pigmentation; the central visual acuity is preserved until the patient's 50s, and retinal blood vessels are not attenuated. The macular area remains intact until late in the disease course, with good visual acuity (Figs. 9.3 and 9.4). Responses on electroretinography (ERG) gradually become extinguished in the late stage. The end stage of choroideremia, however, resembles the end stage of RP.

Early stage choroideremia may mimic gyrate atrophy, in which the margin of the scalloped areas is well demarcated, with hyperpigmentation of the remaining RPE.

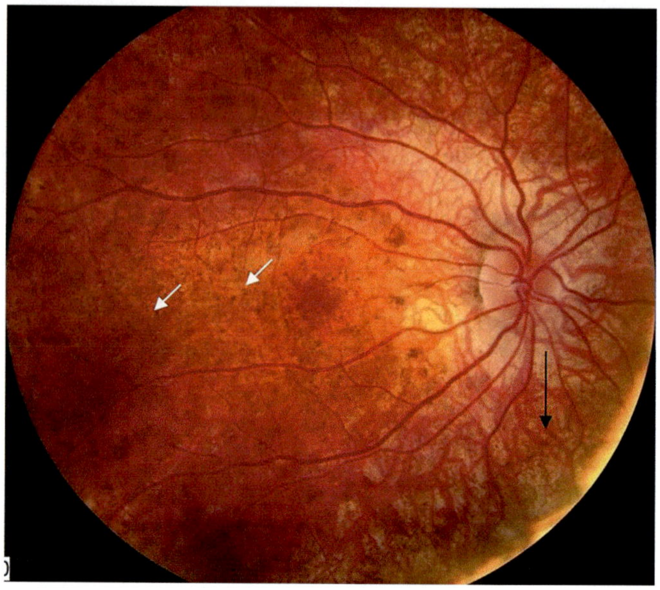

Fig. 9.1 Fundus of a 9-year-old choroideremia patient, showing "salt and pepper" appearance (*white arrows*) temporal to the macula and visualization of large choroidal vessels below the disc (*black arrow*)

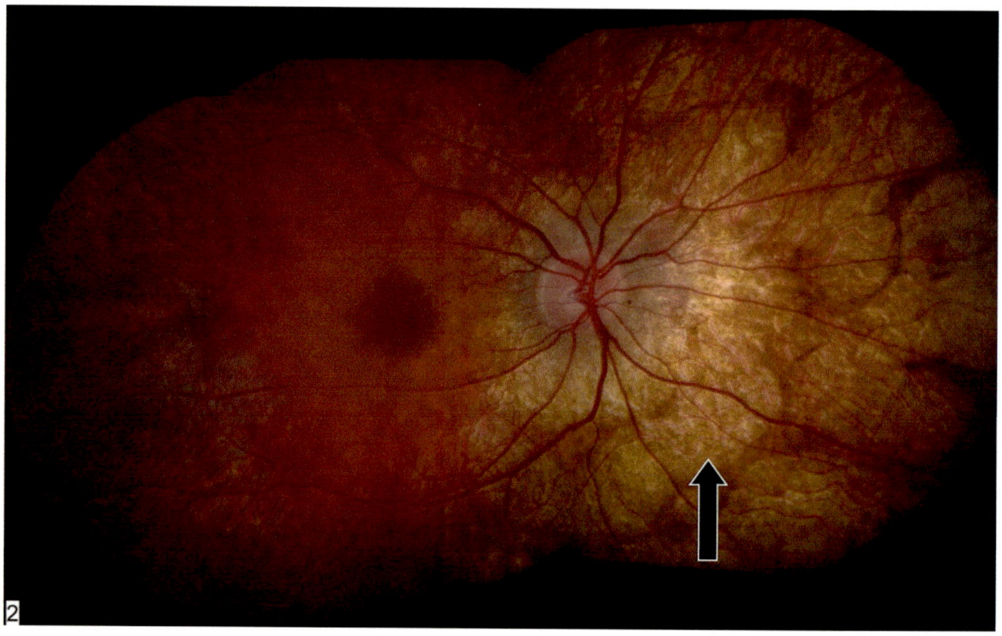

Fig. 9.2 Choroideremia. Right fundus showing patchy atrophy of the retinal pigment epithelium (RPE) (*arrow*)

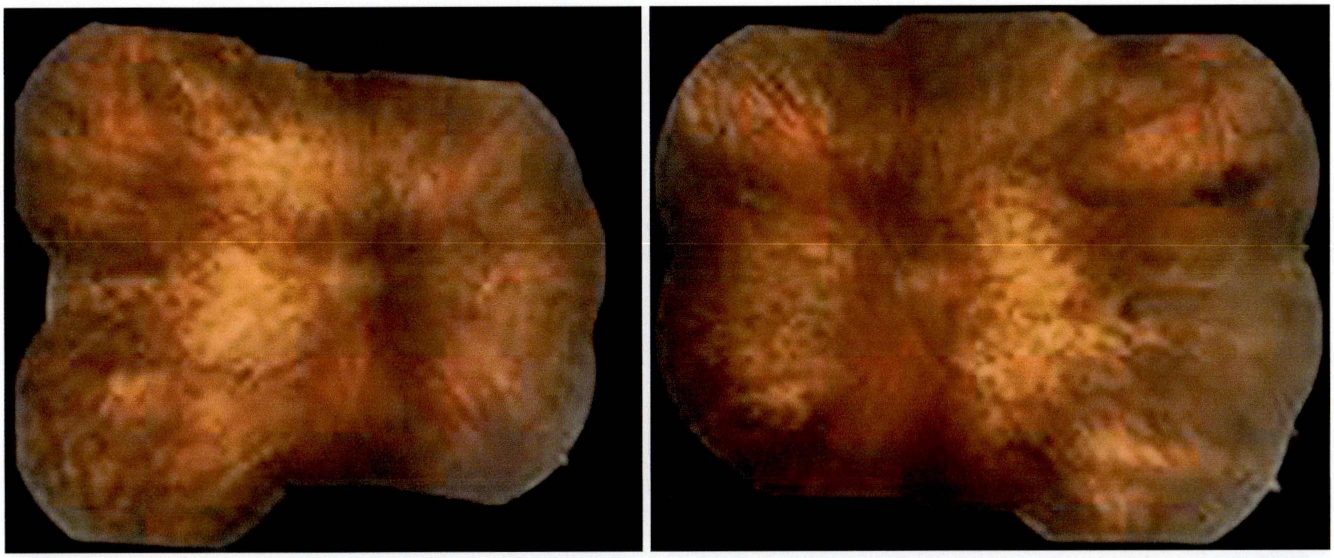

Fig. 9.3 In this older patient with choroideremia, both fundi show an almost symmetrical picture, with scalloped areas of chorioretinal atrophy. The macula is spared until the fifth to sixth decade of life

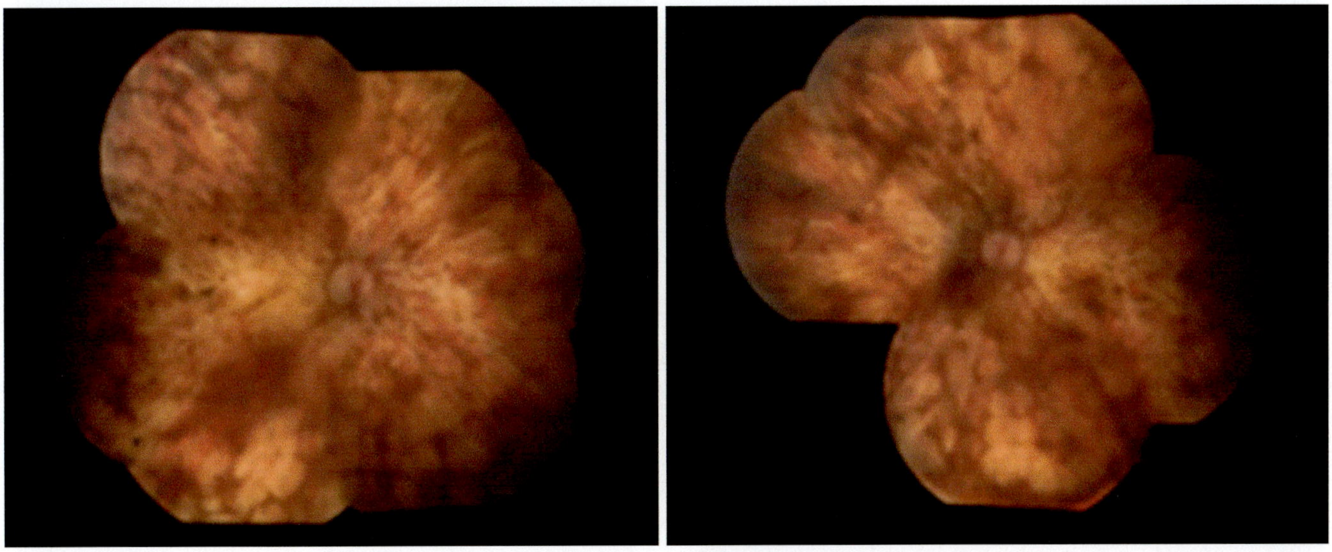

Fig. 9.4 An advanced case of choroideremia, involving the macula in the right eye and just abutting it in the left eye

Imaging and Tests

- **Fluorescein angiogram** (Fig. 9.5) shows diffuse loss of choriocapillaris throughout, except for the macular area. Hypofluorescent scalloped areas appear next to hyperfluorescent areas of perfused choriocapillaris.
- **Visual field** (Fig. 9.6) reveals ring scotoma at the mid-periphery which later progress to complete loss of the peripheral field.
- **Fundus autofluorescence (FAF)** (Fig. 9.7) typically shows loss of fluorescence in areas of chorioretinal atrophy starting in the periphery and an intact RPE/choriocapillaris patch in the macular area.
- **Optical coherence tomography (OCT)** (Fig. 9.8) images showing histopathological changes seen in choroideremia. The affected retina shows areas of RPE and choriocapillaris atrophy and corresponding thinning of outer retinal layers. In some cases, outer retinal tubulations occur, denoting end-stage photoreceptor degeneration.

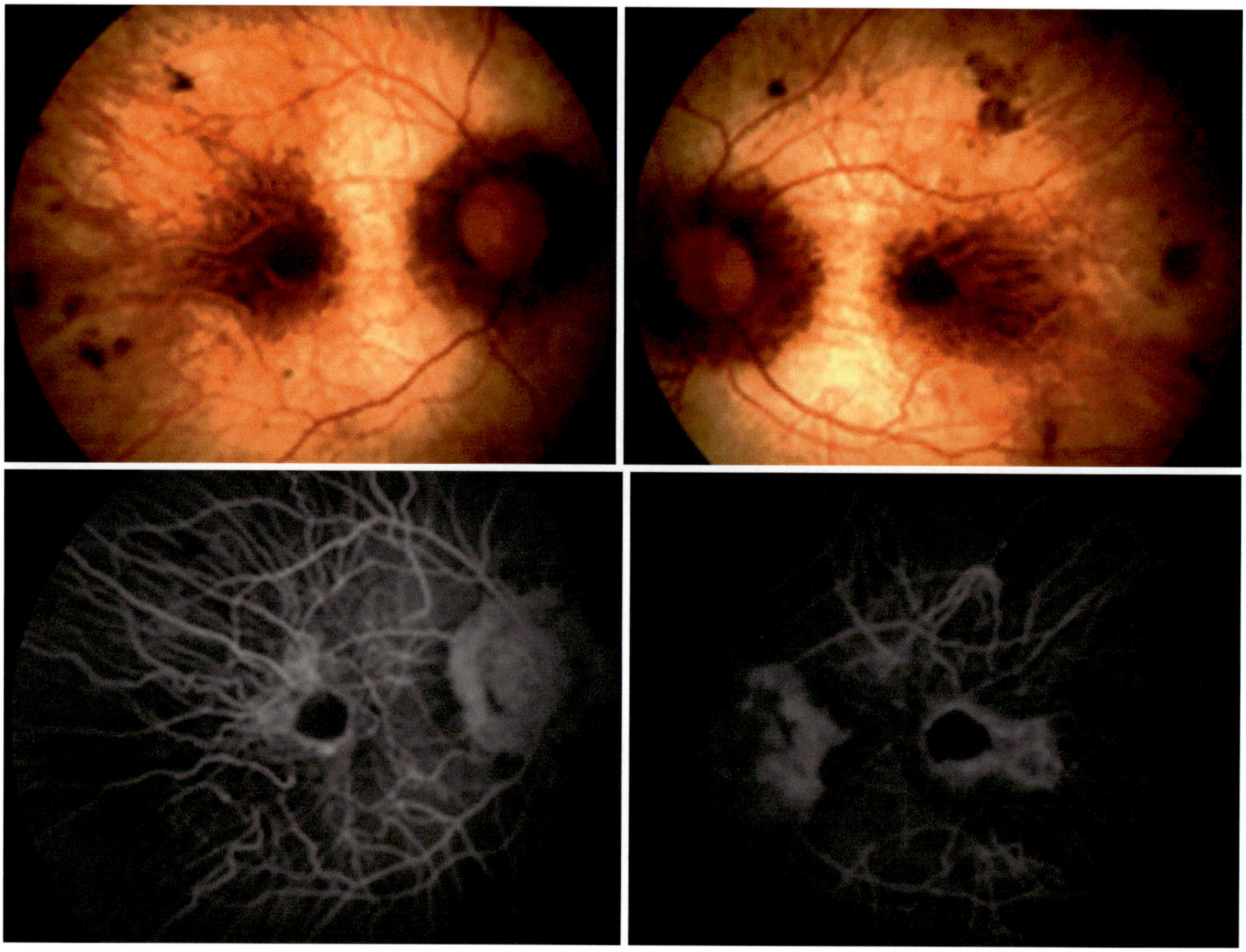

Fig. 9.5 Sparing of the macula in choroideremia; a fluorescein angiogram shows diffuse atrophy of the choriocapillaris throughout, except for the macula

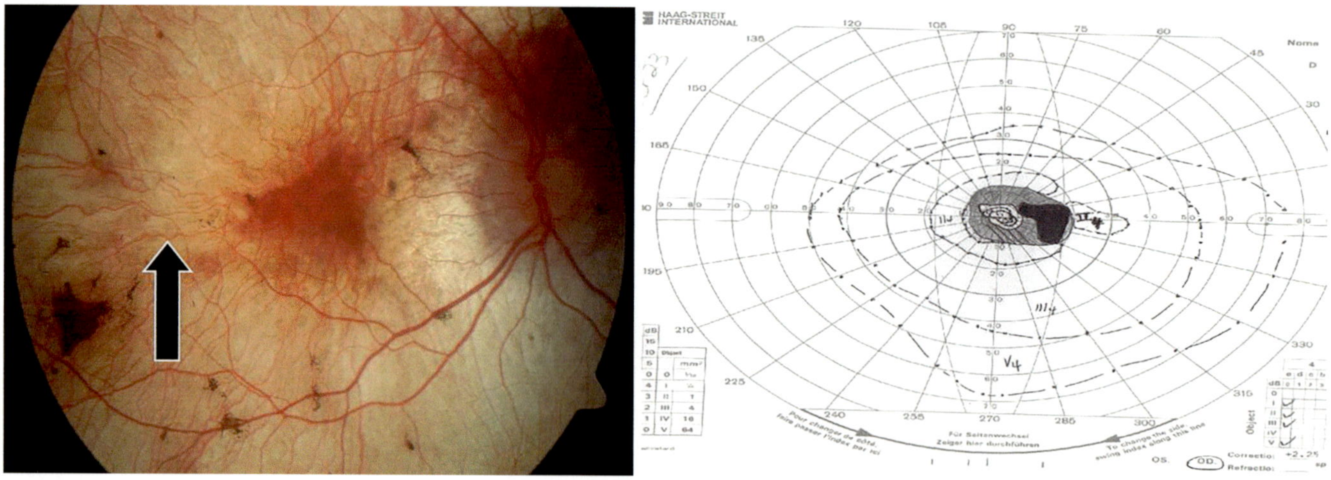

Fig. 9.6 Fundus showing severe atrophy of the RPE and choriocapillaris, allowing white sclera and choroidal blood vessels to be visible (*arrow*); the visual field shows scotomas (*shaded areas*)

Fig. 9.7 Fundus autofluorescence shows an intact RPE patch at the macula

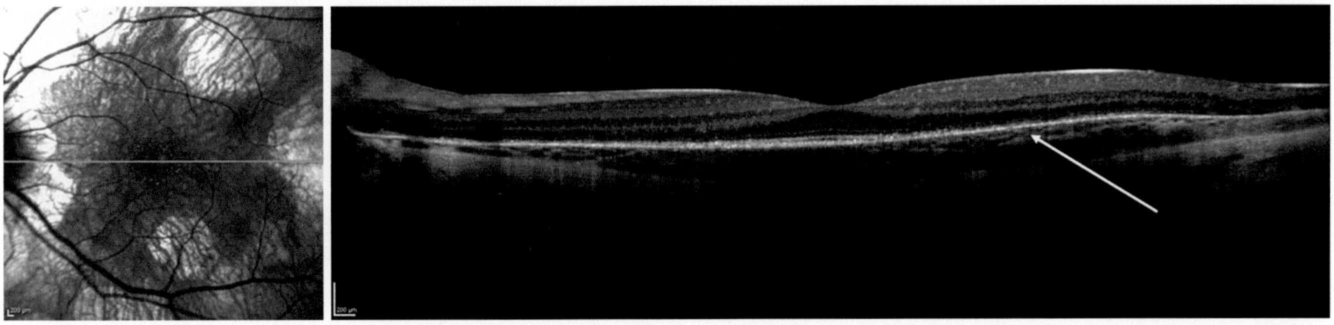

Fig. 9.8 Optical coherence tomogram (OCT) showing atrophy of the RPE and choriocapillaris (*arrow*)

Female Carriers

Heterozygous females may show RPE mottling in the mid-periphery ("moth-eaten" appearance) or radial bands that course from the equatorial retina toward periphery (Figs. 9.9 and 9.10), but their vision and ERG remain normal. Rarely, a female carrier might show changes like those of affected male patients, as a result of random inactivation of the wild type of X chromosome, known as *lyonization*.

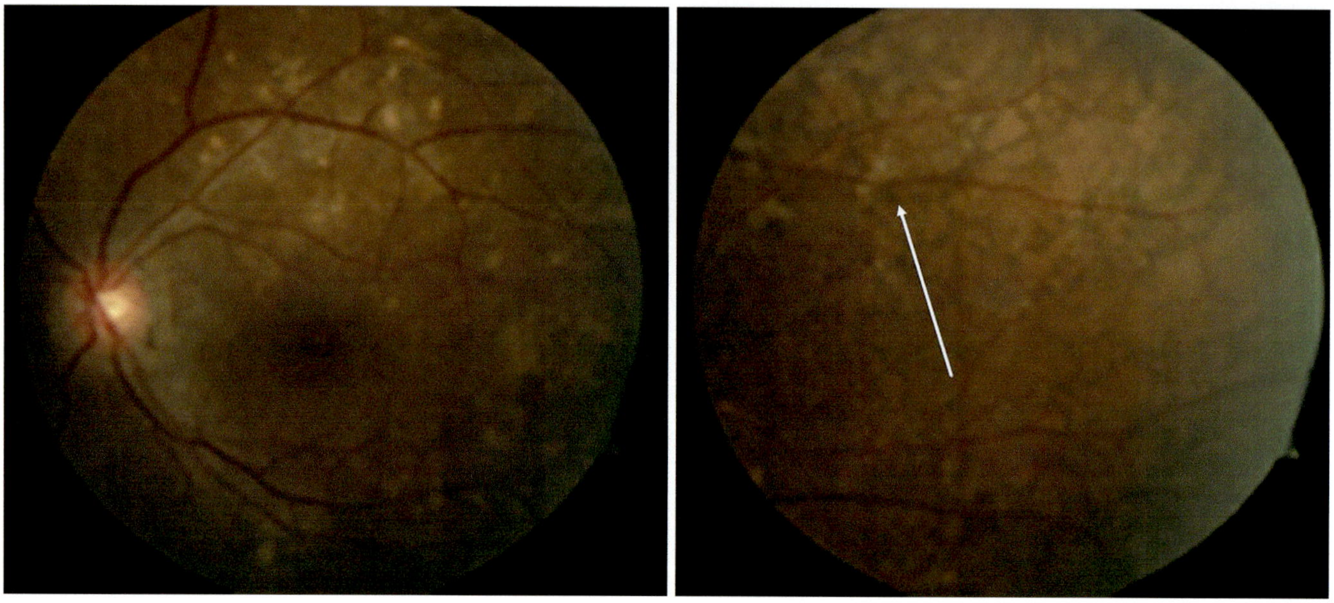

Fig. 9.9 In this female carrier of choroideremia, the fundus shows RPE mottling, predominantly in the periphery (*arrow*)

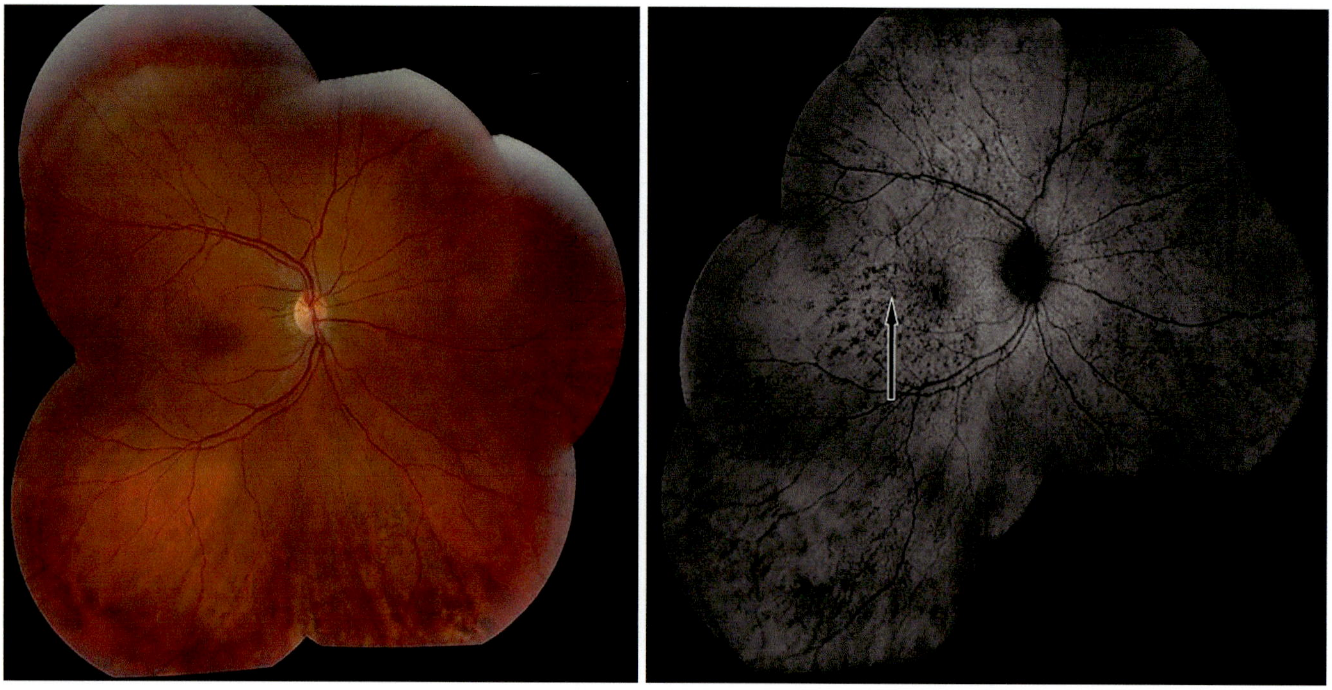

Fig. 9.10 In this female carrier of choroideremia, autofluorescence shows speckling of the RPE (*arrow*)

Molecular Genetics

Choroideremia is caused by a mutation in the *CHM* gene located on chromosome Xq21.2, which encodes for a geranylgeranyl transferase Rab escort protein-1 (REP-1 protein). Rab proteins are involved in intracellular trafficking; as an escort protein, REP-1 attaches to Rab proteins in the cells and directs them to the membranes of various cell organelles. Mutations in the REP-1 protein affect intracellular trafficking of proteins from the Golgi apparatus to the photoreceptor outer segment and RPE phagocytosis of shed outer segments, leading to premature cell death.

More than 280 mutations in the *CHM* gene have been found to cause choroideremia. Most mutations result in protein truncation by replacement of an arginine residue with a stop codon.

Gene Therapy Trials

The *ClinicalTrials.gov* website lists 16 trials of gene therapy for Choroideremia (as of July 8, 2025):

- Natural History of the Progression of Choroideremia Study (NIGHT) (Identifier: NCT03359551)
- "Natural History" Study of Choroideremia (Identifier: NCT02994368)
- REP1 Gene Replacement Therapy for Choroideremia (REGENERATE) (Identifier: NCT02407678)
- Long-term Safety and Efficacy Follow-up of BIIB111 for the Treatment of Choroideremia and BIIB112 for the Treatment of X-Linked Retinitis Pigmentosa (SOLSTICE) (Identifier: NCT03584165)
- Gene Therapy for Blindness Caused by Choroideremia (Identifier: NCT01461213)
- Choroideremia Gene Therapy Clinical Trial (Identifier: NCT02553135)
- A Safety Study of Retinal Gene Therapy for Choroideremia With Administration of BIIB111 (Identifier: NCT03507686)
- An Open Label Clinical Trial of Retinal Gene Therapy for Choroideremia (Identifier: NCT02077361)
- THOR—Tübingen Choroideremia Gene Therapy Trial (Identifier: NCT02671539)
- Dose Escalation Study of Intravitreal 4D-110 in Patients With Choroideremia (Identifier: NCT04483440)
- Efficacy and Safety of BIIB111 for the Treatment of Choroideremia (Identifier: NCT03496012)
- REP1 Gene Replacement Therapy for Choroideremia (Identifier: NCT02407678)
- Safety and Dose Escalation Study of AAV2-hCHM in Subjects With CHM (Choroideremia) Gene Mutations (Identifier: NCT02341807)
- Study to Evaluate Safety of RTx-015 Injection in Retinitis Pigmentosa or Choroideremia Patients (ENVISION) (Identifier: NCT06460844) A Phase I/ II Doseescalating Study of the Safety, Tolerability and Efficacy of KIO-301 Administered Intravitreally to Patients with Retinitis Pigmentosa and Choroideremia (ABACUS) (Identifier: NCT05282953) The Dark-Adapted Retinal Function Response in Choroideremia (DARC) Study (DARC) (Identifier: NCT05045703) Natural Disease Progression in Participants With Choroideremia (Identifier: NCT04795206)

Further Reading

Jacobson SG, Cideciyan AV, Sumaroka A, Aleman TS, Schwartz SB, Windsor EA, *et al.* Remodeling of the human retina in choroideremia: rab escort protein 1 (REP-1) mutations. Invest Ophthalmol Vis Sci. 2006;47:4113–20.

Kärnä J. Choroideremia. A clinical and genetic study of 84 Finnish patients and 126 female carriers. Acta Ophthalmol Suppl. 1986;176:1–68.

MacLaren RE, Groppe M, Barnard AR, Cottriall CL, Tolmachova T, Seymour L, *et al.* Retinal gene therapy in patients with choroideremia: initial findings from a phase 1/2 clinical trial. Lancet. 2014;383:1129–37.

Mura M, Sereda C, Jablonski MM, MacDonald IM, Iannaccone A. Clinical and functional findings in choroideremia due to complete deletion of the CHM gene. Arch Ophthalmol. 2007;125:1107–13.

CHAPTER 10

X-Linked Juvenile Retinoschisis

Leticia Pinheiro, Stephen H. Tsang, and Tarun Sharma

General Features 52
Imaging and Tests 53
Molecular Genetics 55
Treatments 55
Oral Acetazolamide 55
Vitreoretinal Surgery 55
Gene Therapy 56
Further Reading 56

L. Pinheiro
Federal University of Goiás,
Goiânia, Brazil

S. H. Tsang (✉) · T. Sharma
Department of Ophthalmology, Columbia University,
New York, NY, USA
e-mail: sht2@cumc.columbia.edu

S. H. Tsang et al. (eds.), *Atlas of Inherited Retinal Diseases*, Advances in Experimental Medicine and Biology 1467, https://doi.org/10.1007/978-3-031-72230-1_10

General Features

X-linked juvenile retinoschisis (XLRS) occurs exclusively in males and is characterized by visual loss that begins in early childhood; patients are usually school-age and are experiencing visual disturbances, especially in reading. The prevalence is estimated to be 1 in 5000–25,000 men, worldwide. XLRS has complete penetrance but variable expressivity. Carrier females generally remain asymptomatic.

The characteristic finding is foveal schisis: stellate or fine radial striae in a sort of spoke-like pattern (Fig. 10.1) due to small cystoid spaces. Unlike cystoid macular edema, these spaces do not leak on a fluorescein angiogram. The peripheral retina may show retinal pigment epithelial (RPE) alteration (Fig. 10.2). The splitting occurs predominantly within the inner retinal layers (neural retina) (Fig. 10.3). Visual acuity remains stable in most instances to 20/200.

Peripheral retinoschisis is present in about half of patients, and is visible as large holes, which may be so large as to extend from the arcade to the equator. In the inner retina, the retinal blood vessels are freely floating. Rarely, these vessels rupture and patients present with vitreous hemorrhage; retinal detachment occurs infrequently.

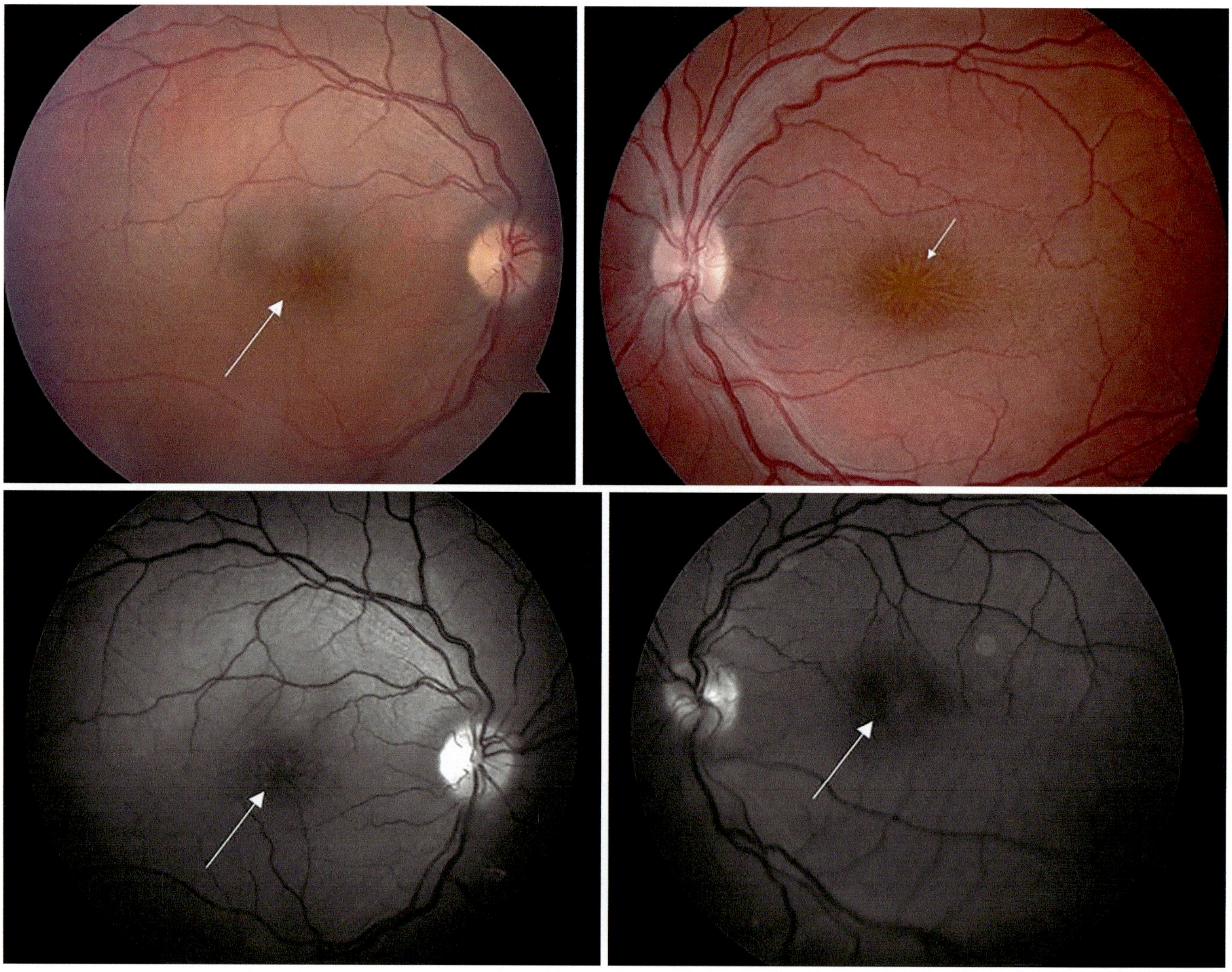

Fig. 10.1 X-linked juvenile retinoschisis. The central macula shows cystic changes around the fovea, in a spoke-like pattern (*arrows*)

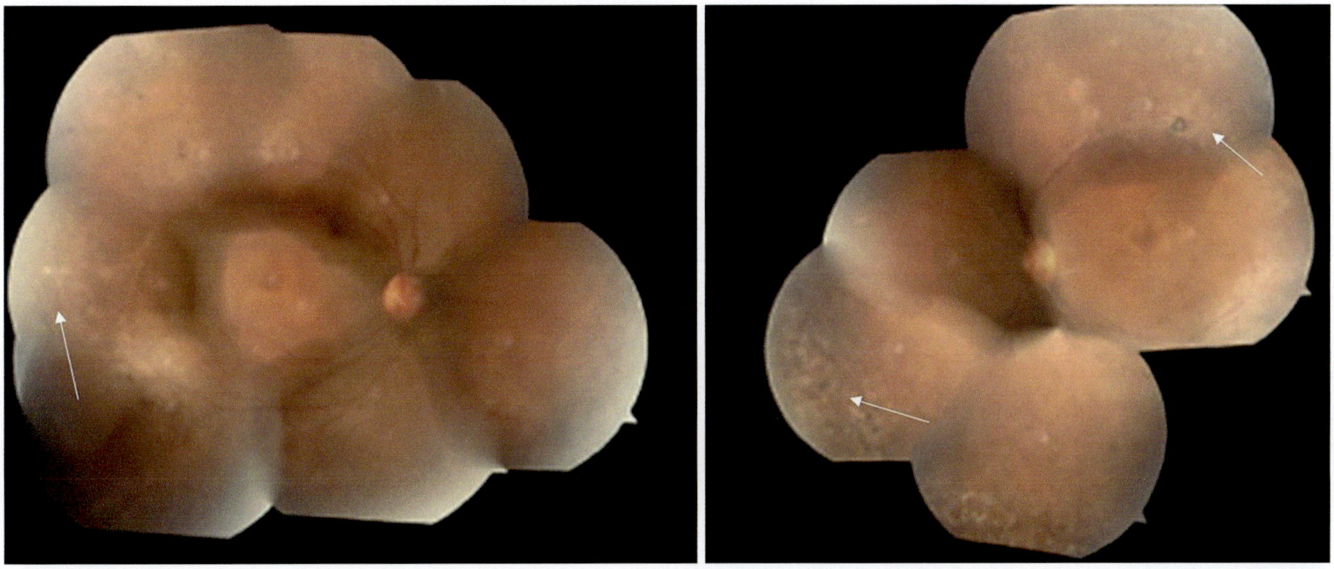

Fig. 10.2 X-linked juvenile retinoschisis. Besides cystic changes at the macula, the peripheral retina shows retinal pigment epithelial (RPE) alteration or mottling (*arrows*)

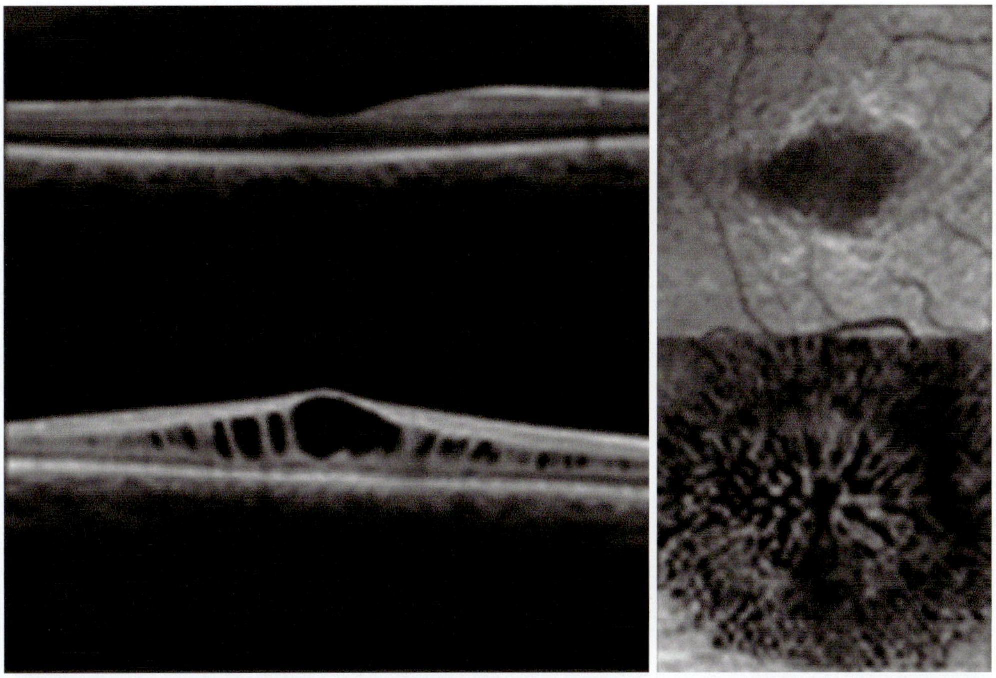

Fig. 10.3 Conventional and en-face optical coherence tomography (OCT) of a normal control (*upper row*) and a patient with (XLRS) (*lower row*). The conventional OCT (*lower left*) shows inner layer schisis, and en-face OCT (*lower right*) shows a cystic labyrinth

Imaging and Tests

- **Fundus autofluorescence (FAF)** shows modification of normal foveal autofluorescence with a radial pattern (Fig. 10.4). Over time, FAF can document progression in terms of RPE changes (Fig. 10.5).
- **Electroretinography (ERG)**: Negative waveform. The a-wave is preserved, but the amplitude of the b-wave is reduced; the alteration of the ratio of these waves (a-wave amplitude exceeding the b-wave amplitude) is an important diagnostic feature (Fig. 10.6). Multifocal ERG also demonstrates reduced amplitude, along with longer implicit time in the central macular area.

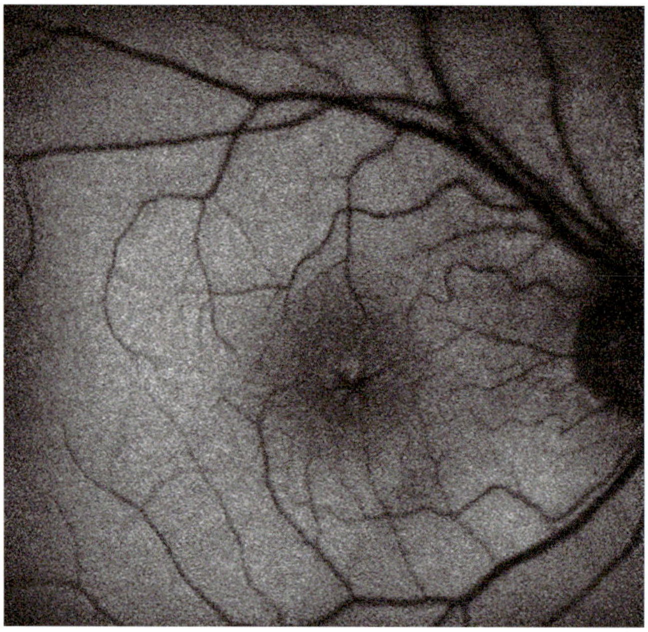

Fig. 10.4 Fundus autofluorescence (FAF) shows a radial pattern of hypofluorescence around the fovea in a patient with XLRS

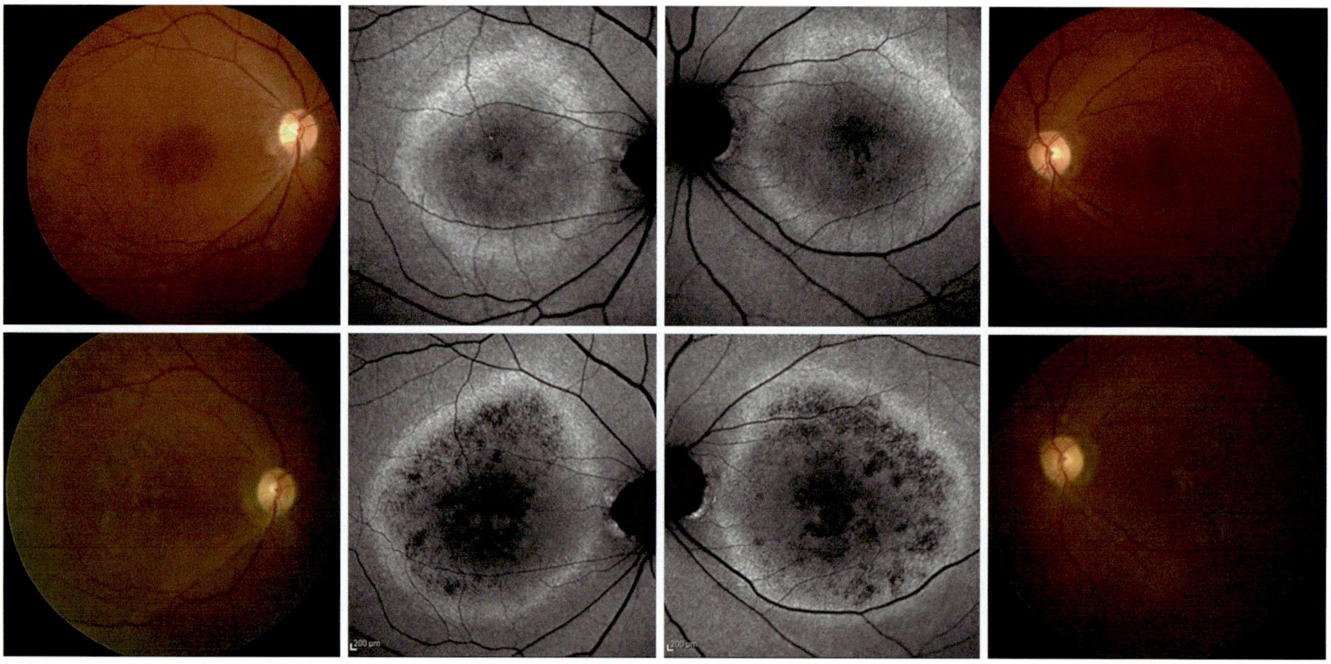

Fig. 10.5 Progression of XLRS as seen from an autofluorescence ring and RPE alteration

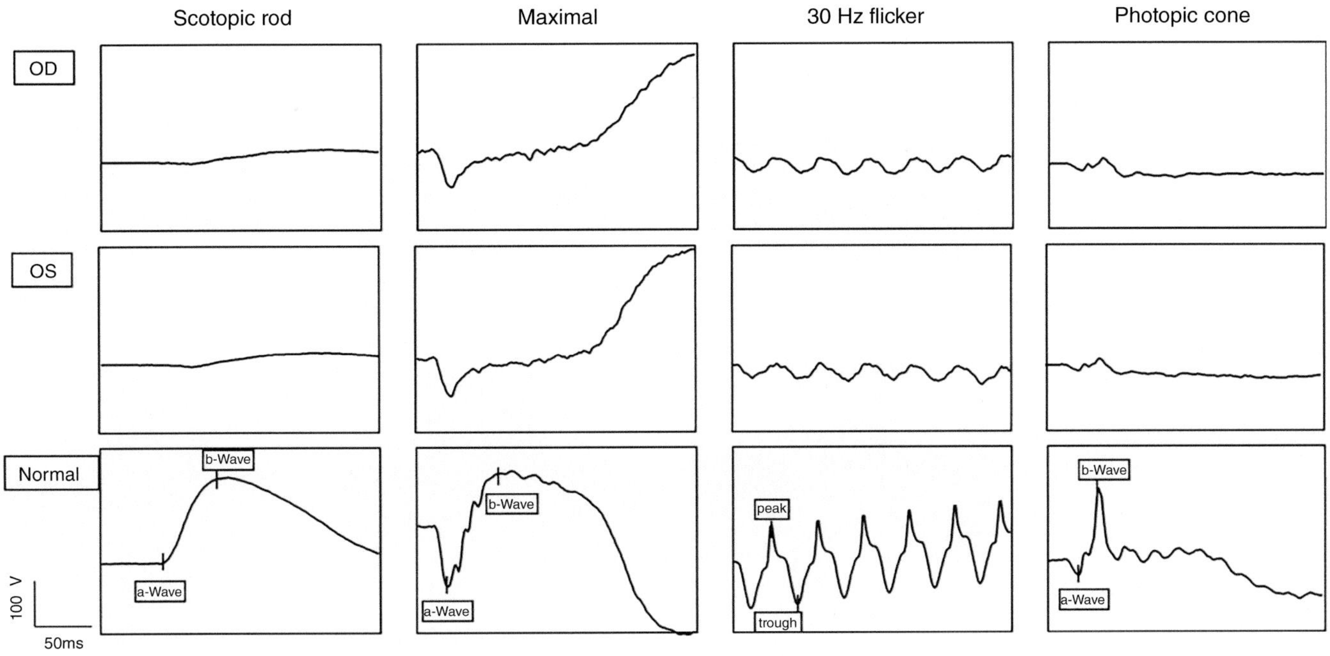

Fig. 10.6 Negative electroretinogram (ERG) waveform in a patient with XLRS: The a-wave amplitude is more than the b-wave amplitude

Molecular Genetics

The gene responsible for X-linked juvenile retinoschisis is *RS1*, which encodes a protein called retinoschisin; this protein is important for cell adhesion among all neuronal cells, including Müller cells. *RS1* gene mutations cause a complete loss of functional retinoschisin protein, so the maintenance and organization of cells in the retina is disrupted. As a result, the retina splits.

About 220 mutations of the *RS1* gene have been reported. Most mutations change one protein building block (amino acid) in the retinoschisin protein. The cytogenetic location is Xp22.13, the short arm of the X chromosome at position 22.13.

Treatments

Oral Acetazolamide

In selected patients, cystic spaces respond to oral acetazolamide and become less in size (Fig. 10.7).

Vitreoretinal Surgery

Patients who develop vitreous hemorrhage or rhegmatogenous retinal detachment may benefit from vitreoretinal surgery.

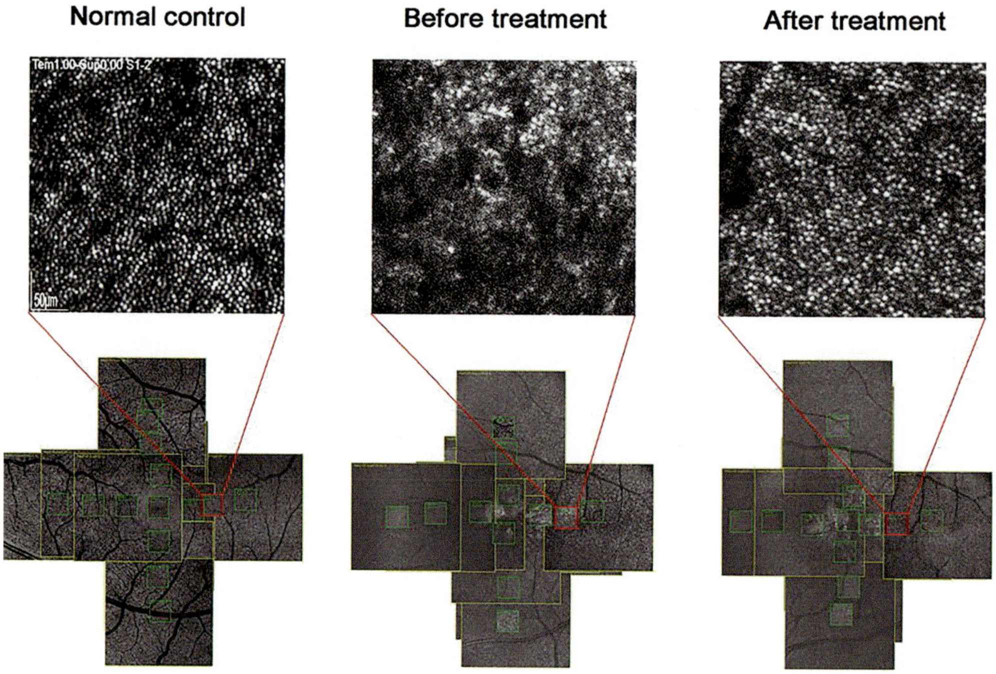

Fig. 10.7 Adaptive optics scanning laser ophthalmoscopy (SLO) showing imaging of a cone mosaic after treatment of XLRS with acetazolamide; the cone mosaic could not be imaged prior to treatment because of cystic cavities in the inner retina

Gene Therapy

The *ClinicalTrials.gov* website lists three trials of gene therapy for XLRS that are currently recruiting (as of January 2, 2018):

- Safety and Efficacy of rAAV-hRS1 in Patients With X-linked Retinoschisis (XLRS) (ClinicalTrials.gov Identifier: NCT02416622)
- Clinical and Genetic Studies of X-Linked Juvenile Retinoschisis (ClinicalTrials.gov Identifier: NCT00055029)
- Study of RS1 Ocular Gene Transfer for X-linked Retinoschisis (ClinicalTrials.gov Identifier: NCT02317887)

Further Reading

Apushkin AM, Fisman GA. Use of dorzolamide for patients with X-linked retinoschisis. Retina. 2006;26:741–5.
Molday RS, Kellner U, Weber BHF. X-linked juvenile retinoschisis: clinical diagnosis, genetic analysis, and molecular mechanisms. Prog Retin Eye Res. 2012;31:195–212.
Tsang SH, Vaclavik V, Bird AC, Robson AG, Holder GE. Novel phenotypic and genotypic findings in X-linked retinoschisis. Arch Ophthalmol. 2007;125:259–67.

CHAPTER 11

X-Linked Ocular Albinism

Benjamin Kuang-Chien Chiang, Stephen H. Tsang, and Tarun Sharma

General Features . . . 58
Female Carriers . . . 58
Oculocutaneous Albinism . . . 60
Molecular Genetics . . . 60
Ocular Albinism . . . 60
Oculocutaneous Albinism . . . 60
Further Reading . . . 60

B. K.-C. Chiang
School of Medicine, College of Medicine,
National Taiwan University, Taipei, Taiwan

S. H. Tsang (✉) · T. Sharma
Department of Ophthalmology, Columbia University,
New York, NY, USA
e-mail: sht2@cumc.columbia.edu

S. H. Tsang et al. (eds.), *Atlas of Inherited Retinal Diseases*, Advances in Experimental Medicine and Biology 1467, https://doi.org/10.1007/978-3-031-72230-1_11

General Features

The prevalence of X-linked ocular albinism (XLOA) is about 1 in 60,000 males. It affects only the eyes; the color of the skin and hairs are normal. Patients usually present with reduced vision, photophobia, nystagmus, and strabismus. Many patients have problem in perceiving depth (stereoscopic vision). The visual loss is permanent, but XLOA is a nonprogressive disorder and visual acuity remains stable throughout life.

Slit-lamp examination reveals a hypopigmented iris showing transillumination defect. The fundus shows hypopigmented fundus revealing underlying large choroidal blood vessels and foveal hypoplasia with no foveal reflex and no luteal pigment (Fig. 11.1). Flash and pattern visual evoked potential (VEP) shows an asymmetric occipital response, in view of the greater number of decussating fibers at the optic chiasma.

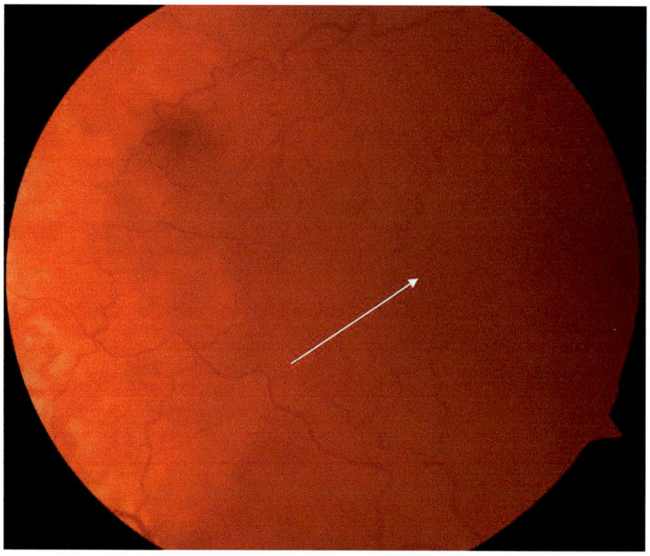

Fig. 11.1 A case of ocular albinism, with absent foveal reflex (*arrow*) and failure of retinal vasculature to wreathe the fovea

Female Carriers

A heterozygous female carrier of ocular albinism may show iris transillumination and a pattern of hypopigmentation and hyperpigmentation in the fundus, in a sort of mosaic pattern (Fig. 11.2).

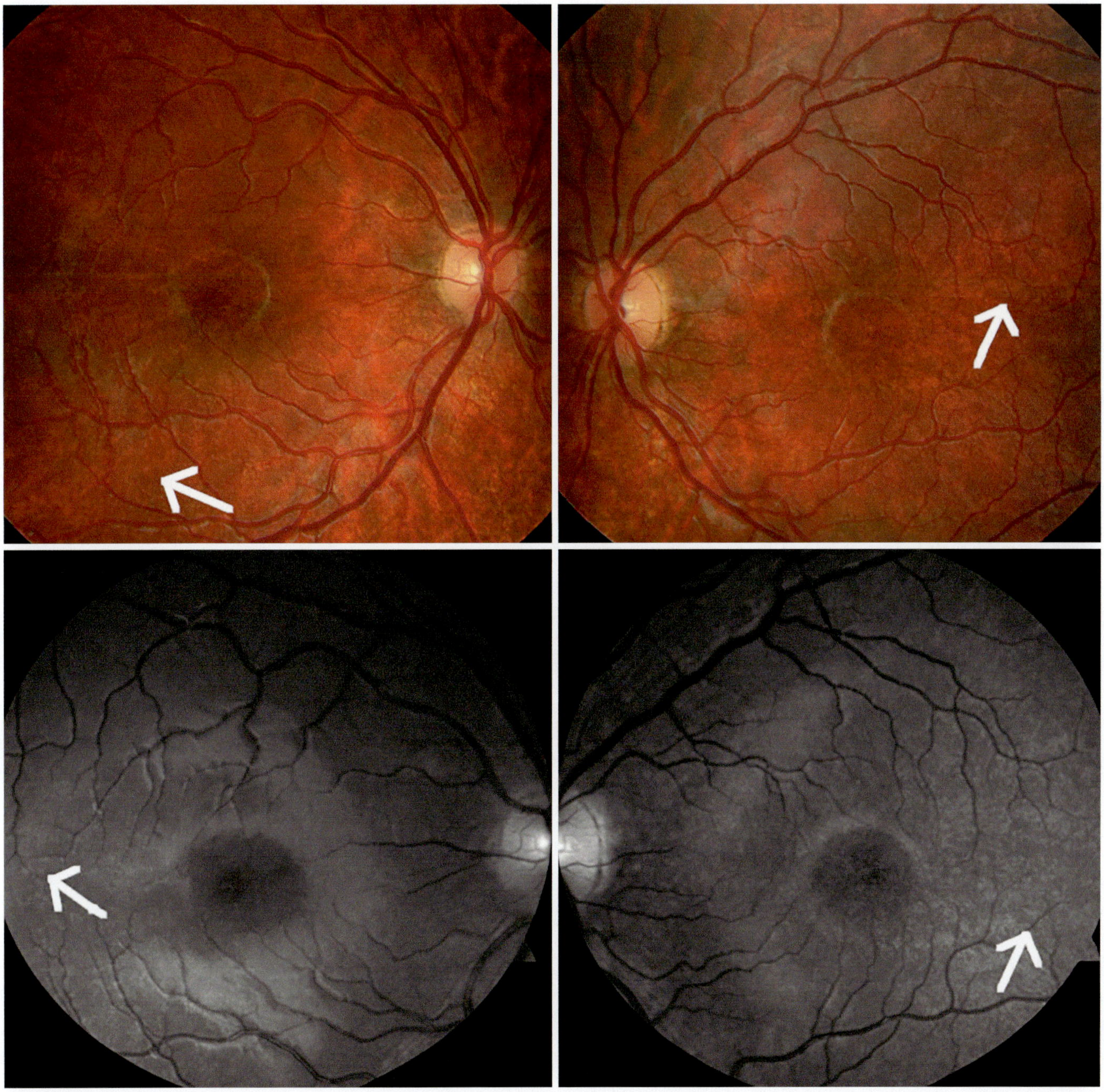

Fig. 11.2 Carrier of ocular albinism, showing areas of retinal pigment epithelial (RPE) mosaicism on color photographs (*arrows*). Autofluorescence shows radially distributed reflex as "mud-splatter" (*arrows*)

Oculocutaneous Albinism

Oculocutaneous albinism (OCA) is inherited in an autosomal recessive fashion; the estimated prevalence is about 1 in 20,000 people. In OCA, there is either a reduction or a complete absence of melanin pigment in the skin, hair, and eyes. Affected individuals have increased risk of skin damage (including skin melanoma) on long-term sun exposure.

Several types of OCA have been identified, depending on the gene and mutation involved (see below):

- Type 1: White hair, pale skin, and light-colored irises
- Type 2: Less severe than type 1; the skin is creamy white, and the hair is light yellow, blond, or light brown
- Type 3: Usually affects dark-skinned individuals and is also known as rufous oculocutaneous albinism; individuals have reddish-brown skin, red hair, and brown irises. Visual problems are milder than in other forms.
- Type 4: Features are similar to Type 2.

OCA is associated with two syndromes:

- **Hermansky-Pudlak syndrome** can feature easy bruising and severe bleeding due to a defect in platelet adhesion.
- **Chédiak–Higashi syndrome** can cause severe infection due to neutropenia or bleeding due to thrombocytopenia.

Molecular Genetics

Ocular Albinism

Ocular albinism is caused by mutations in the G protein-coupled receptor 143 (*GPR143*) gene, also known as *OA1*. Most of the mutations in the *GPR143* gene alter the size or shape of the GPR143 protein; this protein controls the growth of melanosomes, a cellular structure that produces and stores melanin pigment. Without functional GPR143 protein, melanosomes can grow abnormally large (macromelanosomes); how these large melanosomes are related to vision loss and other eye abnormalities is not yet known.

More than 60 *GPR143* mutations have been identified, resulting in the most common form of ocular albinism, called the Nettleship-Falls type or type 1. The cytogenetic location is Xp22.2, the short arm of the X chromosome at position 22.2.

Oculocutaneous Albinism

The enzyme tyrosinase is produced by the *TYR* gene, and this enzyme (located in the melanocytes) is needed for the synthesis of melanin. Tyrosinase converts a protein building block, the amino acid tyrosine, to another compound called dopaquinone, and a series of other reactions converts dopaquinone to melanin. The cytogenetic location is 11q14.3, the long arm of chromosome 11 at position 14.3.

Various mutations in *TYR* are responsible for the types of OCA:

Type 1: *TYR* gene
Type 2: *OCA2* gene
Type 3: *TYRP1* gene
Type 4: *SLC45A2* gene

Further Reading

Charles SJ, Green JS, Grant JW, Yates JR, Moore AT. Clinical features of affected males with X linked ocular albinism. Br J Ophthalmol. 1993;77:222–7.

Rosenberg T, Schwartz M. X-linked ocular albinism: prevalence and mutations—a national study. Eur J Hum Genet 1998;6:570–7.

CHAPTER 12

Progressive Cone Dystrophy and Cone-Rod Dystrophy

Benjamin Kuang-Chien Chiang, Stephen H. Tsang, Tarun Sharma, and Vlad Diaconita

General Features . . . 62
Molecular Genetics . . . 65
Specific Genotype and Phenotype Correlation . . . 65
ABCA4 . . . 65
ADAM9 (CORD9) . . . 66
RPGR1P1 (CORD13, Leber Congenital Amaurosis [LCA6]) . . . 66
KCNV2 (Potassium Voltage-Gated Channel Modifier Subfamily V Member 2) . . . 66
SEMA4A (Semaphoring 4A) (CORD10 or RP35) . . . 66
GUCY2D (CORD6 or LCA1 or RCD2) . . . 66
GUCA1A (Guanylate Cyclase Activator 1A) (COD4 or CORD14) . . . 66
PRPH2 (Peripherin 2 or RDS) . . . 66
CRX (CORD2 or LCA7 or CRD) . . . 66
RIMS1 (Regulating Synaptic Membrane Exocytosis 1, CORD7) . . . 67
PITPNM3 (CORD5) . . . 67
RPGR (COD1 or CORCX1 or RP3 or RP15 or XLRP3) . . . 67
DRAM2 (Damage-Regulated Autophagy Modulator 2, CORD 21) . . . 67
Further Reading . . . 67

B. K.-C. Chiang
School of Medicine, College of Medicine,
National Taiwan University, Taipei, Taiwan

S. H. Tsang (✉) · T. Sharma · V. Diaconita
Department of Ophthalmology,
Columbia University, New York, NY, USA
e-mail: sht2@cumc.columbia.edu

S. H. Tsang et al. (eds.), *Atlas of Inherited Retinal Diseases*, Advances in Experimental Medicine and Biology 1467, https://doi.org/10.1007/978-3-031-72230-1_12

General Features

A heterogenous group of diseases, progressive cone dystrophy usually begins in the mid-teenage years or later in life. The estimated prevalence is 1 in 30,000–40,000 individuals. Patients usually present with decreased central vision and a color vision deficit; the visual loss is progressive and often accompanied by day blindness (hemeralopia) and light intolerance (photophobia). Over time, affected individuals develop night blindness and loss of peripheral field. Visual acuity deteriorates to 20/200 or even counting fingers. There is some association between X-linked cone-rod dystrophy (CORD) and high myopia.

Fundi in the early stage may be normal, but later they may show a symmetric bull's-eye pattern (bull's-eye maculopathy, BEM) of macular dystrophy (Figs. 12.1, 12.2, 12.3, and 12.4). Fundus autofluorescence (FAF) shows alternating areas of hypoautofluorescence and hyperautofluorescence; as atrophy of the retinal pigment epithelium (RPE) sets in, more hypoautofluorescence is evident (Figs. 12.1, 12.2, and 12.3). Optical coherence tomography (OCT) shows foveal thinning and attenuation of the inner segment ellipsoid zone (Figs. 12.2 and 12.3).

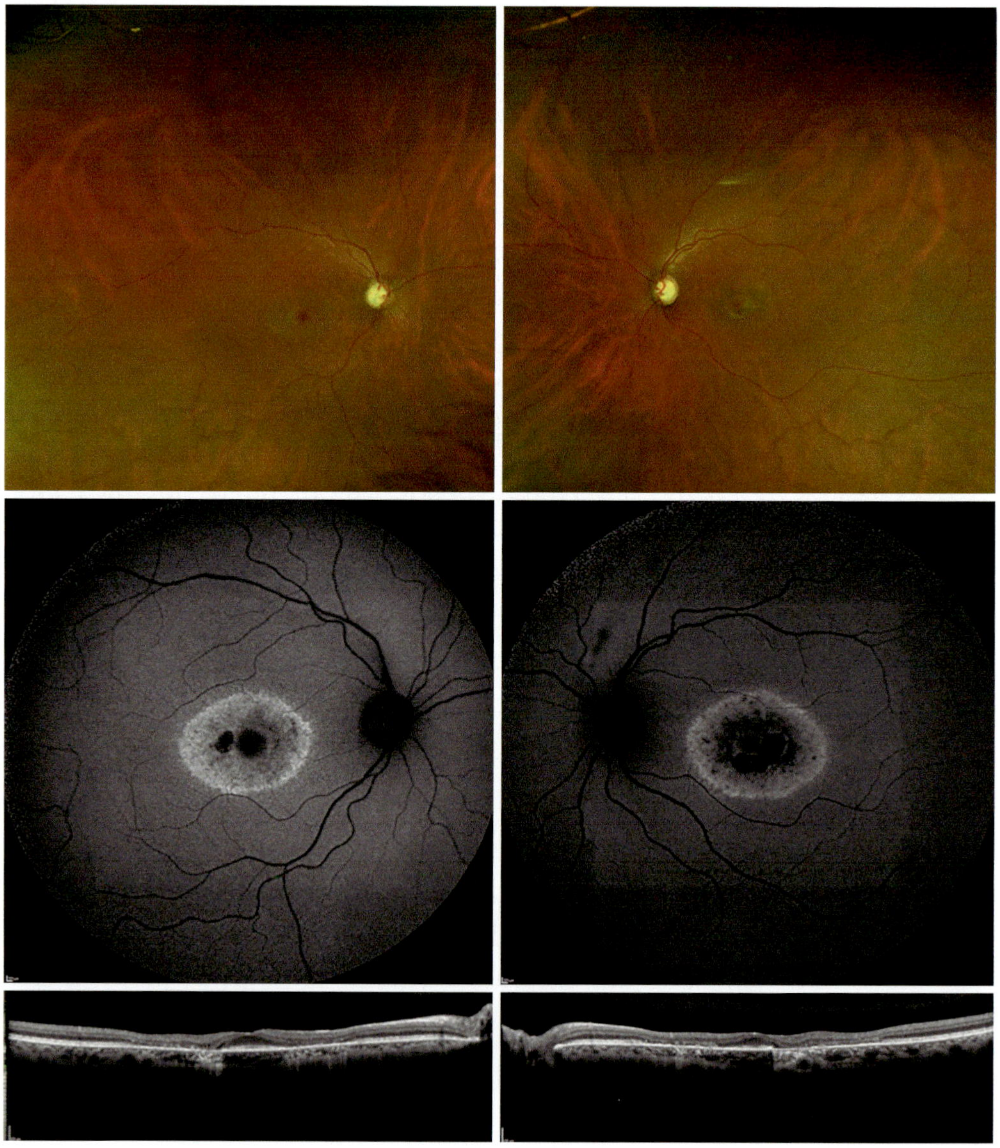

Fig. 12.1 A classic BEM in both eyes on color fundus photograph and corresponding FAF, areas of hypofluorescence surrounded by a ring of hyperfluorescence

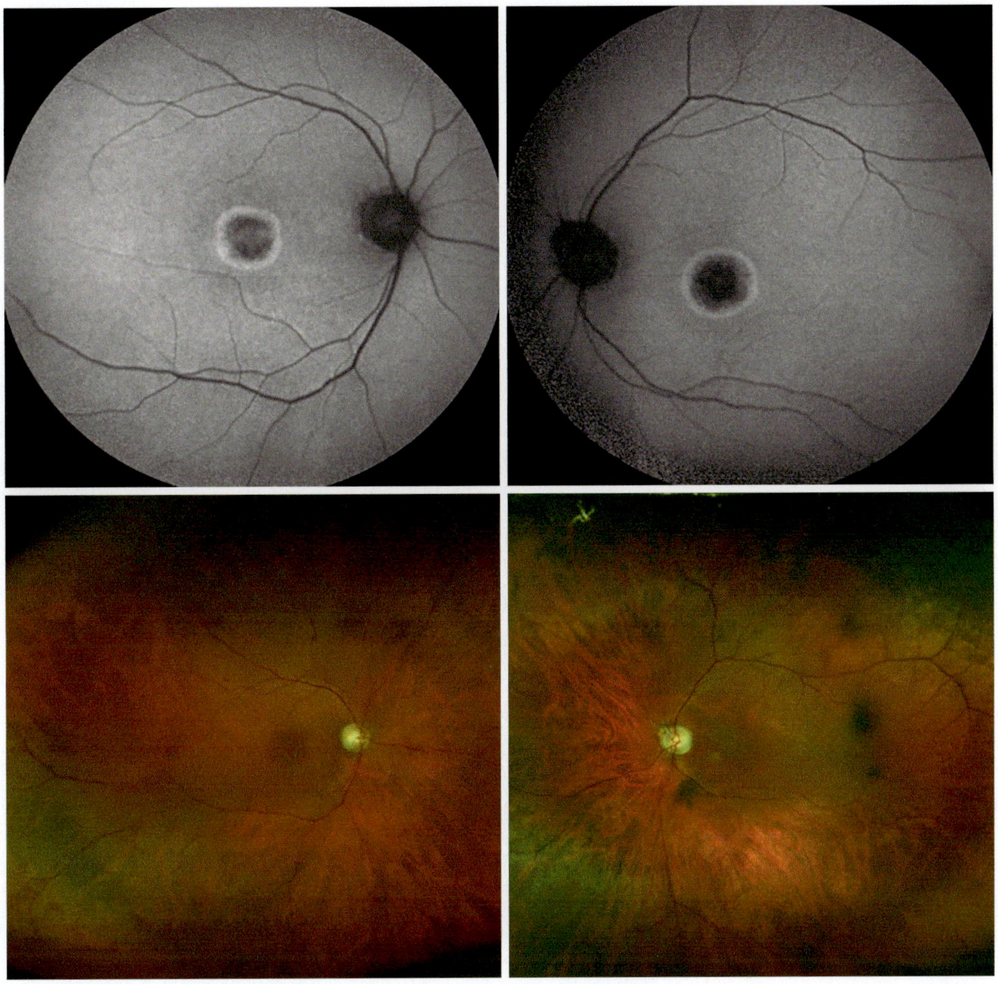

Fig. 12.2 A classic bull's eye maculopathy (BEM) on color fundus photograph (*upper row*). FAF (*middle row*) shows central hypofluorescence surrounded by a ring of hyperfluorescence. OCT (*lower row*) shows foveal thinning with loss of photoreceptors, loss of EZ line, loss of outer nuclear layer, and loss of RPE. A case of RPGR mutation

The electroretinogram (ERG) shows undetectable photopic response and a normal or near-normal scotopic rod response; cone-flicker ERG response is almost invariably delayed.

Peripheral visual fields may remain normal, though central scotoma is present. In some patients, rod photoreceptors are also involved later in life, leading to cone-rod dystrophy (CORD). These patients may show bone spicule-like intraretinal pigmentation and notice defective night vision.

In adults with X-linked cone dystrophy, the fundus may show a tapetal-like reflex and Mizuo-Nakamura phenomenon.

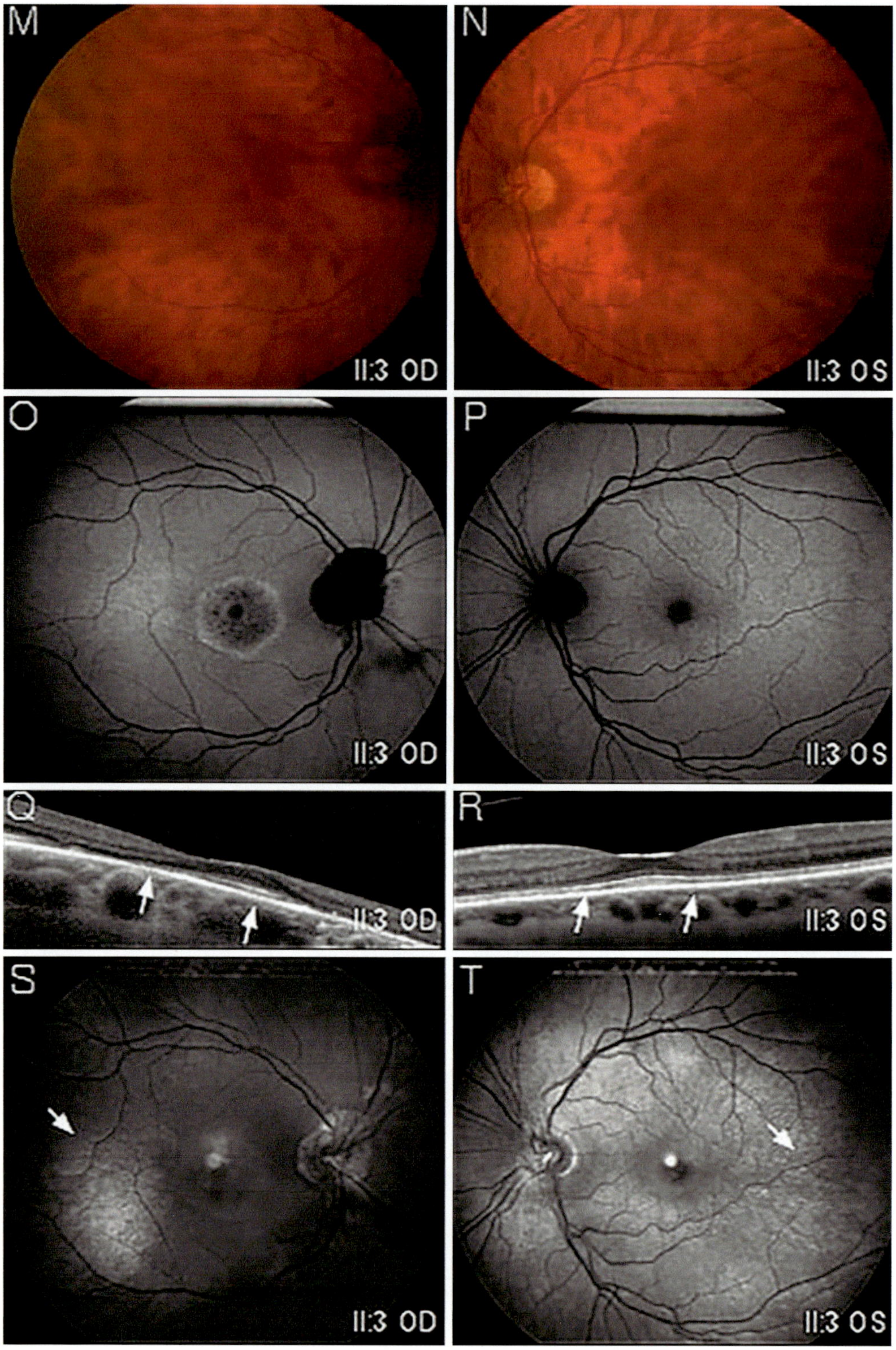

Fig. 12.3 Color photograph (*first row*) of a carrier, minimal RPE changes in the right macula, and the left looks normal. FAF (*second row*) shows bull's eye pattern with peripapillary atrophy in the right eye, and the left eye is normal. OCT (*third row*) shows disruption of EZ line in the perifoveal area in the right eye. The red-free image (*fourth row*) shows a tapetal reflex in an area temporal to macula (*arrow*)

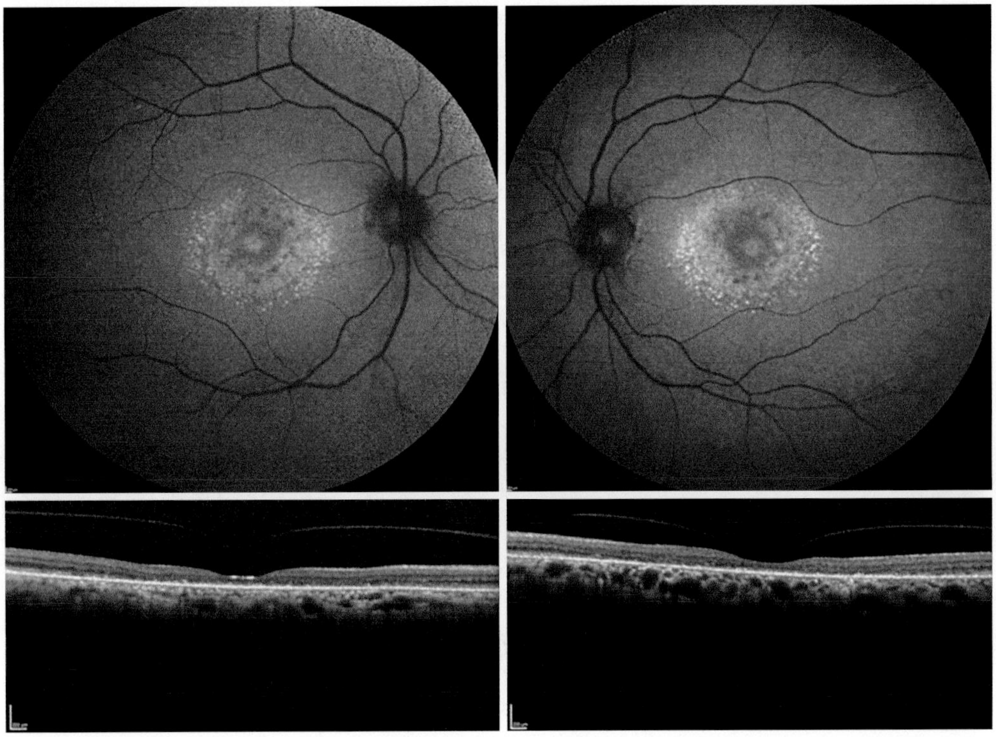

Fig. 12.4 Bull's eye maculopathy on color fundus photographs and corresponding RPE changes on FAF, a speckled appearance

Molecular Genetics

Mutations in about 30 genes are known to cause cone-rod dystrophy. Being genetically heterogenous, all three patterns of Mendelian inheritance are seen: autosomal dominant, autosomal recessive, and X-linked recessive. Mutations in about 20 genes cause autosomal recessive cone-rod dystrophy, and about 10, an autosomal dominant pattern. The autosomal recessive pattern is the commonest (60–70%), followed by autosomal dominant (20–30%) and X-linked (5%). Cone-rod dystrophy has high genetic heterogeneity and also shows great variability in age of onset and associated systemic findings.

Autosomal dominant cone dystrophy is caused by mutations in the gene linked to 6p21.1; this gene encodes for granulate cyclase activator A (GUCA1A). Dominant cone-rod dystrophy is caused by a mutation in the *GUCY2D* gene and *CRX* genes.

For autosomal recessive cone dystrophy, the genes implicated are *ABCA4*, *CNGB3*, *KCNV2*, and *PDE6C*; *ABCA4* accounts for about 30–60% of cases.

For X-linked cases, the affected gene is *RPGR*.

Specific Genotype and Phenotype Correlation

ABCA4

Genotype: Autosomal recessive

Phenotype: This mutation causes both COD and CORD, with early onset (first to third decade) and BEM. FAF shows a ring of increased autofluorescence (AF) surrounding decreased macular AF and sparing of the peripapillary area.

Cytogenetic location: 1p22.1

The mutation in the *ABCA4* gene accounts for 30–60% of cases of CORD. The ABCA4 protein removes from the photoreceptors one of the substances that is produced in phototransduction, called N-retinylidene-PE. With a mutation in *ABCA4*, the altered protein cannot remove the N-retinylidene-PE, and this substance then combines with another substance to produce N-retinylidene-N-retinylethanolamine (A2E). The A2E is toxic to photoreceptors, leading to CORD.

Tip: CORD caused by *ABCA4* gene mutations tends to cause more severe visual loss than CORD caused by other genetic mutations.

ADAM9 (CORD9)

Genotype: Autosomal recessive

Phenotype: Usually CORD, with very early onset (first to second decade); the macula shows RPE atrophy to BEM.

Cytogenetic location: 8p11.22

This gene encodes a member of the ADAM family (disintegrin and metalloprotease domain) and may mediate cell-cell or cell-matrix interactions and regulate the motility of cells via integrins.

RPGR1P1 (CORD13, Leber Congenital Amaurosis [LCA6])

Genotype: Autosomal recessive

Phenotype: Usually CORD, with very early onset (first to second decade); macula shows RPE granularity to atrophy.

Cytogenetic location: 14q11.2

This gene encodes a photoreceptor protein, a scaffolding protein that is needed for normal location of RPGR at the cilium, and for normal disk morphogenesis and disk organization in the outer segment of photoreceptor cells.

KCNV2 (Potassium Voltage-Gated Channel Modifier Subfamily V Member 2)

Genotype: Autosomal recessive

Phenotype: Usually CORD, with very early onset (first to second decade); macula shows RPE alterations to atrophy. FAF shows reduced AF in areas of atrophy.

Cytogenetic location: 9p24.2

This gene encodes a potassium channel subunit.

Tip: ERG shows a supernormal rod response.

SEMA4A (Semaphoring 4A) (CORD10 or RP35)

Genotype: Autosomal recessive

Phenotype: Usually CORD, with very early onset (first to second decade); macula shows RPE granularity to atrophy.

Cytogenetic location: 1q22.

GUCY2D (CORD6 or LCA1 or RCD2)

Genotype: Autosomal dominant

Phenotype: Usually CORD, with very early onset (first to second decade); macula shows RPE atrophy and peripheral changes. FAF shows increased foveal AF, and reduced AF in areas of atrophy. ERG might show an electronegative response; usually reduced cone and rod responses.

Cytogenetic location: 17p13.1.

The GUCY2D protein is involved in a reaction that helps return phototransduction to the dark state after light exposure.

GUCA1A (Guanylate Cyclase Activator 1A) (COD4 or CORD14)

Genotype: Autosomal dominant

Phenotype: Both COD and CORD, with late onset (third to fifth decade); macular RPE changes. FAF shows increased AF in the center of the macula, a perifoveal ring of increased AF, and reduced AF in areas of atrophy. ERG 30-Hz flicker may show normal implicit time.

Cytogenetic location: 6p21.1

The gene encodes an enzyme that promotes the activity of retinal guanylyl cyclase-1 (GC1) at a low calcium concentration and inhibits GC1 at high calcium concentrations. This calcium-sensitive regulation of retinal guanylyl cyclase is an important event in recovery of the dark state of rod photoreceptors following light exposure.

PRPH2 (Peripherin 2 or RDS)

Genotype: Autosomal dominant

Phenotype: Usually CORD, with onset second to third decade; macula shows RPE mottling or atrophy. FAF shows speckled pattern, with areas of increased and decreased AF.

Cytogenetic location: 6p21.1

This gene gives instruction to make peripherin 2 protein, which is involved in stability of the outer segment of photoreceptor cells.

CRX (CORD2 or LCA7 or CRD)

Genotype: Autosomal dominant

Phenotype: Usually CORD, with very early onset (within first decade); macula shows RPE atrophy. FAF shows reduced AF in areas of atrophy. Sometimes, electronegative ERG.

Cytogenetic location: 19q13.33.

The gene provides instruction to make cone-rod homeobox protein, a transcription factor that is necessary for the normal development of photoreceptors.

RIMS1 (Regulating Synaptic Membrane Exocytosis 1, CORD7)

Genotype: Autosomal dominant

Phenotype: Usually CORD, with late onset (third to fifth decade); macula shows RPE mottling or atrophy or BEM, and later peripheral changes along with attenuation of blood vessels. FAF shows speckled pattern. FAF shows reduced AF in the center, surrounded by a ring of increased AF. ERG 30-Hz flicker may show normal implicit time.

Cytogenetic location: 6q13.

The gene plays a role in the regulation of voltage-gated calcium channels during neurotransmission.

PITPNM3 (CORD5)

Genotype: Autosomal dominant

Phenotype: Usually COD, with early onset (first decade); macula shows RPE mottling or atrophy.

Cytogenetic location: 17p13.2–p13.1.

The gene encodes a member of a family of membrane-associated phosphatidylinositol (PI) transfer domain–containing proteins and interacts with the protein tyrosine kinase PTK2B.

RPGR (COD1 or CORCX1 or RP3 or RP15 or XLRP3)

Genotype: X-linked

Phenotype: Both COD and CORD. For COD, late onset (fifth decade); for CORD, onset second to fourth decade. Macula shows RPE atrophy in COD, and peripheral involvement in CORD. FAF shows a perifoveal ring of increased AF.

Cytogenetic location: Xp11.4.

The *RPGR* gene provides instruction for making protein that is essential for the function of a cilia. Several versions or isoforms of RPFG protein are produced; one of the versions is ORF15 exon, expressed predominantly in photoreceptors.

DRAM2 (Damage-Regulated Autophagy Modulator 2, CORD 21)

Genotype: autosomal recessive

Phenotype: CORD, usually with onset in the third decade; macula shows granularity or atrophy. FAF image shows hypo-autofluorescence at the center, surrounded by a hyper-autofluorescence ring. Patients usually have progressive central vision loss, while nyctalopia and light hypersensitivity are inconsistent features of advanced disease.

Cytogenetic location: 1p13.3

The gene encodes a protein that binds microtubule-associated protein 1 light chain 3, which is necessary for autophagy. It is also involved in p53-mediated apoptosis.

Further Reading

Michaelides M, Hardcastle AJ, Hunt DM, Moore AT. Progressive cone and cone-rod dystrophies: phenotypes and underlying molecular genetic basis. Surv Ophthalmol. 2006;51:232–58.

Rahner N, Nuernberg G, Finis D, Nuernberg P, Royer-Pokora B. A novel C8orf37 splice mutation and genotype-phenotype correlation for cone-rod dystrophy. Ophthalmic Genet. 2016;37:294–300.

El-Asrag, M. E., Sergouniotis, P. I., McKibbin, M., Plagnol, V., Sheridan, E., Waseem, N., Abdelhamed, Z., McKeefry, D., Van Schil, K., Poulter, J. A., Johnson, C. A., Carr, I. M., Leroy, B. P., De Baere, E., Inglehearn, C. F., Webster, A. R., Toomes, C., Ali, M., Black, G., … Lord, E. (2015). Biallelic mutations in the autophagy regulator DRAM2 cause retinal dystrophy with early macular involvement. *The American Journal of Human Genetics*, *96*(6), 948–954. https://doi.org/10.1016/j.ajhg.2015.04.006

CHAPTER 13

Congenital Stationary Night Blindness

Justin Cheong, Stephen H. Tsang, and Tarun Sharma

Congenital Stationary Night Blindness with Normal Fundi 70
Types 70
Molecular Genetics 71
Congenital Stationary Night Blindness with Abnormal Fundi. 71
Fundus Albipunctatus. 71
Oguchi Disease 71
Further Reading 72

J. Cheong
Vagelos College of Physicians and Surgeons,
New York, NY, USA

S. H. Tsang (✉) · T. Sharma
Department of Ophthalmology, Columbia University,
New York, NY, USA
e-mail: sht2@cumc.columbia.edu; ts3118@cumc.columbia.edu

S. H. Tsang et al. (eds.), *Atlas of Inherited Retinal Diseases*, Advances in Experimental Medicine and Biology 1467, https://doi.org/10.1007/978-3-031-72230-1_13

Congenital Stationary Night Blindness with Normal Fundi

In congenital stationary night blindness (CSNB), there is a defect in rod photoreceptor signal transmission. This disorder of night vision is non-progressive. The most common inheritance pattern is X-linked, though autosomal recessive and autosomal dominant patterns have been described (Fig. 13.1). There is genetic heterogeneity within these types.

Though these patients usually complain of defective night vision, the fundus is normal. Myopia is an associated common finding. Vision ranges from normal to about 20/200; patients with poor vision also have nystagmus, strabismus, or both.

Dark-adaptometry shows no rod adaptation.

Types

The electroretinogram (ERG) identifies two types of CSNB in these patients:

- **Riggs type**: Absence of a-wave in dark-adapted bright-flash ERG, suggesting a defect in the rod photoreceptor.
- **Schubert-Bornschein type**: Normal a-wave and a markedly decreased rod b-wave with an electronegative waveform under the same stimulus conditions, suggesting abnormal bipolar cell function (Fig. 13.2).

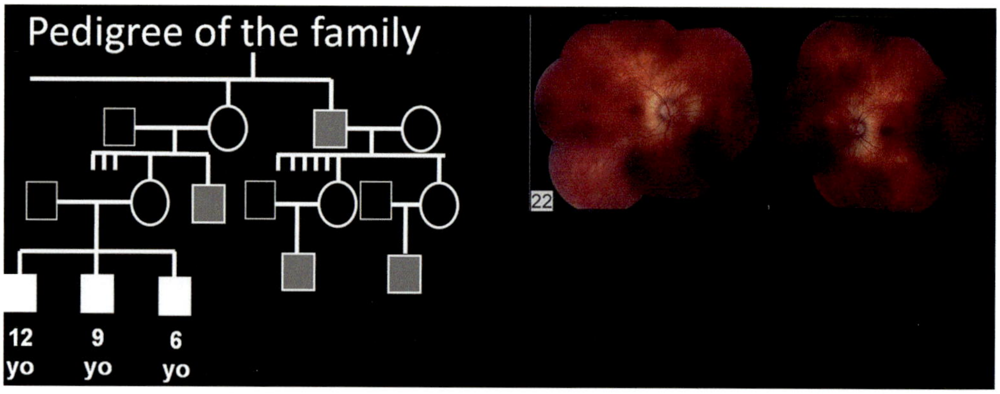

Fig. 13.1 A family with CSNB, only males are affected, and normal fundus

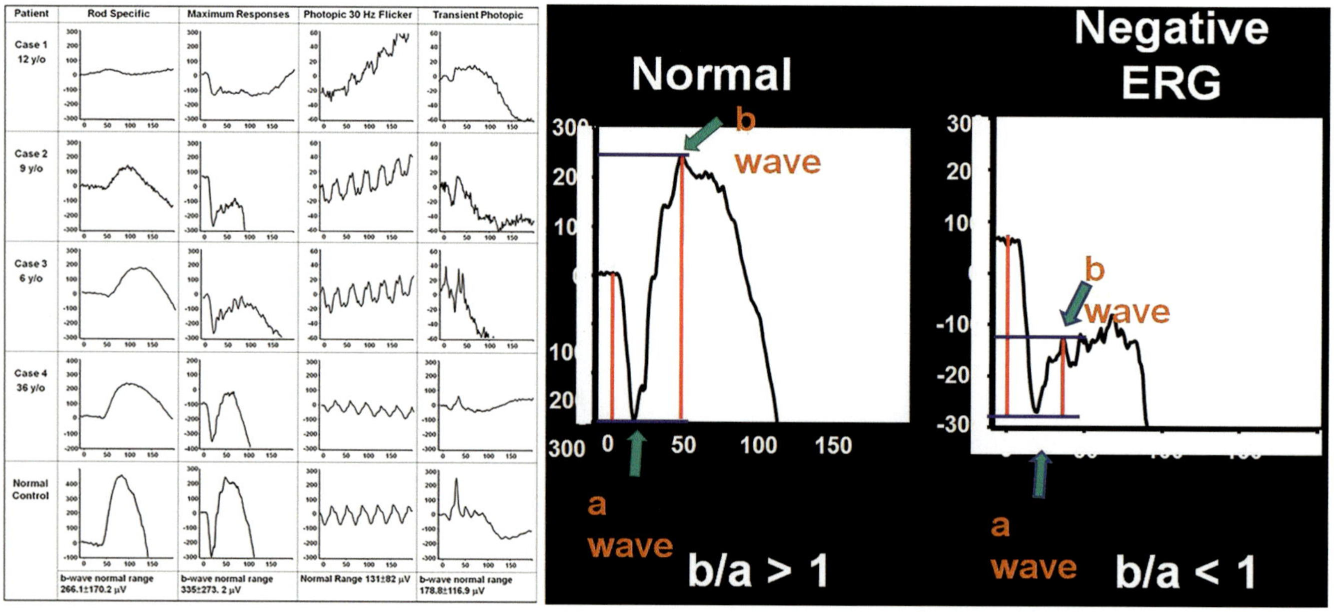

Fig. 13.2 ERG showing electronegative wave pattern

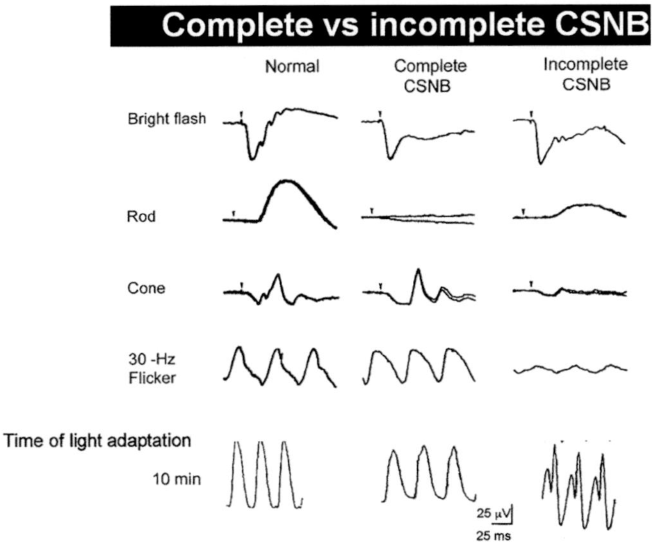

Fig. 13.3 Differentiating features between complete and incomplete CSNB

Two subtypes—complete and incomplete—have been described (Fig. 13.3):

- In the complete type, there is no detectable or recordable rod b-wave (dysfunction in ON-bipolar pathway), and there is a normal-amplitude wave on 30-Hz flicker ERG. Most patients with the complete form have night blindness and usually moderate to high myopia.
- In the incomplete type, there is some detectable rod b-wave (dysfunction in both ON and OFF pathways), and amplitude is markedly subnormal and delayed in the 30-Hz flicker ERG, with a distinctive peak suggestive of greater cone ERG abnormality than in the complete form. Night blindness is less constant in the incomplete form, and refractive error varies from myopia to hyperopia.

Molecular Genetics

- X-linked CSNB is associated with mutations in the *NYX* gene, causing the complete type of CSNB. The gene encodes nyctalopin protein, which plays an important role in the synaptic connection between photoreceptors and ON-bipolar cells. More than 50 mutations have been identified. The cytogenetic location is Xp11.4, the short arm of the X chromosome at position 11.4. The *NYX* mutation accounts for about 45% of cases of X-linked CSNB.
- Another mutation is in *CACNA1F*, causing the incomplete type of CSNB, in which patients have some detectable rod response on ERG. This gene encodes a transmembrane protein that functions as a retina-specific alpha-1 subunit of a voltage-gated L-type calcium channel; it is responsible for regulation of glutamate release from photoreceptors to ON-bipolar and OFF-bipolar cells. The cytogenetic location is Xp11.23, the short arm of the X chromosome at position 11.23. The *CACNA1F* mutation accounts for about 55% of cases of X-linked CSNB.

Congenital Stationary Night Blindness with Abnormal Fundi

Fundus Albipunctatus

Patients with fundus albipunctatus have defective night vision from birth, with multiple white dots visible throughout the fundus, sparing the central fovea. Visual acuity remains good. The inheritance pattern is autosomal recessive.

Imaging and Tests

- *Fundus autofluorescence (FAF)* is diffusely decreased because of less accumulation of lipofuscin pigment.
- *Optical coherence tomography (OCT)* shows hyperreflective lesions in the outer retina.
- *ERG* shows undetectable rod response but normal cone response. The ERG response may become normal after prolonged overnight dark adaptation, however.

Molecular Genetics

There is a mutation in the *RDH5* gene (12q13–12q14). This gene encodes 11-cis-retinol dehydrogenase, which converts 11-cis-retinol into 11-cis-retinal, in the retinal pigment epithelium (RPE). Therefore, regeneration of visual pigment is delayed.

Oguchi Disease

Oguchi disease is most common in Japan. Its inheritance pattern is autosomal recessive. Patients have defective vision since birth.

Imaging and Tests

- *Fundus examination* shows a characteristic yellow tapetal sheen or reflex after a brief light exposure (described as the Mizuo-Nakamura phenomenon). The reflex disappears after 2–3 h of dark adaptation.

- If a *scotopic ERG* is performed after 20 min of dark adaptation, no rod a-wave is detected, but if it is performed after 1–2 h of dark adaptation, the first flash will produce an a-wave that is normal in both amplitude and implicit time. Response to the second flash will be reduced.

Molecular Genetics

Oguchi disease is due to mutation in the *SAG* (S-antigen visual arrestin) gene (2q37.1), which encodes for arrestin, or in the *GRK1* gene (13q34), encoding for rhodopsin kinase.

Further Reading

Marmor MF. Long-term follow-up of the physiologic abnormalities and fundus changes in fundus albipunctatus. Ophthalmology. 1990;97:380–4.

Sergouniotis PI, Sohn EH, Li Z, McBain VA, Wright GA, Moore AT, et al. Phenotypic variability in RDH5 retinopathy (Fundus Albipunctatus). Ophthalmology. 2011;118:1661–70.

Usui T, Tanimoto N, Ueki S, Takagi M, Hasegawa S, Abe H, et al. ERG rod a-wave in Oguchi disease. Vision Res. 2004;44:535–40.

CHAPTER 14

Blue Cone Monochromatism

Kristina J. Hartung, Stephen H. Tsang, Tarun Sharma, and Vlad Diaconita

General Features 74
Imaging and Tests 74
Molecular Genetics 75
Further Reading 75

K. J. Hartung
Department of Ophthalmology, University Medical Centre Ljubljana, Ljubljana, Slovenia

Faculty of Medicine, University of Ljubljana, Ljubljana, Slovenia

S. H. Tsang (✉) · T. Sharma · V. Diaconita
Department of Ophthalmology, Columbia University, New York, NY, USA
e-mail: sht2@cumc.columbia.edu; ts3118@cumc.columbia.edu; vd2374@cumc.columbia.edu

S. H. Tsang et al. (eds.), *Atlas of Inherited Retinal Diseases*, Advances in Experimental Medicine and Biology 1467, https://doi.org/10.1007/978-3-031-72230-1_14

General Features

Blue cone monochromatism (BCM) is a rare X-linked recessive disease characterized by an absent function of L and M cones and a normal function of S cones and rods. The estimated prevalence is 1 in 100,000 individuals, and males are predominantly affected. The patients usually present at birth or in early infancy. The clinical features include impaired color vision (patients only see colors in the blue range of light) and poor visual acuity (between 20/80 and 20/200), photophobia, pendular nystagmus (which may improve over time), and myopia. The condition is typically stationary, but progressive central retinal atrophy may be seen later in life. The presentation is similar to rod monochromatism (achromatopsia), an autosomal recessive disease that affects all three types of cones. However, BCM patients have better visual acuity, preserved tritan discrimination, and myopia (as opposed to hyperopia in achromatopsia). Electrophysiology, psychophysical testing (using Berson plates), and family history (x-linked recessive vs. autosomal recessive) help distinguish the two conditions.

Imaging and Tests

- **Optical coherence tomography (OCT):** may show foveal thinning and disruption of the photoreceptors or outer nuclear layer (Fig. 14.1).
- **Fundus autofluorescence (FAF):** shows a fundus that is usually normal or may show subtle retinal pigment epithelial (RPE) changes or atrophy (in older patients).
- **Electrophysiology:** Full-field ERG shows normal scotopic and non-recordable photopic responses. An S-cone ERG (blue flashlight on yellow background) is preserved and can be used to distinguish BCM from achromatopsia. The 30-Hz flicker ERG is undetectable, as the response is predominantly from L-cones and M-cones, but a low-amplitude single-flash photopic response may be present.

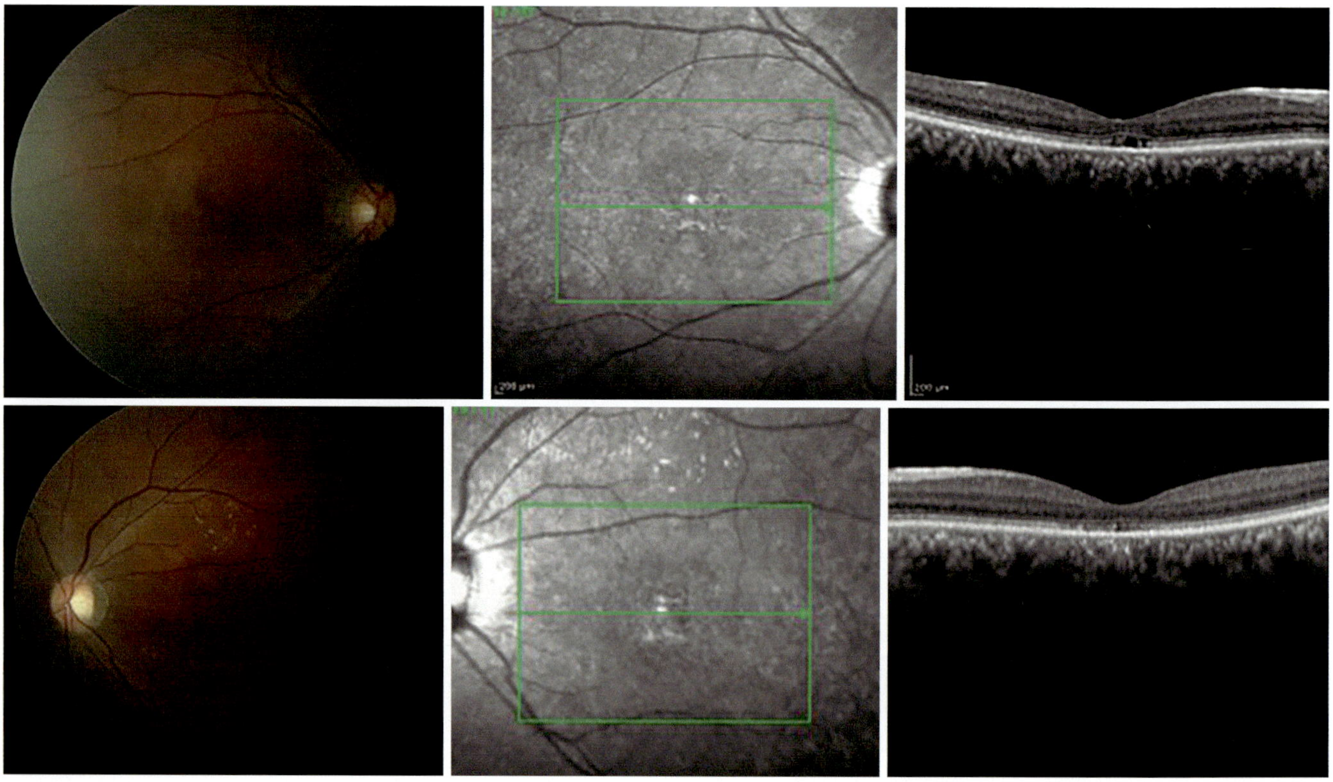

Fig. 14.1 Fundus nearly normal but for a few yellowish dots in the left eye; OCT shows disruption of the ellipsoid zone (EZ) line and optical empty cavities in the center

Molecular Genetics

L-cone and M-cone opsin genes are located on the X-chromosome (Xq28). Their sequences are identical in 98%, which predisposes them to homologous recombination.

- OPN1LW: encodes long-wave sensitive (LWS) type 1 opsin, a G-protein coupled receptor protein located in the disks of the L-cones. It is essential for color vision in the yellow-orange spectrum (peak sensitivity at 565 nm).
- OPN1MW: encodes medium-wave sensitive (MWS) type 1 opsin, a G-protein coupled receptor protein located in the disks of the M-cones. It is essential for color vision in the yellow-green spectrum (peak sensitivity 535 nm).

BCM is usually caused by two types of mutations; either a deletion in the locus control region, which controls the transcription of OPN1LW and OPN1MW or by the presence of a single OPN1LW/OPN1MW hybrid gene inactivated by the p.C203R missense mutation. The resulting disruption in the folding of cone opsin leads to a loss of functional pigment production and inactivation of the corresponding cones.

Further Reading

Aboshiha J, Dubis AM, Carroll J, Hardcastle AJ, Michaelides M. The cone dysfunction syndromes. Br J Ophthalmol. 2016;100(1):115–21.

Michaelides, M., Johnson, S., Simunovic, M. et al. Blue cone monochromatism: a phenotype and genotype assessment with evidence of progressive loss of cone function in older individuals. Eye 19, 2–10 (2005).

Moskowitz A, Hansen RM, Akula JD, Eklund SE, Fulton AB. Rod and rod-driven function in achromatopsia and blue cone monochromatism. Invest Ophthalmol Vis Sci. 2009;50(2): 950–8.

Nathans J, Davenport CM, Maumenee IH, Lewis RA, Hejtmancik JF, Litt M, et al. Molecular genetics of human blue cone monochromacy. Science. 1989; 245:831–8.

Part III
Autosomal Dominant Forms

CHAPTER 15

Autosomal Dominant Retinitis Pigmentosa

Duncan Yi-Te Wang, Stephen H. Tsang, Tarun Sharma, and Vlad Diaconita

General Features 80
Molecular Genetics 81
RHO Gene (RP4, Opsin-2) 81
PRPF3 Gene (RP18, SNRNP90) 82
PRPF8 Gene (RP13, SNRNP220) 82
RP1 Gene (ORP1) 82
Further Reading 85

D. Y.-T. Wang
School of Medicine, College of Medicine, National Taiwan University, Taipei, Taiwan

S. H. Tsang (✉) · T. Sharma
Department of Ophthalmology, Columbia University, New York, NY, USA
e-mail: sht2@cumc.columbia.edu; b08401038@ntu.edu.tw

V. Diaconita
Department of Ophthalmology, Edward S. Harkness Eye Institute
New York-Presbyterian Hospital Columbia University, New York, NY, USA
e-mail: vd2374@cumc.columbia.edu

S. H. Tsang et al. (eds.), *Atlas of Inherited Retinal Diseases*, Advances in Experimental Medicine and Biology 1467, https://doi.org/10.1007/978-3-031-72230-1_15

General Features

More than 70 genes (over 3000 mutations) are known to cause non-syndromic retinitis pigmentosa (RP), including autosomal dominant (AD), autosomal recessive (AR), X-linked, and simplex forms (inheritance not known).

- AD: 15–25% of cases
- AR: 5–20%
- X-linked: 5–15%
- Simplex: 40–50%

As of publication over 26 gene mutations have been identified for the AD form, with over 50 for the AR form. Patients with AD RP usually present later in life, with a less severe and slower progressing presentation as compared to AR RP. The macula may remain spared for a long period, with good central visual acuity (Figs. 15.1 and 15.2).

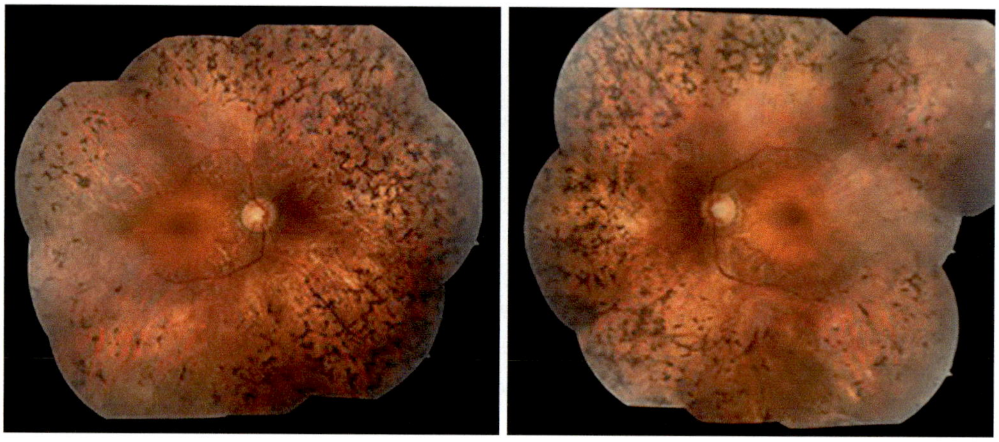

Fig. 15.1 A case of autosomal dominant (AD) retinitis pigmentosa (RP) showing all the classic features, but a healthy macula

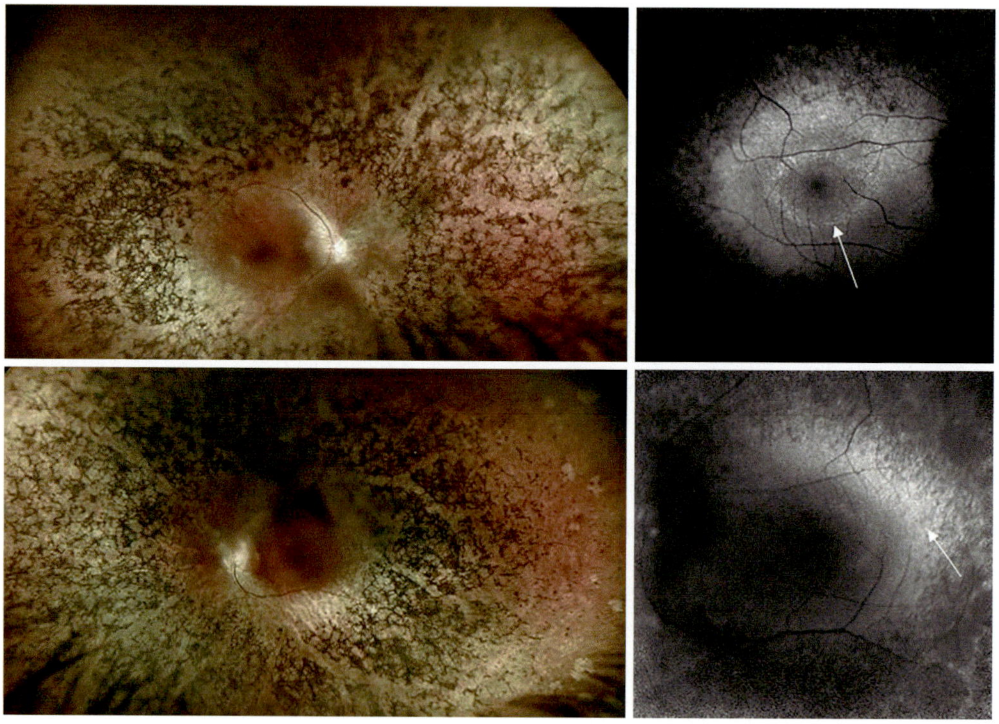

Fig. 15.2 AD RP sparing the macula. Fundus autofluorescence (FAF) shows a characteristic hyperfluorescent ring (*arrow*)

Molecular Genetics

RHO Gene (RP4, Opsin-2)

- Mutations in the *RHO* gene (Fig. 15.3) are the most common cause of AD RP, accounting for 20–30% of all cases.
- This gene encodes for making rhodopsin protein, which is predominantly present in rods.
- Different kinds of mutations in the *RHO* gene may lead to different pigmentation patterns. For example, p.Pro23His (c.68 C>A) mainly shows pigmentary changes in the lower and nasal periphery (Fig. 15.4), while p.Pro347Leu (c.1040 C>T) shows pigmentary changes in the temporal periphery at first (Fig. 15.5) and with diffuse pattern at late stage (Fig. 15.6).
- Cytogenetic location: 3q22.1

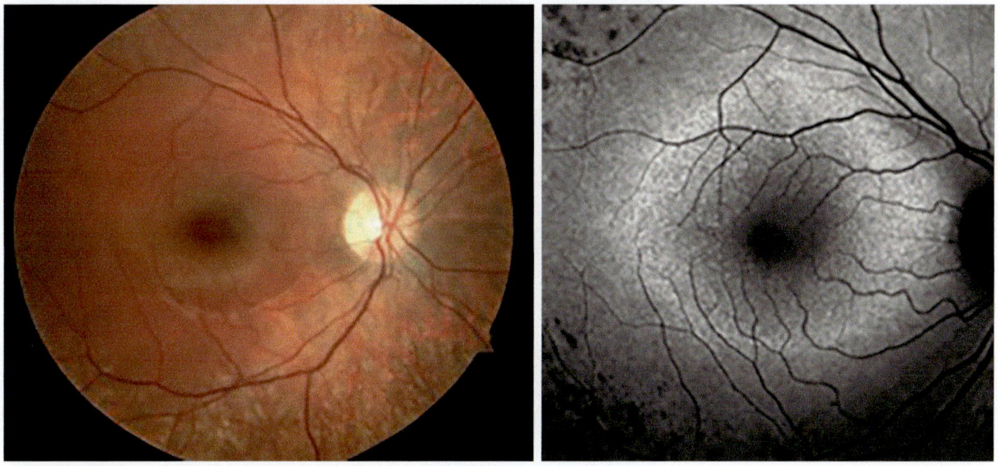

Fig. 15.3 A case of AD RP (*RHO* gene mutation), sparing the macula

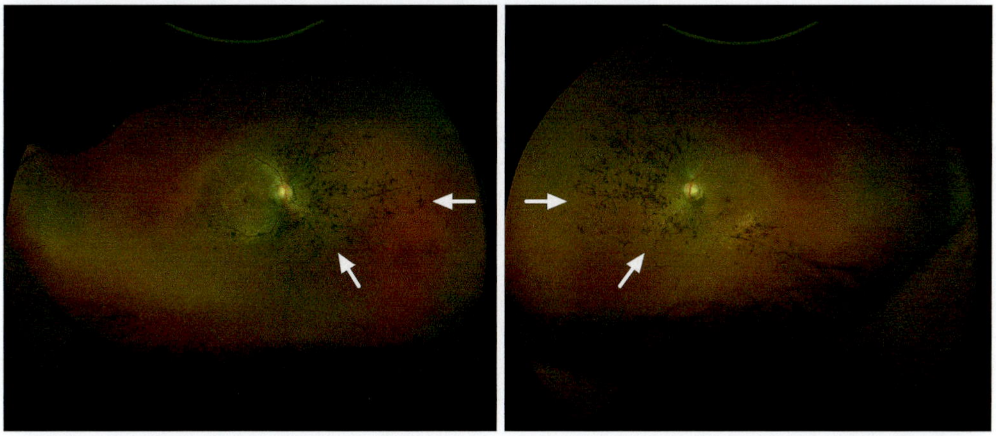

Fig. 15.4 A case of AD RP caused by *RHO* mutation (p.Pro23His), with bony-spicule intraretinal pigmentation both in the nasal and lower periphery (*arrow*)

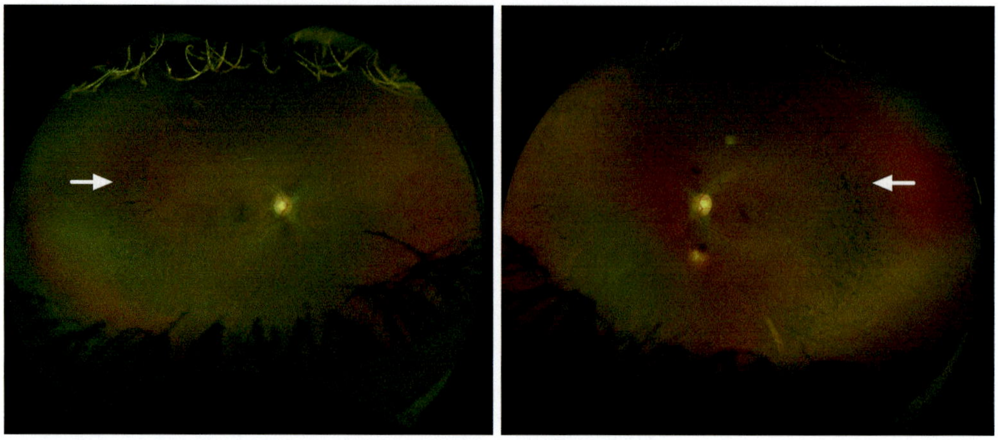

Fig. 15.5 A case of AD RP caused by *RHO* mutation (p. Pro347Leu), with bony-spicule intraretinal pigmentation in the temporal periphery (*arrow*)

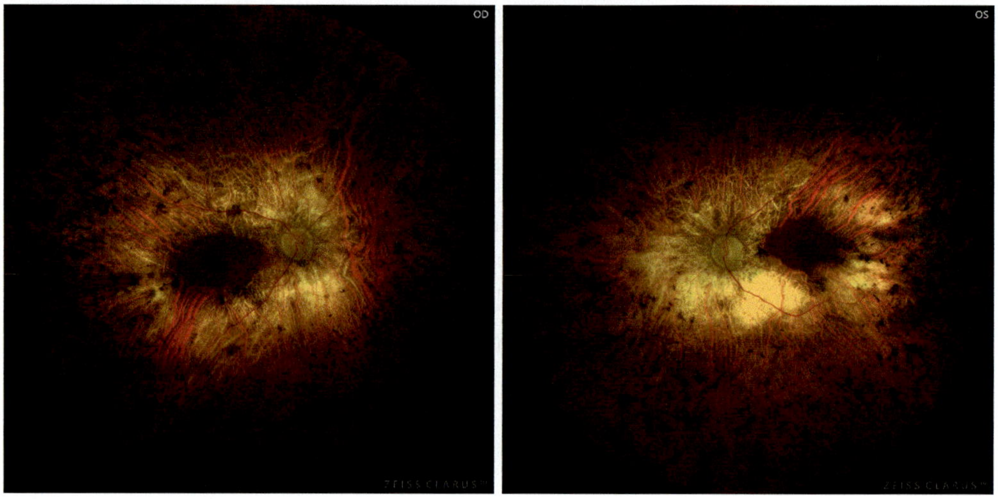

Fig. 15.6 Another AD RP caused by *RHO* mutation (p. Pro347Leu), with diffuse pigmentary changes and retinochoroidal atrophy

PRPF3 Gene (RP18, SNRNP90)

- Figure 15.7 presents an example of AD RP caused by this mutation.
- The removal of introns from nuclear pre-mRNAs occurs on spliceosomes, which are made up of four small nuclear ribonucleoproteins (snRNPs).
- Therefore, this gene participates in pre-mRNA slicing.
- Cytogenetic location: 1q21.2

PRPF8 Gene (RP13, SNRNP220)

- This gene encodes for a scaffold protein helping the assembly of spliceosomes.
- Figure 15.8 presents an example of AD RP caused by this mutation.
- Similar to *PRPF3*, this gene participates in pre-mRNA slicing.
- Cytogenetic location: 17p13.3

RP1 Gene (ORP1)

- The *RP1* gene encodes a doublecortin domain, which binds microtubules and regulates microtubule polymerization.
- This microtubules-associated protein is needed for correct stacking of the outer segment disc (Figs. 15.9 and 15.10)
- Mutation in this gene also causes AR RP.
- Cytogenetic location: 8q11.23–q12.1.

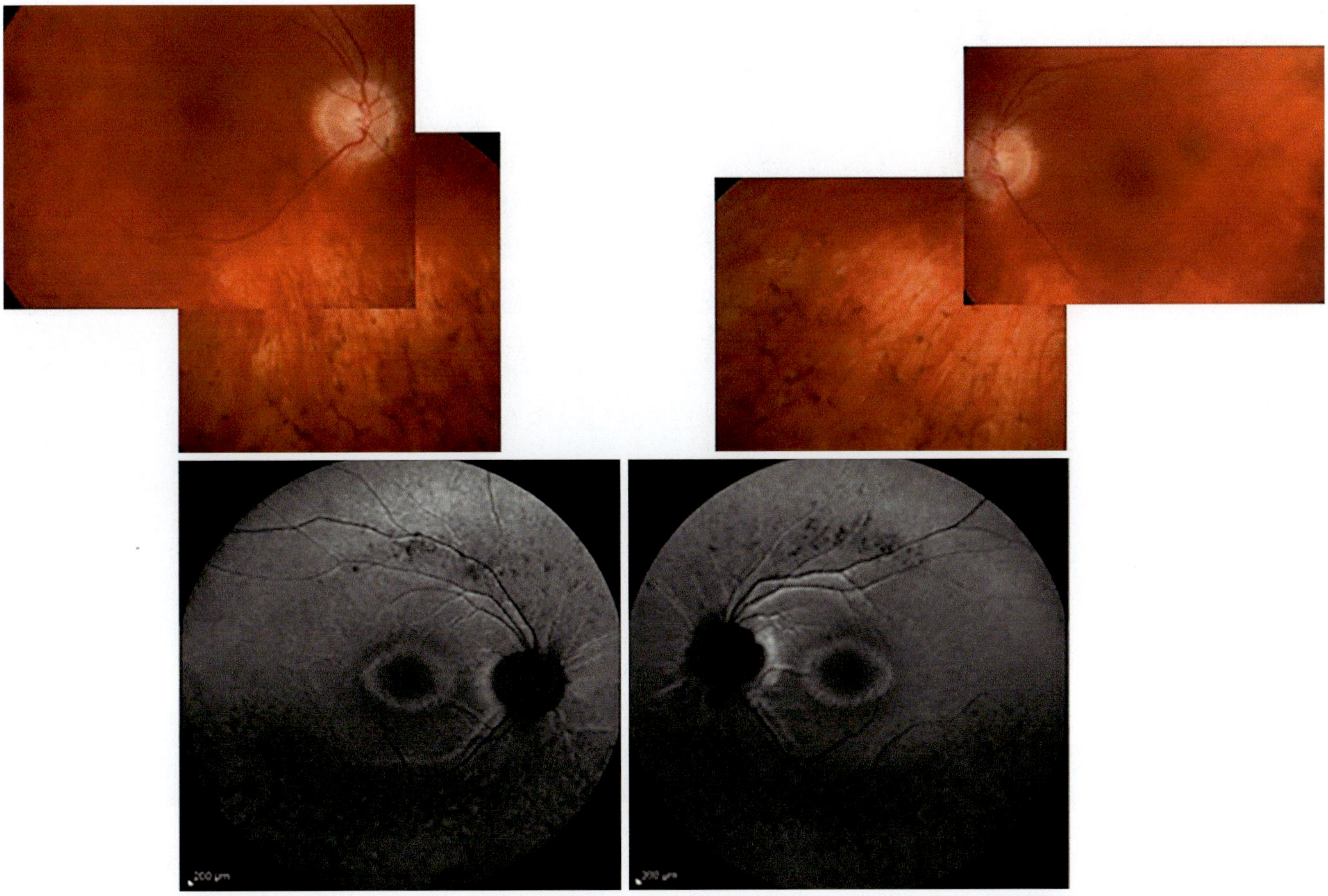

Fig. 15.7 AD RP caused by *PRPF3* mutation, showing pigmentary changes in the lower periphery. FAF shows a ring of hyperfluorescence around the fovea

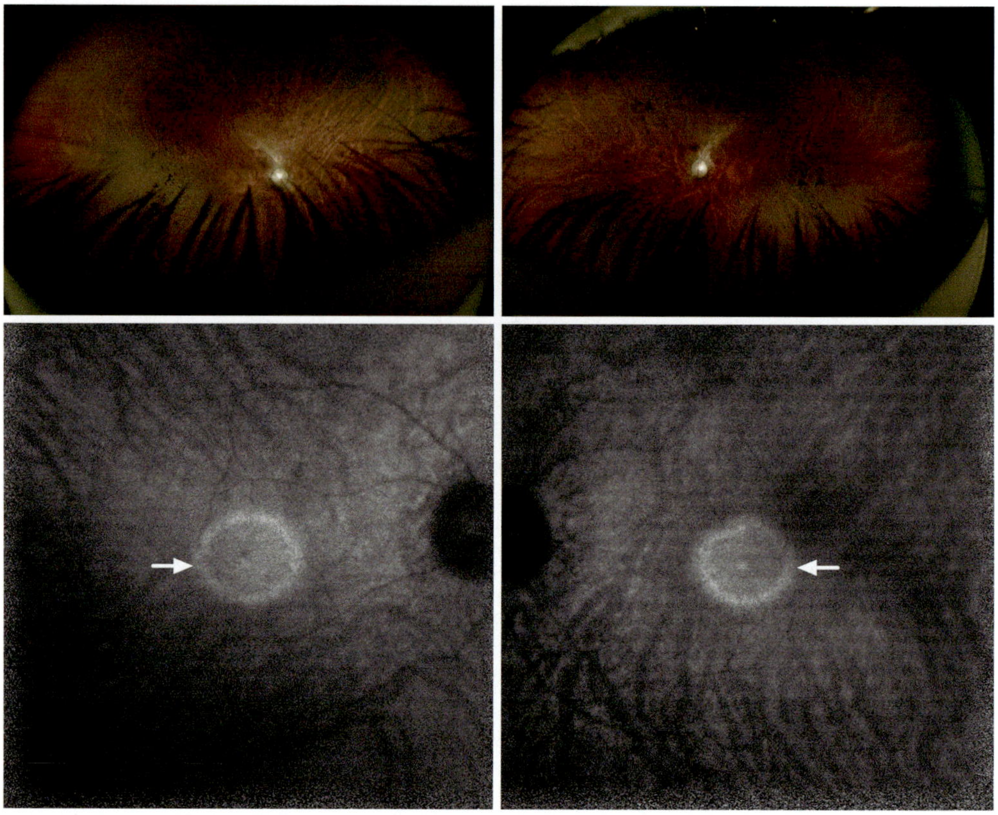

Fig. 15.8 AD RP caused by *PRPF8* mutation, showing pigmentary changes both in the upper and lower periphery. FAF shows a hyperfluorescent ring around the fovea (*arrow*)

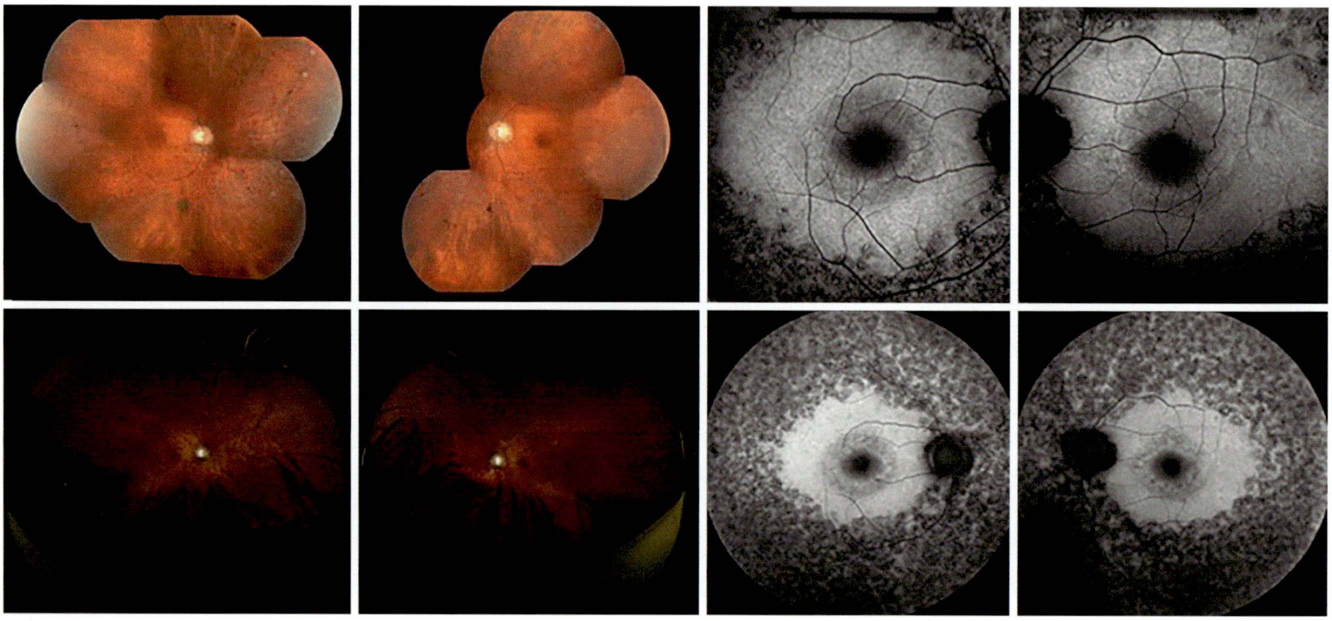

Fig. 15.9 A case of *RP1* mutation AD RP. The macula remains spared despite a 5-year follow-up

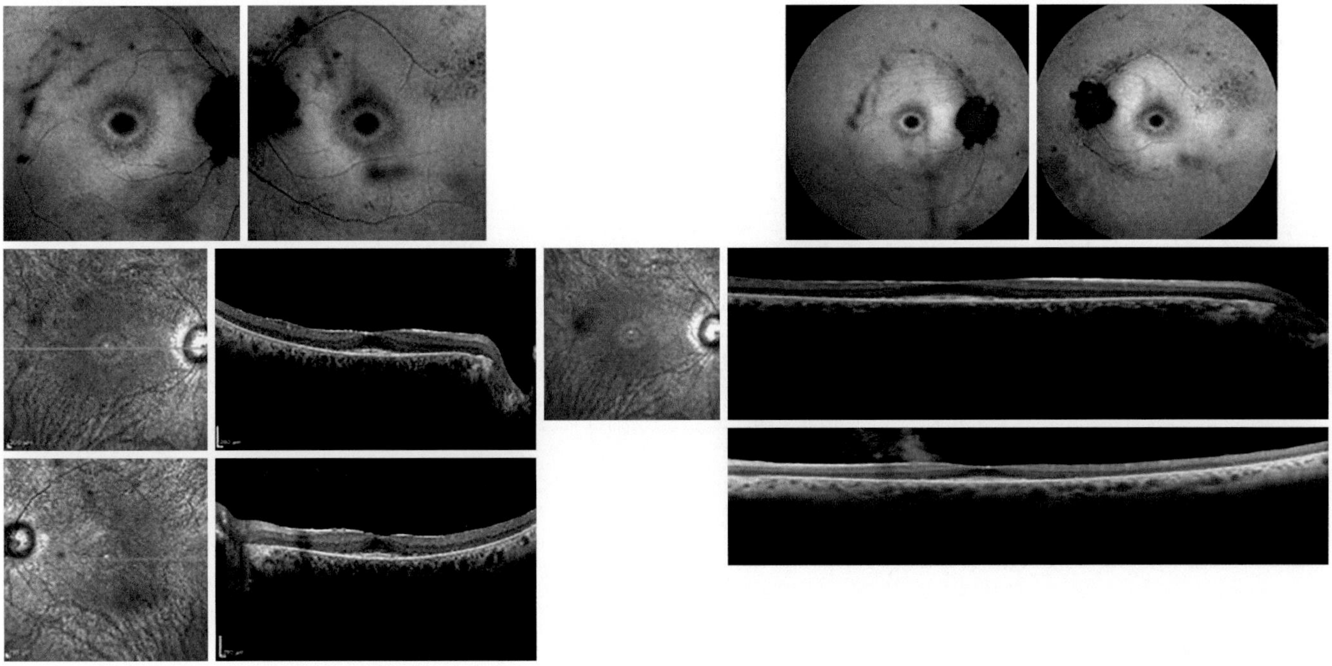

Fig. 15.10 AD RP caused by *RP1* mutation, showing a ring of hyperfluorescence on FAF and loss of the outer nuclear layer on optical coherence tomography (OCT). A 2-year follow-up showed no significant progression

Further Reading

Borràs E, de Sousa Dias M, Hernan I, Pascual B, Mañé B, Gamundi MJ, et al. Detection of novel genetic variation in autosomal dominant retinitis pigmentosa. Clin Genet. 2013;84:441–52.

Daiger SP, Sullivan LS, Bowne SJ. Genes and mutations causing retinitis pigmentosa. Clin Genet. 2013;84:132–41.

CHAPTER 16

Best Vitelliform Macular Dystrophy

Laura A. Jenny, Stephen H. Tsang, and Tarun Sharma

General Features 88
Imaging and Tests 90
Subtypes 90
Adult-Onset Foveomacular Vitelliform Dystrophy (AFMD) 90
Autosomal Recessive Bestrophinopathy (ARB) 90
Autosomal Dominant Vitreoretinochoroidopathy (ADVIRC) 90
Molecular Genetics 91
BEST1 Gene (VMD2 Gene or RP50 or BMD) 91
Further Reading 91

L. A. Jenny · S. H. Tsang (✉) · T. Sharma
Department of Ophthalmology, Columbia University, New York, NY, USA
e-mail: lj2539@cumc.columbia.edu; sht2@cumc.columbia.edu; ts3118@cumc.columbia.edu

S. H. Tsang et al. (eds.), *Atlas of Inherited Retinal Diseases*, Advances in Experimental Medicine and Biology 1467, https://doi.org/10.1007/978-3-031-72230-1_16

General Features

Best vitelliform macular dystrophy (VMD or BVMD) is one of the most common macular dystrophies, affecting 1 in 10,000 individuals. The clinical presentation varies, depending on the stage of the disease at which the patient presents, usually one of these five stages:

- Previtelliform
- Vitelliform (Figs. 16.1, 16.2, 16.3, and 16.4)
- Pseudohypopyon
- Vitelliruptive (Fig. 16.5)
- Atrophic

The vitelliform stage is characterized by an egg yolk–like appearance in the central macula, due to deposition of yellowish material in the subretinal space. As the material breaks down, there is clear fluid in the upper part and yellowish material in the lower part, giving a picture of pseudohypopyon. As the condition progress, the yellow material breaks down and resembles a scrambled egg, and eventually appears as an atrophic scar. Surprisingly, vision remains good until the last stage of geographic atrophic appearance. The choroidal neovascular membrane develops in about 20% of cases and may account for decreased visual acuity.

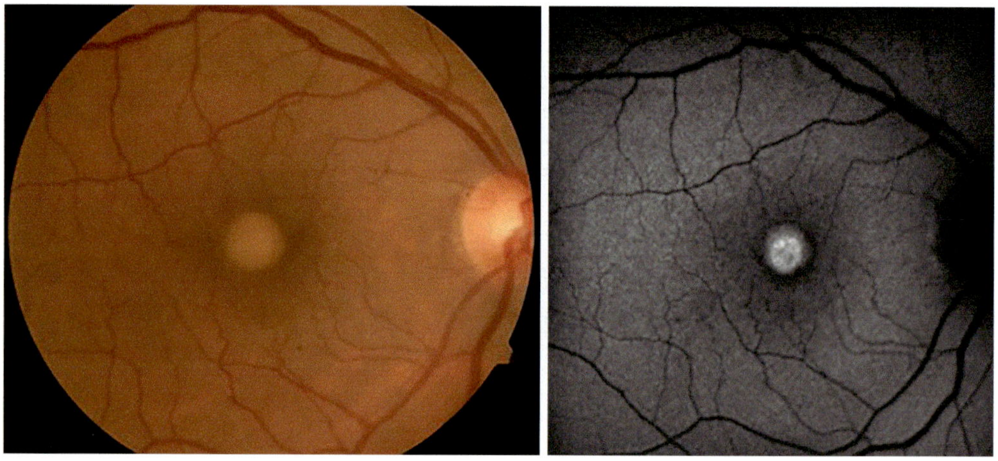

Fig. 16.1 Egg-yolk appearance on color fundus photograph, with corresponding intense hyperfluorescence on fundus autofluorescence (FAF)

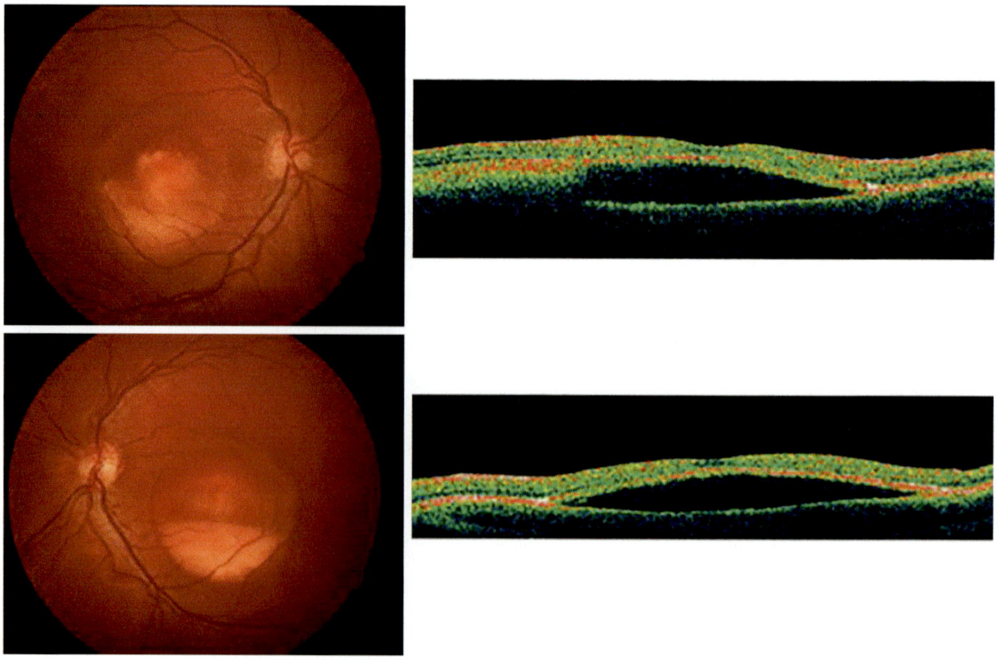

Fig. 16.2 Partially absorbed yolk on color fundus photograph, with neurosensory macular detachment

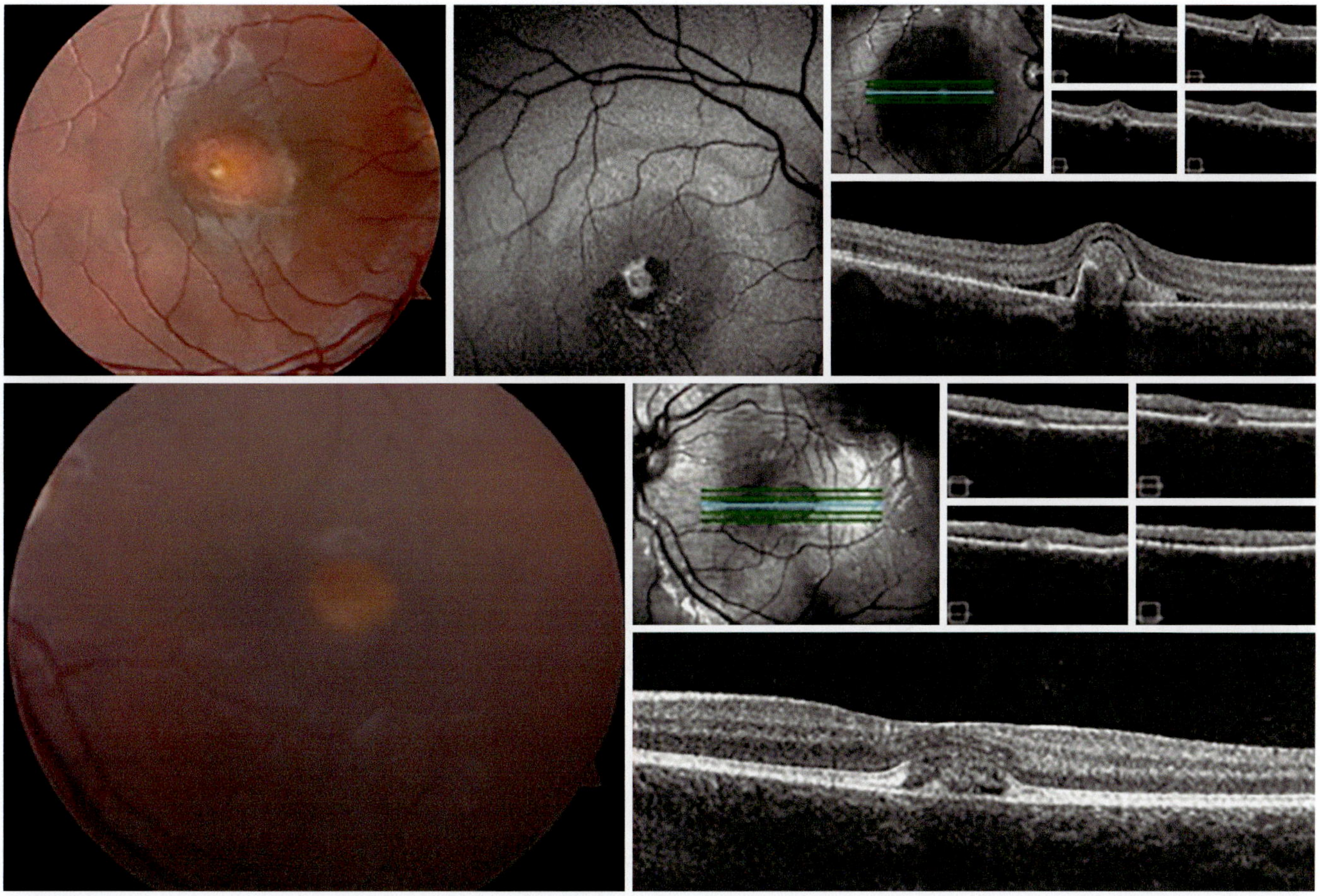

Fig. 16.3 Vitelliform lesion is getting absorbed. Autofluorescence shows retinal pigment epithelium (RPE) dysfunction in the right eye, and the left eye shows neurosensory detachment at the macula

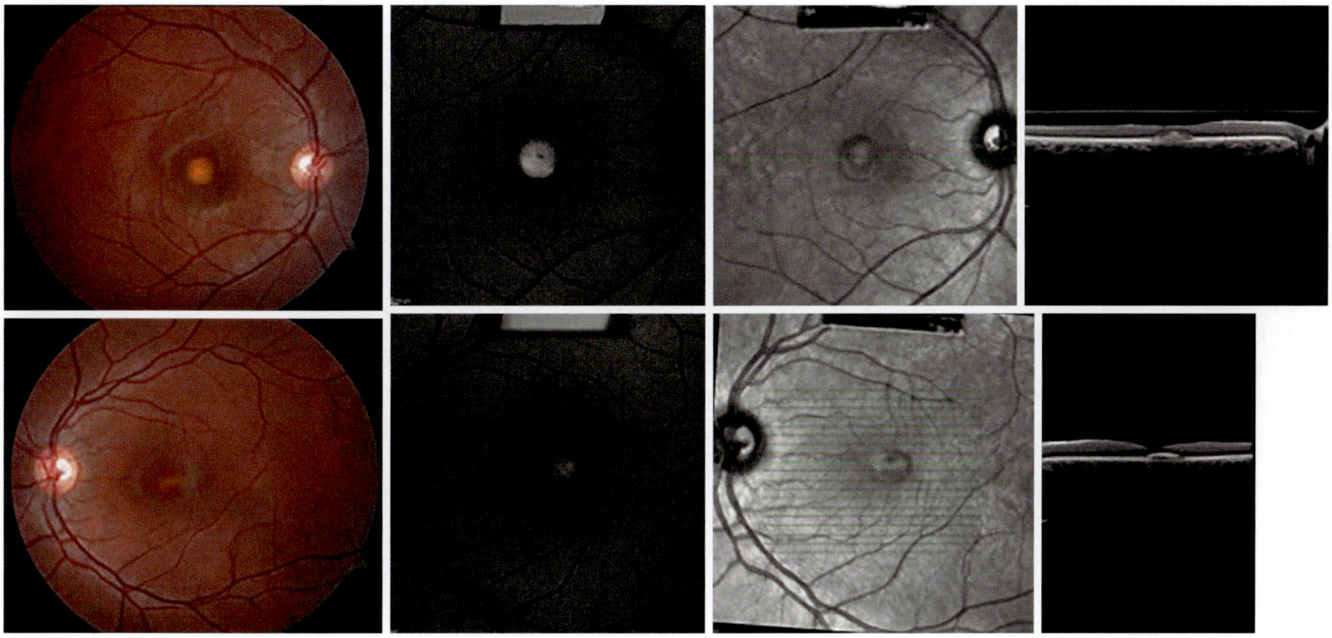

Fig. 16.4 In this asymmetrical picture, the yellowish yolk is less absorbed in the right eye than in the left one, with corresponding intense hyperfluorescence. Optical coherence tomography (OCT) shows hyperreflective material in the subretinal space

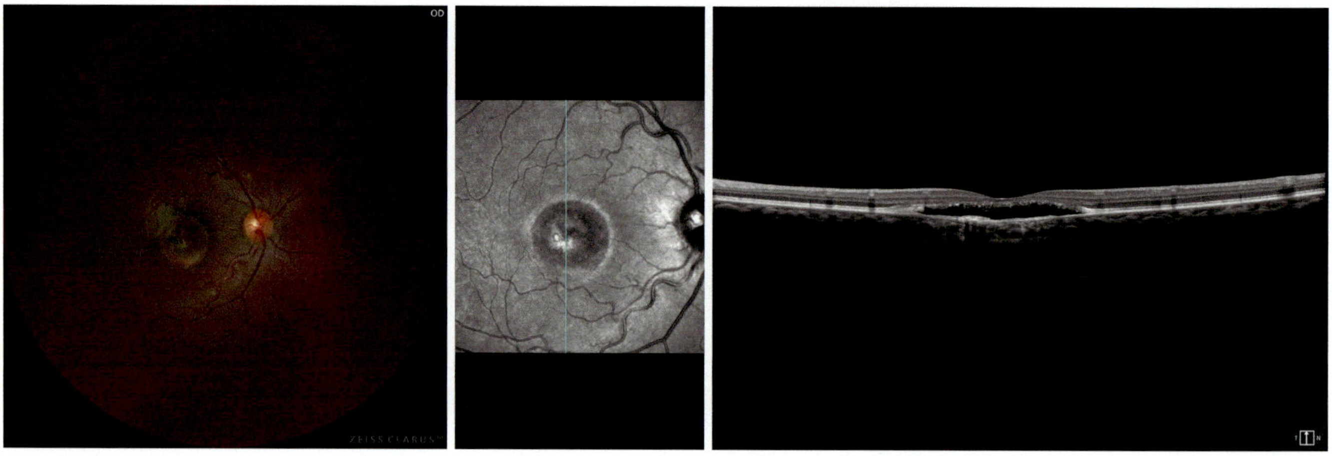

Fig. 16.5 Vitelliruptive stage; OCT shows persistence of neurosensory elevation

Imaging and Tests

- **Fundus autofluorescence (FAF):** The yellowish material (photoreceptor outer segment and lipofuscin-like material) shows intense hyperfluorescence, but the pattern may change as the lesion passes through various stages.
- **Optical coherence tomography (OCT)** shows neurosensory detachment at the macula; rarely, choroidal neovascular membrane and scarring are also visible. The material involved is shed photoreceptor outer segments.
- **Electroretinography (ERG)** is normal, but **electrooculography (EOG)** is always abnormal, with an Arden ratio of about 1.0 or less than 1.5. This is an important marker in those with a normal fundus or poorly defined lesion.
- Some *BEST1* mutation carriers (previtelliform stage) do show an abnormal EOG; the fundus may be normal or just show some RPE alterations.

Subtypes

Adult-Onset Foveomacular Vitelliform Dystrophy (AFMD)

- This disorder resembles VMD, but its onset is at a higher age (30–60 years) and patients have normal EOG.
- The gene involved is *PRPH-2*/*RDS*.

Autosomal Recessive Bestrophinopathy (ARB)

- In this rare entity (Fig. 16.6), lesions are multifocal and may extend into the mid-periphery and be associated with flecks. ERG may show progressive photoreceptor damage affecting central vision.
- The macula may show cystoid intraretinal fluid.
- Hyperopia is common and patients do have abnormal EOG.
- Subretinal fibrosis is more common.

Autosomal Dominant Vitreoretinochoroidopathy (ADVIRC)

- These patients show a characteristic hyperpigmented band from equator to ora serrata (almost 360°), along with peripapillary atrophy.
- Usually, patients are hyperopic.

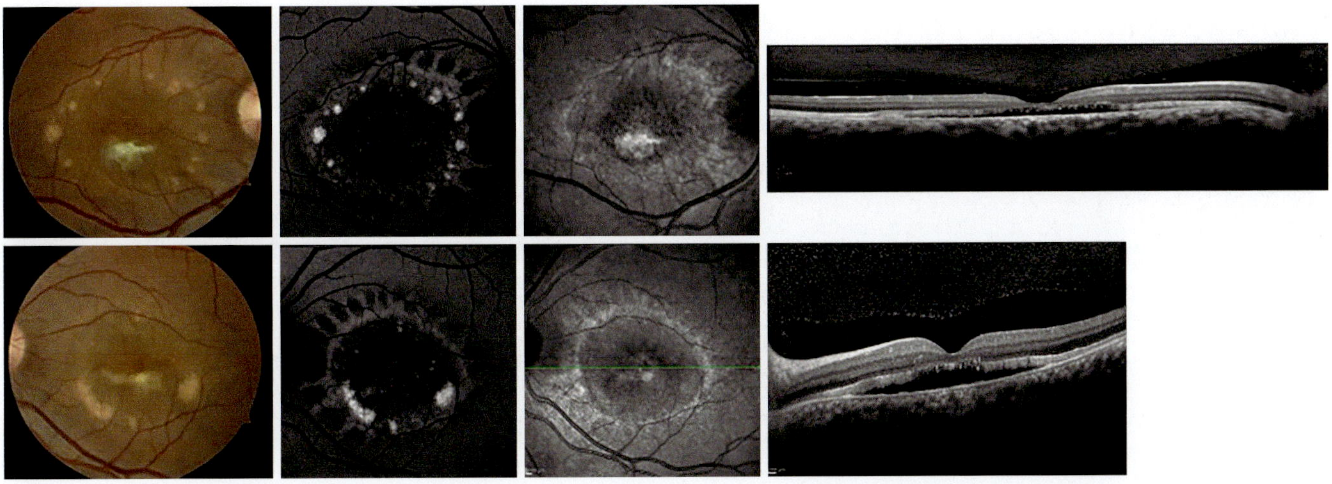

Fig. 16.6 Multifocal yellowish lesions surrounding the macula, with extensive RPE damage on FAF, suggestive of AR bestrophinopathy (ARB)

Molecular Genetics

BEST1 Gene (*VMD2* Gene or *RP50* or *BMD*)

- Mutations in the *BEST1* gene cause Best VMD.
- The *BEST1* gene encodes for a multifunctional protein called bestrophin-1.
- Bestrophin-1, a chloride channel protein, is located on the basolateral membrane of the RPE; abnormal function results in accumulation of lipofuscin in the RPE and subretinal space. Prolonged neurosensory retinal detachment and overloading of the RPE with lipofuscin eventually result in photoreceptor and RPE dysfunction.
- About 150 mutations have been described in *BEST1*, with a broad range of phenotypes. The phenotypes most often associated with *BEST1* mutations are Best VMD and adult-onset foveomacular vitelliform dystrophy, confined to the macular area.
- On the other hand, autosomal recessive bestrophinopathy, autosomal dominant vitreoretinochoroidopathy (ADVIRC), and autosomal dominant MRCS (MRCS syndrome: microcornea, rod-cone dystrophy, cataract, and posterior staphyloma) have widespread ocular changes.
- Cytogenetic location: 11q12.3.

Further Reading

Boon CJ, Klevering BJ, Leroy BP, Hoyng CB, Keunen JE, den Hollander AI. The spectrum of ocular phenotypes caused by mutations in the *BEST1* gene. Prog Retin Eye Res. 2009;28:187–205.

Guerriero S, Preising MN, Ciccolella N, Causio F, Lorenz B, Fischetto R. Autosomal recessive bestrophinopathy: new observations on the retinal phenotype—clinical and molecular report of an Italian family. Ophthalmologica. 2011;225:228–35.

Kolesnikova M, Oh JK, Wang J, Lee W, Zernat J, Su PY, Kim AH, Jenny LA, Yang T, Allikmets R, and Tsang SH. A pathogenic in-fram deletion-insertion variant in *BEST1* phenocopies Stargardt disease. JCI Insight. 2022;7(23):e162687.

Pfister TA, Zein WM, Cukras CA, Sen HN, Maldonado RS, Huryn LA, Hufnagel RS. Recessive Bestrophinopathy and Best Vitelliform Macular Dystrophy. IOVS. 2023;62(6):22.

CHAPTER 17

Pattern Dystrophy

Benjamin Kuang-Chien Chiang, Stephen H. Tsang, and Tarun Sharma

General Features 94
Butterfly-Type Pattern Dystrophy 94
Adult-Onset Foveomacular Vitelliform Dystrophy 97
Sjogren's Reticular-Type Pattern Dystrophy 97
Fundus Pulverulentus 97
Molecular Genetics 97
PRPH2 Gene (Peripherin/RDS or rd2 or Peripherin-2 or RDS) 97
Further Reading 98

B. K.-C. Chiang
School of Medicine, College of Medicine, National Taiwan University, Taipei, Taiwan

S. H. Tsang (✉) · T. Sharma
Department of Ophthalmology, Columbia University, New York, NY, USA
e-mail: sht2@cumc.columbia.edu; ts3118@cumc.columbia.edu

S. H. Tsang et al. (eds.), *Atlas of Inherited Retinal Diseases*, Advances in Experimental Medicine and Biology 1467, https://doi.org/10.1007/978-3-031-72230-1_17

General Features

The pattern dystrophies of the retinal pigment epithelium (RPE) include a group of disorders:

- Butterfly-type pattern dystrophy (Figs. 17.1, 17.2, 17.3, 17.4, and 17.5)
- Adult-onset foveomacular vitelliform dystrophy
- Sjögren's reticular-type pattern dystrophy
- Fundus pulverulentus

Fundus details can vary considerably among affected family members or between the two eyes of an individual. Sometimes one pattern may evolve into another pattern over a period of time.

Visual loss is the result of either geographic atrophy or choroidal neovascular membrane. Another common complaint is delayed recovery from exposure to bright light.

Butterfly-Type Pattern Dystrophy

- The fundus shows a butterfly-shaped pigmentary lesion having three to five wings or arms, surrounded by an area of depigmentation.
- The lesion may begin as just pigmented dots or a faint, yellowish lesion that eventually evolves into a butterfly pattern over years.
- Some patients develop yellowish flecks in the macular region or beyond the arcades.
- Fundus autofluorescence (FAF) shows areas of hyperfluorescence or hypofluorescence corresponding to lipofuscin contents in the RPE.
- Most patients retain good vision for many years or decades, until the development of choroidal neovascularization or marked atrophy of the RPE.

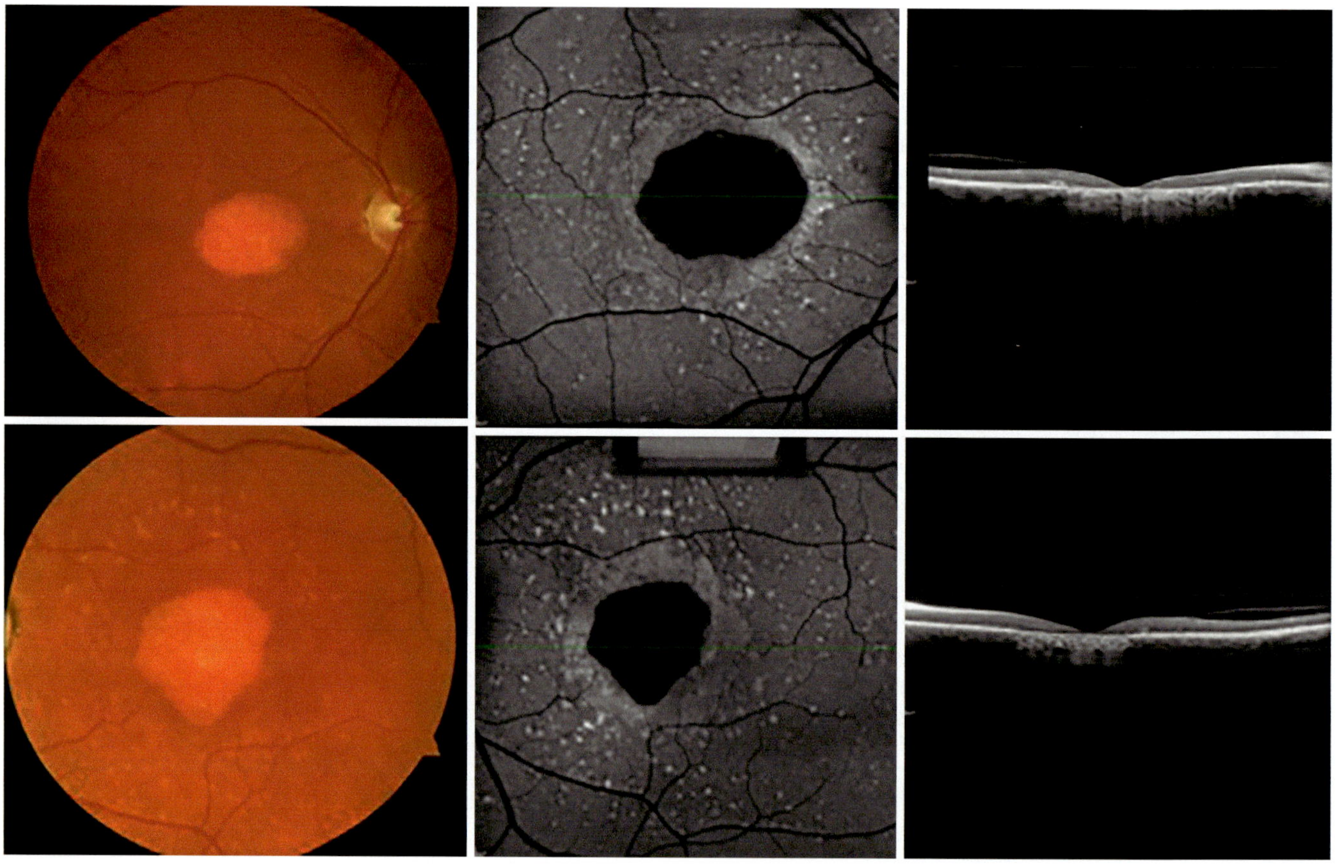

Fig. 17.1 In this patient with *PRPH2* mutation (late stage), an area of retinal pigment epithelial (RPE) atrophy is seen in the macular area, with multiple faint flecks. Fundus autofluorescence (FAF) shows a central area of RPE atrophy surrounded by hyperfluorescent dots. Optical coherence tomography (OCT) shows central RPE atrophy with increased transmission

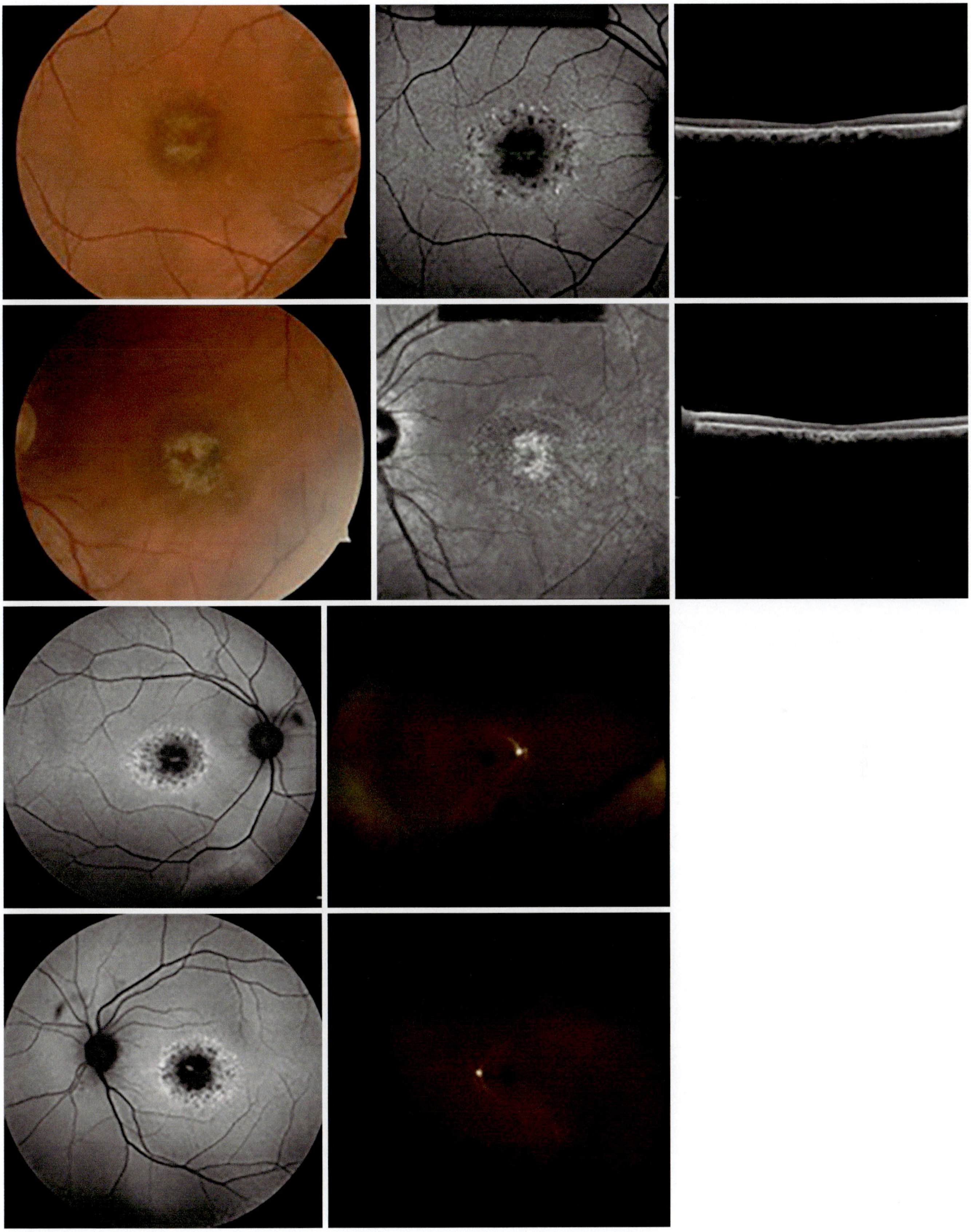

Fig. 17.2 A 5-year follow-up of a case of advanced-stage butterfly-type pattern dystrophy. The right eye remains stable, but the left macula shows increasing RPE dysfunction

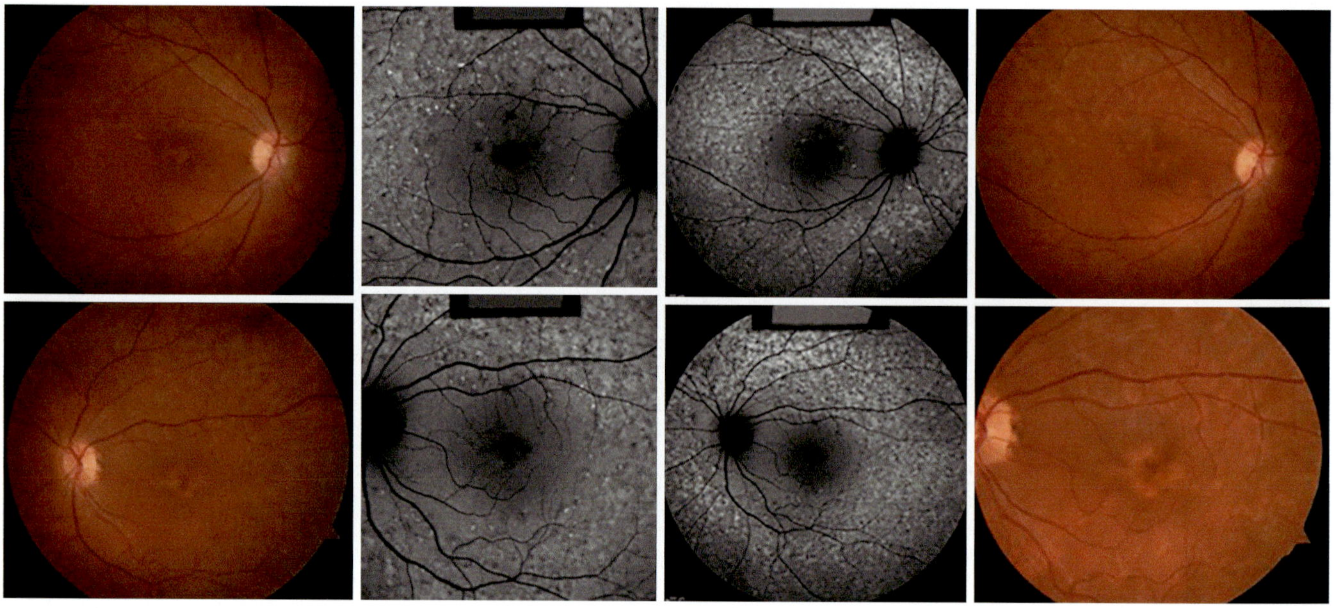

Fig. 17.3 A 2-year follow-up of a case of butterfly-type pattern dystrophy with *PRPH2* mutation. No change was observed in RPE dysfunction on FAF

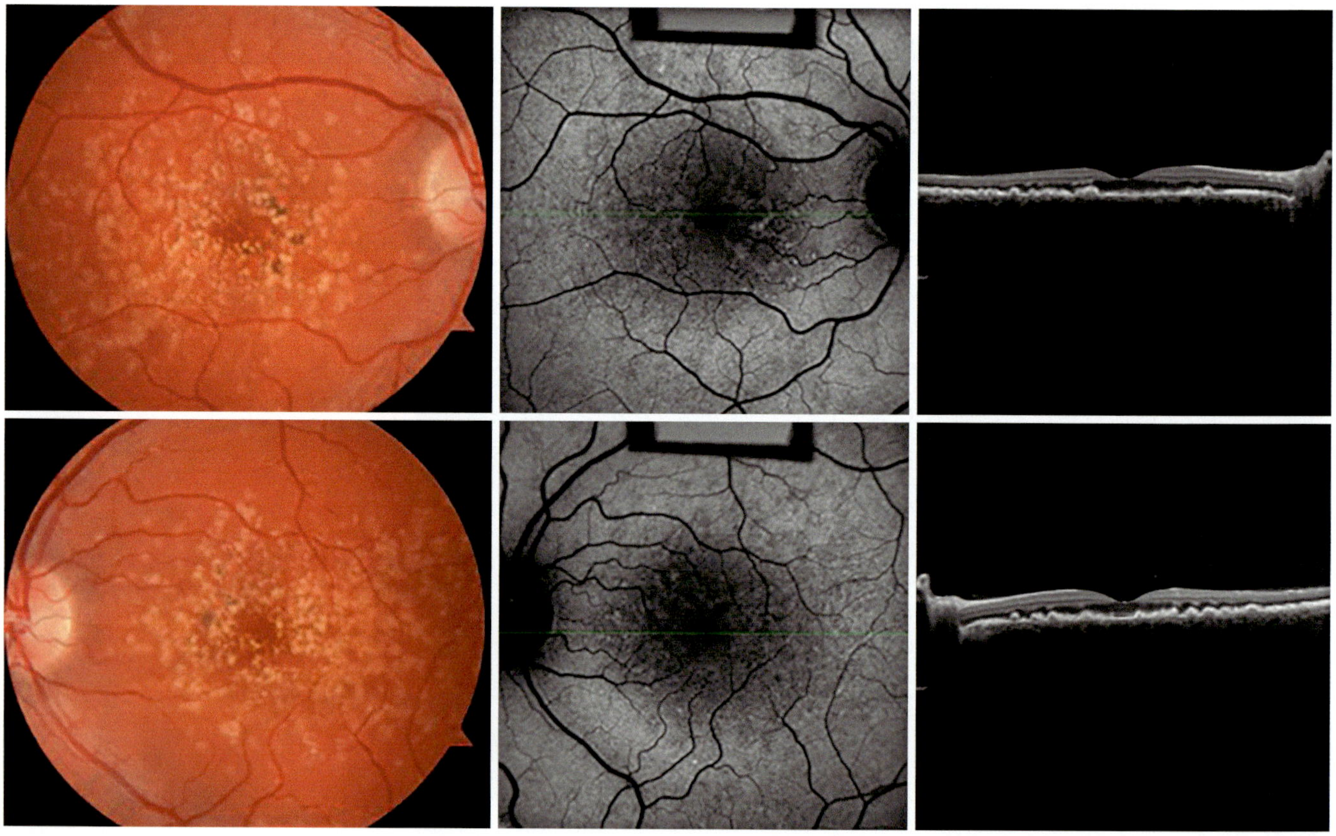

Fig. 17.4 A case of pattern dystrophy with drusen and *PRPH2* mutation. Color fundus shows pigmented dots surrounded by a halo and numerous drusen. FAF shows relatively good RPE function; OCT shows drusen excrescences

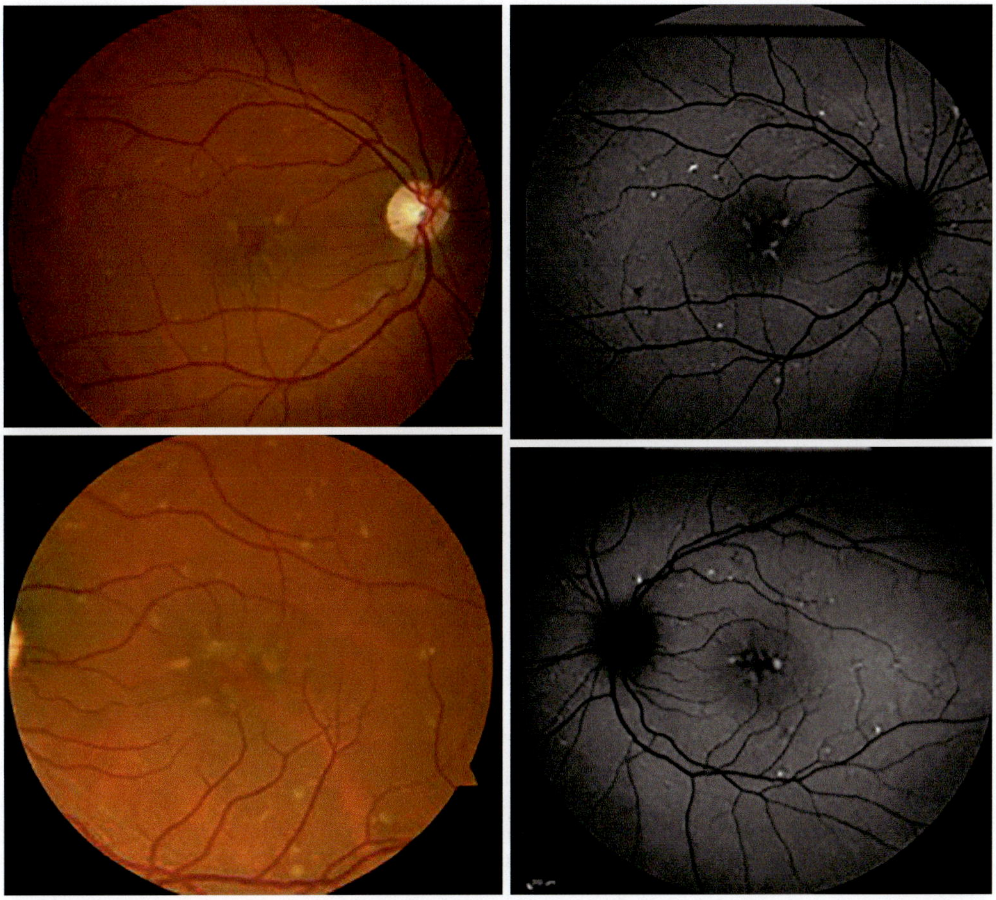

Fig. 17.5 A case of pattern dystrophy. A stellate pattern is more discernible on FAF

Adult-Onset Foveomacular Vitelliform Dystrophy

- The lesion mimics Best disease (*see* Chap. 16), but the yellow lesion is smaller and has an associated central pigment clump. Unlike Best disease, patients present late in life (fourth or fifth decade) with mild visual impairment, which progresses slowly. The Arden ratio is normal or slightly low. FAF shows hyperautofluorescence.
- With time, some lesions may evolve to butterfly-shaped pattern dystrophy or may develop flecks (as in Stargardt disease).

Sjogren's Reticular-Type Pattern Dystrophy

- As the name suggests, the lesion is a network of black-pigmented lines, crisscrossing like fishnet with knots, which may even extend to the periphery. The pattern is well delineated on fluorescein and FAF.

Fundus Pulverulentus

- In this rare entity, RPE stippling at the macula occurs with fine pigment clumping interspersed with areas of RPE depigmentation.

Molecular Genetics

PRPH2 Gene (*Peripherin/RDS* or *rd2* or *Peripherin-2* or *RDS*)

- Most of the pattern dystrophies are associated with mutations in *PRPH2*. Some cases of adult-onset vitelliform dystrophy may be associated with mutations in *BEST1* or *IMPG1* and *IMP*.
- The *PRPH2* gene is responsible for making a protein called peripherin 2.
- The peripherin 2 protein is located in the rim region of rods and cones' outer segment (OS) discs and lamellae; it plays a critical role in OS morphogenesis.

- *PRPH2* mutations result in an autosomal dominant phenotype in most situations, except for digenic retinitis pigmentosa (RP), which requires a mutation in the *ROM1* gene.
- *PRPH2* mutation results in vitelliform macular dystrophy, cone-rod dystrophy, RP, and butterfly-type pattern dystrophy. At least 100 mutations have been identified.
- Phenotypic variations are common even in the same family with identical *PRPH2* mutations, because of decreased penetrance and variable expression.
- Cytogenetic location: 6p21.1.

Further Reading

Boon CJ, den Hollander AI, Hoyng CB, Cremers FP, Klevering BJ, Keunen JE. The spectrum of retinal dystrophies caused by mutations in the *peripherin/RDS* gene. Prog Retin Eye Res. 2008;27:213–35.

Manes G, Meunier I, Avila-Fernández A, Banfi S, Le Meur G, Zanlonghi X, et al. Mutations in *IMPG1* cause vitelliform macular dystrophies. Am J Hum Genet. 2013;93:571–8.

Marmor MF. The pattern dystrophies. In: Heckenlively JR, Arden GB, eds. Principles and practice of clinical electrophysiology of vision. 2nd ed. Cambridge, MA: MIT Press; 2006. p. 757–61.

CHAPTER 18

Doyne Honeycomb Retinal Dystrophy (Malattia Leventinese, Autosomal Dominant Drusen)

Kristina J. Hartung, Stephen H. Tsang, Tarun Sharma, and Vlad Diaconita

General Features 100
Molecular Genetics 105
EFEMP1 Gene (DHRD or DRAD or FBLN3 or FIBL-3) 105
Further Reading 105

K. J. Hartung
Department of Ophthalmology, University Medical Centre Ljubljana, Ljubljana, Slovenia

Faculty of Medicine, University of Ljubljana, Ljubljana, Slovenia

S. H. Tsang (✉) · T. Sharma
Department of Ophthalmology, Columbia University, New York, NY, USA
e-mail: sht2@cumc.columbia.edu; ts3118@cumc.columbia.edu

V. Diaconita
Department of Ophthalmology, Columbia University, New York, NY, USA

Department of Ophthalmology, Edward S. Harkness Eye Institute
New York-Presbyterian Hospital Columbia University, New York, NY, USA
e-mail: vd2374@cumc.columbia.edu

S. H. Tsang et al. (eds.), *Atlas of Inherited Retinal Diseases*, Advances in Experimental Medicine and Biology 1467, https://doi.org/10.1007/978-3-031-72230-1_18

General Features

Doyne Honeycomb Retinal Dystrophy, also known as Malattia Leventinese or Familial Dominant Drusen, is an autosomal dominant condition with a wide range of clinical presentations; variability in the clinical picture is common even within families.

- Symptoms: Patients are often asymptomatic until the fourth to fifth decade, and the drusen are usually first discovered incidentally on routine exams. As the disease progresses, blurry vision, metamorphopsia, dyschromatopsia, and scotomas may be noticed. Poor visual acuity (less than 20/200) is common in the seventh and eighth decades.
- Fundus: Early onset drusen are present on the posterior pole, in the center of the macula, peripapillary, especially nasally of the optic disc (a characteristic finding), and along the vascular arcades. While small, often radial drusen are present at first, they increase in size and number with age. The periphery is usually spared. Over time central atrophy, pigment hyperplasia, irregular subretinal fibrosis, and rarely choroidal neovascular membrane may appear, resembling Sorsby fundus dystrophy and AMD.
- Imaging: Drusen show areas of increased hyperautofluorescence (whereas hypoautofluorescent drusen are present in AMD), but hypoautofluorescence may be seen in areas of RPE atrophy. OCT shows the accumulation of the drusen material under the RPE, seen as pigment epithelial detachments and sub-RPE hyperreflectivity. Diffuse ellipsoid zone loss, outer and inner segment disruption, and retinal atrophy may be seen in later stages. FA and OCTA are useful in CNV detection (Figs. 18.1, 18.2, 18.3, 18.4, 18.5, 18.6, and 18.7).
- Electrophysiology: The full-field ERG is usually normal, and pattern ERG is abnormal in most patients.

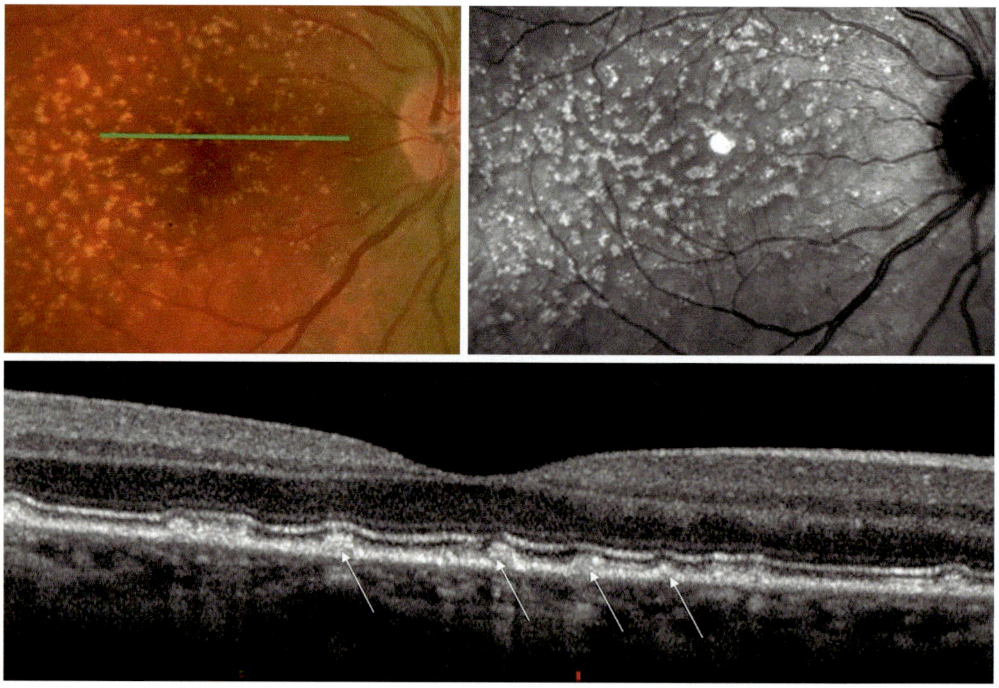

Fig. 18.1 Radial pattern of autosomal dominant drusen (upper row). Optical coherence tomography (OCT) shows drusen located under the retinal pigment epithelium (RPE) (lower row, *arrows*)

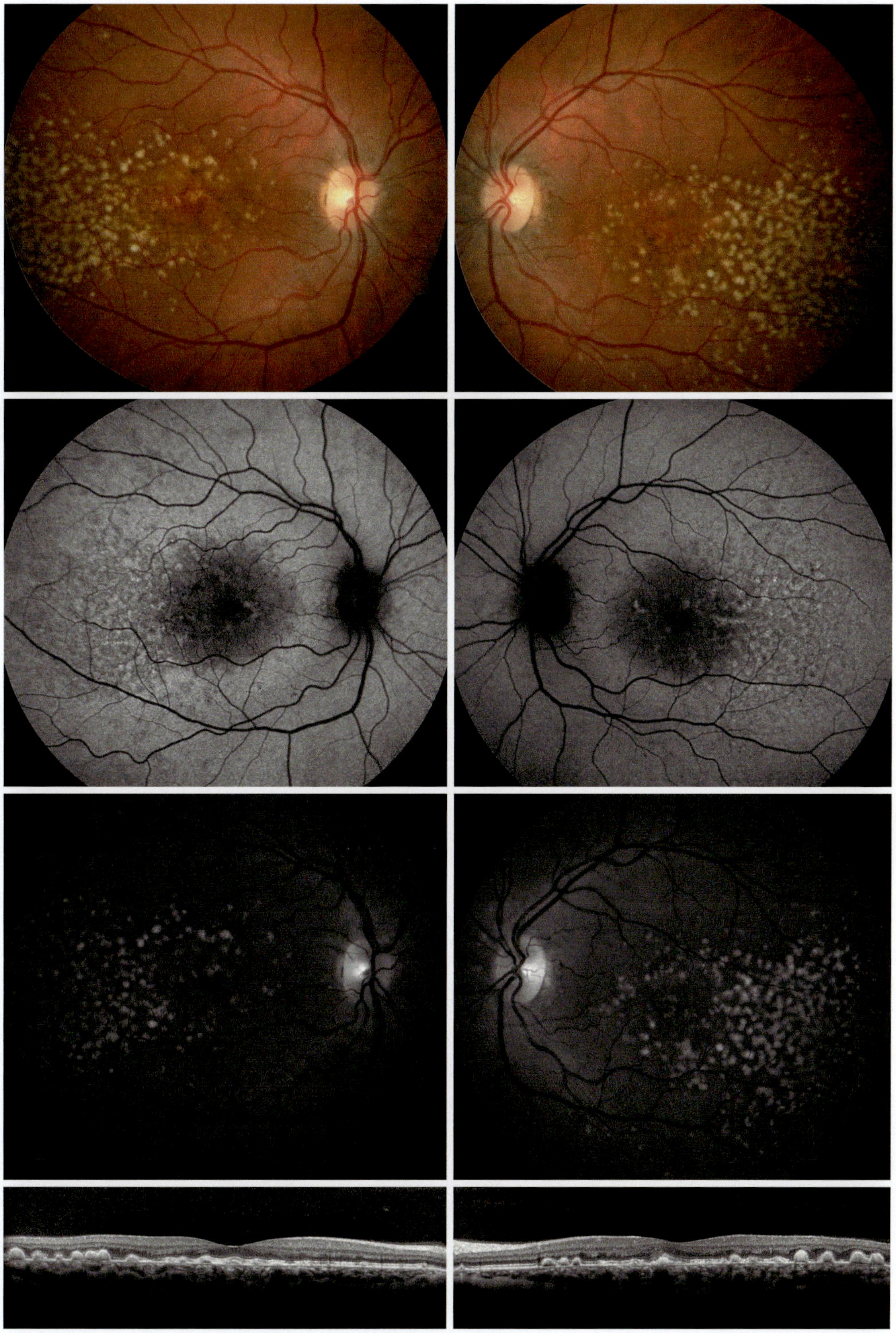

Fig. 18.2 Multiple drusen along the horizontal raphe; fundus autofluorescence (FAF) shows hyperautofluorescent drusen and good RPE function. OCT shows multiple bumps beneath the RPE

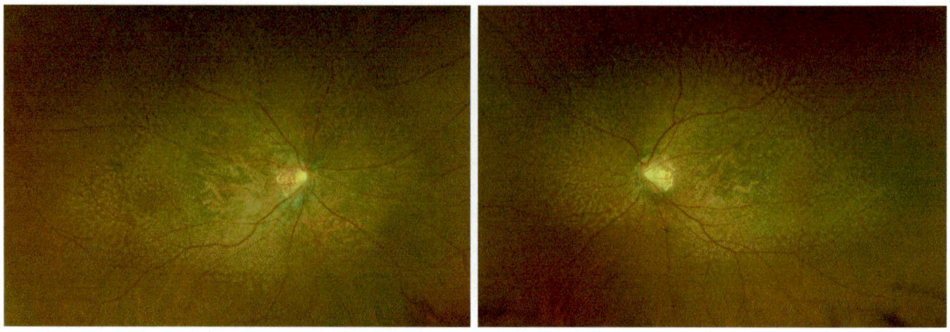

Fig. 18.3 Drusen at the macula and nasal to the optic disc (a characteristic finding)

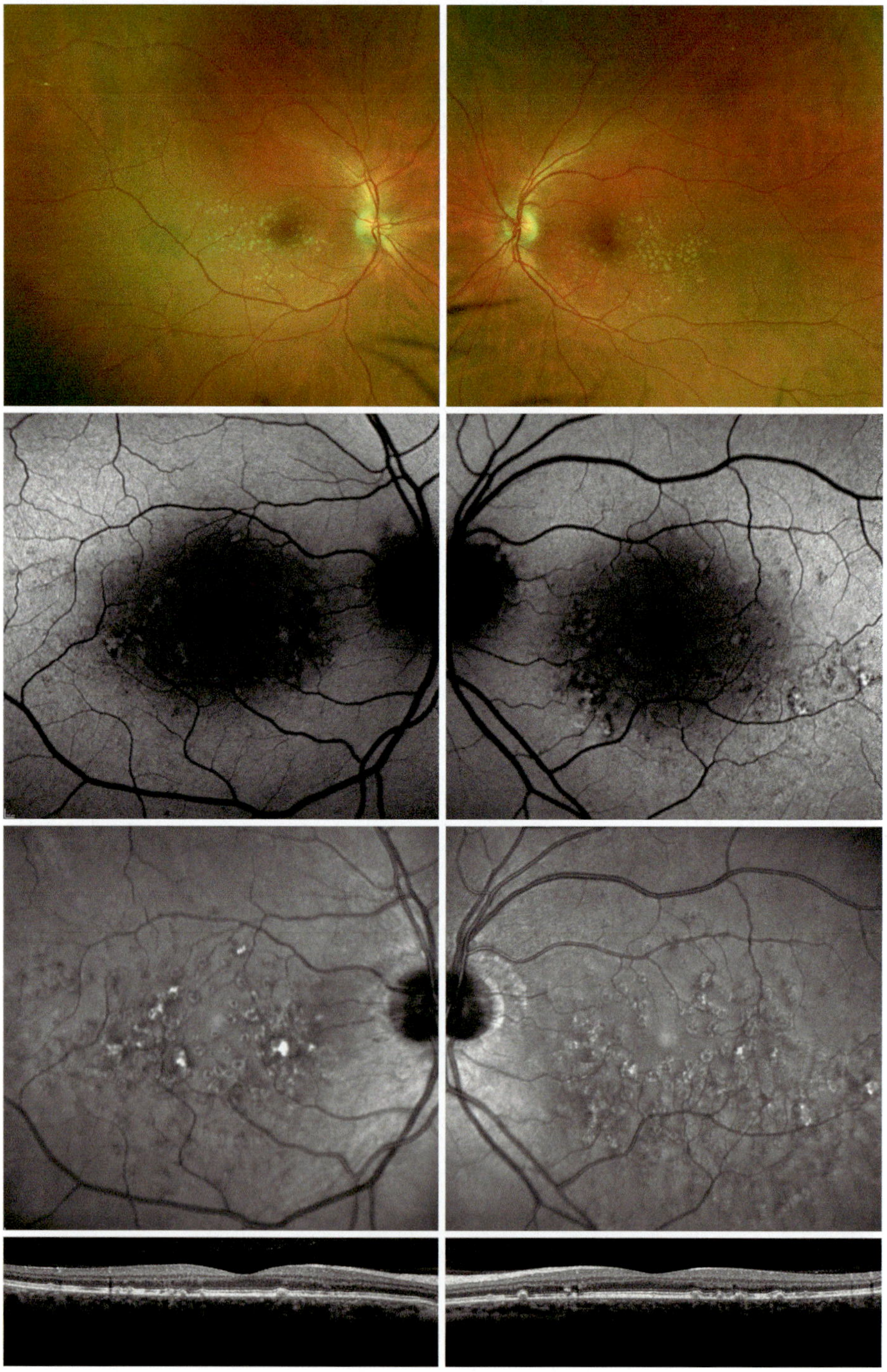

Fig. 18.4 Color fundus photographs show early-stage bilateral radial drusen in the macula. Note the subtle hyperfluorescence in the corresponding FAF and IR, as well as small drusenoid deposits on the OCT with preserved EZ line subfoveally

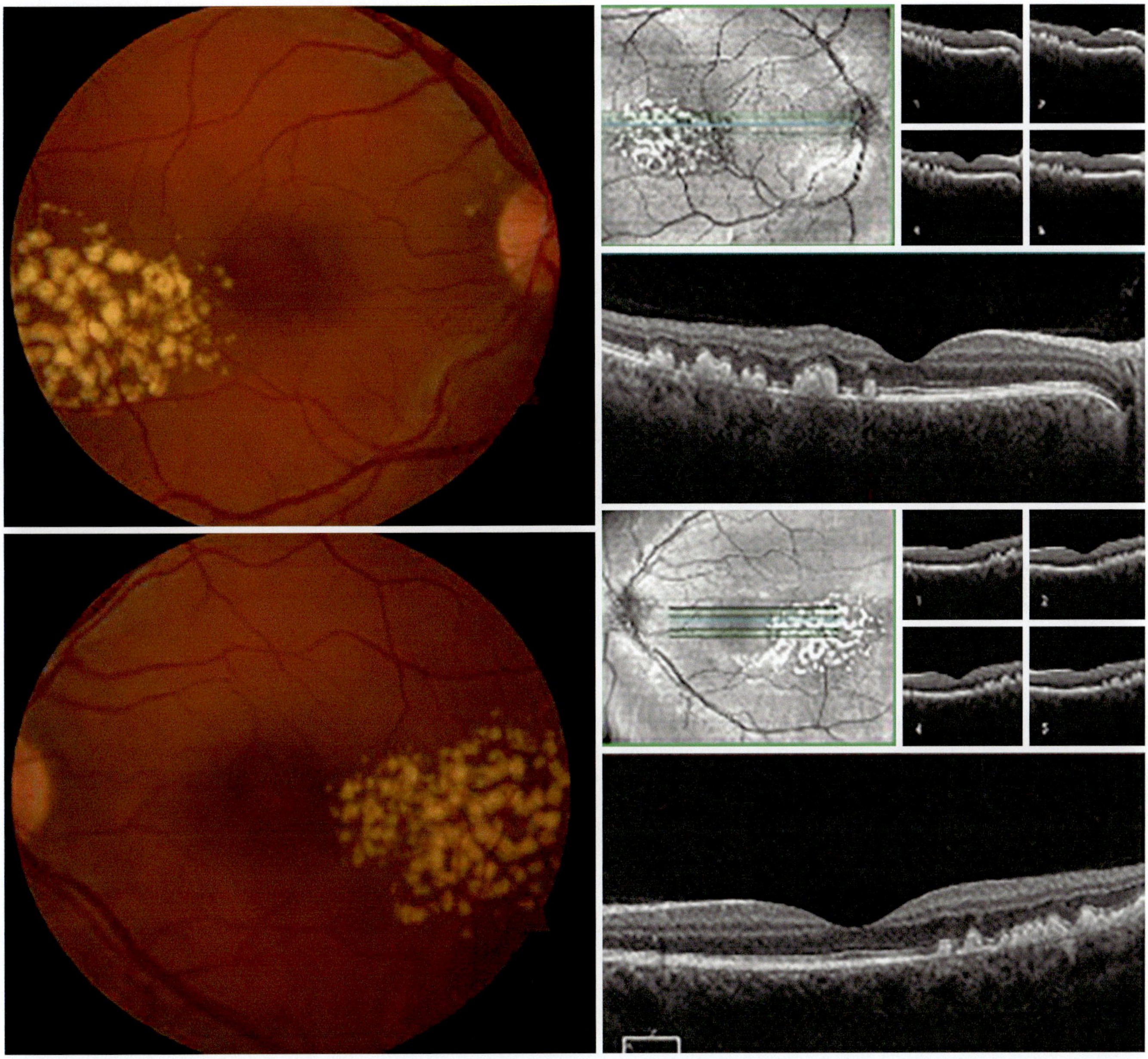

Fig. 18.5 Bilateral Doyne (arrows) almost encircling the fovea in a young boy (p.Arg345Trp)

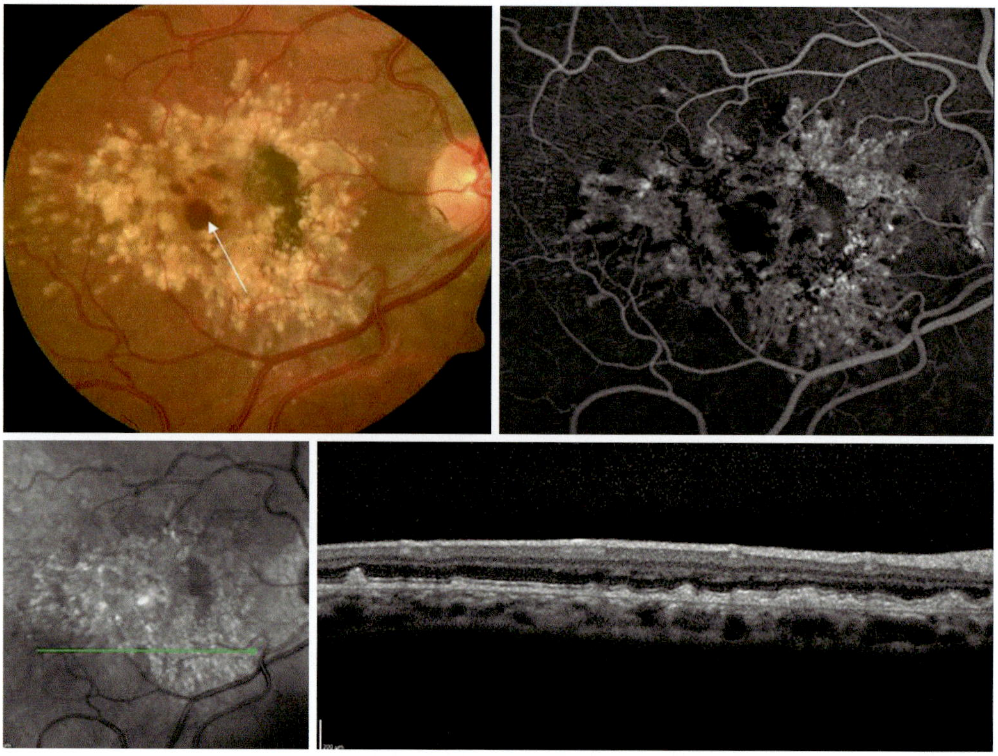

Fig. 18.6 Confluent drusen and radial pattern are discernible; the presence of hemorrhage (*arrow*) suggests a choroidal neovascular membrane

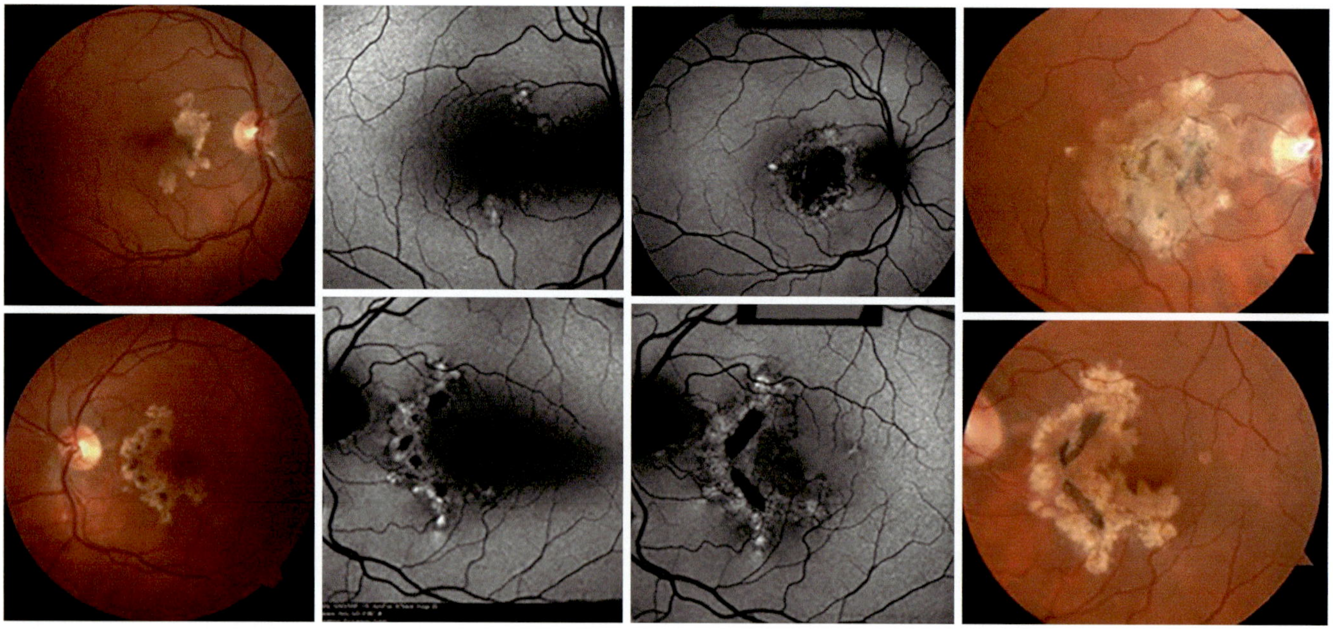

Fig. 18.7 Ten-year follow-up shows the progression of drusen (p.Arg345Trp), which are becoming confluent. FAF shows progressive RPE dysfunction

Molecular Genetics

EFEMP1 Gene (*DHRD* or *DRAD* or *FBLN3* or *FIBL-3*)

- A single heterozygous missense mutation (p.Arg345Trp) in the *EFEMP1* (epidermal growth factor [EGF]–containing fibulin-like extracellular matrix protein 1) gene is responsible for this condition.
- This gene is a member of the fibulin family of extracellular matrix glycoproteins.
- Cytogenetic location: 2p16.1.

Further Reading

Sohn EH, Wang K, Thompson S, Riker MJ, Hoffmann JM, Stone EM, Mullins RF. Comparison of drusen and modifying genes in autosomal dominant radial drusen and age-related macular degeneration. Retina. 2015 Jan;35(1):48–57.

Stone EM, Lotery AJ, Munier FL, Héon E, Piguet B, Guymer RH, et al. A single *EFEMP1* mutation associated with both Malattia Leventinese and Doyne honeycomb retinal dystrophy. Nat Genet. 1999;22:199–202.

Takeuchi T, Hayashi T, Bedell M, Zhang K, Yamada H, Tsuneoka H. A novel haplotype with the R345W mutation in the *EFEMP1* gene associated with autosomal dominant drusen in a Japanese family. Invest Ophthalmol Vis Sci. 2010;51:1643–50.

CHAPTER 19

Occult Macular Dystrophy

Kristina J. Hartung, Stephen H. Tsang,
Tarun Sharma, and Vlad Diaconita

General Features 108
Molecular Genetics 109
Further Reading 109

K. J. Hartung
Department of Ophthalmology, University Medical Centre
Ljubljana, Ljubljana, Slovenia

Faculty of Medicine, University of Ljubljana,
Ljubljana, Slovenia

S. H. Tsang (✉) · T. Sharma · V. Diaconita
Department of Ophthalmology, Columbia University,
New York, NY, USA

Department of Ophthalmology, Edward S. Harkness Eye Institute
New York-Presbyterian Hospital Columbia University,
New York, NY, USA
e-mail: sht2@cumc.columbia.edu; ts3118@cumc.columbia.edu;
vd2374@cumc.columbia.edu

S. H. Tsang et al. (eds.), *Atlas of Inherited Retinal Diseases*, Advances in Experimental Medicine and Biology 1467, https://doi.org/10.1007/978-3-031-72230-1_19

General Features

Occult macular dystrophy (OMD), also known as Miyake disease, is a rare, autosomal dominant disease. Symptom onset is variable; the majority of patients present in their late twenties with a progressive bilateral decline of central vision, photophobia, and central scotoma. The mean visual acuity is 20/80 but can vary from 20/30 to 20/200 or worse.

No significant abnormality is seen in the fundus or fluorescein angiography. Optical coherence tomography (OCT) findings range from blurring of the interdigitation zone (IZ) and the ellipsoid zone (EZ) with thinning of the outer nuclear layer to extinguished IZ and disrupted EZ resulting in the optical gap in the fovea (Fig. 19.1). Full-field electroretinography (ERG) is normal; multifocal ERG and pattern ERG reveal diminished amplitudes, suggestive of macular dysfunction.

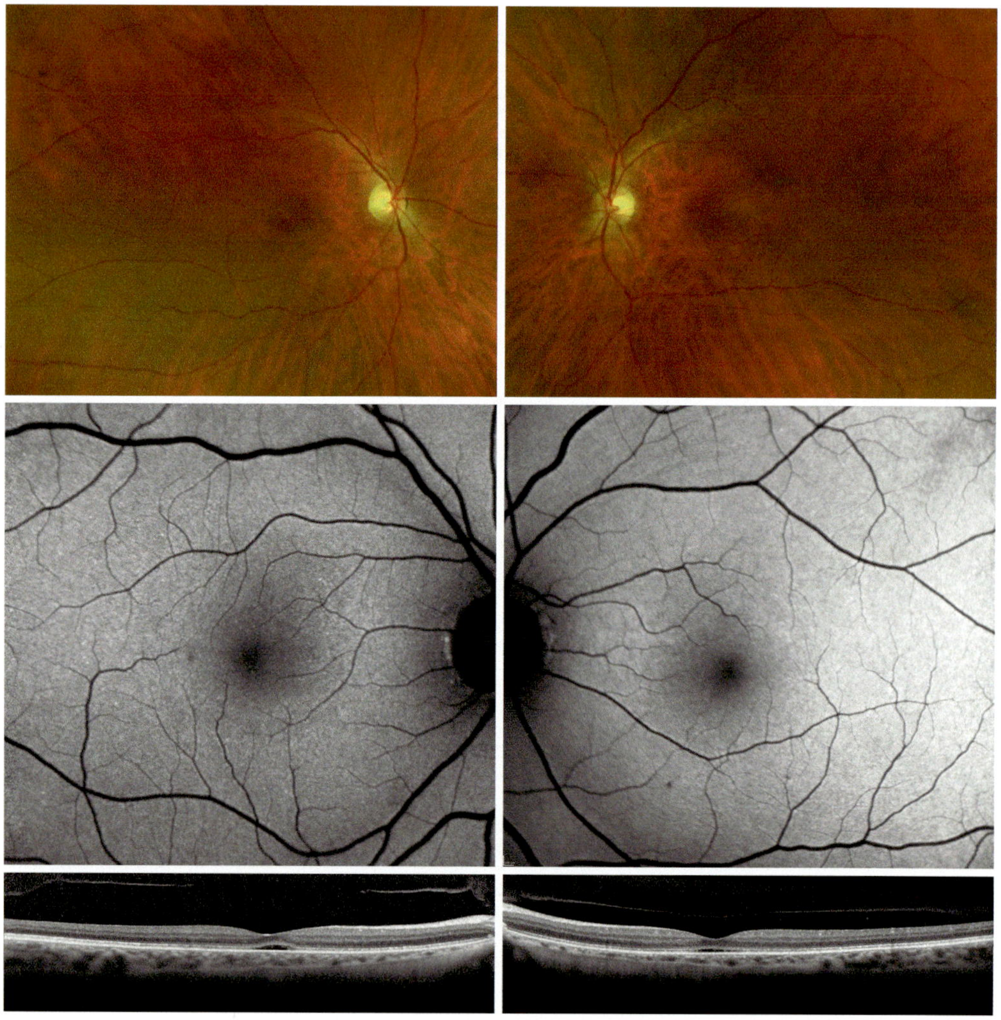

Fig. 19.1 Normal fundus and fundus autofluorescence (FAF) in a patient with occult macular dystrophy (RP1L1; V1: C.133C>T:P.Arg45Trp; V2: c.449C>T:p.Thr150Ilw). Optical coherence tomography (OCT) shows an optical gap—the absence of the interdigitation zone (IZ) and ellipsoid zone (EZ) disruption in the fovea

Molecular Genetics

- The RP1L1 (retinitis pigmentosa 1-like 1) mutation can result in varied patterns of retinal diseases, including autosomal dominant OMD, autosomal recessive retinitis pigmentosa, and autosomal recessive cone dystrophy.
- Cytogenetic location: 8p23.1.
- Thus far, 24 variants of RP1L1 have been reported in association with OMD, and the most common one is the c.133C > T, p.Arg45Trp mutation in exon 2. Another mutation hot spot includes six amino acids (1196–1201) in exon 4.
- The gene (also known as *DCDC4B*, doublecortin domain-containing 4B) encodes a member of the doublecortin family and encodes a retina-specific protein. This protein and the RP1 protein have synergistic roles in outer segment morphogenesis.

Further Reading

Ahn SJ, Cho SI, Ahn J, Park SS, Park KH, Woo SJ. Clinical and genetic characteristics of Korean occult macular dystrophy patients. Invest Ophthalmol Vis Sci. 2013;54:4856–63.

Hayashi T, Gekka T, Kozaki K, Ohkuma Y, Tanaka I, et al. Autosomal dominant occult macular dystrophy with an RP1L1 mutation (R45W). Optom Vis Sci. 2012;89:684–91.

Nakamura N, Tsunoda K, Mizuno Y, Usui T, Hatase T, et al. Clinical stages of occult macular dystrophy based on optical coherence tomographic findings. Invest Ophthalmol Vis Sci. 2019;60(14):4691–4700.

Wang DD, Gao FJ, Li JK, Chen F, Hu FY, et al. Clinical and genetic characteristics of chinese patients with occult macular dystrophy. Invest Ophthalmol Vis Sci. 2020;61(3):10.

Yang L, Joo K, Tsunoda K, Kondo M, Fujinami-Yokokawa Y, et al. Spatial functional characteristics of east asian patients with occult macular dystrophy (miyake disease); EAOMD Report No. 2. Am J Ophthalmol 2021;221:169–180.

CHAPTER 20

Sorsby Pseudoinflammatory Fundus Dystrophy

Kristina J. Hartung, Stephen H. Tsang, Tarun Sharma, and Vlad Diaconita

General Features 112
Molecular Genetics 114
TIMP3 Gene (K222 or SFD) 114
Further Reading 114

K. J. Hartung
Department of Ophthalmology, University Medical Centre Ljubljana, Ljubljana, Slovenia

Faculty of Medicine, University of Ljubljana, Ljubljana, Slovenia

S. H. Tsang (✉) · T. Sharma
Department of Ophthalmology, Columbia University, New York, NY, USA
e-mail: sht2@cumc.columbia.edu; ts3118@cumc.columbia.edu

V. Diaconita
Department of Ophthalmology, Edward S. Harkness Eye Institute New York-Presbyterian Hospital Columbia University, New York, NY, USA
e-mail: vd2374@cumc.columbia.edu

S. H. Tsang et al. (eds.), *Atlas of Inherited Retinal Diseases*, Advances in Experimental Medicine and Biology 1467, https://doi.org/10.1007/978-3-031-72230-1_20

General Features

Sorsby fundus dystrophy (SFD) is a rare autosomal dominant disease with complete penetrance and great variability even within the same mutation.

- Symptoms: Night blindness usually begins in the third decade; however, patients present to the ophthalmologist in the fourth to the sixth decades of life with bilateral loss of central vision (20/200 or less), central scotomas, reduced color vision, and metamorphopsia in the setting of bilateral choroidal neovascularization. The choroidal neovascular membranes mimic age-related macular degeneration (AMD) but have an earlier onset.
- Fundus: Initially, symmetrical, bilateral, pale, drusen-like deposits or confluent faint yellow 20–30 μm thick deposits at the level of Bruch's membrane are seen. They eventually progress to either geographic atrophy with pigmentary clumps or a scar due to choroidal neovascular membrane (Figs. 20.1, 20.2, and 20.3). As opposed to AMD, the lesions often extend beyond the temporal vascular arcades.
- Differential diagnosis: AMD, familial dominant drusen, pattern dystrophy (PRPH2), and other dominantly inherited macular dystrophies, including late-stage Best disease (BEST1).

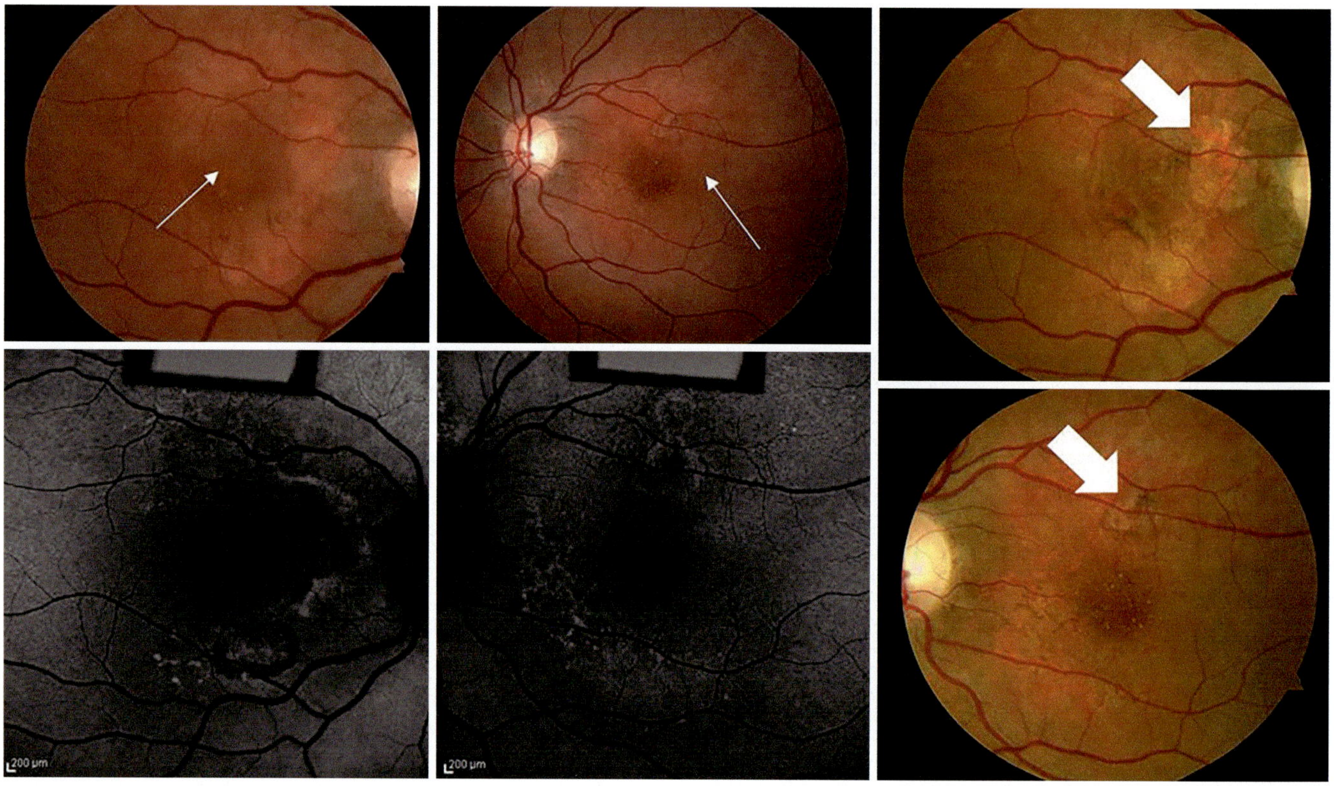

Fig. 20.1 At the initial visit (left), the fundus showed fine, yellowish flecks at the macula (*arrows*) and corresponding retinal pigment epithelial (RPE) dysfunction on fundus autofluorescence (FAF); 6 months later (right), it has progressed to RPE atrophy, more in the right eye than in the left (*thick arrows*)

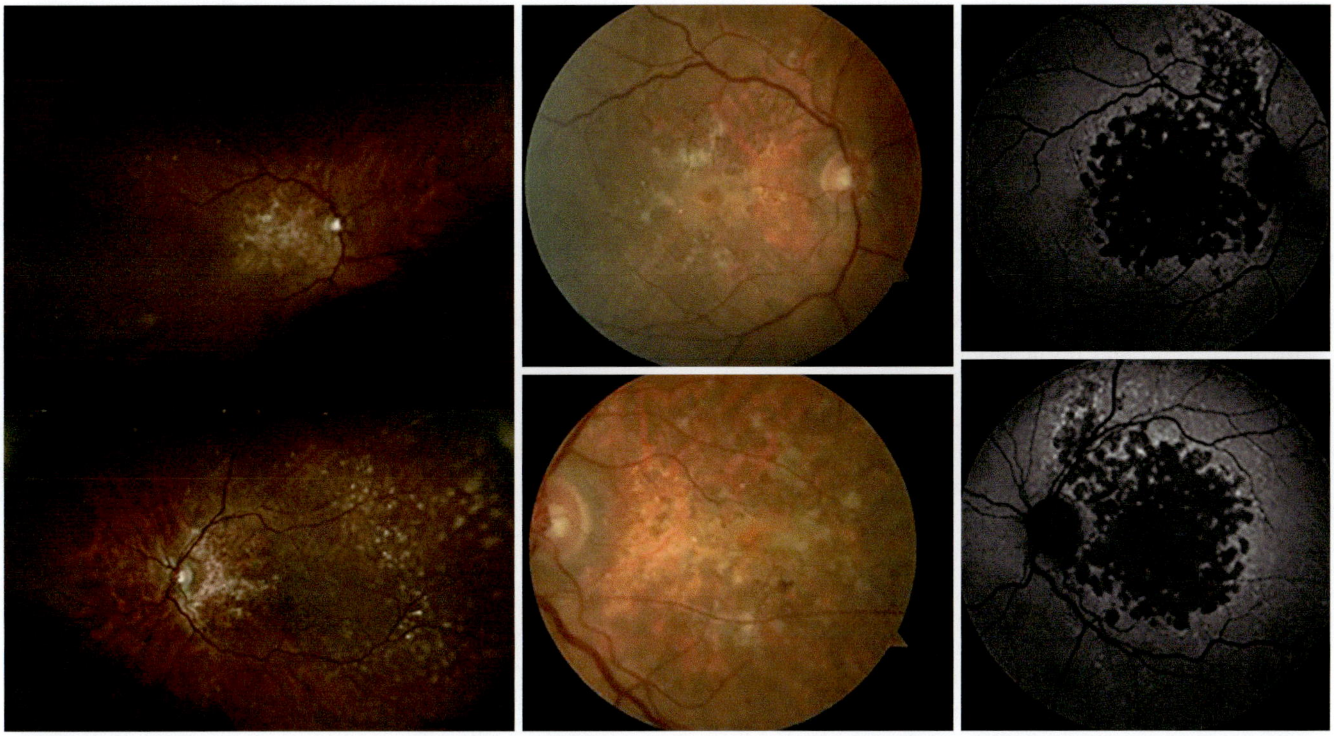

Fig. 20.2 Atrophic changes at the macula along with some peripheral yellowish deposit and severe RPE dysfunction on FAF

Fig. 20.3 *TIMP3* gene mutation (c.610A>T p. Ser204Cys) causing Sorsby macular dystrophy with advanced RPE and choriocapillaris (CC) atrophy on color fundus and FAF (greater in the left eye)

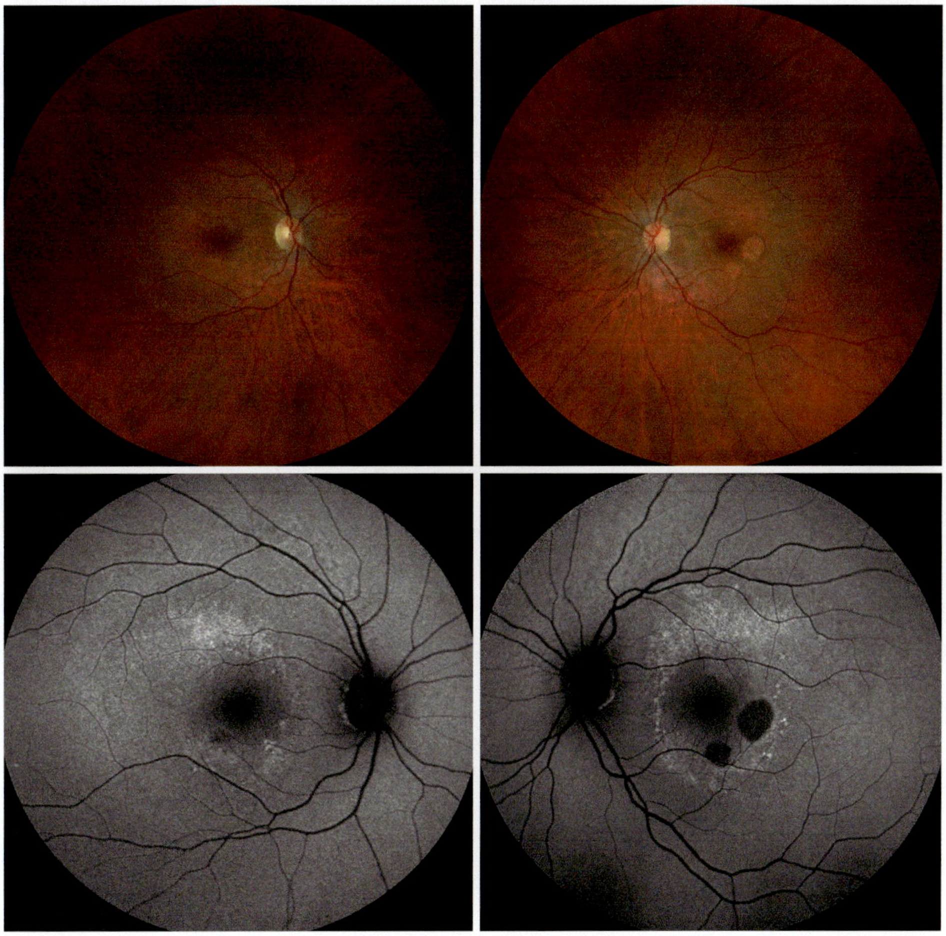

Molecular Genetics

TIMP3 Gene (K222 or SFD)

The TIMP3 gene encodes a tissue inhibitor of metalloproteinase-3, which catalyzes the breakdown of the extracellular matrix.

- Cytogenetic location: 22q12.3.
- Most mutations are located in exon 5 (C-terminal portion of the protein) and result in either a gain or a loss of a cysteine residue.
- There are currently 16 known mutations: stop mutation (p.(Glu162*)), mutation at the intron 4/exon 5 boundary (c.439–2dupA), and 14 missense mutations that result in amino acid substitutions.
- Mutations cause altered extracellular matrix turnover, which results in the thickening of the Bruch's membrane and the accumulation of lipidic and proteinaceous material beneath the RPE.

Further Reading

Guan B, Huryn LA, Hughes AB, Li Z, Bender C, et al. Early-onset TIMP3-related retinopathy associated with impaired signal peptide. JAMA Ophthalmol. 2022;140(7):730–733.

Polkinghorne PJ, Capon MR, Berninger T, Lyness AL, Sehmi K, Bird AC. Sorsby's fundus dystrophy. A clinical study. Ophthalmology. 1989;96:1763–8.

Sorsby A, Mason ME. A fundus dystrophy with unusual features. Br J Ophthalmol. 1949;33:67–97.

CHAPTER 21

North Carolina Macular Dystrophy

Amber Diaz, Kristina J. Hartung, and Kent Small

General Features 116
Molecular Genetics 117
References 117

A. Diaz · K. Small (✉)
Macula & Retina Institute, Glendale and Los Angeles, CA, USA

Molecular Insight Research Foundation, Glendale and Los Angeles, CA, USA

K. J. Hartung
Department of Ophthalmology, University Medical Centre Ljubljana, Ljubljana, Slovenia

Faculty of Medicine, University of Ljubljana, Ljubljana, Slovenia

S. H. Tsang et al. (eds.), *Atlas of Inherited Retinal Diseases*, Advances in Experimental Medicine and Biology 1467, https://doi.org/10.1007/978-3-031-72230-1_21

General Features

North Carolina Macular Dystrophy (NCMD) is an autosomal dominant, congenital, completely penetrant non-progressive macular malformation. The NCMD phenotype is highly variable even within the same family (Fig. 21.1). Grade 1 individuals have few drusen centrally while grade 2 can appear with confluent drusen to central vitelliform lesions. Grade 3 appears as dramatic choroidal excavations and coloboma like lesions (Small 1989; Small et al. 1991, 2022a). There can be some vision decline secondary to the development of choroidal neovascular membranes (CNVMs) (Bakall et al. 2018). Patients who develop

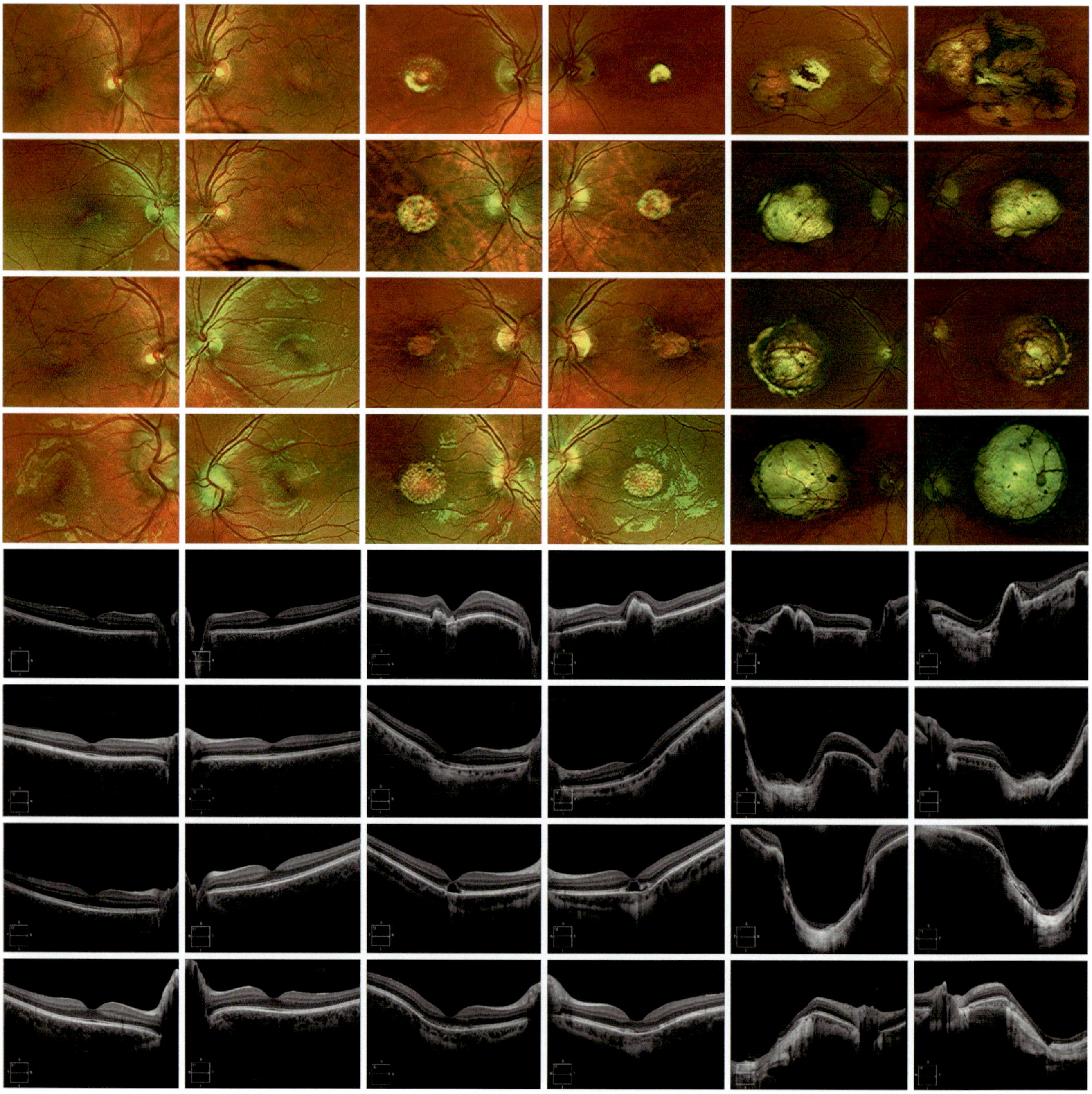

Fig. 21.1 Color fundus photography and corresponding SD-OCT images of the original NCMD family. The first column is grade 1 subjects with few central drusen. The second column is grade 2 subjects with confluent drusen and central vitelliform lesions. The third column shows grade 3 subjects with large central defects with atrophy that bow back into the choroid and sclera with surrounding subretinal fibrosis

CNVMs are the subjects who experience progressive moderate to severe vision impairment, typically confined to one eye. In grade 3 lesions, the CNVMs and resulting fibrosis typically occur along the temporal edge of the "coloboma" and do not affect the visual acuity. Peripheral drusen are variably reported.

Molecular Genetics

The causative genetic variants in NCMD have been mapped to the MCDR1 locus on chromosome 6 in 1991 (Small et al. 1992). In 2016 Small et al found 3-point mutations (i.e., single nucleotide variants [SNVs]) 12 Kb from the closest gene (PRDM13 a retinal transcription factor) in a DNASE1 hypersensitive binding site (Small et al. 2016). Small et al also found at that time large tandem duplications involving the DNASE1 site, PRDM13 at the MCDR1 locus, and a large duplication in a Danish family at the chromosome 5 MCDR3 locus involving another DNASE1 site and the IRX1 gene. To date there are 19 identified mutations causing NCMD (Small et al. 2016, 2020, 2022b, 2023; Bowne et al. 2016; Manes et al. 2017; Silva et al. 2019; Namburi et al. 2020; Wu et al. 2022; Sompele et al. 2022; Seo et al. 2024; Cipriani et al. 2017); Chacon-Camacho (2024).

References

Bakall B, Bryan J S, Stone E M, et al. Choroidal neovascularization in North Carolina macular dystrophy responsive to anti–vascular endothelial growth factor therapy[j]. Retinal Cases Brief Rep. 2018.

Bowne SJ, Sullivan LS, Wheaton DK, et al. North Carolina macular dystrophy (MCDR1) caused by a novel tandem duplication of the PRDM13 gene. Mol Vis. 2016;22:1239-1247.

Cipriani V, Silva RS, Arno G, et al. Duplication events downstream of IRX1 cause North Carolina macular dystrophy at the MCDR3 locus. Sci Rep. 2017;7(1):7512.

Chacon-Camacho OF, Flores-Lagunes LL, Small KW, Udar N, Udar U, Diaz A, Arce-González R, Molina-Garay C, Martínez-Aguilar A, Montes-Almanza L, Garcia-Martinez F, Gudiño A, Matsui-Serrano R, Fest-Parra S, Alaez-Verson C, Shaya F, Zenteno JC. A novel PRDM13 gene duplication causing congenital North Carolina macular dystrophy phenotype in a Mexican family. Mol Vis. 2024;30:400-408.

Manes G, Joly T, Smirnov S, et al. A novel duplication of PRMD13 causes North Carolina macular dystrophy: overexpression of PRDM13 orthologue in Drosophila eye reproduces the human phenotype. Hum Mol Genet. 2017;26(22):4367-4374.

Namburi P, Khateb S, Meyer S, et al. A unique PRDM13-associated variant in a Georgian Jewish family with probable North Carolina macular dystrophy and the possible contribution of a unique CFH variant. Molecular Vision 2020; 26: 299-310.

Seo Y, Joo K, Lee J, Diaz A, Jang S, Cherry TJ, Bujakowska KM, Han J, Woo SJ, & Small KW. Two novel non-coding single nucleotide variants in the DNase1 hypersensitivity site of PRDM13 causing North Carolina macular dystrophy in Korea. Molecular Vision. 2024;30, 58–66. http://www.molvis.org/molvis/v30/58/.

Silva RS, Arno G, Cipriani V, et al. Unique noncoding variants upstream of PRDM13 are associated with a spectrum of developmental retinal dystrophies including progressive bifocal chorioretinal atrophy. Hum Mutat. 2019;40(5):578–587.

Small KW. North Carolina macular dystrophy, revisited. Ophthalmology 1989;96:1747–54.

Small KW, Killian J, McLean WC. North Carolina's dominant progressive foveal dystrophy: how progressive is it? Br J Ophthalmol 1991;75:401–6.

Small KW, Weber JL, Roses A, Lennon F, Vance JM, Pericak-Vance MA. North Carolina macular dystrophy is assigned to chromosome 6. Genomics. 1992;13(3):681-685. doi:https://doi.org/10.1016/0888-7543(92)90141-e

Small KW, DeLuca AP, Whitmore SS, Rosenberg T, Silva-Garcia R, Udar N, Puech B, Garcia CA, Rice TA, Fishman GA, Héon E, Folk JC, Streb LM, Haas CM, Wiley LA, Scheetz TE, Fingert JH, Mullins RF, Tucker BA, Stone EM. North Carolina macular dystrophy is caused by dysregulation of the retinal transcription factor PRDM13. Ophthalmology 2016;123(1):9–18. doi: https://doi.org/10.1016/j.ophtha.2015.10.006. Epub 2015 Oct 24.

Small KW, Sompele SV, Nuytemans K, Vincent A, Yuregir O, Ciloglu E, et al. A novel duplication involving PRDM13 in a Turkish family supports its role in North Carolina macular dystrophy (NCMD/MCDR1). Mol Vis. 2021;27:518–527. http://www.molvis.org/molvis/v27/518. e contribution of a unique CFH variant. Mol Vis. 2020;26:299–310.

Small KW, Wiggins R, Udar N, et al. North Carolina macular dystrophy: long-term follow-up of the original family. Ophthalmol Retina. 2022a;6(6):512-519. doi:https://doi.org/10.1016/j.oret.2022.02.003

Small KW, Tawfik CA, Udar N, et al. CONGENITAL POSTERIOR POLAR CHORIORETINAL HYPOPLASIA: expansion of the clinical spectrum, mutation, and its association with PRDM13. Retina (Philadelphia, PA.). 2022b;42(12):2379–2387. DOI: https://doi.org/10.1097/iae.0000000000003611.

Small KW, Van de Sompele S, Avetisjan J, Udar N, Agemy S, De Baere E, Shaya FS. New noncoding base pair mutation at the identical locus as the original NCMD/MCDR1 in a Mexican family, suggesting a mutational hotspot. J Vitreoretin Dis. 2023;7(1):33-42. doi: https://doi.org/10.1177/24741264221129432.

Sompele SV, Small KW, Cicekdal MB, López V, D'haene E, Shaya F, Agemy S, Van der Snickt T, Dueñas Rey A, Rosseel T, Van Heetvelde M, Vergul S, Balikova I, Bergen AA, Boon CJF, De Zaeytijd J, Inglehearn CF, Kousal B, Leroy BP, Rivolta C, Vaclavik V, van den Ende J, van Schooneveld MJ, Gómez-Skarmeta JL, Tena JJ, Martinez-Morales JR, Liskoval P, Vleminck K, De Baere E. Multi-omics profiling, in vitro and in vivo enhancer assays dissect the cis-regulatory mechanisms underlying North Carolina macular dystrophy, a retinal enhanceropathy. Am J Hum Genet. 2022;109:2029–2048. doi:https://doi.org/10.1016/j.ajhg.2022.09.013

Wu, S., Yuan, Z., Sun, Z. et al. A novel tandem duplication of PRDM13 in a Chinese family with North Carolina macular dystrophy. Graefes Arch Clin Exp Ophthalmol 260, 645–653 (2022). doi:https://doi.org/10.1007/s00417-021-05376-w

Further Reading

Small KW, DeLuca AP, Whitmore SS, Rosenberg T, Silva-Garcia R, Udar N, Puech B, Garcia CA, Rice TA, Fishman GA, Héon E, Folk JC, Streb LM, Haas CM, Wiley LA, Scheetz TE, Fingert JH, Mullins RF, Tucker BA, Stone EM. North Carolina macular dystrophy is caused by dysregulation of the retinal transcription factor PRDM13. Ophthalmology. 2016;123(1):9–18. doi: https://doi.org/10.1016/j.ophtha.2015.10.006. Epub 2015 Oct 24.

CHAPTER 22

Pigmented Paravenous Chorioretinal Atrophy (PPCRA)

Justin Cheong, Stephen H. Tsang, and Tarun Sharma

General Features 120
Molecular Genetics 121
Further Reading 121

J. Cheong
Vagelos College of Physicians and Surgeons,
New York, NY, USA

S. H. Tsang (✉) · T. Sharma
Department of Ophthalmology, Columbia University,
New York, NY, USA
e-mail: sht2@cumc.columbia.edu; ts3118@cumc.columbia.edu

S. H. Tsang et al. (eds.), *Atlas of Inherited Retinal Diseases*, Advances in Experimental Medicine and Biology 1467, https://doi.org/10.1007/978-3-031-72230-1_22

General Features

As the name suggests, in pigmented paravenous chorioretinal atrophy (PPCRA), patches of chorioretinal atrophy and pigment clumping are distributed along the veins (Figs. 22.1 and 22.2). In most cases, the retinal vessels, macula, and optic discs are normal, and the disease is usually nonprogressive.

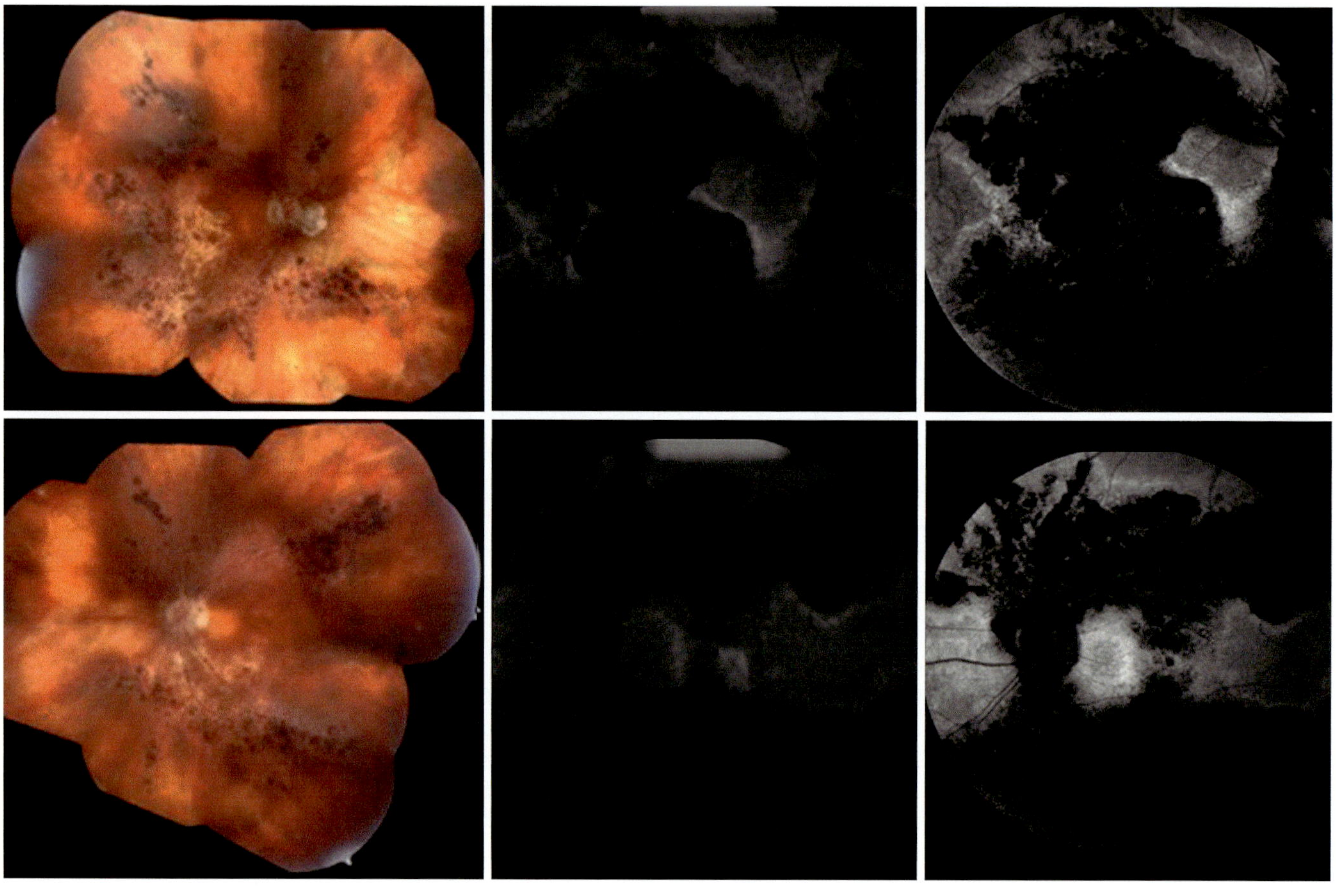

Fig. 22.1 Five-year follow-up of a case of pigmented paravenous chorioretinal atrophy (PPCRA), showing little progression

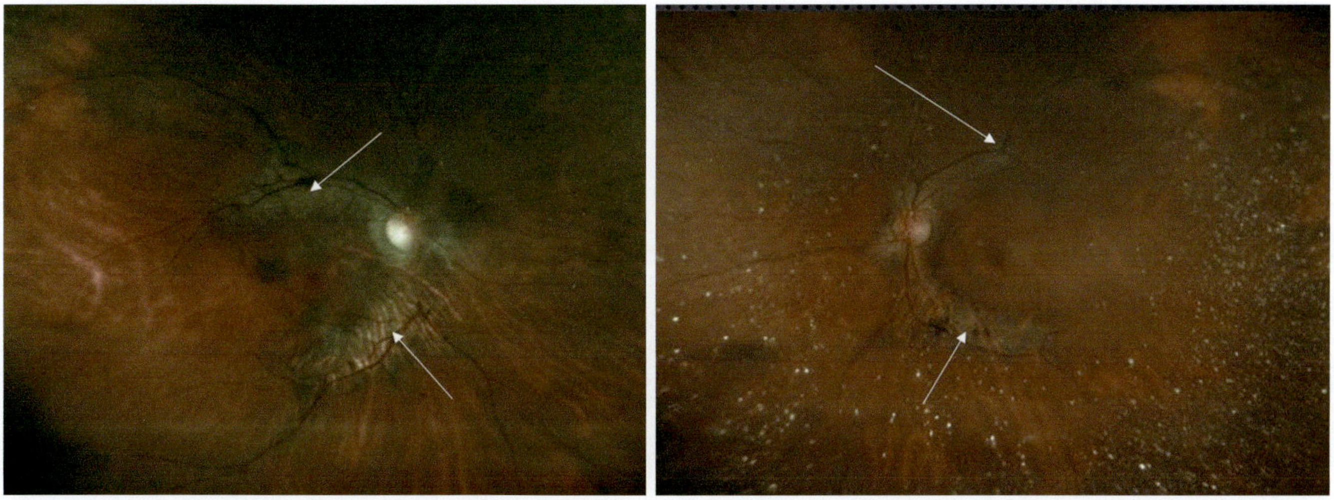

Fig. 22.2 Note the distribution of pigmentary changes along the veins (*arrows*)

Molecular Genetics

The mutation is in the gene *CRB1*, located at 1q31.3.

Further Reading

Barteselli G. Fundus autofluorescence and optical coherence tomography findings in pigmented paravenous retinochoroidal atrophy. Can J Ophthalmol. 2014;49:144–6.

CHAPTER 23

Late-Onset Retinal Degeneration

Stephen H. Tsang and Tarun Sharma

General Features 124
Molecular Genetics 124
C1QTNF5 Mutation (CTRP5) 124
Further Reading 124

S. H. Tsang (✉) · T. Sharma
Department of Ophthalmology, Columbia University,
New York, NY, USA
e-mail: sht2@cumc.columbia.edu; ts3118@cumc.columbia.edu

S. H. Tsang et al. (eds.), *Atlas of Inherited Retinal Diseases*, Advances in Experimental Medicine and Biology 1467, https://doi.org/10.1007/978-3-031-72230-1_23

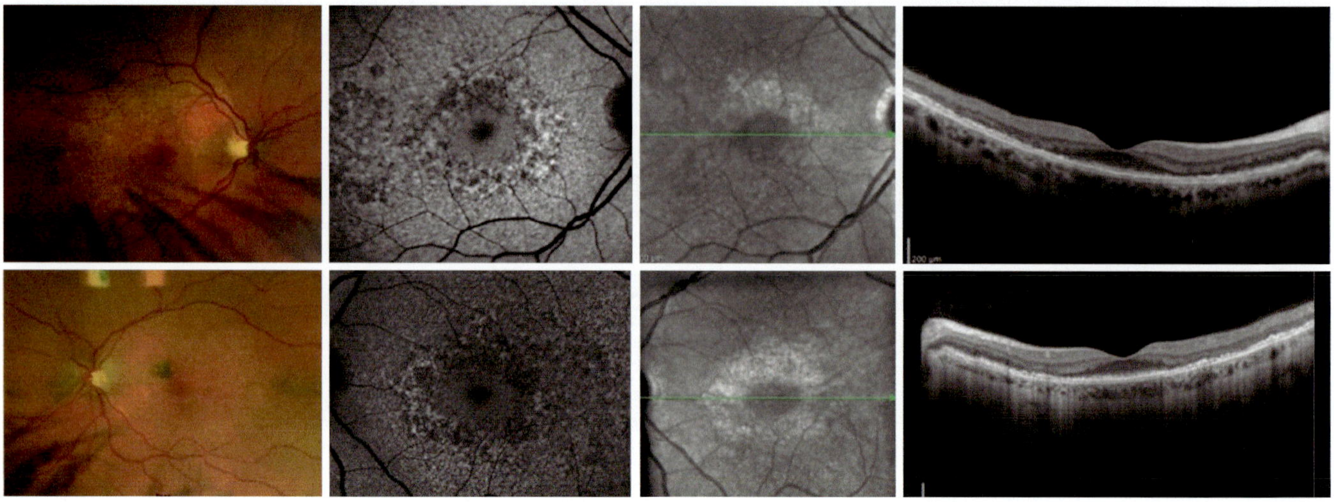

Fig. 23.1 Late-onset retinal degeneration (LORD) caused by *CTRP5* mutation. Fundus autofluorescence (FAF) shows a speckled pattern

General Features

Patients with late-onset retinal degeneration (LORD) usually present with nyctalopia in the fifth or sixth decade of life. The fundus shows yellowish-white, punctate deposits, usually progressive and giving rise to scalloped areas of retinal pigment epithelium (RPE) atrophy in the mid-periphery and posterior pole (Fig. 23.1). The anterior segment shows elongated zonules with central lens insertion and transillumination defect in the iris, due to iris atrophy.

Molecular Genetics

C1QTNF5 Mutation (*CTRP5*)

- LORD is related to mutation in the *C1QTNF5* gene, which is expressed in the RPE, lens, and ciliary epithelium.
- The gene encodes a family of protein that functions as components of basement membranes and may play a role in cell adhesion.
- Cytogenetic location: 11q23.3.

Further Reading

Cukras C, Flamendorf J, Wong WT, Ayyagari R, Cunningham D, Sieving PA. Longitudinal structural changes in late-onset retinal degeneration. Retina. 2016;36:2348–56.

Part IV
Autosomal Recessive Forms

CHAPTER 24

Achromatopsia

Kristina J. Hartung, Stephen H. Tsang, Tarun Sharma, and Vlad Diaconita

General Features . . . 128
Stages . . . 128
Molecular Genetics . . . 132
Clinical Trials . . . 133
Further Reading . . . 133

K. J. Hartung
Department of Ophthalmology, University Medical Centre Ljubljana, Ljubljana, Slovenia

Faculty of Medicine, University of Ljubljana, Ljubljana, Slovenia

S. H. Tsang (✉) · T. Sharma
Department of Ophthalmology, Columbia University, New York, NY, USA
e-mail: sht2@cumc.columbia.edu; ts3118@cumc.columbia.edu

V. Diaconita
Department of Ophthalmology, Edward S. Harkness Eye Institute New York-Presbyterian Hospital Columbia University, New York, NY, USA
e-mail: vd2374@cumc.columbia.edu

S. H. Tsang et al. (eds.), *Atlas of Inherited Retinal Diseases*, Advances in Experimental Medicine and Biology 1467, https://doi.org/10.1007/978-3-031-72230-1_24

General Features

- Achromatopsia is a rare autosomal recessive congenital disease affecting 1 in 30,000 individuals. The prevalence is much higher on the Island of Pingelap in Micronesia, affecting 4–10%, due to a founder effect and high homozygosity of a mutation in CNGA3.
- All three types of cones are defective, which distinguishes it from common forms of color blindness.
- Symptoms: Patients usually present in infancy with pendular nystagmus (which is usually the first symptom noted by the pediatrician and can improve with age) and photophobia. Some patients show paradoxical pupillary response; that is, the pupils dilate in bright light. Other symptoms include decreased visual acuity, absent or diminished color vision, hemeralopia, and glare. It is typically stationary but can be progressive; however, the progression is very slow, and patients usually maintain the vision that develops during childhood.
- There are two forms:
  - **Complete achromatopsia (rod monochromatism):**
    Visual acuity is 20/2000 or worse.
    Complete color blindness, patients see the world in shades of gray. They can identify several of the pseudochromatic (Ishihara) color plates (they distinguish colors as shades of gray); Farnsworth D-15 (may reveal a scotopic axis between the deutan and tritan Axes) or the Sloan achromatopsia test may be used.
  - **Incomplete achromatopsia:**
    Milder.
    Visual acuity is 20/80 to 20/120.
    Residual color vision.
- Fundus: can appear normal early in the course; later retinal pigment epithelial mottling, macular atrophy, and temporal pallor of the optic disc may be seen. Foveal hypoplasia is very common (83%).

Stages

Figure 24.1 illustrates the stages of achromatopsia on the basis of spectral domain optical coherence tomography (SD-OCT):

- Stage 1: Intact outer retina.
- Stage 2: Inner segment ellipsoid line disruption.
- Stage 3: Presence of an optically empty space.
- Stage 4: Optically empty space with partial retinal pigment epithelium (RPE) disruption. It is weighted to the nasal side of the fovea due to the higher density of cones.
- Stage 5: Complete RPE disruption and/or loss of outer nuclear layer.

On fundus autofluorescence (FAF) imaging, the presence of hypoautofluorescence is suggestive of progressive retinal degeneration (Fig. 24.2).

Fundus autofluorescence (FAF) corresponds to and usually precedes the structural changes seen on OCT. Patients

Fig. 24.1 (**a**) Spectral domain optical coherence tomography (SD-OCT) shows subtle discontinuity of the inner segment ellipsoid (ISe) line posterior to the foveola. The cone outer segment tip (COST) layer (*left arrow*) is thinned toward the foveal center, and the external limiting membrane (*right arrow*) is hyperreflective. (**b**) Fundus autofluorescence (FAF) imaging shows decreased macular pigment contrast and fine pigment hyperautofluorescent dots scattered across the macula. (**c**) SD-OCT shows disruption of the ISe line, interruption of the COST layer, and hyperreflectivity of photoreceptor inner segments. (**d**) FAF shows centrally reduced autofluorescence with subtle hyperautofluorescence around the inferotemporal fovea (*arrow*). (**e**) SD-OCT shows the classic optically empty space (OES), but the retinal pigment epithelium (RPE) is intact. (**f**) FAF shows hyperautofluorescence, with a stippled appearance. (**g**) SD-OCT shows an OES and partial RPE disruption (choroidal hyperreflectance, *arrows*). The roof of the OES shows reflective material, possibly from photoreceptor debris. (**h**) FAF shows a central area of reduced autofluorescence suggestive of some RPE dysfunction. (**i**) SD-OCT shows complete disruption of the RPE and loss of the outer nuclear layer. (**j**) FAF shows an area of hypofluorescence in the center (absent autofluorescence, due to RPE atrophy) with a surrounding hyperfluorescent ring

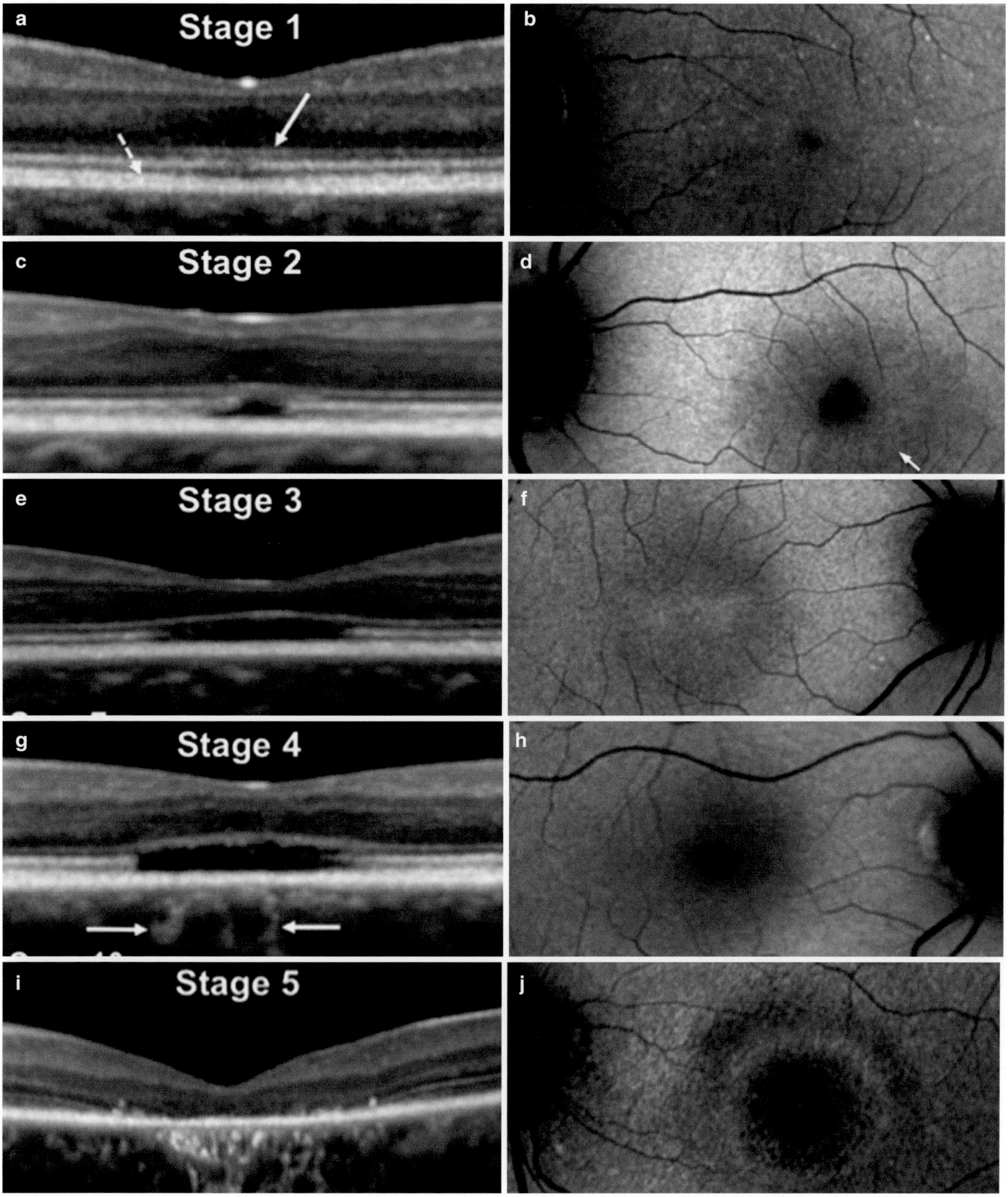
a
Stage 1
b
c
Stage 2
d
e
Stage 3
f
g
Stage 4
h
i
Stage 5
j

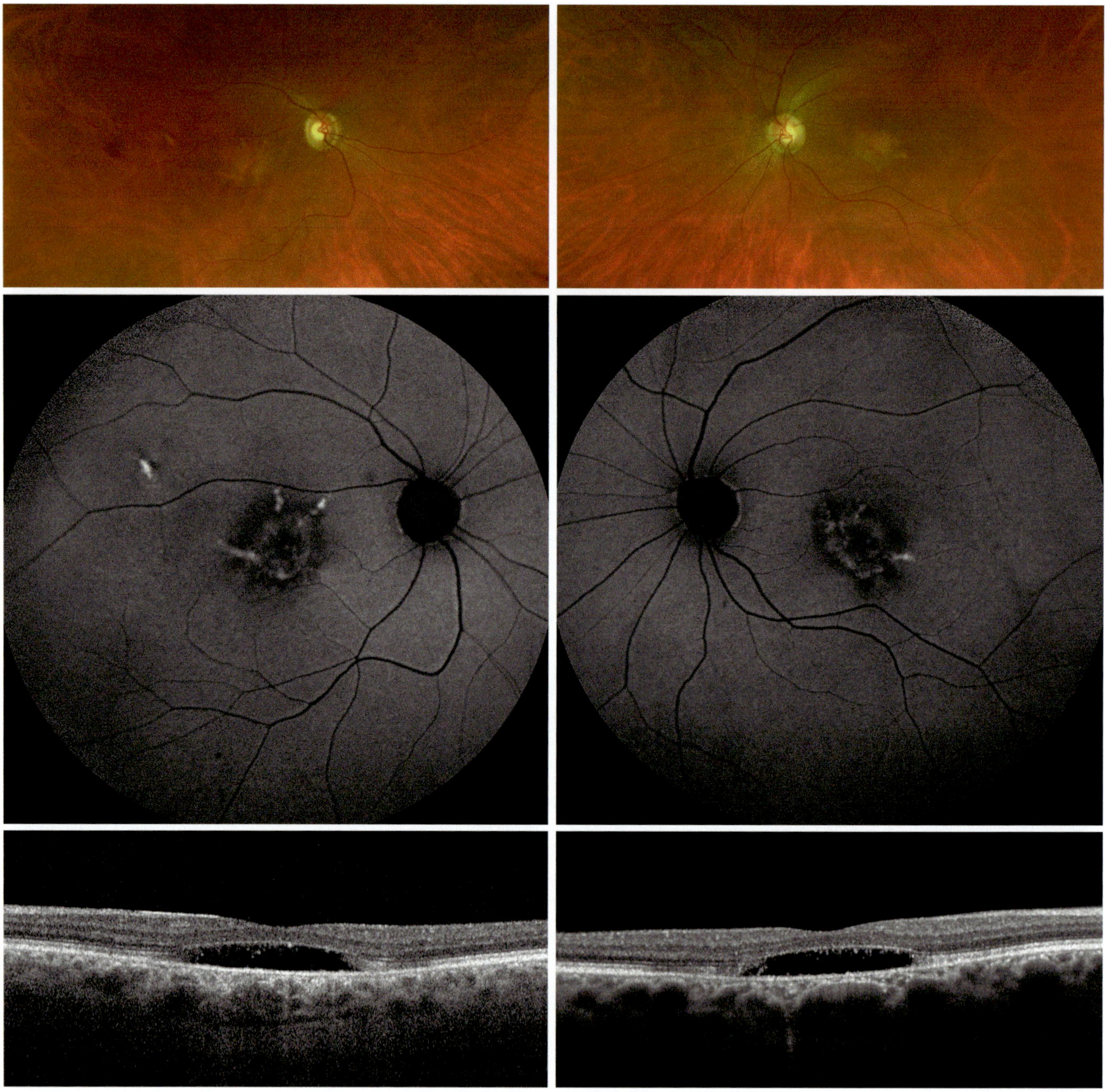

Fig. 24.2 Multimodal imaging and electroretinography (ERG) of a patient with achromatopsia (CNGA3: V1: c.1669G>A:p.Gly557Arg; V2: c.667C>G:p.Arg223Gly). Fundus photographs show retinal pigment mottling. FAF shows punched-out hypoautofluorescence with surrounding hyperautofluorescent ring. SD-OCT shows an optically empty space, disruption of the inner/outer-segment (IS–OS) junction, and marked thinning of the outer nuclear layer. ERG shows non-recordable cone responses and normal rod responses

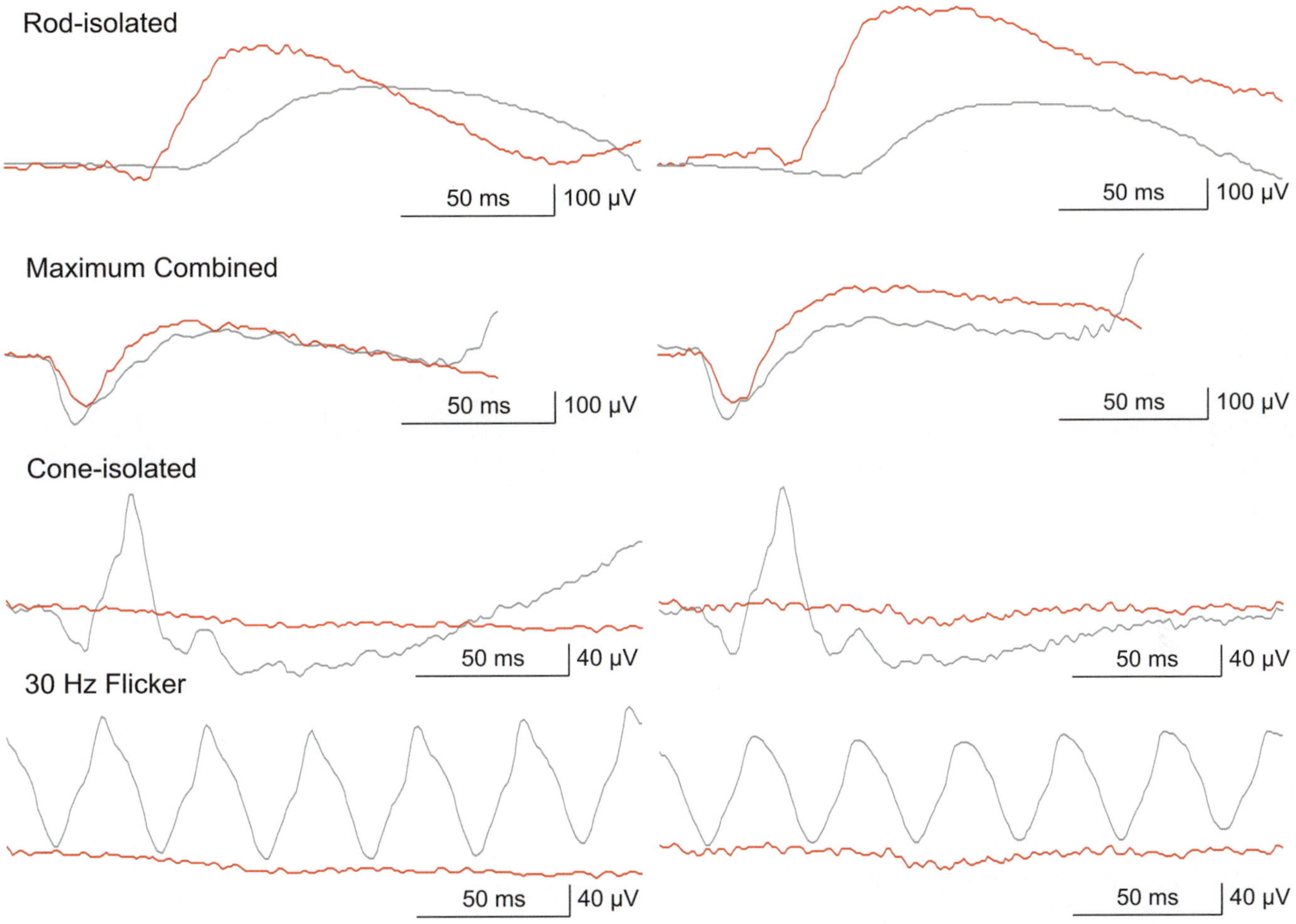

Fig. 24.2 (continued)

with normal OCT exhibit foveal hyper autofluorescence, whereas punched-out hypo autofluorescence with variable degrees of surrounding hyper autofluorescence is seen in patients with optically empty space OCT (Fig. 24.3).

Electrophysiology: A full-field electroretinography (ERG) shows nearly or completely non-recordable cone responses and normal or near-normal rod responses (Fig. 24.2). Mf-ERG is severely diminished. ERG is crucial for diagnosis as it helps to differentiate from other conditions that share similar clinical features, e.g., LCA (extinguished photopic and scotopic ERGs) and blue cone monochromatism (recordable cone responses when stimulated with the blue light on a yellow background).

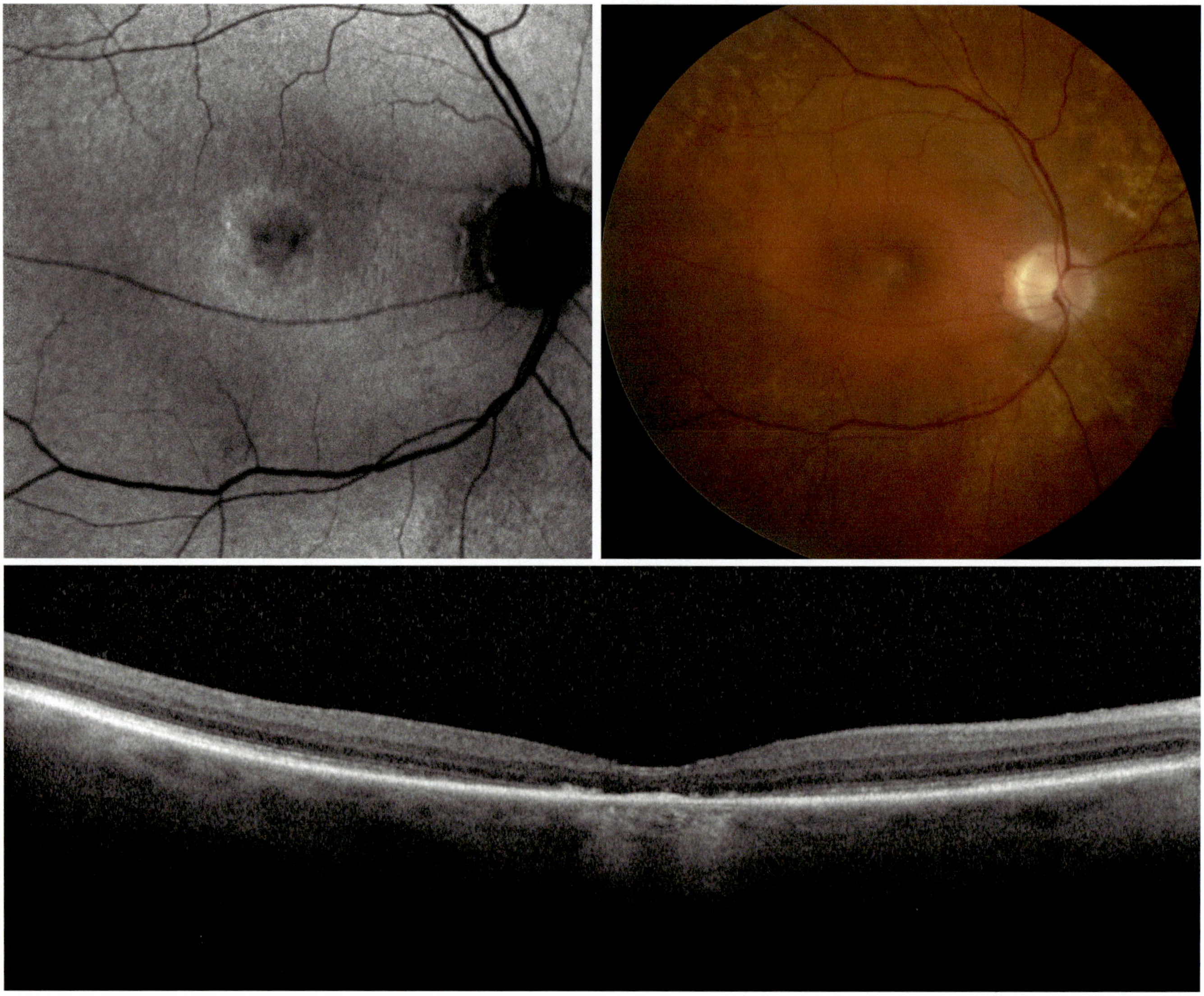

Fig. 24.3 Multimodal imaging of a patient (CNGA3; V1: c.8484G>A:p.Arg283Gln; V2: c.1279C>T:p.Arg427Cys). Color fundus photograph shows RPE mottling and atrophy in the macula. FAF shows corresponding central hypoautofluorescence, surrounded by a hyperautofluorescent ring. OCT shows a complete loss of RPE in the fovea and marked thinning of the outer nuclear layer

Molecular Genetics

Currently, variants in six genes are linked to achromatopsia; all but ATF6 encode essential components of phototransduction.

1. **CNGB3 (cyclic nucleotide gated channel subunit beta 3).** Accounts for most cases (40–50%). It encodes beta 3 subunit of the cGMP-gated cation channel. Cytogenetic location: 8q21.3
2. **CNGA3 (cyclic nucleotide gated channel alpha 3):** Accounts for **20–30% of cases.** It encodes for alpha subunit of cone cyclic guanosine monophosphate (cGMP)–gated cation channel. Cytogenetic location: 2q11.2.
3. **GNAT2 (G protein subunit alpha transducin 2):** encodes for the alpha subunit of cone-related transducin. Cytogenetic location: 1p13.3.
4. **PDE6C (Phosphodiesterase 6C):** encodes for cGMP-specific cone phosphodiesterase 6C alpha protein. Cytogenetic location: 10q23.33.

5. **PDE6H:** on chromosome 12, encodes for inhibitory Y subunit of the cone photoreceptor cGMP phosphodiesterase. Cytogenic location: 12p12.3.
6. **ATF6 (activating transcription factor 6):** encodes an endoplasmatic stress-regulated transmembrane protein that activates the transcription of molecular chaperones. Cytogenic location: 1q23.3.

The latter four genes, GNAT2, PDE6C, PDE6H, and ATF6, account for just 5–8% of all cases.

Mutations in CNGA3 and PDE6H have been associated with incomplete achromatopsia.

Clinical Trials

The ClinicalTrials.gov website lists five trials investigating the treatment of achromatopsia that are currently active (as of October 16th, 2022):

- Evaluation of Glycerol Phenylbutyrate (PBA) Use in Endoplasmic Reticulum Stress Reduction in ATF6-/- Patients (Columbia University) (ClinicalTrials.gov identifier: NCT04041232)
- Safety and Efficacy of rAAV.hCNGA3 Gene Therapy in Patients With CNGA3-linked Achromatopsia (Colourbridge) (STZ eyetrial) (NCT02610582)
- Long-Term Follow-Up Gene Therapy Study for Achromatopsia CNGB3 (MeiraGTx UK II Ltd) (ClinicalTrials.gov Identifier: NCT03278873)
- Safety and Efficacy Trial of AAV Gene Therapy in Patients With CNGA3 Achromatopsia (A Clarity Clinical Trial) (AGTC) (ClinicalTrials.gov identifier: NCT02935517)
- Safety and Efficacy Trial of AAV Gene Therapy in Patients With CNGB3 Achromatopsia (AGTC) (ClinicalTrials.gov identifier: NCT02599922)

Further Reading

Fischer MD, Michalakis S, Wilhelm B, et al. Safety and vision outcomes of subretinal gene therapy targeting cone photoreceptors in achromatopsia: a nonrandomized controlled trial. JAMA Ophthalmol. 2020;138(6):643–651.

Hirji N, Aboshiha J, Georgiou M, Bainbridge J, Michaelides M. Achromatopsia: clinical features, molecular genetics, animal models and therapeutic options. Ophthalmic Genet. 2018;39(2):149–157.

Greenberg JP, Sherman J, Zweifel SA, Chen RW, Duncker T, Kohl S, et al. Spectral-domain optical coherence tomography staging and autofluorescence imaging in achromatopsia. JAMA Ophthalmol. 2014;132:437–45.

Sundaram V, Wilde C, Aboshiha J, Cowing J, Han C, Langlo CS, et al. Retinal structure and function in achromatopsia: implications for gene therapy. Ophthalmology. 2014;121:234–45.

CHAPTER 25

Retinitis Pigmentosa (Non-syndromic)

Ahmet Hondur, Stephen Tsang, Tarun Sharma, and Vlad Diaconita

General Features 136
Genes and Mutations. 138
PDE6 (Phosphodiesterase) 138
RP25 or EYS (Eyes Shut Homolog) 138
TULP1 (Tubby-Like Protein 1) Gene or RP14 or TUBL1 or LCA15. 138
CERKL (Ceramide Kinase-Like) Gene or RP26 140
RPE65 Gene 140
Other Genes Associated with AR-RP. 140
Further Reading 141

A. Hondur · S. Tsang (✉) · T. Sharma · V. Diaconita
Department of Ophthalmology, Columbia University,
New York, NY, USA
e-mail: ahondur@gazi.edu.tr; sht2@cumc.columbia.edu

S. H. Tsang et al. (eds.), *Atlas of Inherited Retinal Diseases*, Advances in Experimental Medicine and Biology 1467, https://doi.org/10.1007/978-3-031-72230-1_25

General Features

- Most of the genes causing autosomal recessive retinitis pigmentosa (AR-RP) are rare and cause 1% of all cases.
- Some of the genes like *PDE6* (*PDE6A*, *PDE6B*, *PDE6G*), *RP25*, and *RPE65* have higher prevalence, about 2–5% of all cases.
- Overall, autosomal recessive RP accounts for about 15–20% of all cases of RP.
- Clinically, it shows all the classic features of RP, such as attenuated retinal blood vessels, intraretinal pigmentation, waxy pallor of the optic disc, and hyperfluorescent rings on fundus autofluorescence (FAF) (Figs. 25.1, 25.2, and 25.3). The ring is suggestive of increased metabolic burden of the corresponding retinal pigment epithelium (RPE).

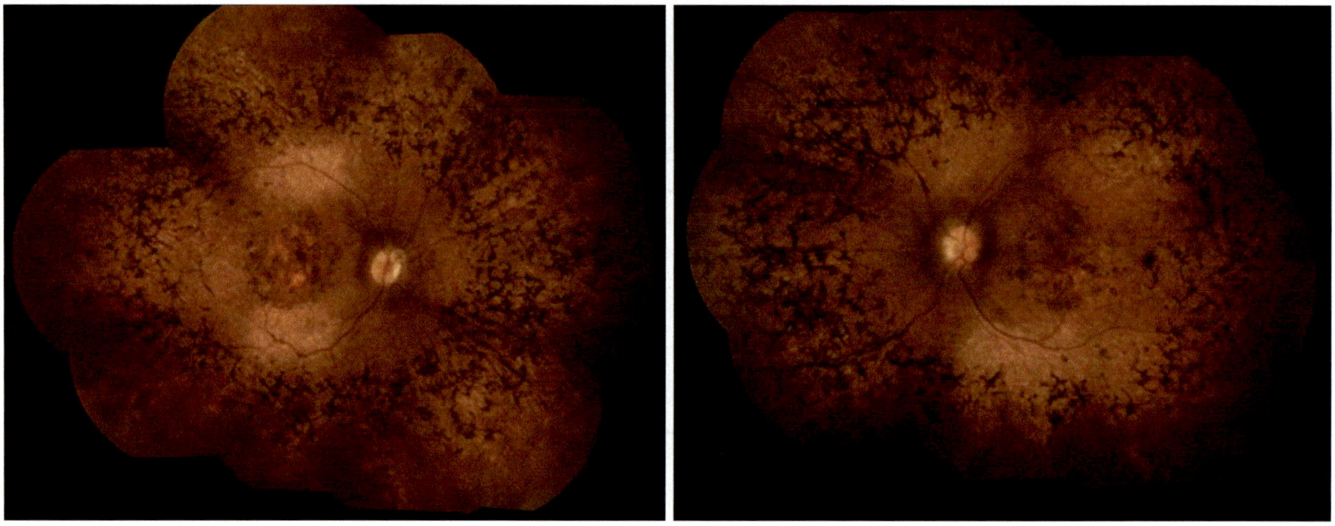

Fig. 25.1 Retinitis pigmentosa with attenuated vessels, intraretinal pigmentation, and macular involvement; mutation *RDH12*

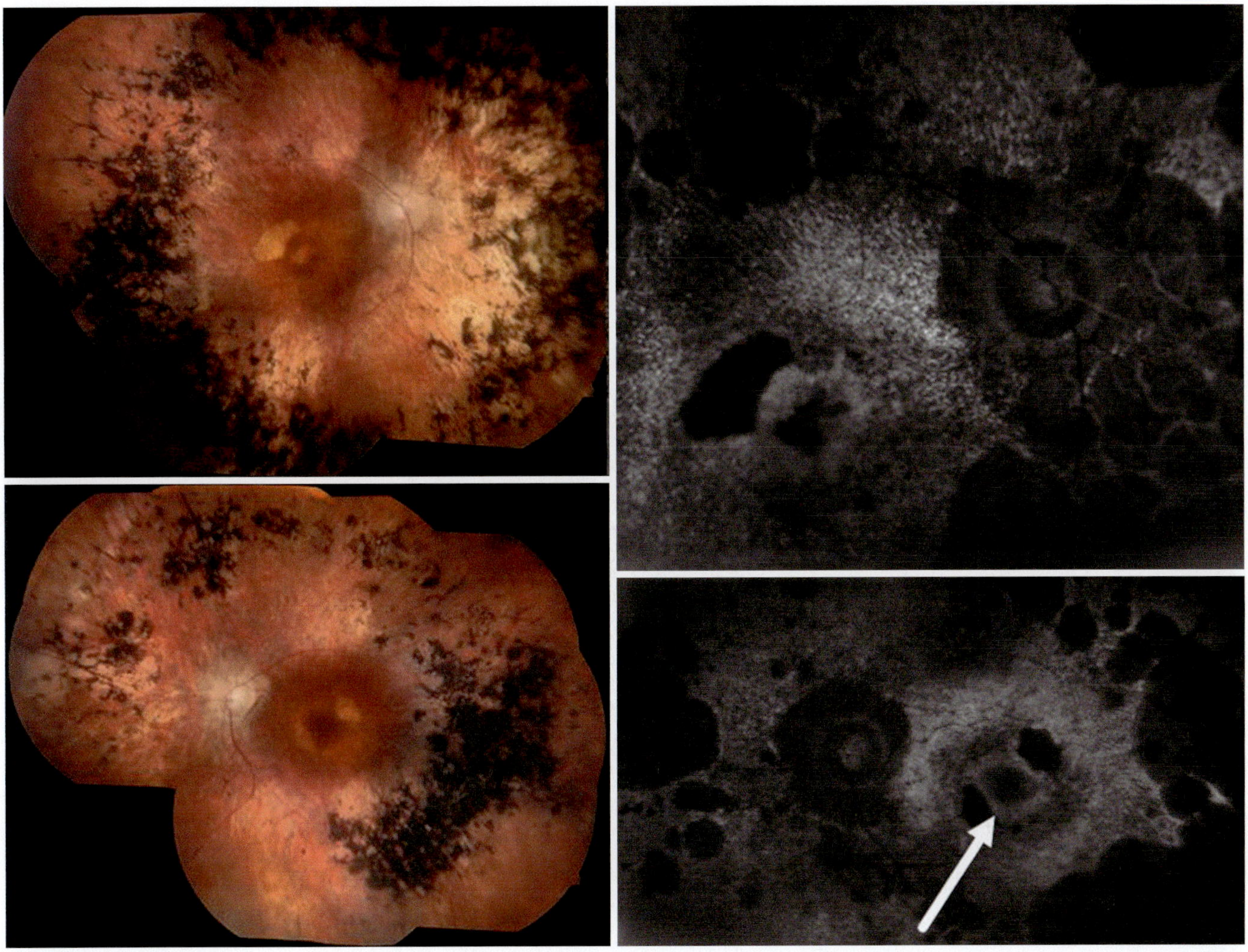

Fig. 25.2 Retinitis pigmentosa with attenuated vessels, intraretinal pigmentation, fundus autofluorescence (FAF) showing a hyperfluorescent ring around the fovea (*arrow*); mutation *PDE6B*

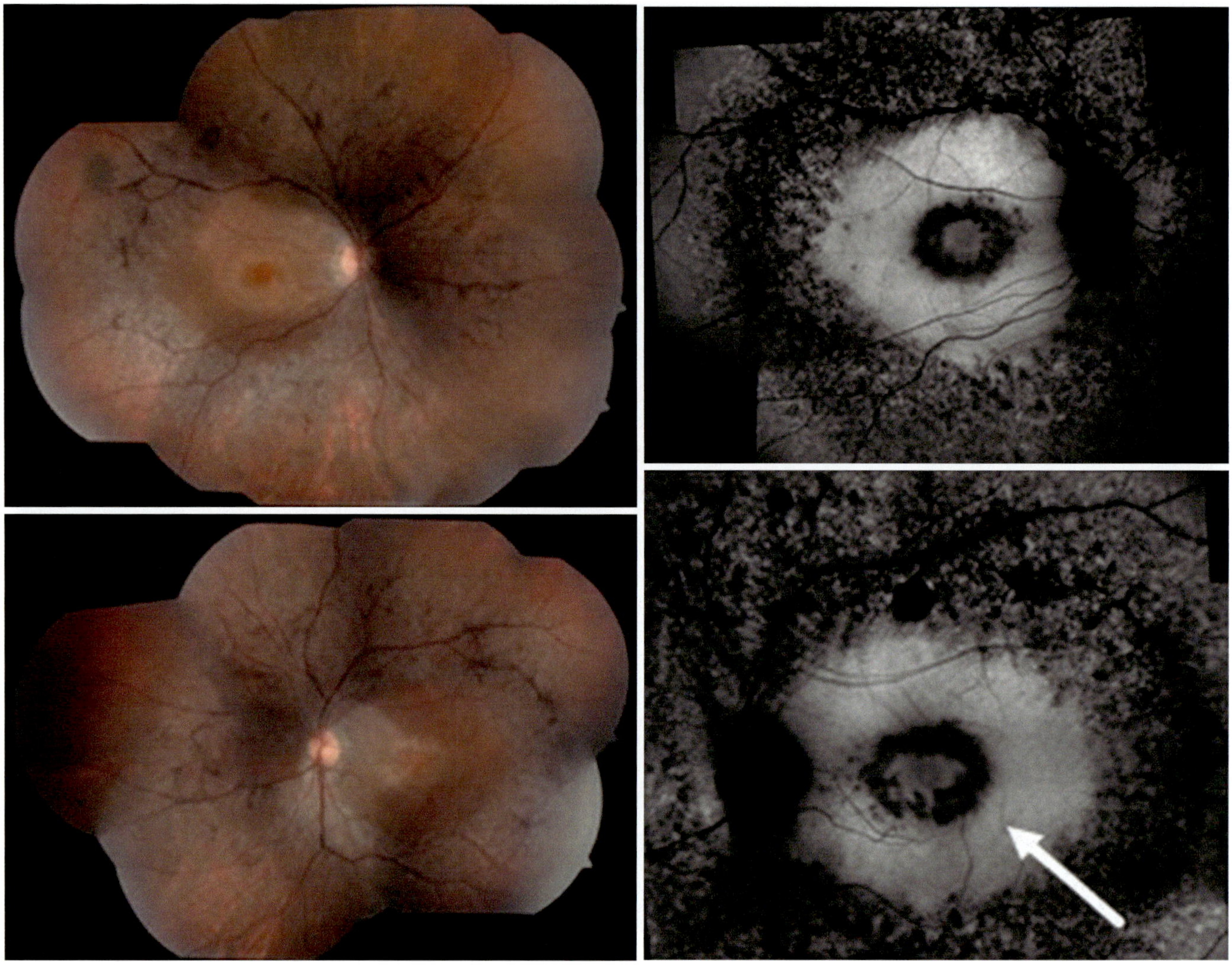

Fig. 25.3 Retinitis pigmentosa with attenuated vessels, intraretinal pigmentation, FAF showing a broad hyperfluorescent ring around the fovea (*arrow*); mutation *USH2A*

Genes and Mutations

PDE6 (Phosphodiesterase) (Fig. 25.4)

The PDE complex has one α, one β, and two γ subunits. Mutations in each subunit can lead to AR-RP. The complex encodes a protein that play an important role in rod phototransduction and maintenance of the intracellular cGMP level by activating cGMP in response to photons. This closes the cGMP-gated ion channels and hyperpolarizes the photoreceptor cells to transduce the signal. *PDE6G* might be associated with early-onset RP.

- *PDE6A* or *RP43*
- *PDE6B* or *RP40*
- *PDE6G* or *RP57*

RP25 or *EYS* (Eyes Shut Homolog) (Fig. 25.5)

This gene accounts for 10–20% of AR-RP in Spanish and 15–33% of AR-RP in Japanese populations. Cytogenetic location: 6q12.

TULP1 (Tubby-Like Protein 1) Gene or *RP14* or *TUBL1* or *LCA15* (Fig. 25.6)

The protein encoded by the *TULP1* gene is thought to play a role in the physiology of photoreceptors. Cytogenetic location: 6p21.31.

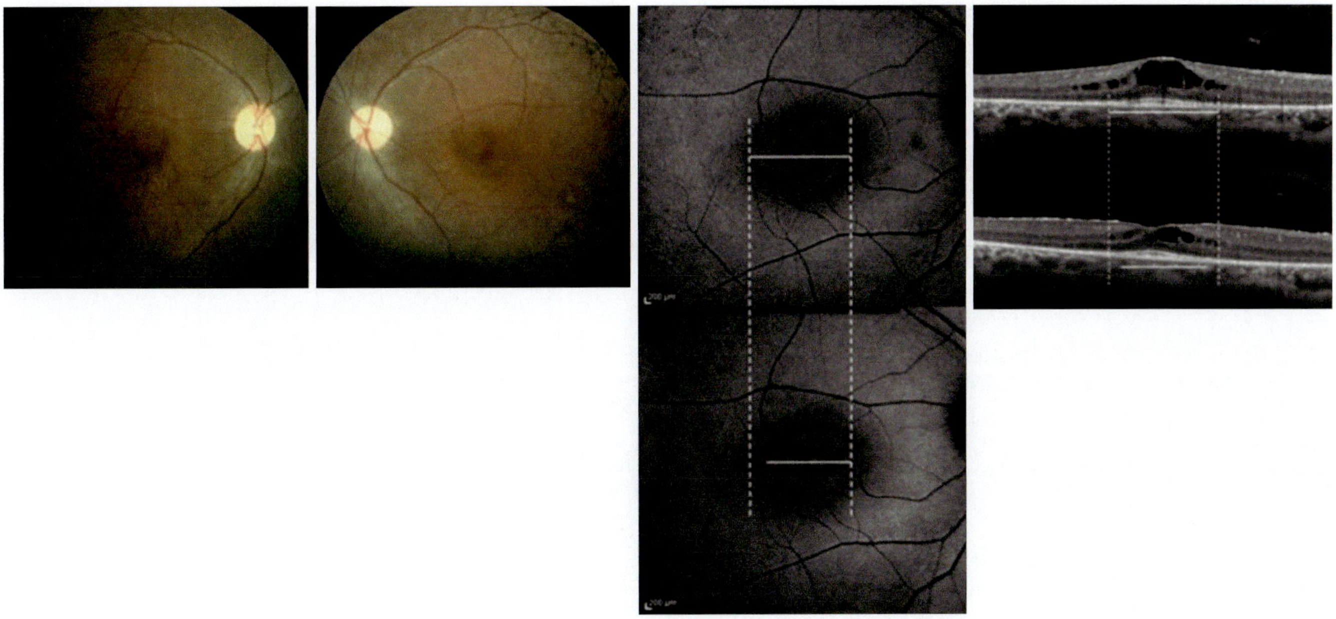

Fig. 25.4 Autosomal recessive retinitis pigmentosa (AR-RP). Rod-cone dystrophy secondary to PDE6B mutations. The size of the hyperfluorescent ring on AF and the length of the ellipsoid zone (EZ) decreased over a 2-year follow-up, suggestive of progressive rod photoreceptor damage

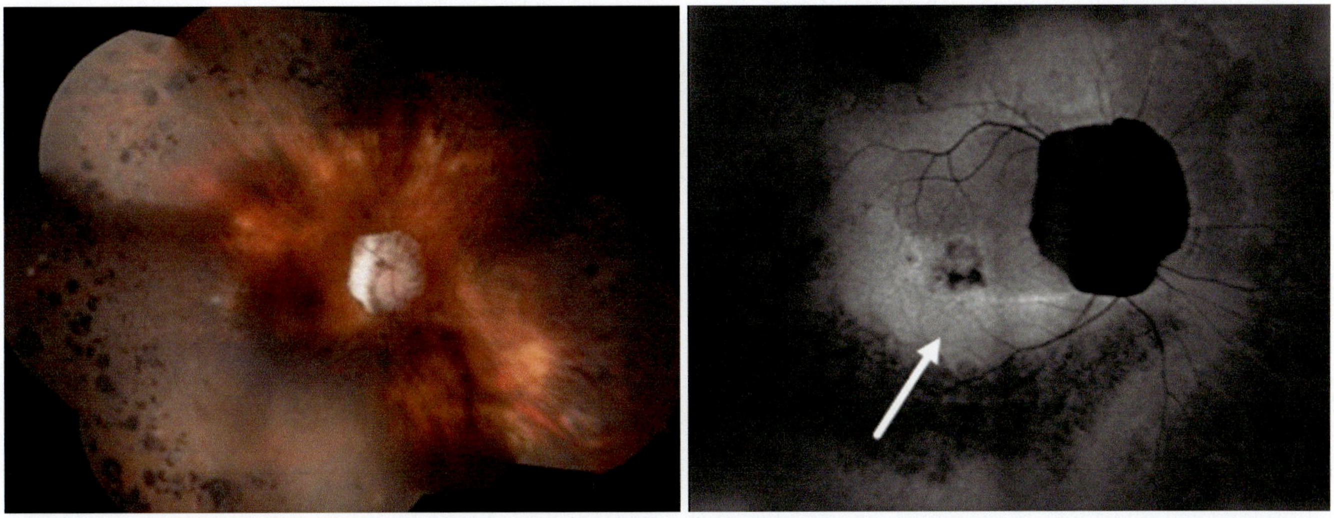

Fig. 25.5 AR-RP. Rod-cone dystrophy secondary to eyes shut homolog (*EYS*) mutations; FAF shows an atypical crescent-shaped hyperAF ring (*arrow*)

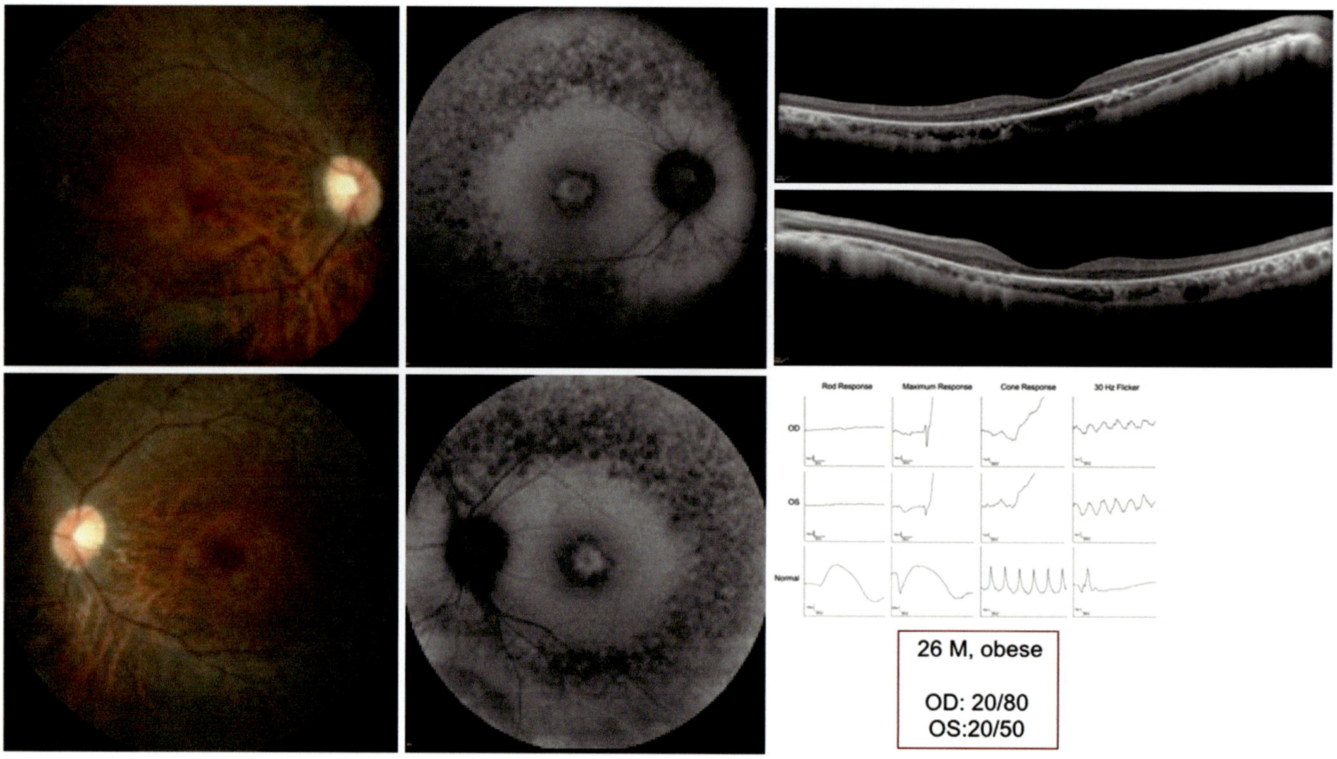

Fig. 25.6 AR-RP. Rod-cone dystrophy secondary to homozygous *TULP1* mutations and heterozygous *ABCA4* mutations; bulls-eye maculopathy (BEM) pattern along with peripheral involvement

CERKL (Ceramide Kinase-Like) Gene or *RP26* (Fig. 25.7)

This gene encodes a protein with ceramide kinase-like domains; the protein may be a negative regulator of apoptosis in photoreceptor cells. Cytogenetic location: 2q31.3.

RPE65 Gene

This gene is expressed in the RPE and is vital for converting all-*trans*-retinyl esters into 11-*cis*-retinal, an important step in regeneration of visual pigment. The *RPE* gene (about 60 mutations) is responsible for around 2% of AR-RP and 16% of Leber congenital amaurosis. Cytogenetic location: 1p31.3.

Subretinal gene therapy with Voretigene Neparvovec (VN, Luxturna®, Spark Therapeutics, Philadelphia, PA) for eyes with biallelic RPE65 mutation has been the first ever FDA-approved gene therapy.

Other Genes Associated with AR-RP

Among other genes are *ABCA4* (1p22.1, 680 mutations), *BEST1* (11q12.3, 232 mutations), *CRB1* (1q31.3, 183 mutations), *CRX* (19q13.32, 51 mutations), *MAK* (6p24.2, 9 mutations), and *MERTK* (2q13, 27 mutations).

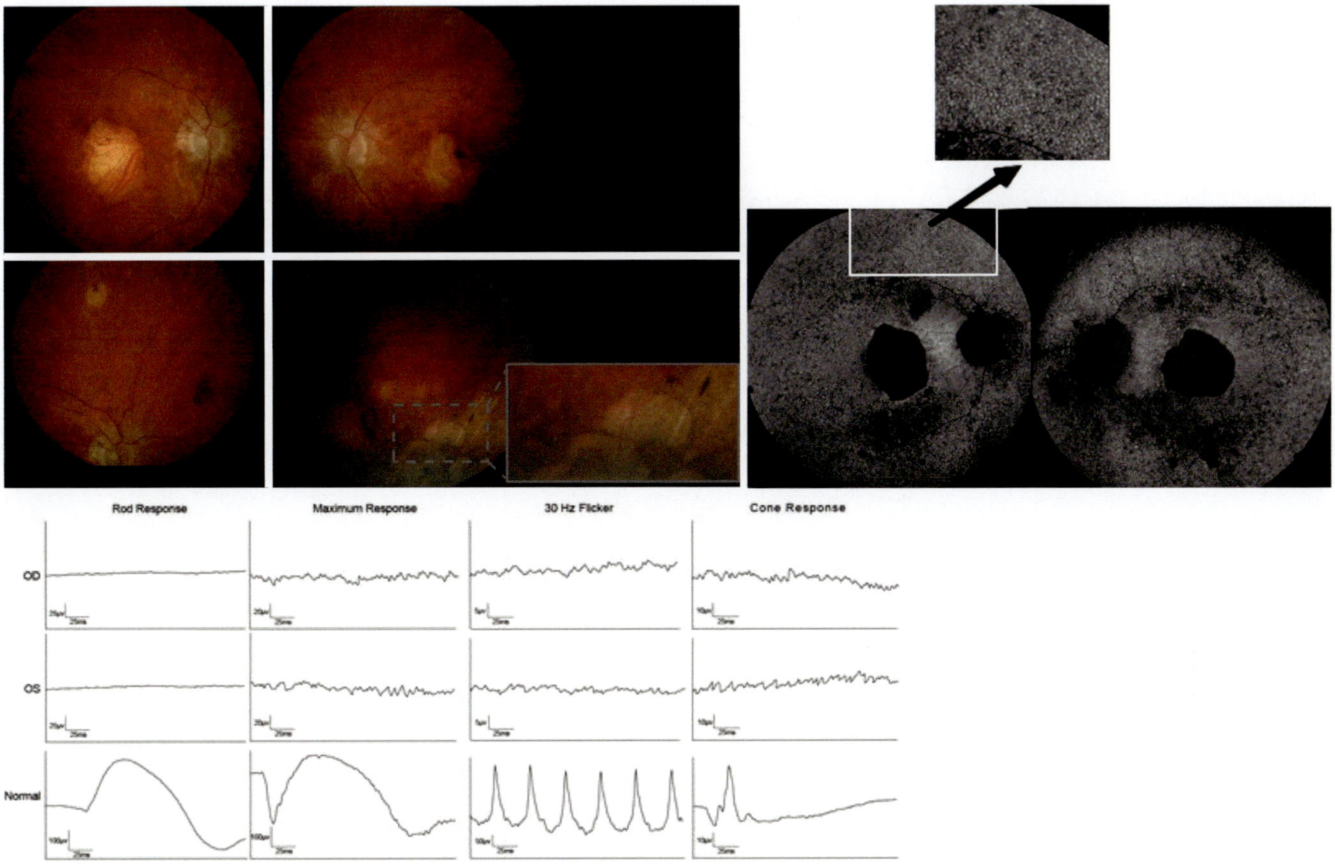

Fig. 25.7 AR-RP. Rod-cone dystrophy secondary to mutation in the ceramide kinase-like (*CERKL*) gene, with early-onset maculopathy, peripheral lacunae, intraretinal pigment migration, and white hyperAF dots on FAF (*arrow*)

Further Reading

Bravo-Gil N, González-Del Pozo M, Martín-Sánchez M, Méndez-Vidal C, Rodríguez-de la Rúa E, Borrego S, et al. Unravelling the genetic basis of simplex retinitis pigmentosa cases. Sci Rep. 2017;7:41937.

Iwanami M, Oishi A, Ogino K, Seko Y, Nishida-Shimizu T, Yoshimura N, Kato S. Five major sequence variants and copy number variants in the EYS gene account for one-third of Japanese patients with autosomal recessive and simplex retinitis pigmentosa. Mol Vis. 2019;25:766–779

Martin-Merida I, Avila-Fernandez A, Del Pozo-Valero M, Blanco-Kelly F, Zurita O, Perez-Carro R, Aguilera-Garcia D, et al. Genomic Landscape of Sporadic Retinitis Pigmentosa: Findings from 877 Spanish Cases. Ophthalmology. 2019;126:1181–1188

McGuigan DB, Heon E, Cideciyan AV, Ratnapriya R, Lu M, Sumaroka A, et al. EYS mutations causing autosomal recessive retinitis pigmentosa: changes of retinal structure and function with disease progression. Genes (Basel). 2017;8:E178.

Perea-Romero I, Gordo G, Iancu IF, Del Pozo-Valero M, Almoguera B, Blanco-Kelly F, Carreño E, Jimenez-Rolando B, et al. Genetic landscape of 6089 inherited retinal dystrophies affected cases in Spain and their therapeutic and extended epidemiological implications. Sci Rep. 2021;11:1526

Russell S, Bennett J, Wellman JA, Chung DC, Yu ZF, Tillman A, et al. Efficacy and safety of voretigene neparvovec (AAV2-hRPE65v2) in patients with RPE65-mediated inherited retinal dystrophy: a randomised, controlled, open-label, phase 3 trial. Lancet. 2017;390:849–860.

Sharon D, Banin E. Nonsyndromic retinitis pigmentosa is highly prevalent in the Jerusalem region with a high frequency of founder mutations. Mol Vis. 2015;21:783–92.

CHAPTER 26

Leber Congenital Amaurosis

Stephen H. Tsang and Tarun Sharma

General Features 144
Molecular Genetics 144
Specific Genotype and Phenotype Correlations 144
LCA1. 144
LCA2. 144
LCA3. 144
LCA4. 146
LCA5. 146
LCA6. 146
LCA7. 146
LCA8. 146
LCA9. 147
LCA10 147
LCA11 147
LCA12 147
LCA14 148
LCA15 148
LCA16 149
RP38 149
LCA57 149
SLSN5 149

S. H. Tsang (✉) · T. Sharma
Department of Ophthalmology, Columbia University,
New York, NY, USA
e-mail: sht2@cumc.columbia.edu

S. H. Tsang et al. (eds.), *Atlas of Inherited Retinal Diseases*, Advances in Experimental Medicine and Biology 1467, https://doi.org/10.1007/978-3-031-72230-1_26

ABCA-4 149
Further Reading 149

General Features

- Leber congenital amaurosis (LCA) is a part of the spectrum of early-onset retinal dystrophy (EORD). It usually presents in the first few years of life, most often before the age of 1 year. The prevalence is about 1:80,000.
- Also known as congenital retinitis pigmentosa (RP), patients have wandering nystagmus, with reduced vision from birth. The fundus is almost normal to start with. Later, pigmentary disturbances develop.
- The best corrected visual acuity (BCVA) ranges from no light perception (in nearly one third of cases) to no better than 20/400, often with a hyperopic refraction (≥5.0D). Patients have absent pupillary reflex and some of them have keratoconus (*CRB1* and *AIPL1*).
- Oculodigital reflex (eye rubbing or poking) is a common association.
- Electroretinography (ERG) responses are almost undetectable.
- Patients may have normal intelligence, but some studies suggest that as many as 20% of children with LCA without associated anomalies develop intellectual disability.
- Patients with *GUCY2D* mutations have a stable clinical course; those with *RPE65* mutations show a period of limited improvement followed by progressive deterioration; and those with *AIPL1*, *CRB1*, *CEP290*, and *NMNAT1* mutations show gradual progression over several decades.

Molecular Genetics

Those with *CEP290* or *ICQB1* mutations should have a renal evaluation and neurological evaluation for Joubert syndrome or Senior Loken syndrome.

In December 2017, the U.S. Food and Drug Administration (FDA) approved a gene therapy, Luxturna (voretigene neparvovec-rzyl) for the treatment of patients with confirmed biallelic *RPE65* mutation–associated retinal dystrophy, which affects about 1000–2000 patients in the United States. Luxturna works by delivering a normal copy of the *RPE65* gene directly to retinal cells.

Specific Genotype and Phenotype Correlations

Most often LCA is inherited in an autosomal recessive manner. Rarely, it can be inherited in an autosomal dominant mode, particularly due to mutations in *IMPDH1*, *OTX2*, or *CRX* gene.

LCA1

- **Genotype**: *GUCY2D* gene located at 17p13.1, is expressed in photoreceptors, and encodes for guanylate cyclase; needed to convert GTP into cGMP (recovery of photoreceptors after phototransduction). Accounts for 6–21% cases of LCA.
- **Phenotype**: Severe cone-rod dystrophy, poor stable vision, and no visual improvement.

LCA2 (Fig. 26.1)

- **Genotype**: *RPE65* located at 1p31, expressed in RPE and encodes for 65 KD protein; needed to convert all-*trans*-retinyl ester to 11-*cis*-retinol. Accounts for 3–16% cases of LCA.
- **Phenotype**: Rod-cone dystrophy. Early RPE mottling and small white intraretinal spots, with later intraretinal pigmentation. Some patients developed peculiar star-shaped maculopathy or have translucent RPE. Some have transient improvement in vision and eventual progressive visual loss.

LCA3 (Fig. 26.2)

- **Genotype**: *SPATA7* located at 14q13.3, expressed in photoreceptors (IS) and testis; needed for spermatogenesis.
- **Phenotype**: Rod-cone dystrophy, intraretinal pigmentation.

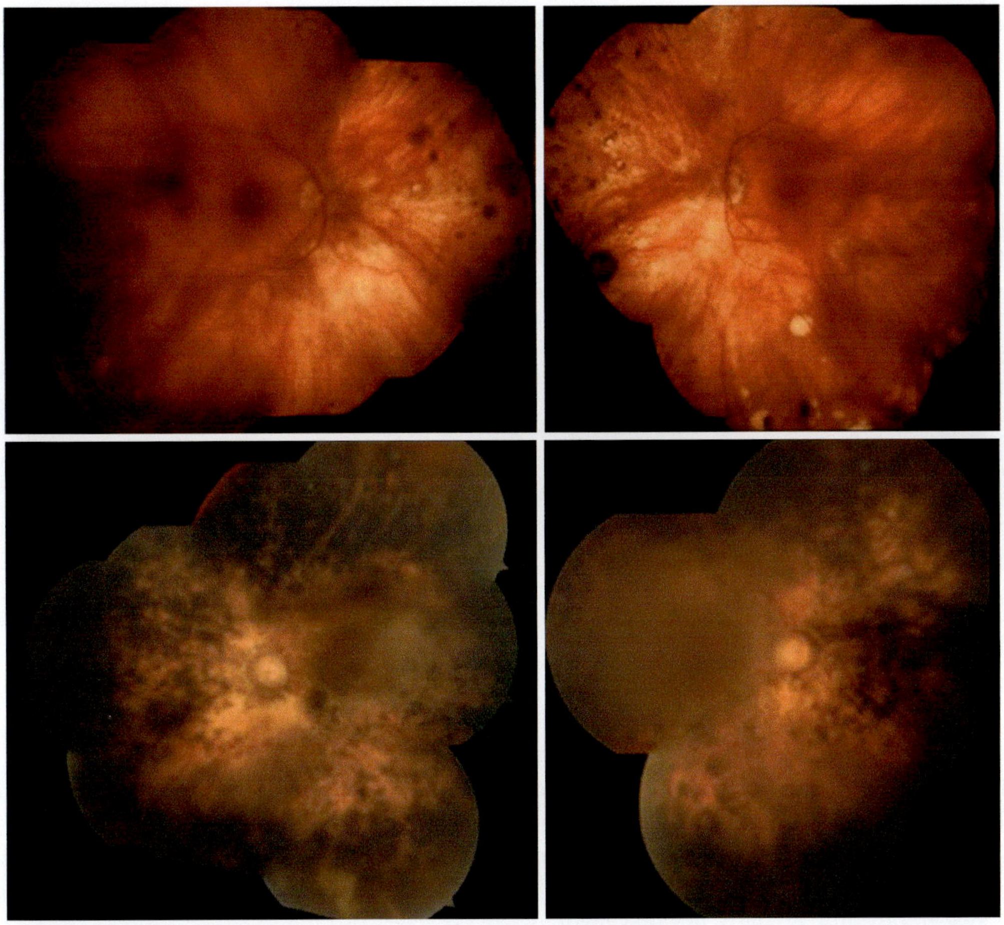

Fig. 26.1 Leber congenital amaurosis (LCA) caused by *RPE65* mutation (A132T); note extensive RPE mottling and advanced pigmentary changes (*lower row*)

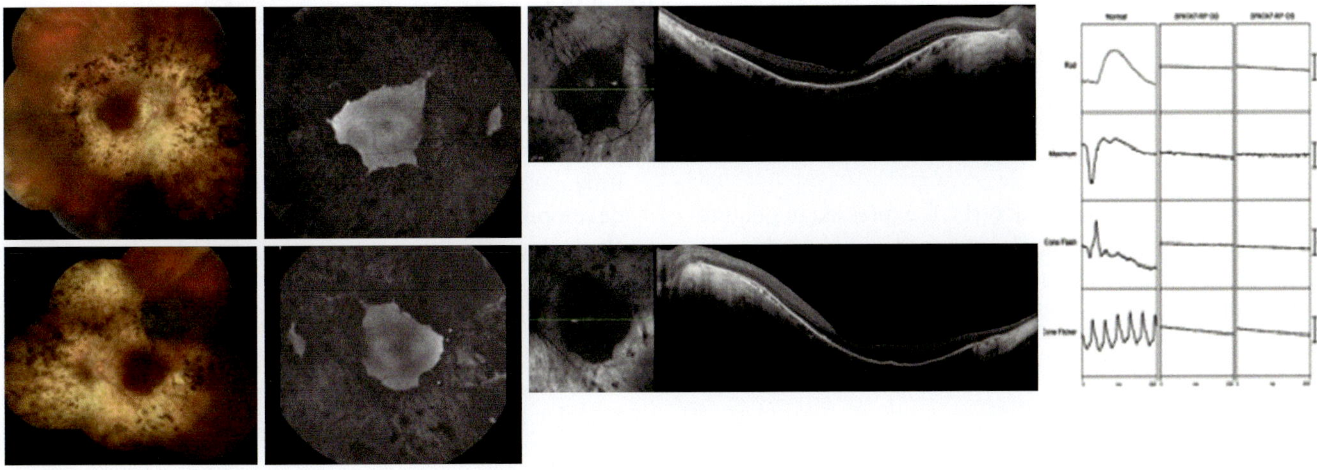

Fig. 26.2 LCA caused by *SPATA7* mutation. Note an unusual choroideremia-like phenotype in this 64-year-old man, with evidence of extensive RPE atrophy and choroidal sclerosis, extensive intraretinal pigment migration, and a centrally spared RPE island on fundus autofluorescence (FAF)

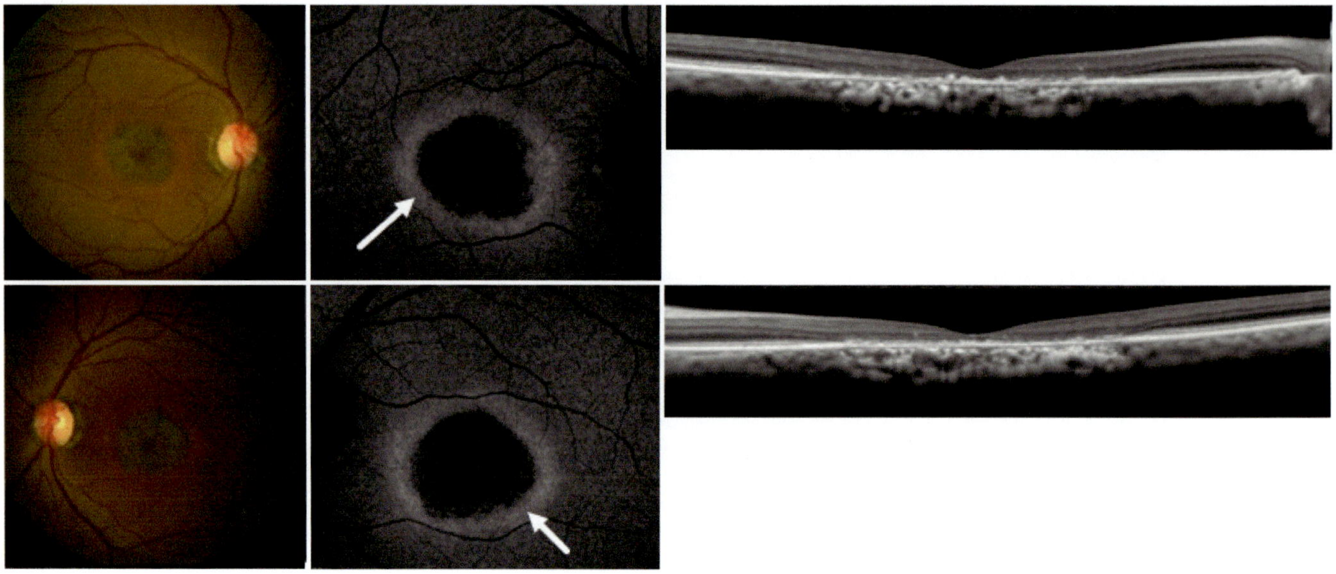

Fig. 26.3 LCA caused by *CRX* mutation; only macular involvement, with periphery spared. FAF shows annular hyperAF (*arrows*) surrounding the atrophic RPE

LCA4

- **Genotype**: *AIPL1* located at 17p13.1, expressed in photoreceptors and pineal gland; needed for normal development of rods and cones. Accounts for 4–8% of cases of LCA.
- **Phenotype**: Cone-rod dystrophy; early macular involvement, progressive atrophic lesion with a well-defined margin, at times called a "macular coloboma." These patients have severe, marked pigmentary retinopathy, with maculopathy. A large subset have keratoconus and cataract.

LCA5

- **Genotype**: *LCA5* located at 6q14.1, expressed in photoreceptor cilia and microtubules; lebercilin, needed for various ciliary proteins. Accounts for 1–2% of cases of LCA.
- **Phenotype**: Rod-cone dystrophy.

LCA6

- **Genotype**: *RPGRIP1* located at 14q11, expressed in photoreceptors and other organs including the heart, liver, spleen, kidney, testis, and brain; RP GTPase regulator protein. Accounts for 5% cases of LCA.
- **Phenotype**: Cone-rod dystrophy.

LCA7 (Fig. 26.3)

- **Genotype**: *CRX* located at 19q13.33, expressed in photoreceptors, inner nuclear layer, and pineal gland; encodes cone-rod homeobox–containing transcription factor needed for photoreceptor development. Accounts for 3% of cases of LCA.
- **Phenotype**: Cone-rod dystrophy.

LCA8 (Figs. 26.4 and 26.5)

- **Genotype**: *CRB1* located at 1q31.3, expressed in photoreceptors (IS) and other organs like testis and brain; encodes crumbs homolog 1 and is needed for outer limiting membrane (OLM) integrity and photoreceptor development.
- **Phenotype**: Rod-cone dystrophy, occasionally cone-rod dystrophy. Nummular intraretinal pigmentation and preservation of para-arteriolar RPE, pseudopapilledema, and prepapillary paravascular fibrosis. Coats-like vascular anomalies. Also progressive macular atrophic lesion and macular atrophic lesion with sharp border (like macular coloboma).

LCA9

- **Genotype**: *NMNAT1* located at 1p36.22, expressed ubiquitously; involved in retinal neuroprotection.
- **Phenotype**: Cone-rod dystrophy, early optic atrophy, macular atrophic lesion with sharp border (like macular coloboma).

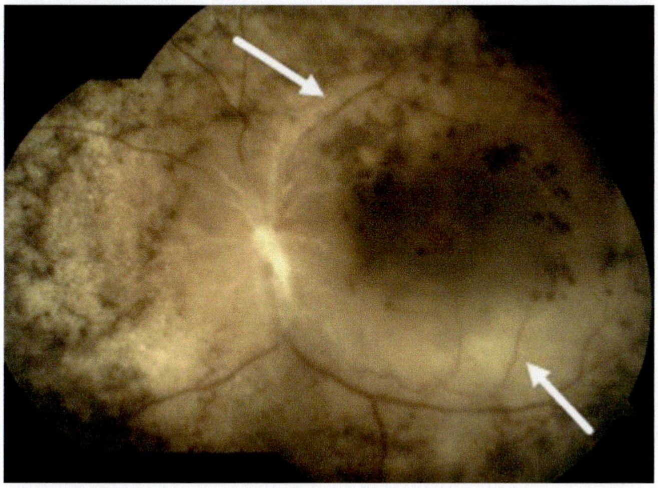

Fig. 26.4 LCA caused by *CRB1* mutation. Note the relative sparing of the para-arteriolar RPE (*arrows*)

LCA10 (Fig. 26.6)

- **Genotype**: *CEP290* located at 12q21.32, expressed in photoreceptor cilium and also in ciliated and nonciliated cells of multiple organs, including the brain; encodes for centrosomal protein 290 KD. Accounts for ≤20% of cases of LCA.
- **Phenotype**: Cone-rod dystrophy. These patients may also have hypotonia, ataxia, or intellectual disability and autistic behavior.

LCA11

- **Genotype**: *IMPDH1* located at 7q31.1, expressed in photoreceptors (IS) and several other organs, ubiquitously; it catalyzes the rate-limiting step in de novo guanine synthesis.
- **Phenotype**: Rod-cone dystrophy.

LCA12 (Figs. 26.7 and 26.8)

- **Genotype**: *RDH12* located at 14q24.1, expressed in photoreceptors (OS); involved in visual cycle with dual specificity for all-*trans*-retinols and all-*cis*-retinols. Accounts for 2.7% of cases of LCA.
- **Phenotype**: Rod-cone dystrophy, peripapillary sparing.

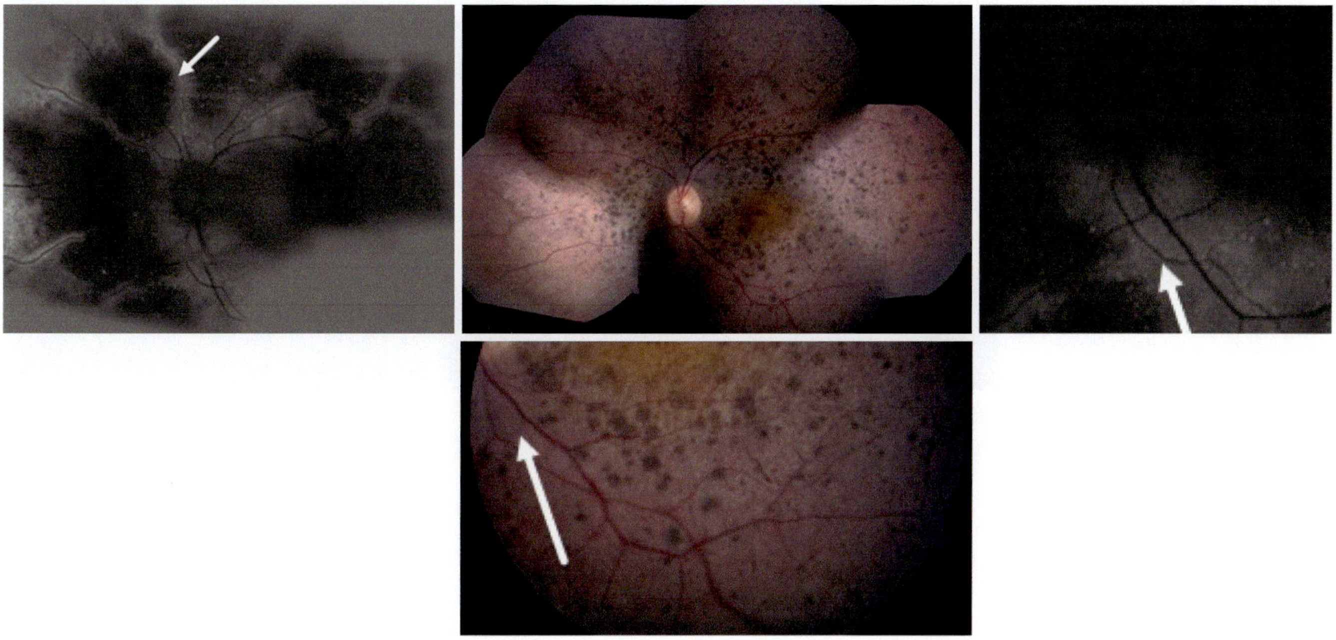

Fig. 26.5 LCA caused by *CRB1* mutation. Note the relative sparing of the para-arteriolar RPE (*arrows*)

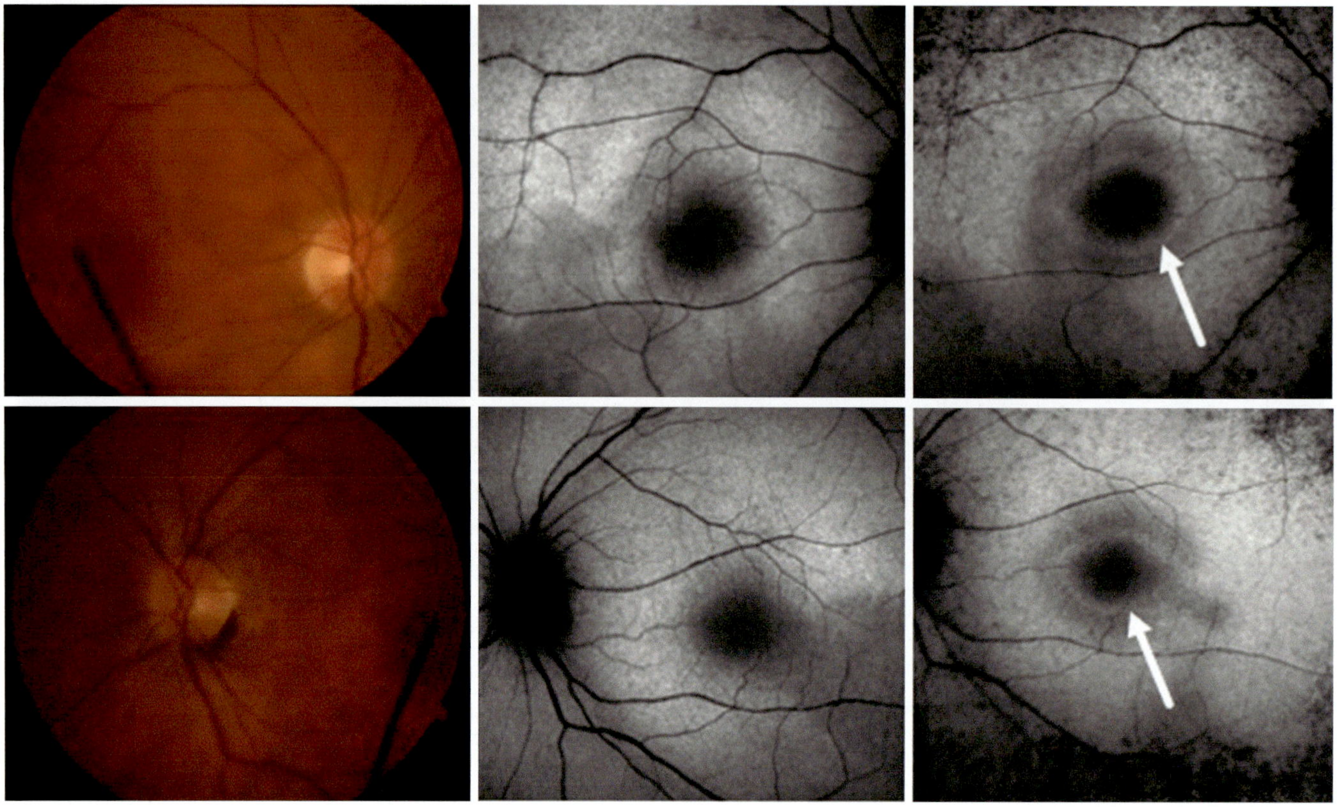

Fig. 26.6 LCA caused by *CEP290* mutation. Cone-rod dystrophy. Six-year follow-up progression of macular changes (hyperAF ring, *arrow*)

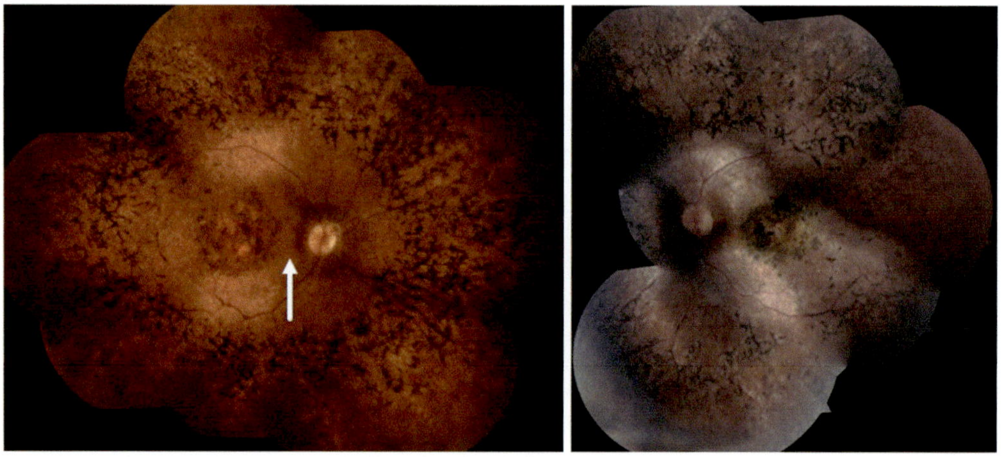

Fig. 26.7 LCA caused by *RDH12* mutation; note peripapillary sparing (*arrow*). Fundus shows involvement of both periphery and macula

LCA14

- **Genotype**: *LRAT* located at 4q32.1, expressed in RPE and involved in retinoid cycle.
- **Phenotype**: Rod-cone dystrophy.

LCA15

- **Genotype**: *TULP1* located at 6p21.31, expressed in photoreceptors and involved in tubby-like protein 1.
- **Phenotype**: Rod-cone dystrophy; pinpoint yellowish deposits at the macula, leading to atrophy and peripheral intraretinal pigmentation.

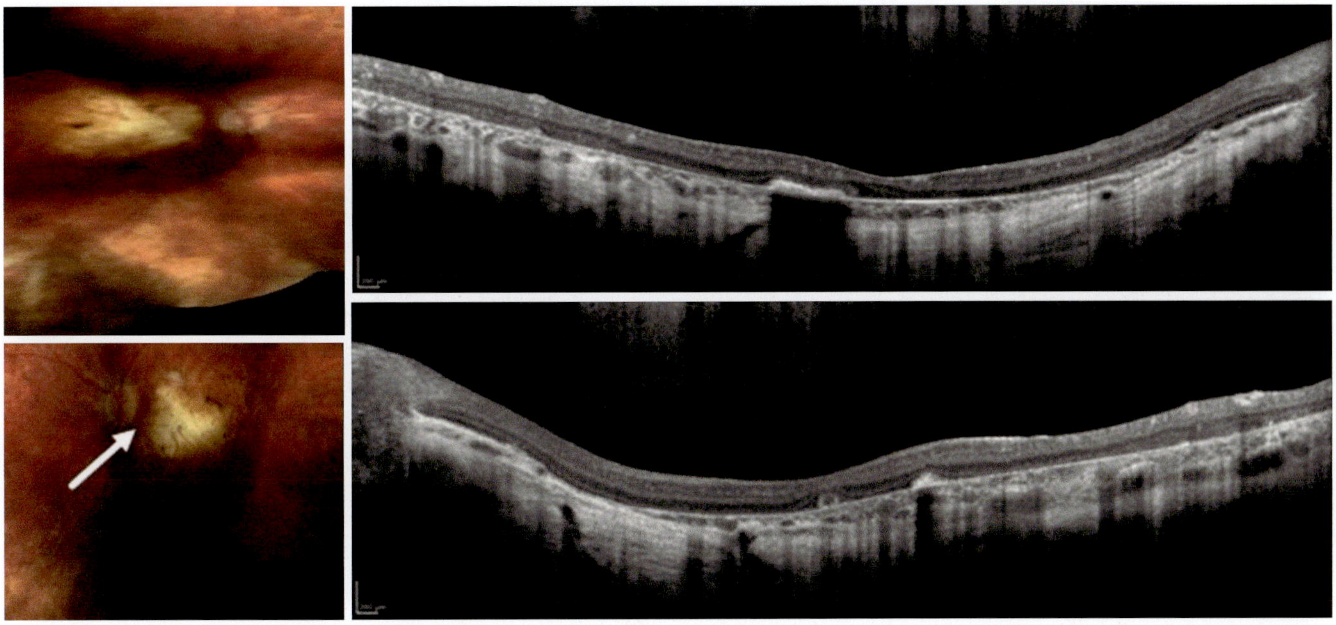

Fig. 26.8 LCA caused by *RDH12* mutation; note peripapillary sparing (*arrow*). There is a macular atrophy, with no intraretinal pigment migration

LCA16

- **Genotype**: *KCNJ13* located at 2q37.1, expressed in RPE apical membrane and other ion-transporting epithelia of the choroidal plexus, thyroid, prostate, kidney, and intestine; involved in potassium channel transport.
- **Phenotype**: Rod-cone dystrophy; extensive nummular intraretinal pigmentation.

RP38

- **Genotype**: *MERTK* located at 2q14.1, expressed in RPE and sclera; needed for phagocytosis of outer segments by RPE.
- **Phenotype**: Rod-cone dystrophy; dot-like whitish deposits at macula, leading to atrophy and intraretinal pigmentation.

LCA57

- **Genotype**: *PDE6G* located at 17q25.3, expressed in photoreceptors, in particular rod cGMP-specific phosphodiesterase 6.
- **Phenotype**: Rod-cone dystrophy, intraretinal pigmentation, and cystoid macular edema.

SLSN5

- **Genotype**: *IQCB1* located at 3q21.1, expressed in photoreceptor cilium; needed for transport of cargo molecules photoreceptor outer segment.
- **Phenotype**: Rod-cone dystrophy.

ABCA-4

- **Genotype**: *ABCA4* located at 1p22.1, expressed in photoreceptor (OS); needed for clearance of visual cycle intermediates from the intradiscal lumen of OS.
- **Phenotype:** Cone-rod dystrophy.

Further Reading

Kumaran N, Moore AT, Weleber RG, Michaelides M. Leber congenital amaurosis/early-onset severe retinal dystrophy: clinical features, molecular genetics and therapeutic interventions. Br J Ophthalmol. 2017;101:1147–54.

Sheck L, Davies WIL, Moradi P, Robson AG, Kumaran N, Liasis AC, et al. Leber congenital amaurosis associated with mutations in *CEP290*, clinical phenotype, and natural history in preparation for trials of novel therapies. Ophthalmology. 2018. https://doi.org/10.1016/j.ophtha.2017.12.013. [Epub ahead of print]

Takkar B, Bansal P, Venkatesh P. Leber's congenital amaurosis and gene therapy. Indian J Pediatr. 2018;85:237–42.

CHAPTER 27

Stargardt Disease

Leticia Pinheiro, Stephen H. Tsang, and Tarun Sharma

General Features 152
Clinical Stages and Groupings 152
Stages, Based on Fundus Appearance. 152
Groupings, Based on Electroretinography (ERG) 155
Types, Based on FAF 155
Classes, Based on Optical Coherence Tomography (OCT) 155
Molecular Genetics 158
ABCA4 Gene (RIM Protein or ABCR) 158
Other Mutations 158
Genotype-Phenotype Correlation 158
Further Reading 161

L. Pinheiro
Federal University of Goiás, Goiânia, Brazil

S. H. Tsang (✉) · T. Sharma
Department of Ophthalmology, Columbia University,
New York, NY, USA
e-mail: sht2@cumc.columbia.edu

S. H. Tsang et al. (eds.), *Atlas of Inherited Retinal Diseases*, Advances in Experimental Medicine and Biology 1467, https://doi.org/10.1007/978-3-031-72230-1_27

General Features

- Stargardt disease (STGD) is one of the most common macular dystrophies in young adults. It progresses slowly. Its prevalence is about 1:8000–10,000.
- Age of onset is a surrogate marker: The earlier the onset, the more severe the disease course. Onset usually occurs in childhood or early adolescence, at about 10–15 years of age.
- Vision is between about 20/70 and 20/200.
- The fundus shows a bull's eye pattern or beaten-bronze appearance, with or without yellowish flecks (fundus flavimaculatus).
- Fluorescein angiography may show dark choroid in about 80% of cases.
- On fundus autofluorescence (FAF), newer flecks appear hyperautofluorescent (hyperAF); older ones become progressively more hypoAF with time. Some flecks are surrounded by a ring of decreased AF.
- Peripapillary sparing is one the characteristics of Stargardt disease, but this area can be involved in about 2–7% of cases. The reason for this sparing is unclear; this area may be more resilient to the deleterious effect of *ABCA4* gene mutation, and there might be a more favorable RPE photoreceptor ratio, resulting in less lipofuscin build-up, in the presence of a thicker overlying peripapillary retinal nerve fiber layer.
- Patients with Stargardt disease should avoid bright light and excessive vitamin A.

Clinical Stages and Groupings

Stages, Based on Fundus Appearance (Figs. 27.1, 27.2, 27.3, 27.4, 27.5, and 27.6)

STAGE I: Confined to macula, with beaten-metal appearance (discontinuous, as opposed to classic appearance); a discontinuous ring of flecks around fovea, about 1 disc diameter (DD). The electrooculogram (EOG) and electroretinogram (ERG) are normal.

STAGE II: Widespread flecks beyond temporal arcades and/or nasal to disc. Subnormal cone and rod response.

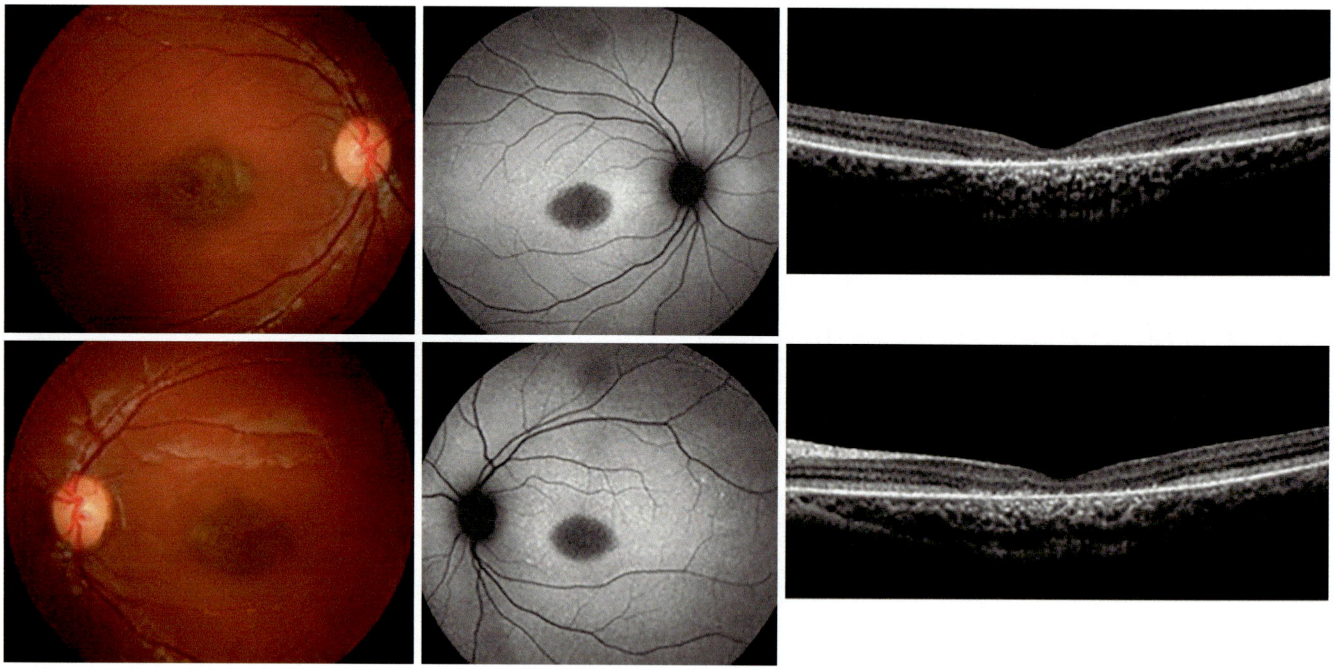

Fig. 27.1 Stargardt disease (STGD) presenting as bull's eye maculopathy (BEM), with no flecks. HypoAF on FAF. Optical coherence tomography (OCT) shows marked damage to photoreceptor layer and RPE (C4537delC)

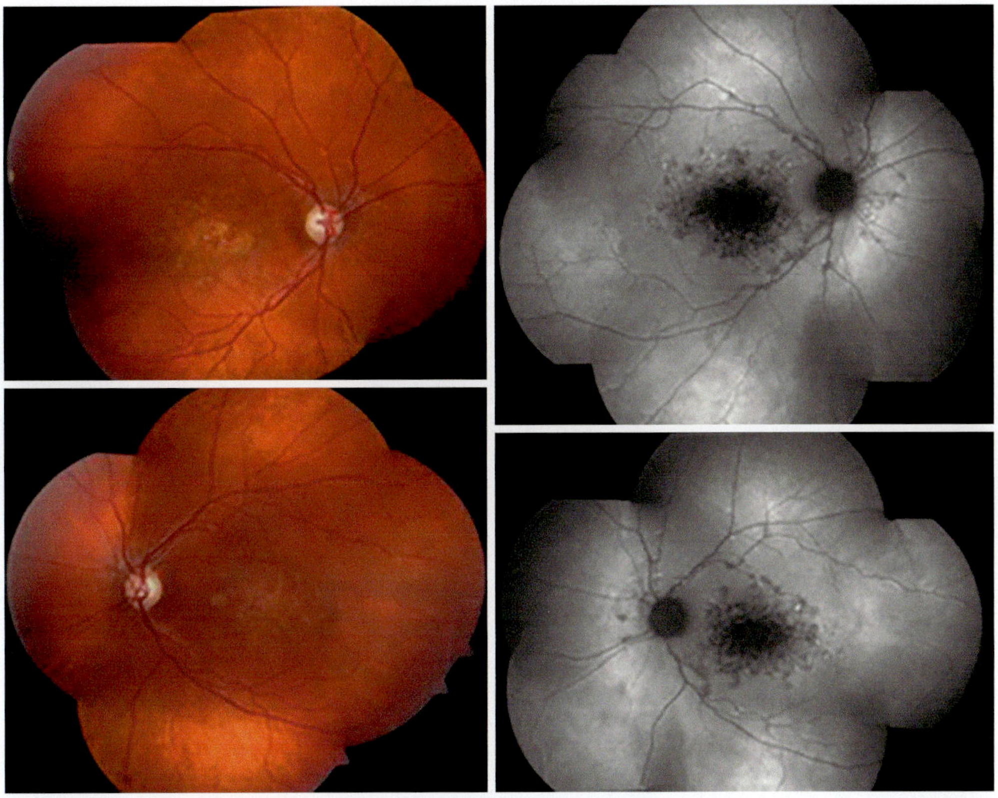

Fig. 27.2 STGD presenting as atrophic patch at the macula and flecks. Note sparing of the peripapillary area

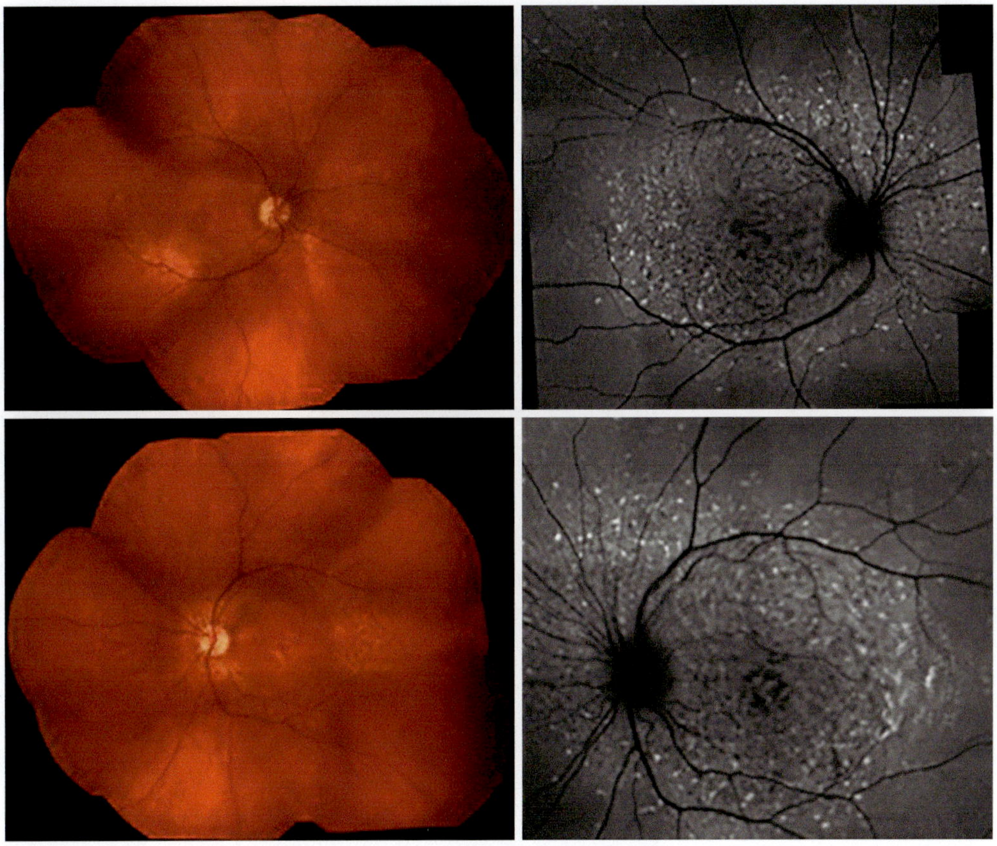

Fig. 27.3 STGD presenting as multiple flecks at the posterior pole, with a few atrophic patches at the macula

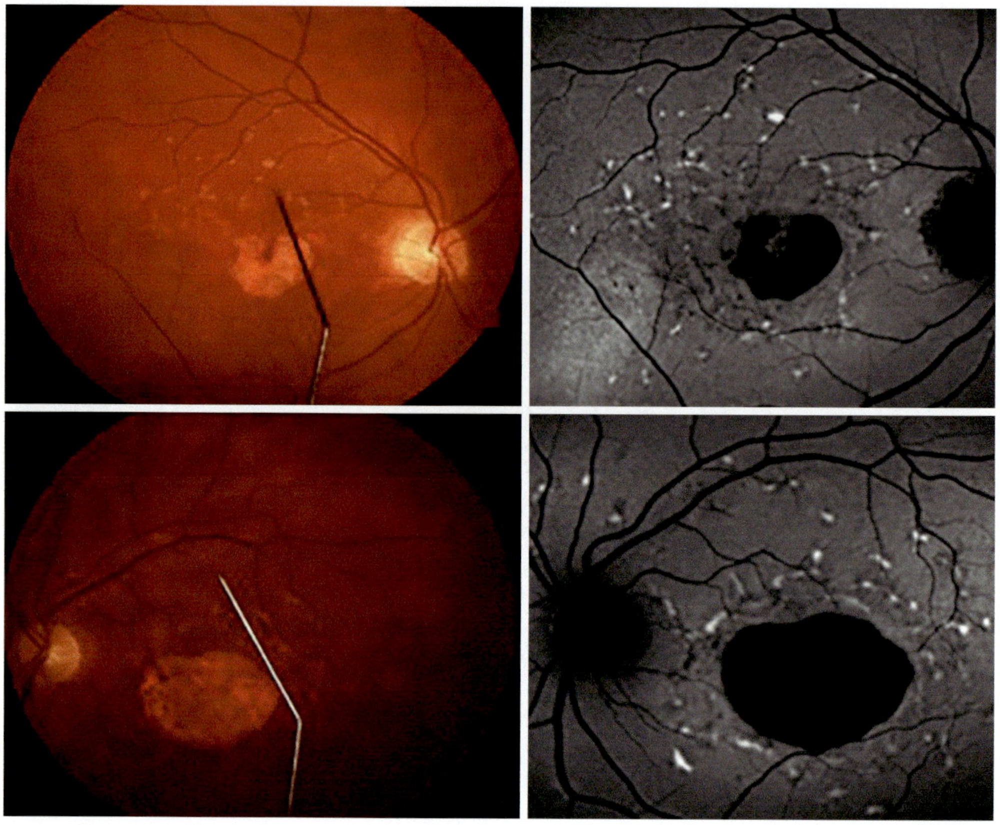

Fig. 27.4 STGD presenting as a large atrophic patch at the macula with a few flecks, some of them encroaching on the peripapillary area (28:p.P1380L, 36:p.S1696N)

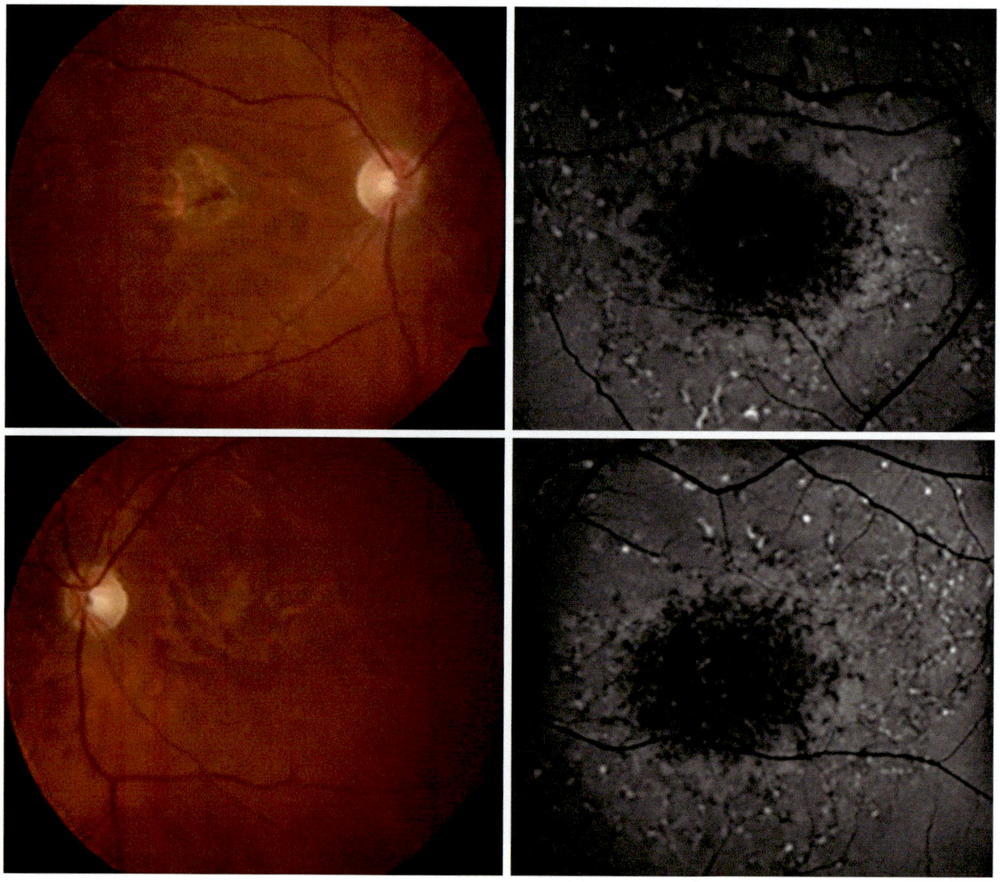

Fig. 27.5 STGD presenting as a large RPE atrophic area in the macula with sparse flecks, some of them showing RPE atrophy as well

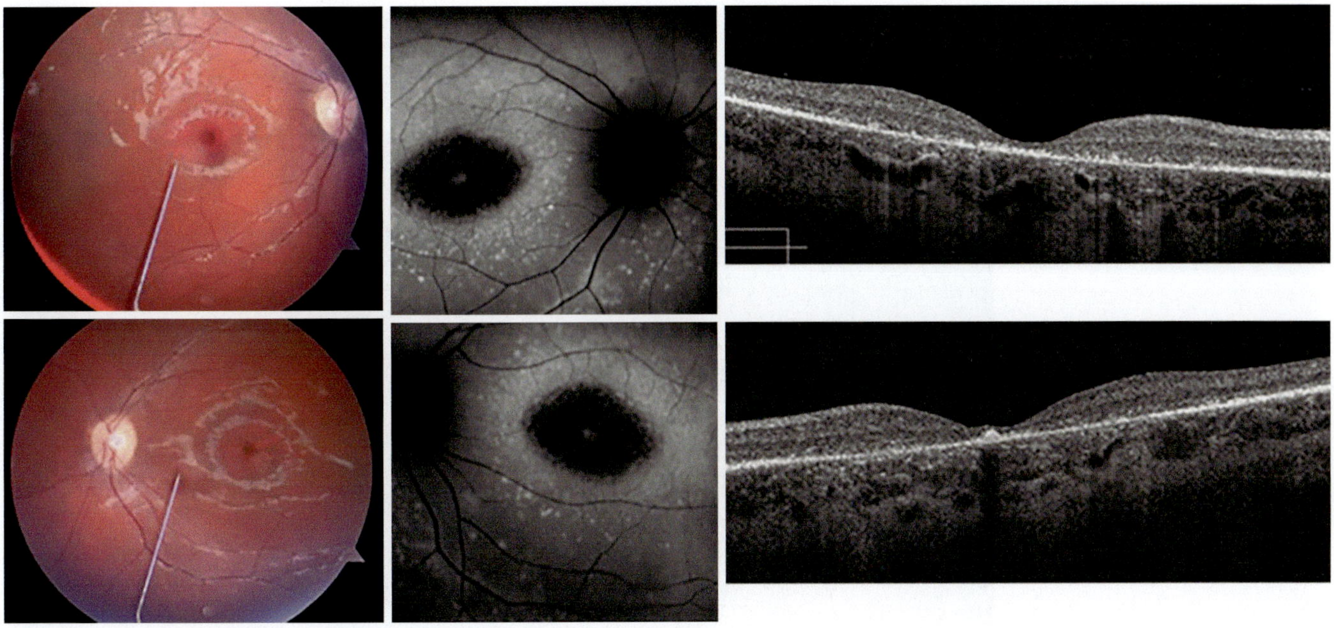

Fig. 27.6 STGD presenting as BEM with minimal flecks; OCT shows marked thinning of the central retina. Note peripapillary sparing

STAGE III: Re-absorption of flecks and widespread choriocapillaris (CC) atrophy. EOG shows subnormal ratio, and ERG shows cone/rod dysfunction.

STAGE IV: Further resorption of flecks, with extensive CC and RPE atrophy.

Groupings, Based on Electroretinography (ERG)

GROUP I: Isolated abnormalities in the flash ERG (FERG) and pattern ERG (PERG); normal photopic/scotopic full-field (FF) ERG.

GROUP II: Abnormal FERG, PERG, and photopic FF ERG; normal scotopic FF ERG.

GROUP III: Abnormal FERG, PERG, and photopic/scotopic FF ERG.

Types, Based on FAF (Figs. 27.7, 27.8, 27.9, 27.10, and 27.11)

TYPE A: Central hypofluorescence, surrounded by a hyperfluorescent ring.

TYPE B: Only central hypofluorescence without surrounding hyperfluorescence.

TYPE C: No central hypofluorescence; instead, a speckled pattern of alternating hypofluorescence and hyperfluorescence.

Classes, Based on Optical Coherence Tomography (OCT)

CLASS A: Flecks confined to outer segment (OS) layer.

CLASS B: Flecks extend through junction of OS and inner segment (IS).

CLASS C: Protrude into outer nuclear layer (ONL).

CLASS D: Only in ONL.

CLASS E: Drusen-like pigment epithelial detachment (PED). (Familial drusen shows echolucency in the center; flecks are echodense.)

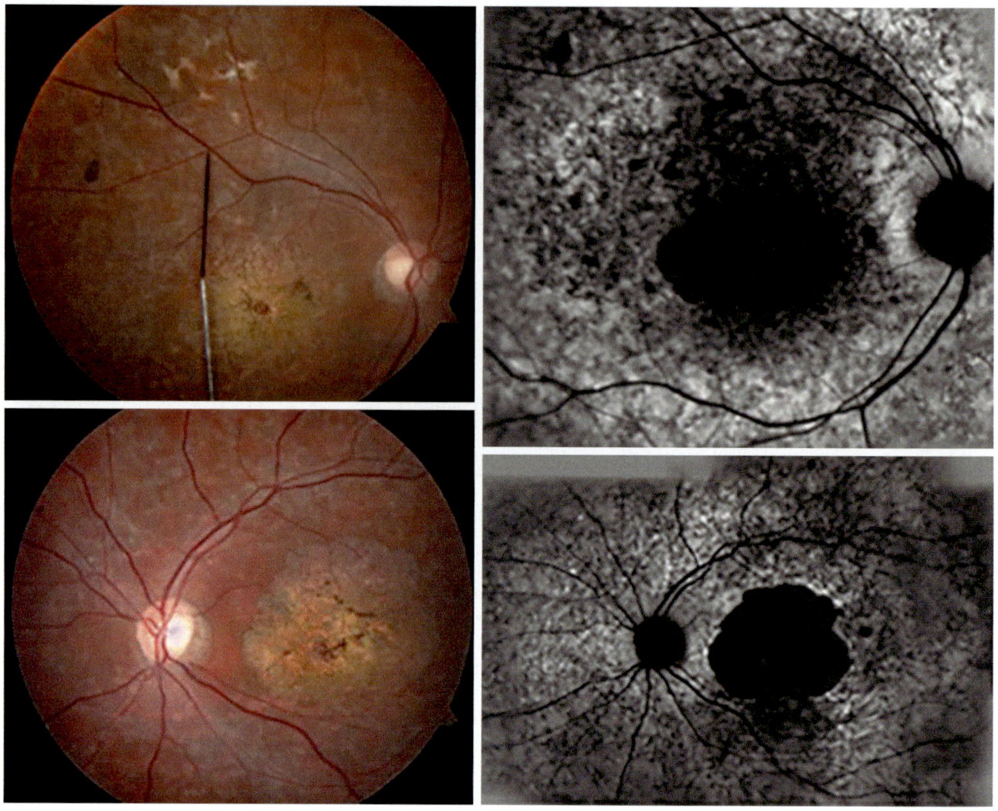

Fig. 27.7 An advanced case of STGD showing RPE atrophy over the macula and over all of the flecks as well. p.R18W and p.D2102E

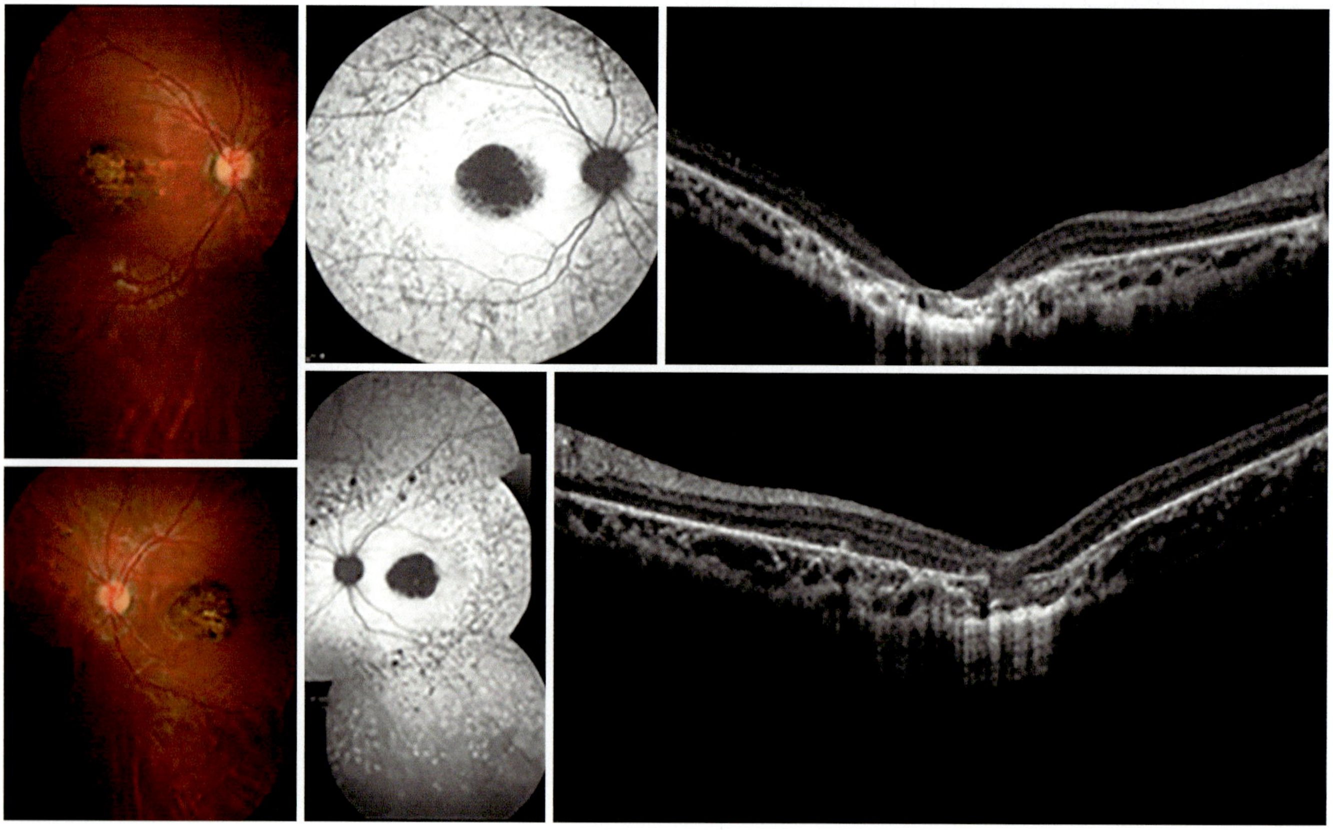

Fig. 27.8 STGD involving predominantly the macula, with sparse flecks in the periphery; OCT shows marked thinning of central retinal layers, including disruption of RPE layer (OS)

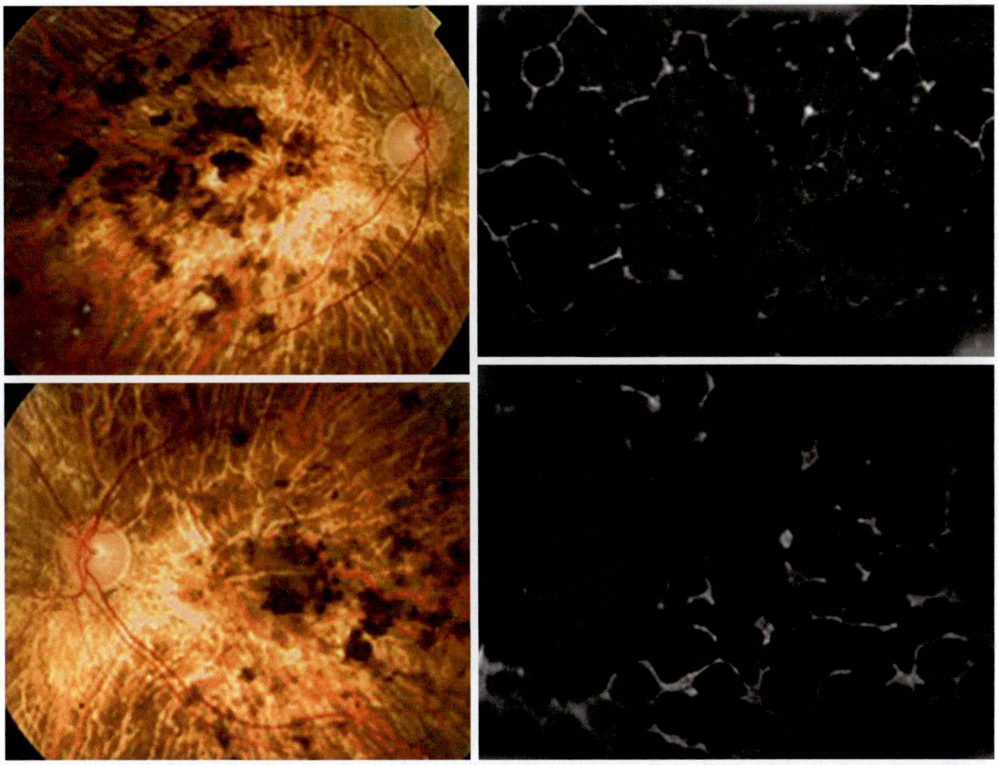

Fig. 27.9 An advanced stage of STGD, with marked RPE atrophy, confluent at the posterior pole (3:p. C54Y,IVS14:c.2160+1G>C)

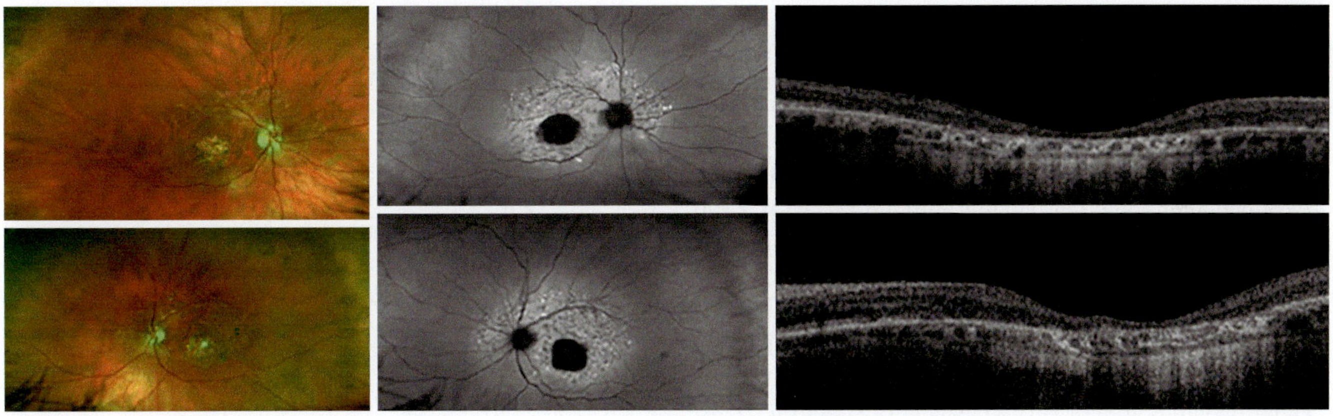

Fig. 27.10 STGD with just posterior involvement; the periphery is healthy. OCT shows marked thinning of the central retina. (Compare with Fig. 27.11, mother.) Two genetic variations: Cys54Tyr and Ala1038Val

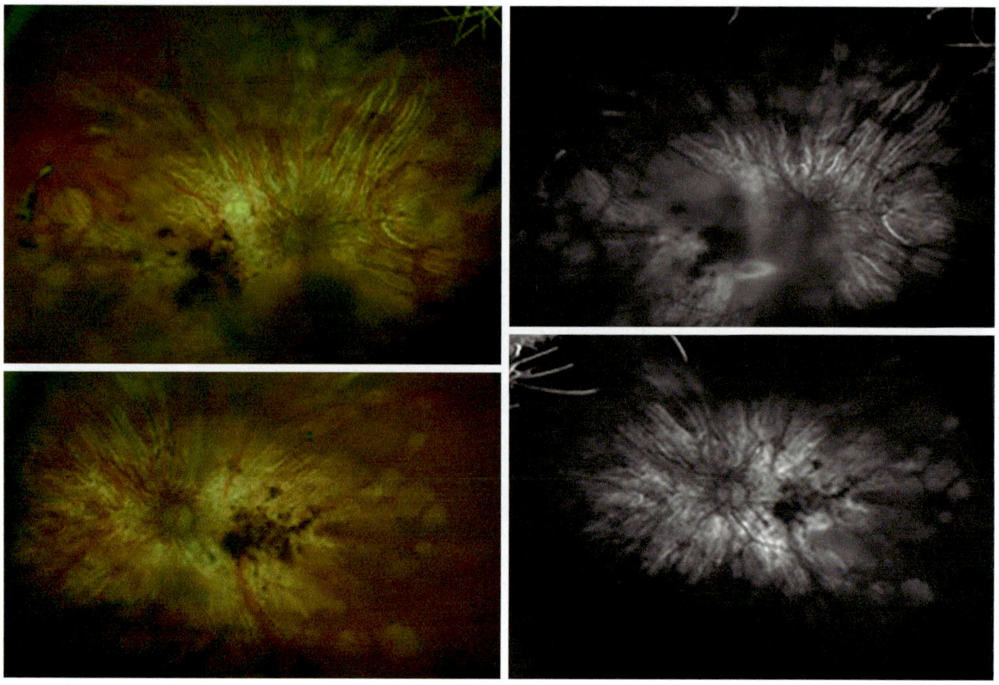

Fig. 27.11 STGD with involvement of the posterior pole and the periphery and marked atrophy of the RPE layer. (Compare with Fig. 27.10, son.) Two genetic variations: Cys54Tyr and Ala1038Val

Molecular Genetics

ABCA4 Gene (RIM Protein or ABCR)

Mutation in *ABCA4* gene, which encodes an ATP-binding cassette (ABC) transporter protein, expressed in the outer segment (OS) of rods. ABCA4 protein transports potentially toxic substances out of photoreceptor cells. These substances are formed after phototransduction. Buildup of the toxic substance—lipofuscin—in the photoreceptor cells and surrounding cells causes cell death. Cytogenetic location: 1p22.1.

The genetics is extremely heterogenous and may be responsible for wide variations in clinical presentations. Mild reduction in ABCA4 activity results in about 95% of cases of STGD; moderate loss of activity results in cone-rod dystrophy (about 30–50% of cases); and complete loss results in retinitis pigmentosa (RP) with complete loss of rod and cone function (about 8% of cases of autosomal recessive RP).

Other Mutations

Other mutations, accounting for the remaining 5% of STGD cases, are in the dominant genes *STGD4* and *ELOVL4* and *PRPH2*. The ELOVL4 protein plays a role in making a group of fats called very long-chain fatty acids. Mutations in the *ELOVL4* gene lead to the formation of ELOVL4 protein clumps in the cells, interfering with their activity and eventually leading to cell death. Cytogenetic location:6q14.1.

Genotype-Phenotype Correlation

c.5603A>T (p.Asn1868Ile) (Figs. 27.12, 27.13, and 27.14)

- This is a hypomorphic allele that is expressed only when in *trans* with a deleterious mutation.
- Phenotypes are late-onset (age about 36 years or later). About 85% of cases have foveal sparing and can masquerade as age-related macular degeneration (AMD).
- Very few flecks are seen, which are somewhat larger, well-defined (a kind of "pizza sign"), and sparsely distributed.

c.5882G>A (p.Gly1961Glu) (Fig. 27.15)

- These patients exhibit mild disease expression, few specks, no confluence of flecks, and not much disease progression.
- Age of onset is about 23 years, with foveal sparing in about 40% of cases.
- Flecks are relatively large, well-defined, and sparsely distributed.
- On OCT, these patients show focal loss of the ellipsoid zone (EZ) (Stage I), an optical gap (Stage II), and eventual collapse of the optical gap.

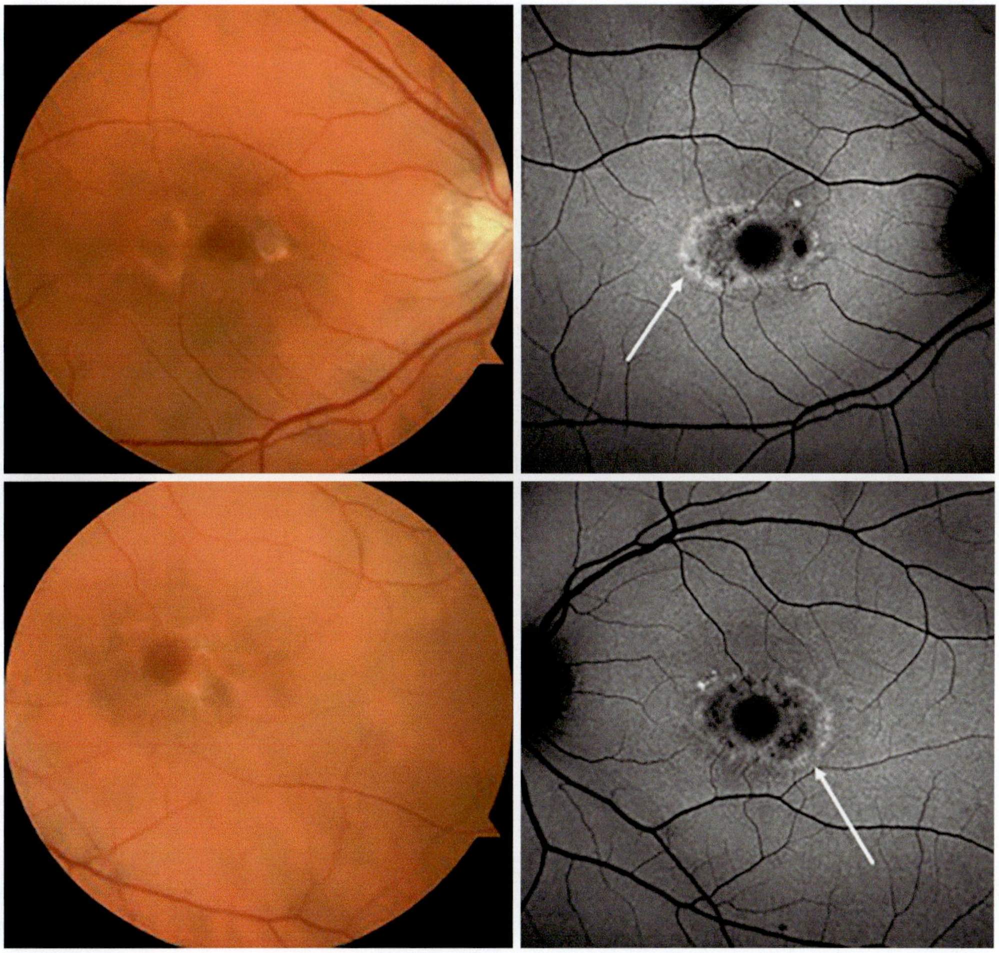

Fig. 27.12 STGD with predominant macular involvement; FAF shows central hypoAF surrounded by stippled AF ("pizza sign") (*arrows*)

Other *ABCA4* Alleles

- These patients show innumerable flecks, which are confluent across the posterior pole and progressive; onset is about 18 years of age.
- Foveal sparing, about 30%.

Biallelic Null *ABCA4*

- Age of onset about 10 years or earlier.
- Confluent flecks appear early and progress more rapidly, leading to end-stage atrophy.

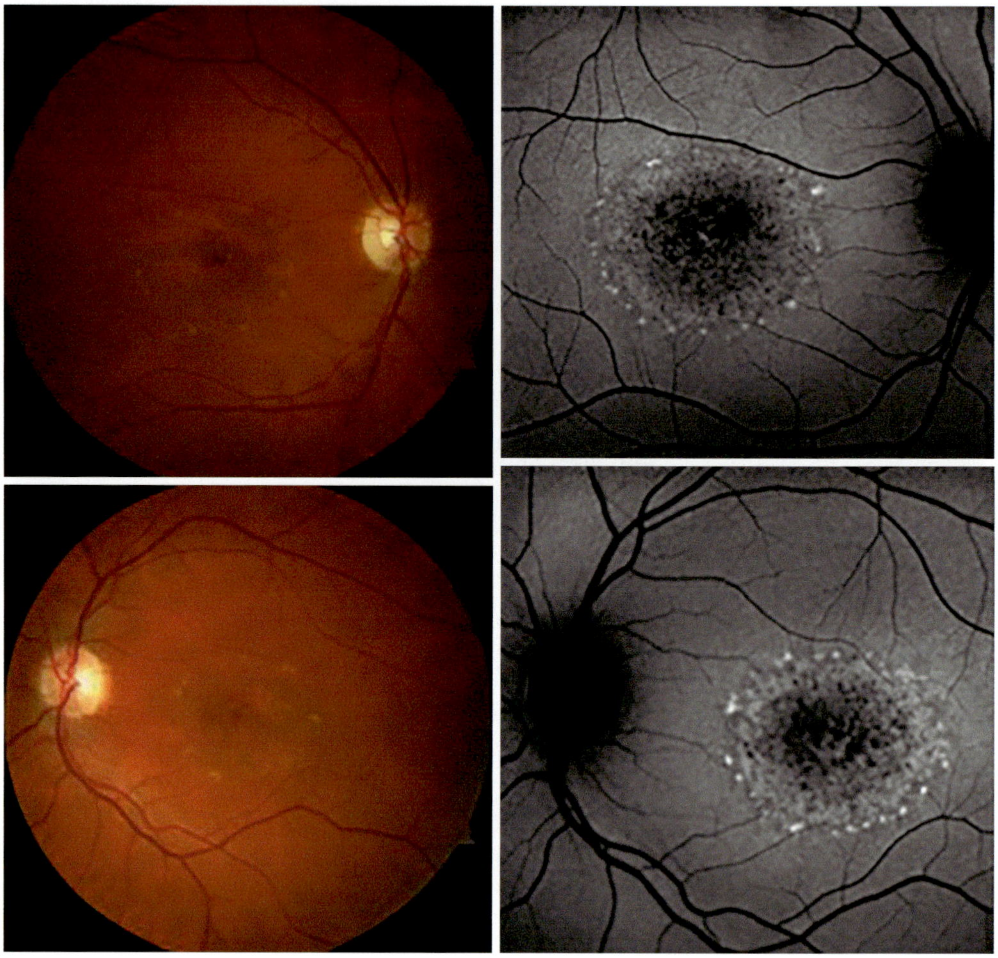

Fig. 27.13 STGD caused by c.5603A>T (p.N1868I). Note the characteristic "pizza sign"

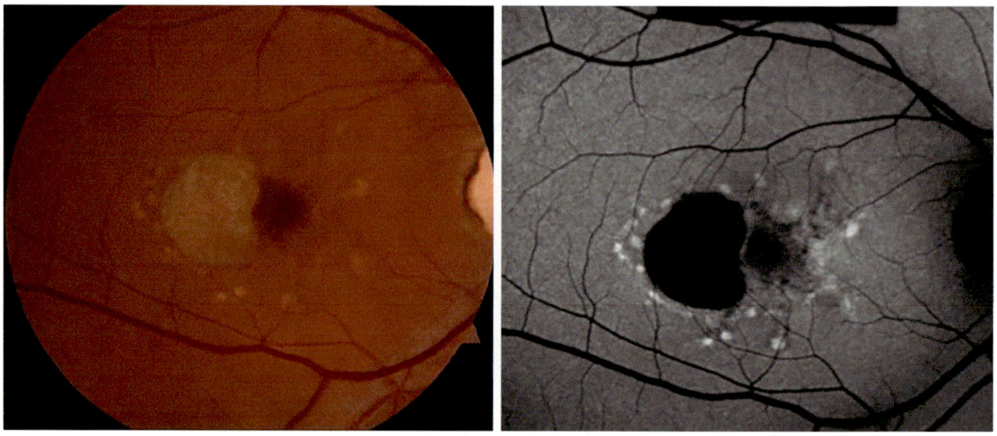

Fig. 27.14 STGD caused by c.5603A>T (p.N1868I). Note foveal sparing on AF. This mutation accounts for foveal sparing in about 80% of cases, along with late onset and mild progression

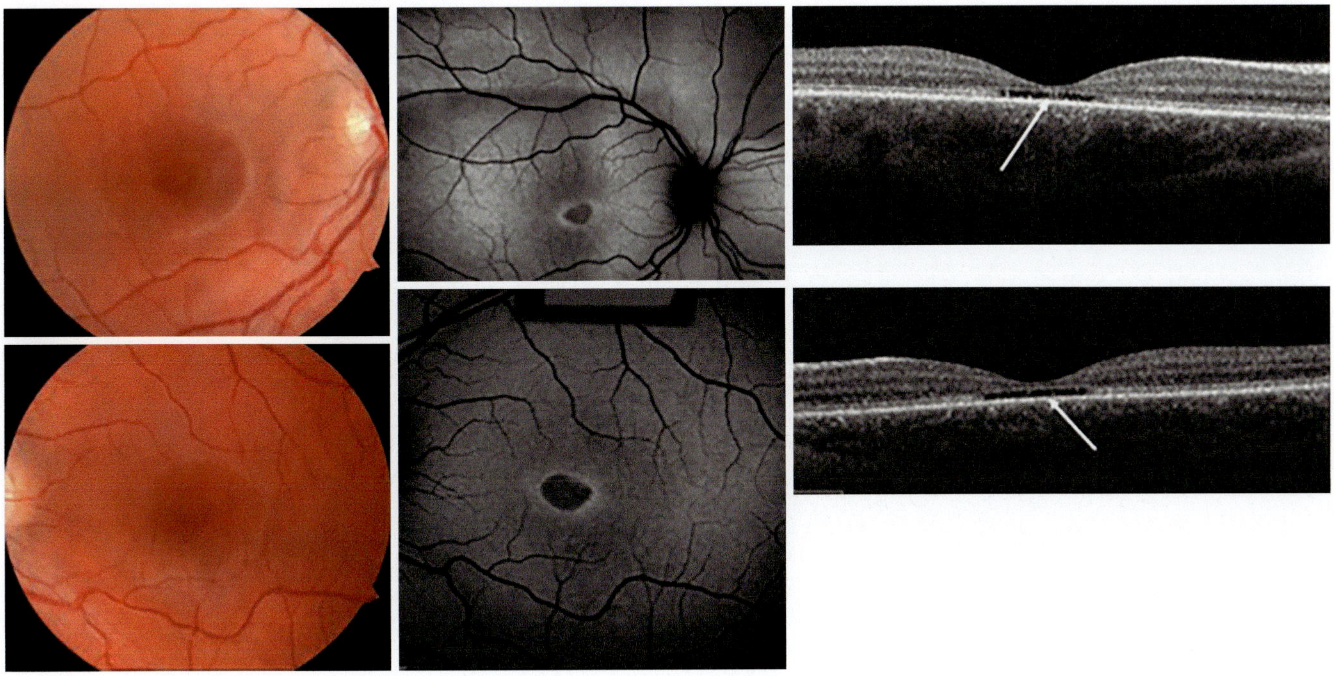

Fig. 27.15 STGD caused by c.5882G>A (p.G1961E) mutation. Note loss of foveal reflex, and central hypoAF surrounded by a hyperAF ring. OCT shows loss of the ellipsoid zone (EZ), causing an optical gap (*arrow*)

Further Reading

Burke TR, Tsang SH. Allelic and phenotypic heterogeneity in ABCA4 mutations. Ophthalmic Genet. 2011;32:165–74.

Gelman R, Smith RT, Tsang SH. Diagnostic accuracy evaluation of visual acuity and fundus autofluorescence macular geographic atrophy area for the discrimination of Stargardt groups. Retina. 2016;36:159–601.

Greenstein VC, Nunez J, Lee W, Schuerch K, Fortune B, Tsang SH, et al. A comparison of en face optical coherence tomography and fundus autofluorescence in Stargardt disease. Invest Ophthalmol Vis Sci. 2017;58:5227–36.

Nõupuu K, Lee W, Zernant J, Tsang SH, Allikmets R. Structural and genetic assessment of the ABCA4-associated optical gap phenotype. Invest Ophthalmol Vis Sci. 2014;55:7217–26.

Tanna P, Strauss RW, Fujinami K, Michaelides M. Stargardt disease: clinical features, molecular genetics, animal models and therapeutic options. Br J Ophthalmol. 2017;101:25–30.

Zernant J, Lee W, Collison FT, Fishman GA, Sergeev YV, Schuerch K, et al. Frequent hypomorphic alleles account for a significant fraction of ABCA4 disease and distinguish it from age-related macular degeneration. J Med Genet. 2017;54:404–12.

CHAPTER 28

Enhanced S-Cone Syndrome (Goldmann-Favre Syndrome)

Laura A. Jenny, Stephen H. Tsang, and Tarun Sharma

General Features 164
Molecular Genetics 166
Further Reading 167

L. A. Jenny · S. H. Tsang (✉) · T. Sharma
Department of Ophthalmology, Columbia University,
New York, NY, USA
e-mail: lj2539@cumc.columbia.edu; sht2@cumc.columbia.edu

S. H. Tsang et al. (eds.), *Atlas of Inherited Retinal Diseases*, Advances in Experimental Medicine and Biology 1467, https://doi.org/10.1007/978-3-031-72230-1_28

General Features

- In enhanced S-cone syndrome (ESCS), rods and red and green cone receptors are degenerated, but S-cones are enhanced (increased in number).
- Patients have increased sensitivity to blue light, with night blindness from an early age (from birth) and impaired central vision.
- The fundus shows characteristic nummular pigmentary changes along the vascular arcades at the level of the retinal pigment epithelium (RPE). (In retinitis pigmentosa [RP], pigment is deposited within the retina.) Sometimes, there are white-yellow dots at the level of the RPE along the vascular arcades, along with focal hyperpigmentation within the arcades (Figs. 28.1, 28.2, and Fig. 28.3).
- Patients may also have foveal schisis. (Goldmann-Favre syndrome consists of foveal schisis plus peripheral degeneration.)

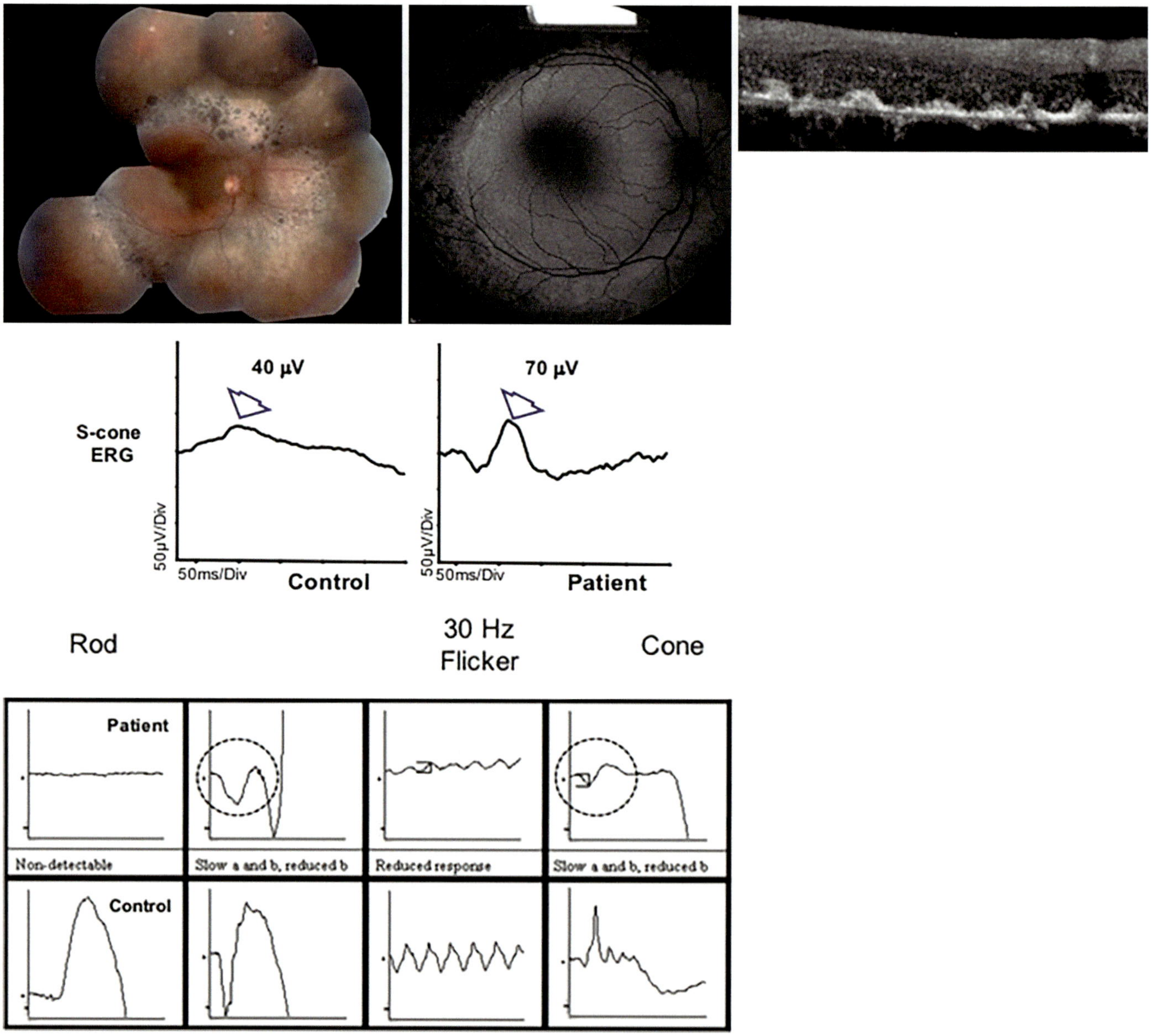

Fig. 28.1 A case of enhanced S-cone syndrome (ESCS) with *NR2E3* mutation. Note the nummular pigmentation pattern along the temporal arcades and nasal to the disc, as well as hyperAF in the involved area with hypoAF in the areas of pigmentation. Optical coherence tomography (OCT) shows pigmentation at the RPE level, and electroretinography (ERG) shows reduced response in both photopic and scotopic conditions, with decreased amplitude on 30 Hz flicker

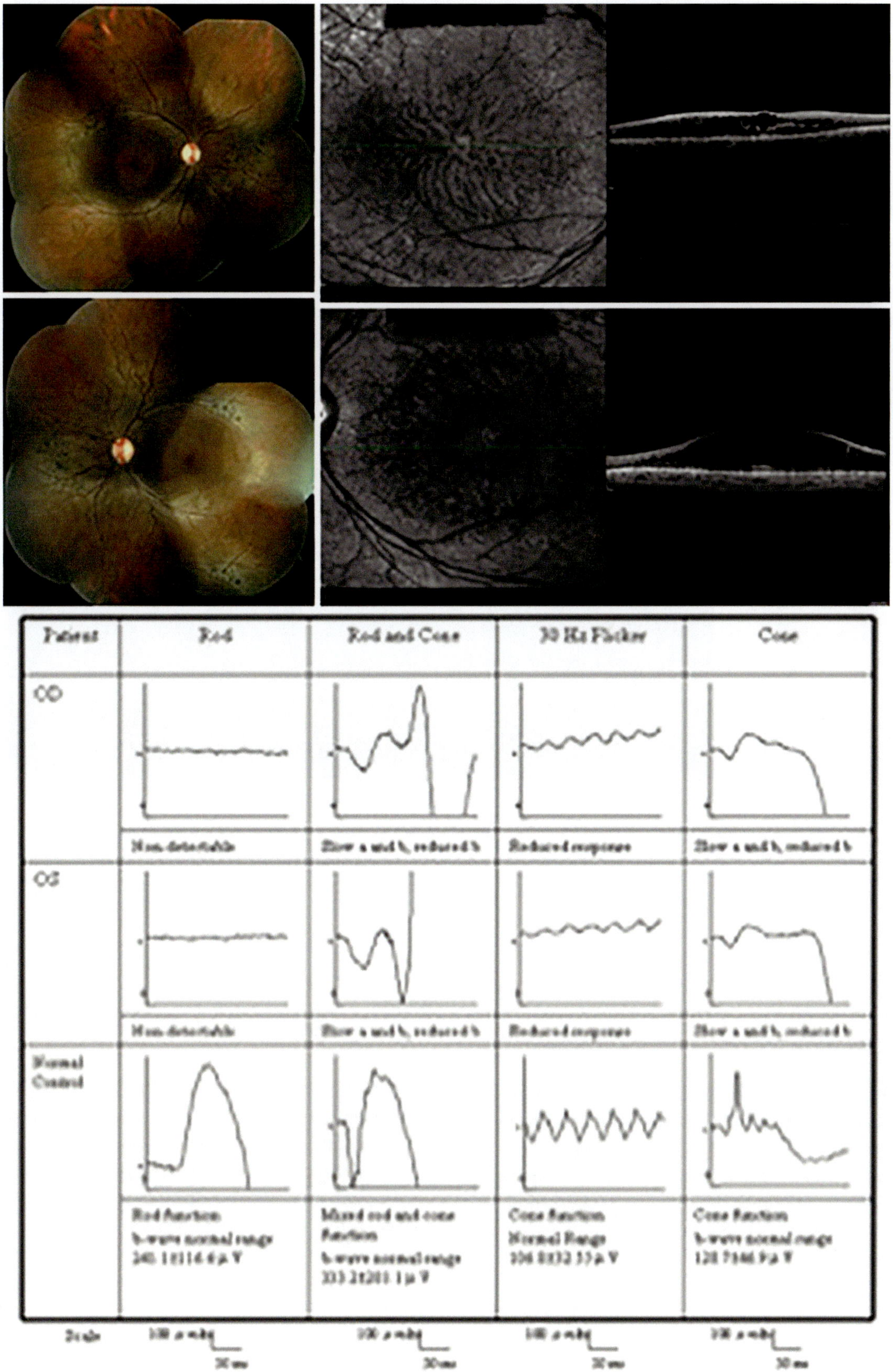

Fig. 28.2 *NR2E3* (homozygous R311Q) mutation, showing characteristic nummular pigmentation, macular schisis on OCT, and typical ERG findings (similar waveform in the maximum scotopic and photopic), with disproportionately reduced amplitude on 30 Hz flicker compared to single-flash cone amplitude

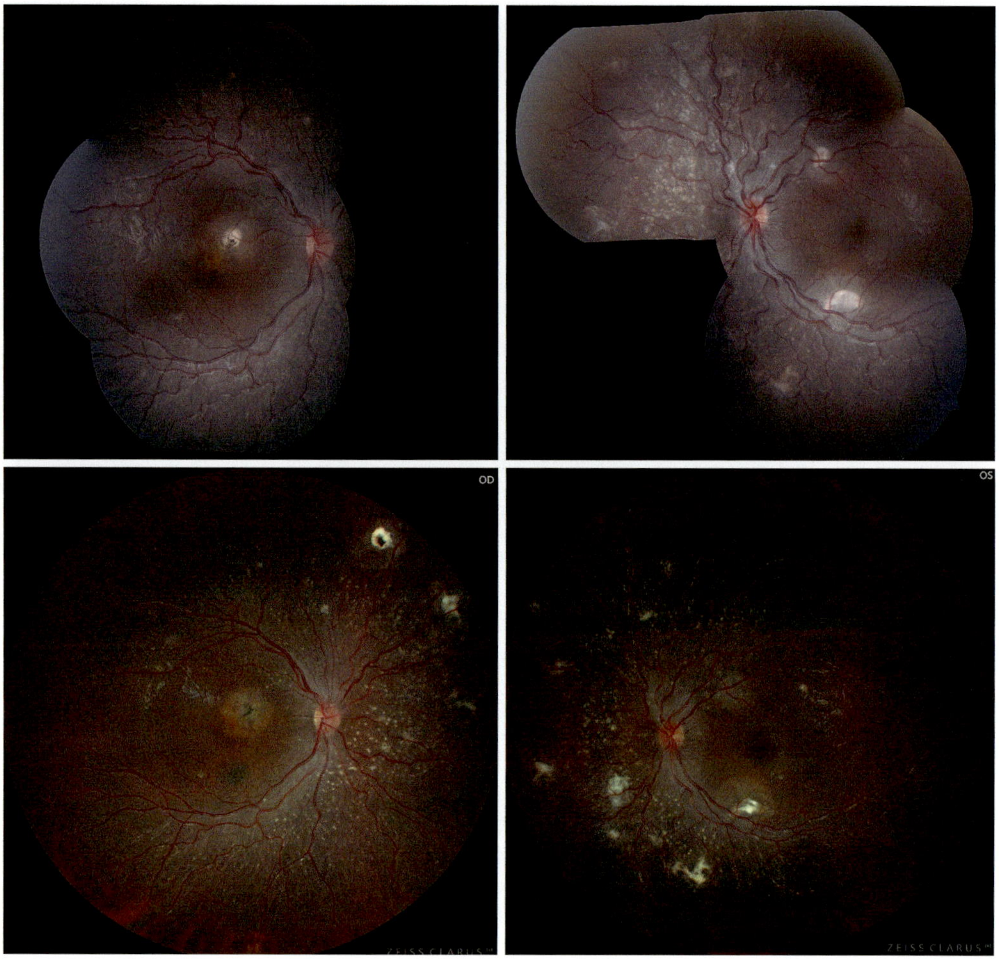

Fig. 28.3 A case of ESCS showing areas of chalky white preretinal reticular deposits outside of the vascular arcades nasally and inferiorly

- Fundus autofluorescence (FAF) shows a decrease or lack of AF outside the arcades, possibly due to loss of photoreceptors in this region; a ring of hyperautofluorescence (hyperAF) is seen in the transition zone between the region of absent AF beyond the arcade and the central zone of spoke-like, relatively increased AF, centered on the fovea; increased AF may be related to lipofuscin accumulation secondary to RPE-photoreceptor dysfunction in that area.
- Electroretinograms (ERGs) in ESCS are pathognomonic and show extinguished rod response and a similar wave form in both scotopic and photopic conditions.
  - The 30 Hz flicker responses are markedly delayed and are of lower amplitude. (Normally, flicker amplitude lies between that of the photopic a- and b-waves, but in ESCS, it is less than that of the photopic a-wave.)
  - With orange background (suppressing red and green cones), increased response to short wavelength (blue) is elicited, suggestive of functioning of S-cones. The multifocal ERG (mfERG) shows preservation of central responses, though somewhat delayed.
- ESCS is a slowly progressive disorder that often leads to severe visual loss in adults.

Molecular Genetics

- Mutation in *NR2E3* (located on chromosome 15q23); this gene encodes for photoreceptor nuclear transcription factor, which controls retinal progenitor cell fate.
- The *NR2E3* gene promotes differentiation and survival of rod photoreceptors by differentially regulating transcription of rod-specific and cone-specific genes.
- Mutation in *NR2E3* results in improper photoreceptor cell differentiation, possibly by encouraging S-cone pathway from rod photoreceptor pathway.

Further Reading

Carvalho ER, Robson AG, Arno G, Boon CJF, Webster AA, Michaelides M. Enhanced S-Cone Syndrome: Spectrum of Clinical, Imaging, Electrophysiologic, and Genetic Findings in a Retrospective Case Series of 56 Patients. Ophthalmology Retina. 2021;5(2):195–214.

Garafalo AV, Calzetti G, Cideciyan AV, Roman AJ, Saxena S, Sumaroka A, Choi W, Wright AF, Jacobson SG. Cone Vision Changes in the Enhanced S-Cone Syndrome Caused by *NR2E3* Gene Mutations. IOVS. 2018;59(8):3209–3219.

Gelman R, Greenberg JP, Duncker T, Nguyen HV, Yannuzzi LA, Tsang SH. Hyperautofluorescent macular ring in a series of patients with enhanced S-cone syndrome. Ophthalmic Surg Lasers Imaging Retina. 2014;45:592–5.

Hull S, Arno G, Sergouniotis PI, Tiffin P, Borman AD, Chandra A, et al. Clinical and molecular characterization of enhanced S-cone syndrome in children. JAMA Ophthalmol. 2014;132:1341–9.

CHAPTER 29

Best Vitelliform Macular Dystrophy

Justin Cheong, Stephen H. Tsang, Tarun Sharma, and Vlad Diaconita

General Features 170
Further Reading 170

J. Cheong
Vagelos College of Physicians and Surgeons,
New York, NY, USA

S. H. Tsang (✉) · T. Sharma · V. Diaconita
Department of Ophthalmology, Columbia University,
New York, NY, USA
e-mail: sht2@cumc.columbia.edu

S. H. Tsang et al. (eds.), *Atlas of Inherited Retinal Diseases*, Advances in Experimental Medicine and Biology 1467, https://doi.org/10.1007/978-3-031-72230-1_29

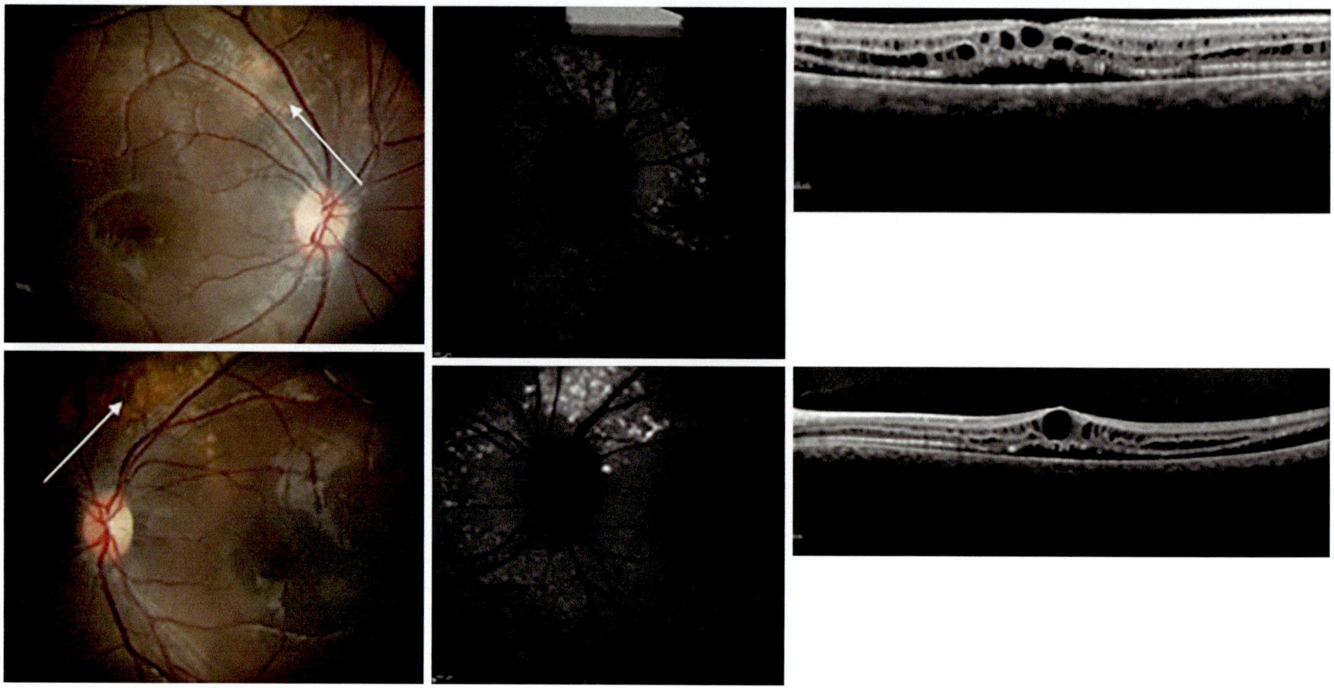

Fig. 29.1 Autosomal recessive Best macular dystrophy caused by mutation in the *BEST1* gene (homozygous truncation at F283 amino acid). Note the yellowish subretinal deposits (*arrows*), which show hyperautofluorescence. Optical coherence tomography (OCT) shows cystoid macular edema

General Features

- Autosomal recessive bestrophinopathy (ARB) results from a total absence of functional bestrophin-1 protein owing to two *BEST1* mutations, one on each of the chromosomes.
- If present at an early age, the presenting feature could be decreased vision due to amblyopia.
- Refractive error is hyperopia, predisposing these eyes for acute angle-closure glaucoma.
- The yellowish lesions are larger and more extensive—extending beyond the arcades—than in the typical autosomal dominant Best disease. Some of the eyes also show numerous yellowish subretinal dots. Lesions are multifocal (Fig. 29.1).
- Subretinal fibrosis in the macular area is a common feature.
- Optical coherence tomography (OCT) may show cystoid changes in the neurosensory retina.
- Fundus autofluorescence (FAF): Increased AF reflects lipofuscin accumulation in the RPE; decreased AF reflects RPE atrophy.
- Electroretinography (ERG): As panretinal photoreceptor dysfunction progresses with advancing age, full-field (FF) ERG shows delayed rod and cone responses.
- Electrooculography (EOG): Abnormal Arden ratio.

Further Reading

Fung A, Yzer S, Allikmets R. Clinical and genetic misdiagnosis of autosomal recessive bestrophinopathy. JAMA Ophthalmol. 2013;131:1651.

Silva RA, Berrocal AM, Lam BL, Albini TA. Novel mutation in *BEST1* associated with retinoschisis. JAMA Ophthalmol. 2013;131:794–8.

Tian L, Sun T, Xu K, Zhang X, Peng X, Li Y. Screening of *BEST1* gene in a Chinese cohort with Best vitelliform macular dystrophy or autosomal recessive bestrophinopathy. Invest Ophthalmol Vis Sci. 2017;58:3366–75.

Part V
Systemic Disorders

CHAPTER 30

Mitochondrial Disorder: Kearns-Sayre Syndrome

Benjamin Kuang-Chien Chiang, Stephen H. Tsang, Alicia R. P. Aycinena, and Tarun Sharma

General Features 174
Molecular Genetics 175
Further Reading 175

B. K.-C. Chiang
School of Medicine, College of Medicine,
National Taiwan University, Taipei, Taiwan

S. H. Tsang (✉) · T. Sharma
Department of Ophthalmology, Columbia University,
New York, NY, USA
e-mail: sht2@cumc.columbia.edu

A. R. P. Aycinena
Kaiser Permanente, San Francisco, CA, USA

S. H. Tsang et al. (eds.), *Atlas of Inherited Retinal Diseases*, Advances in Experimental Medicine and Biology 1467, https://doi.org/10.1007/978-3-031-72230-1_30

General Features

- Mitochondrial diseases are multisystem disorders: anemia, myopathy, lactic acidosis, CNS abnormality, endocrine abnormalities, renal disease, sensorineural deafness, and retinal involvement. The clinical abnormalities are heterogeneous, and they usually begin in childhood. Premature death occurs because of cardiac conduction defects.
- The onset is usually before 20 years of age. The fundus shows pigmentary retinopathy, with a salt-and-pepper appearance (Fig. 30.1), but vision remains good in most patients.

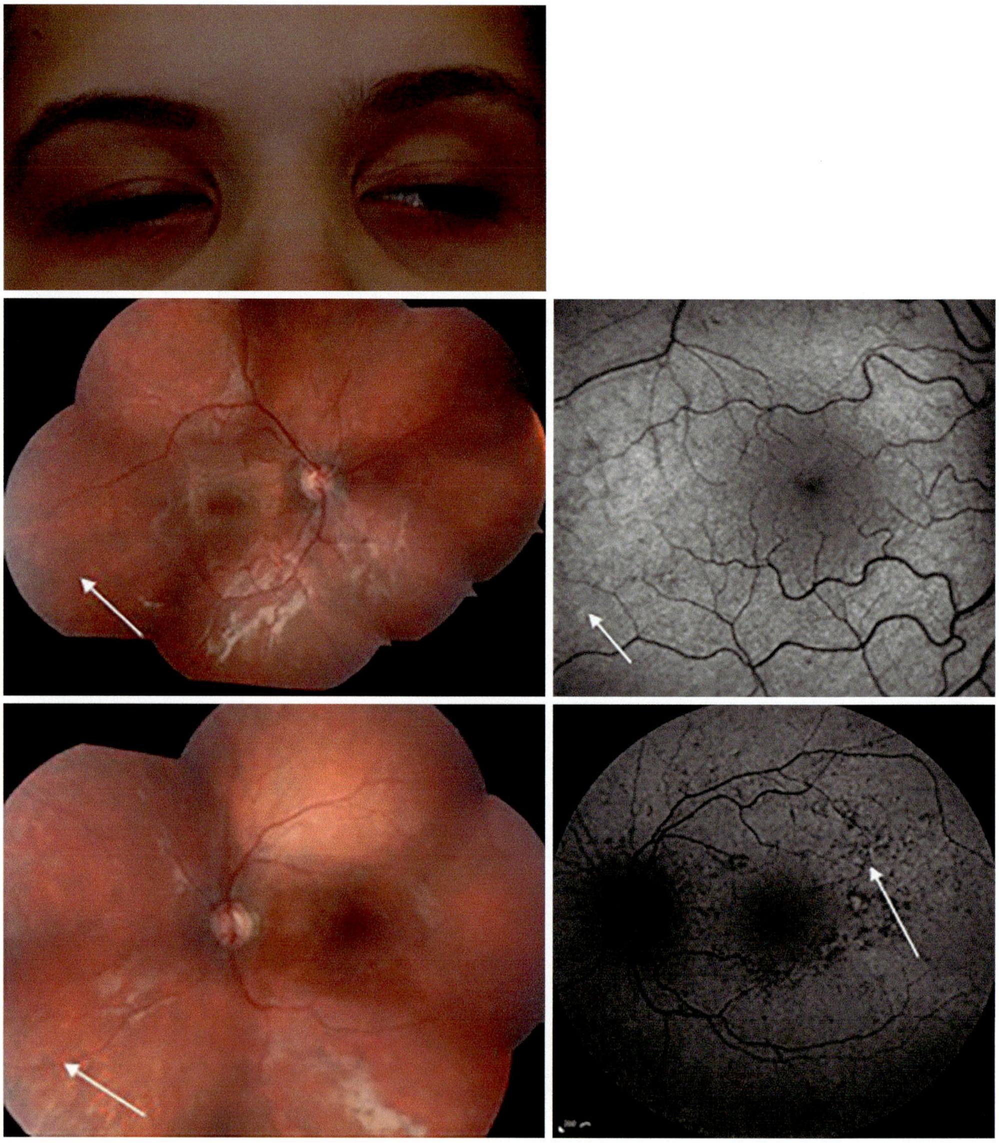

Fig. 30.1 Kearns-Sayre syndrome in this patient is evidenced by ptosis, more marked in the right eye; pigmentary retinopathy (*arrows*), and a stippled autofluorescence pattern (*arrows*)

- Systemic involvement includes chronic progressive external ophthalmoplegia (CPEO), with ptosis being the most common complaint, and cardiomyopathy.
- Other variable features are short stature; cerebellar symptoms; weakness of muscles of the face, pharynx, trunk, or extremities; and progressive hearing loss.
- Full-field ERG does show evidence of generalized retinal dysfunction, involving both rods and cones.
- Skeletal muscle biopsy shows ragged red fibers and abnormal mitochondria.

Molecular Genetics

Mitochondrial DNA (mtDNA) deletions can be identified, and rarely point mutation.

Further Reading

Ortiz A, Arias J, Cárdenas P, Villamil J, Peralta M, Escaf LC, Ortiz J. Macular findings in spectral domain optical coherence tomography and OCT angiography in a patient with Kearns-Sayre syndrome. Int J Retina Vitreous. 2017;3:24.

Shemesh A, Margolin E. Kearns Sayre syndrome. Treasure Island, FL: StatPearls Publishing; 2018.

CHAPTER 31

Mitochondrial Disorder: Maternally Inherited Diabetes and Deafness

Fabiha Ahmed, Stephen H. Tsang, Alicia R. P. Aycinena, and Tarun Sharma

General Features 178
Molecular Genetics 180
Further Reading 180

F. Ahmed
Columbia University, New York, NY, USA
e-mail: fa2558@barnard.edu

S. H. Tsang (✉) · T. Sharma
Department of Ophthalmology, Columbia University, New York, NY, USA
e-mail: sht2@cumc.columbia.edu

A. R. P. Aycinena
Kaiser Permanente, San Francisco, CA, USA

S. H. Tsang et al. (eds.), *Atlas of Inherited Retinal Diseases*, Advances in Experimental Medicine and Biology 1467, https://doi.org/10.1007/978-3-031-72230-1_31

General Features

- Patients with maternally inherited diabetes and deafness (MIDD) have insulin-dependent diabetes with relatively low BMI; usually the onset of the diabetes is during the third or fourth decade of life and it is associated with progressive neurosensory deafness.
- The fundus shows circumferentially oriented but discontinuous patches of RPE and choriocapillaris (CC) atrophy around the macula, within the arcades (Figs. 31.1 and 31.2). Sometimes even hyperpigmentation can be seen, also around the optic nerve, or pattern-like dystrophy may occur.
- Vision is usually good, about 20/40 or better.
- Fundus autofluorescence (FAF) shows decreased AF in the areas of RPE atrophy, surrounded by a zone of speckled AF.
- No generalized retinal dysfunction is seen on full-field electroretinography (ERG), but pattern ERG or multifocal ERG shows abnormal function.
- Asymptomatic maternal relatives harboring the mutation may show pigmentary changes, hearing loss, and in some, diabetes.
- Another disorder associated with the same A3243G mitochondrial DNA (mtDNA) mutation as MIDD is Mitochondrial Encephalomyopathy with Lactic Acidosis and Stroke-like episodes (MELAS). The stroke-like episode (such as hemiplegia) occurs at about 5–15 years of age, and severe encephalopathy may cause death at a young age. Patients may also experience headache, vomiting, visual field defects, or cortical blindness.

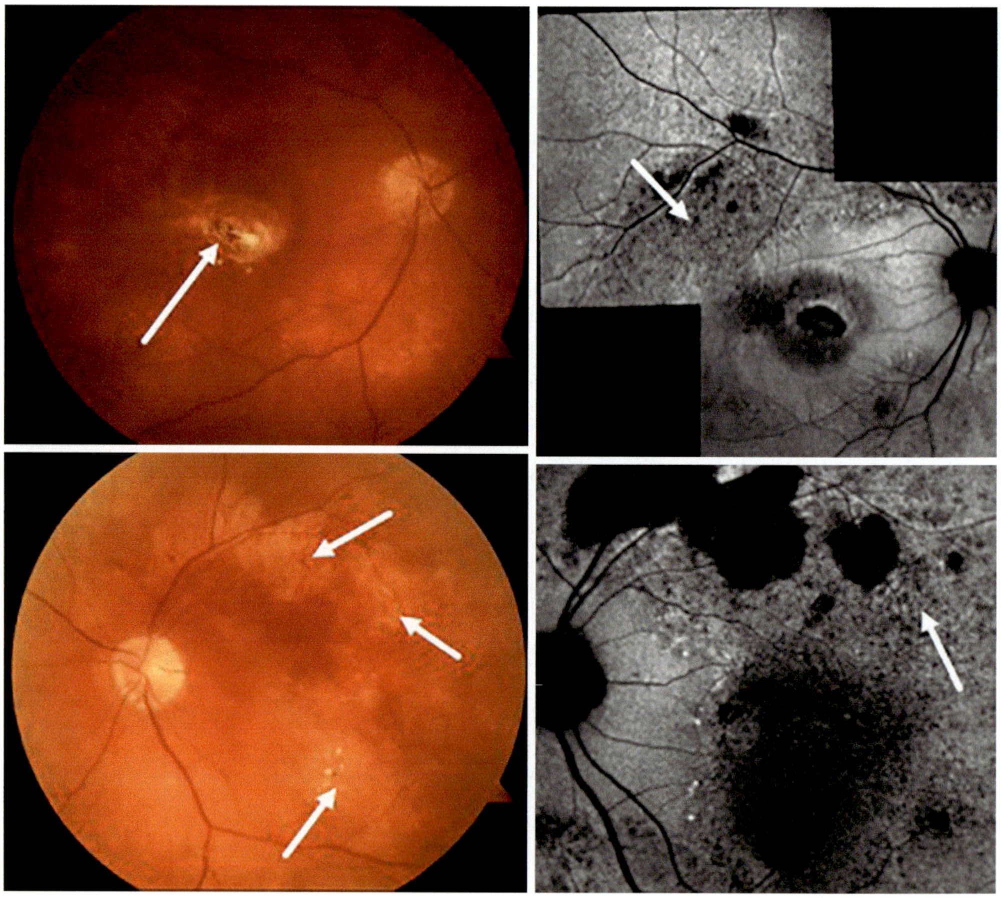

Fig. 31.1 The fundus of a patient with maternally inherited diabetes and deafness (MIDD) shows circular areas of RPE depigmentation (*left, arrows*) with corresponding hypoautofluorescence (hypoAF) and stippled AF (*right, arrows*)

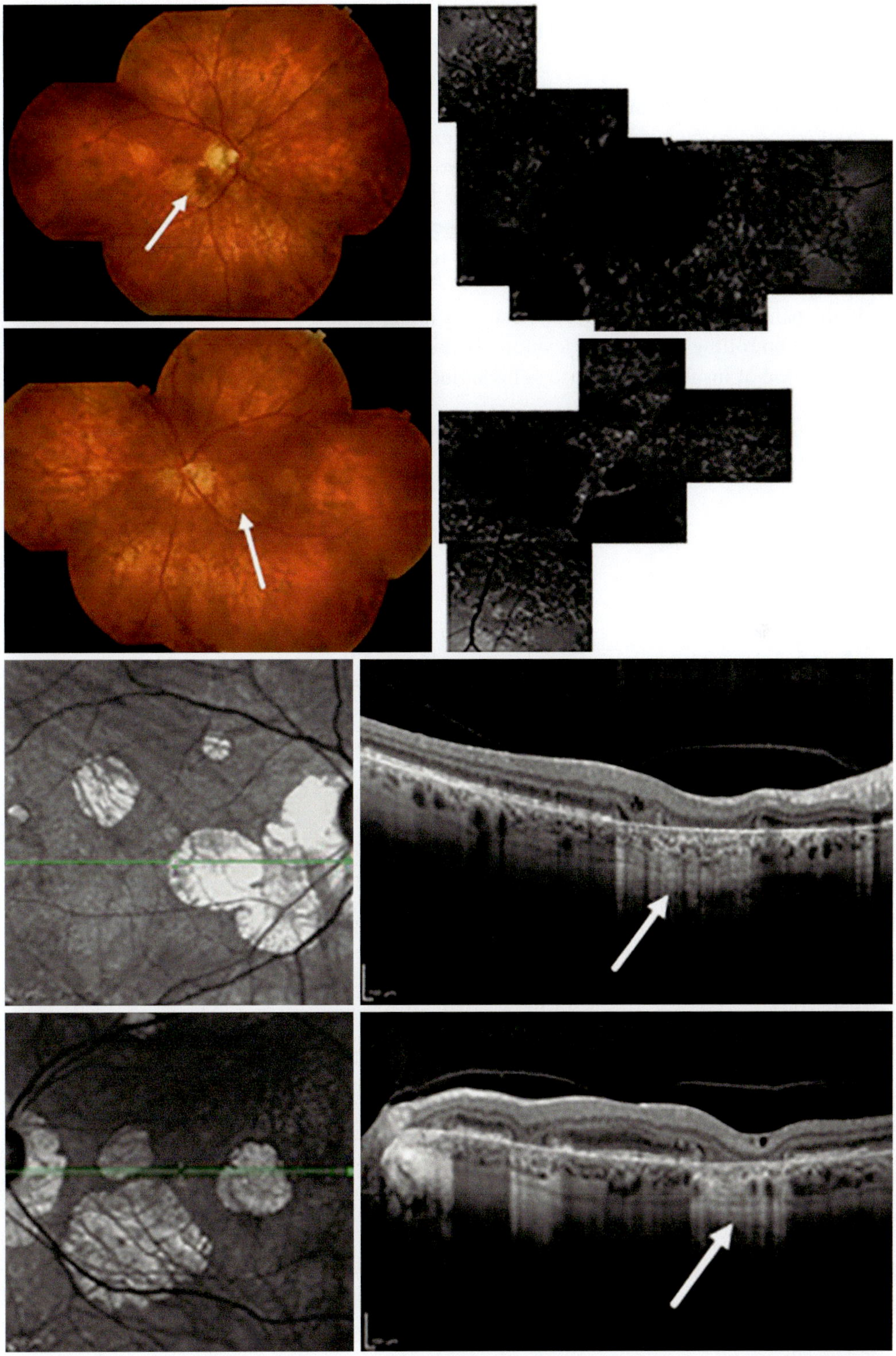

Fig. 31.2 MIDD. The fundus shows circular areas of RPE and choriocapillaris atrophy (*arrows*) at the posterior pole, with corresponding hypoAF and increased RPE transmission on optical coherence tomography (OCT)

Molecular Genetics

- The associated mutation is A3243G mtDNA point mutation (adenine-to-guanine transition at position 3243).
- The *MT-TL1*, *MT-TK*, and *MT-TE* genes provide instructions for making molecules called transfer RNAs; these tRNA molecules are present only in mitochondria and help to assemble proteins that are needed for producing energy for cells.
- In the pancreas, there are beta cells that play an important role in controlling the glucose level; mitochondria trigger the release of insulin in response to high glucose levels.

Further Reading

Kameoka K, Isotani H, Tanaka K, Kitaoka H, Ohsawa N. Impaired insulin secretion in Japanese diabetic subjects with an A-to-G mutation at nucleotide 8296 of the mitochondrial DNA in tRNA(Lys). Diabetes Care. 1998;21:2034–5.

van den Ouweland JM, Lemkes HH, Ruitenbeek W, Sandkuijl LA, de Vijlder MF, Struyvenberg PA, et al. Mutation in mitochondrial $tRNA^{Leu(UUR)}$ gene in a large pedigree with maternally transmitted type II diabetes mellitus and deafness. Nat Genet. 1992;1:368–71.

CHAPTER 32

Ciliopathy: Usher Syndrome

Benjamin Kuang-Chien Chiang, Stephen H. Tsang, Alicia R. P. Aycinena, and Tarun Sharma

General Features 182
Molecular Genetics 183
USH1 183
USH2 183
USH3 183
Therapy Undergoing Clinical Trials 184
Further Reading 184

B. K.-C. Chiang
School of Medicine, College of Medicine,
National Taiwan University, Taipei, Taiwan

S. H. Tsang (✉) · T. Sharma
Department of Ophthalmology, Columbia University,
New York, NY, USA
e-mail: sht2@cumc.columbia.edu

A. R. P. Aycinena
Kaiser Permanente, San Francisco, CA, USA

S. H. Tsang et al. (eds.), *Atlas of Inherited Retinal Diseases*, Advances in Experimental Medicine and Biology 1467, https://doi.org/10.1007/978-3-031-72230-1_32

General Features

- Ciliopathies are a group of disorders caused by a defect in ciliogenesis, ciliary protein trafficking. Because nearly every cell in the body (including the photoreceptors) contains cilia, defects in ciliary proteins typically affect multiple organ systems.
- Usher syndrome is the most common syndromic cause of retinitis pigmentosa (RP) and accounts for 10–20% of cases of RP.
- Inheritance is autosomal recessive, and the retinal dystrophy is usually rod-cone dystrophy (Figs. 32.1 and 32.2).
- These patients have RP with sensorineural hearing loss (partial or complete) since birth; some may have vestibular dysfunction.
- Most patients retain central vision of about 20/40 until about age 40.
- **Usher Syndrome 1 (USH1):** Profound congenital sensorineural hearing loss on audiometry, absent vestibular function, and typical RP (onset by 10 years of age); accounts for about 70% of all Usher cases. Patient may benefit from a cochlear implant. The retinitis pigmentosa occurs at an early age (childhood onset) and progress slowly.
- **Usher Syndrome 2 (USH2):** Moderate to severe congenital sensorineural hearing loss on audiometry (predominantly for higher frequencies), normal vestibular function, and typical RP (onset by 20 years of age); accounts for about 26% of all Usher cases.
- **Usher Syndrome 3 (USH3):** Progressive sensorineural hearing loss and typical RP (onset in second decade); accounts for about 4% of all Usher cases. Vestibular function is normal in about half of patients, but abnormal in the other half.

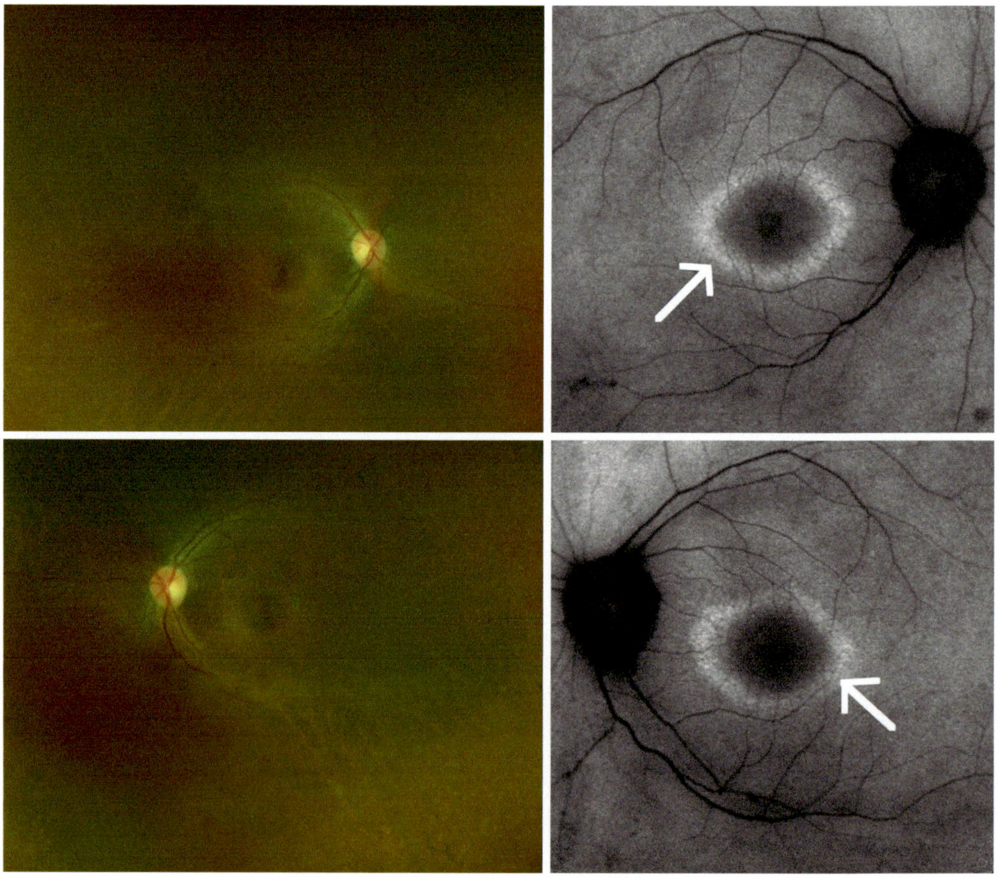

Fig. 32.1 A case of Usher showing a hyperautofluorescent ring around the fovea (*arrows*)

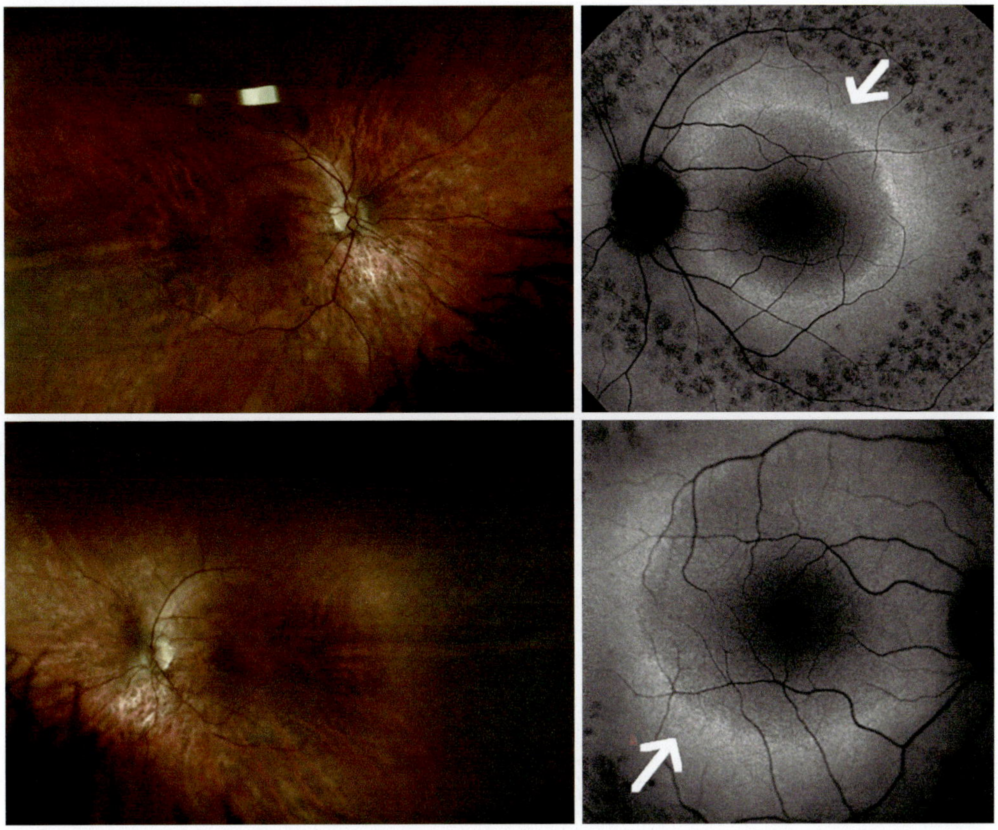

Fig. 32.2 A case of Usher showing a broad hyperautofluorescent ring along the arcades (*arrows*)

Molecular Genetics

USH1

There are nine known loci, USH1B to USH1K (no USH1A or USH1I).

- The most commonly mutated genes are *MYO7A* (USH1B, 53–63%), *CDH23* (USH1D, 7–20%, cadherin protein), *USH1C* (1–15%, harmonin protein), and *PCDH15* (USH1F, 7–12%, protocadherin 15).
- Almost all patients with *USH1C* mutations (specifically c.497-79_497-35 (VNTR exp) or p.V72=) have Acadian ancestry, whereas the p.R245X nonsense mutation in *PCDH15* is more common in Ashkenazi Jewish individuals.
- *MYO7A* produces unconventional myosin VIIa, a motor protein that is found in retinal pigment epithelium (RPE), photoreceptor cells, and cochlear and vestibular neuroepithelia.

USH2

There are three known genes associated with type II:

- *USH2A* (57–79%), Usherin protein, cytogenetic location 1q41.
- *ADGRV1* (USH2C, 6.6–19%), G-protein coupled receptor 98, cytogenetic location 5q14.3.
- *WHRN* (USH2D, 0–9.5%), Whirlin protein, cytogenetic location 9q32.

USH3

There are two known genes associated with this type:

- USH3A: Mutations in *CLRN1* (Clarin 1); cytogenetic location 3q25.1.
- USH3B: Mutations in *HARS* (histidyl-tRNA synthetase); cytogenetic location 5q31.3.

Therapy Undergoing Clinical Trials

The ClinicalTrials.gov website lists five ongoing trials for Usher syndromes (as of August 5, 2022). These include RNA and cell therapies:

- Study to Evaluate Safety and Tolerability of QR-421a in Subjects With RP Due to Mutations in Exon 13 of the USH2A Gene (Stellar) (Completed, ClinicalTrials.gov Identifier: NCT03780257). The result is not yet publicly available.
- An Open-Label Extension Study to Evaluate Safety & Tolerability of QR-421a in Subjects With Retinitis Pigmentosa (HELIA) (Enrolling by invitation, ClinicalTrials.gov Identifier: NCT0508596).
- A Study to Determine the Long-Term Safety, Tolerability, and Biological Activity of SAR421869 in Patients With Usher Syndrome Type 1B (Not recruiting, ClinicalTrials.gov Identifier: NCT02065011).
- Retinal Imaging of Subjects Implanted With Ciliary Neurotrophic Factor (CNTF)-Releasing Encapsulated Cell Implant for Early-stage Retinitis Pigmentosa (Completed, ClinicalTrials.gov Identifier: NCT01530659). The result is not yet publicly available.
- Stem Cell Ophthalmology Treatment Study II (SCOTS2) (Recruiting, ClinicalTrials.gov Identifier: NCT03011541).

Further Reading

Lentz J, Keats BJB. Usher syndrome Type I. 1999 Dec 10 [updated 2016 May 19]. In: Adam MP, Ardinger HH, Pagon RA, Wallace SE, Bean LJH, Stephens K, Amemiya A, editors. GeneReviews® [Internet]. Seattle (WA): University of Washington, Seattle; 1993–2018. Available from http://www.ncbi.nlm.nih.gov/books/NBK1265/

Lentz J, Keats B. Usher syndrome Type II. 1999 Dec 10 [updated 2016 Jul 21]. In: Adam MP, Ardinger HH, Pagon RA, Wallace SE, Bean LJH, Stephens K, Amemiya A, editors. GeneReviews® [Internet]. Seattle (WA): University of Washington, Seattle; 1993–2018. Available from http://www.ncbi.nlm.nih.gov/books/NBK1341/

Lopes VS, Williams DS. Gene therapy for the retinal degeneration of Usher syndrome caused by mutations in *MYO7A*. Cold Spring Harb Perspect Med. 2015;5:a017319.

CHAPTER 33

Ciliopathy: Bardet-Biedl Syndrome

Ahmet Hondur, Stephen Tsang, Alicia R. P. Aycinena, Tarun Sharma, and Vlad Diaconita

General Features 186
Molecular Genetics 187
Management 188
Further Reading 188

A. Hondur · S. Tsang (✉) · T. Sharma · V. Diaconita
Department of Ophthalmology, Columbia University,
New York, NY, USA
e-mail: ahondur@gazi.edu.tr; sht2@cumc.columbia.edu

A. R. P. Aycinena
Kaiser Permanente, San Francisco, CA, USA

S. H. Tsang et al. (eds.), *Atlas of Inherited Retinal Diseases*, Advances in Experimental Medicine and Biology 1467, https://doi.org/10.1007/978-3-031-72230-1_33

General Features

- Bardet-Biedl syndrome (BBS) is an autosomal recessive disease with a prevalence of about 1/125,000.
- It has heterogeneous clinical manifestations. Mixed rod-cone dystrophy (which becomes obvious by 5–10 years of age) is the most consistent (94–100%) and key manifestation for diagnosis. About two thirds of patients have postaxial polydactyly. Sometimes syndactyly, brachydactyly, and/or clinodactyly may be present. Hypogonadism and renal involvement occur in about 40%, mental retardation in about 50%, and truncal obesity in about 70%; it is present early, along with insulin resistance, type 2 diabetes, dyslipidemia, and hypertension. Vision becomes markedly impaired by about age 30 years. The BBS is a genetically heterogeneous entity with considerable phenotypic variability.
- Other associated problems include CNS-related ataxia, abnormal gait, and facial hypotonia, as well as anomalies such as high palate, hearing loss, and cardiac malformations. In males, there may be oligospermia, leading to infertility.
- Around 50–80% of BBS patients have renal malformations (like cyst, agenesis or scarring) and renal dysfunction leading to end-stage renal disease.
- There are no pigmentary changes before the age of 1–2 years. Later, subtle pigmentary changes appear in the macula or peripapillary area. Several years later, pigments appear in the equatorial region, along with attenuation of retinal blood vessels and waxy pallor of the optic disc. Eventually, the macula may show atrophic changes (Figs. 33.1, 33.2, and 33.3).
- Electroretinography (ERG) shows involvement of rods and cones and is abnormal even before the fundus shows changes.

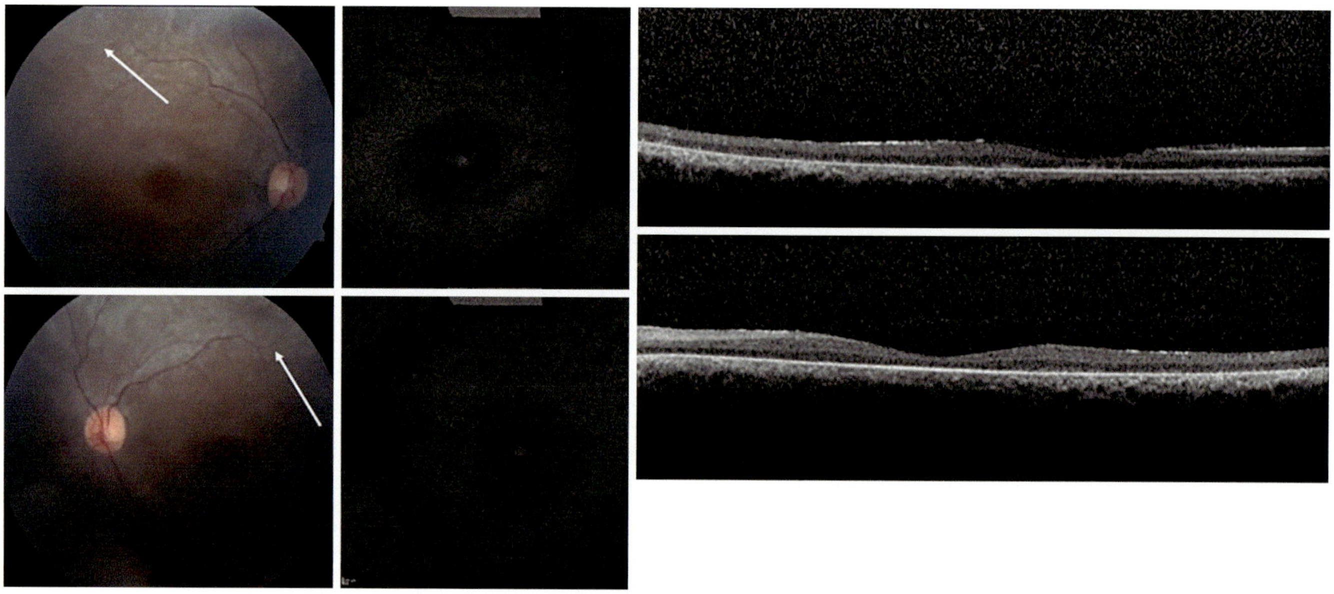

Fig. 33.1 Early in this case of Bardet-Biedl syndrome (BBS), subtle pigment mottling (*arrows*) is more prominently seen on fundus autofluorescence (FAF), with relative thinning of the fovea

Fig. 33.2 An advanced case of BBS shows marked pigmentary changes, more In the OS>OD (*arrows*)

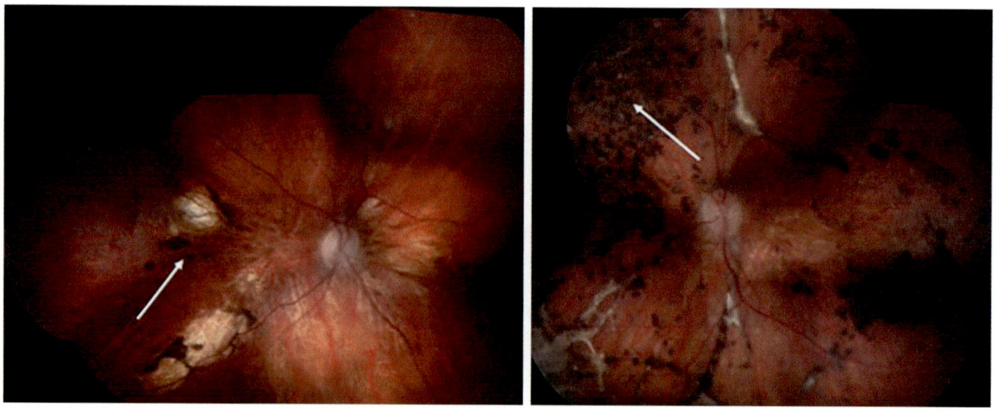

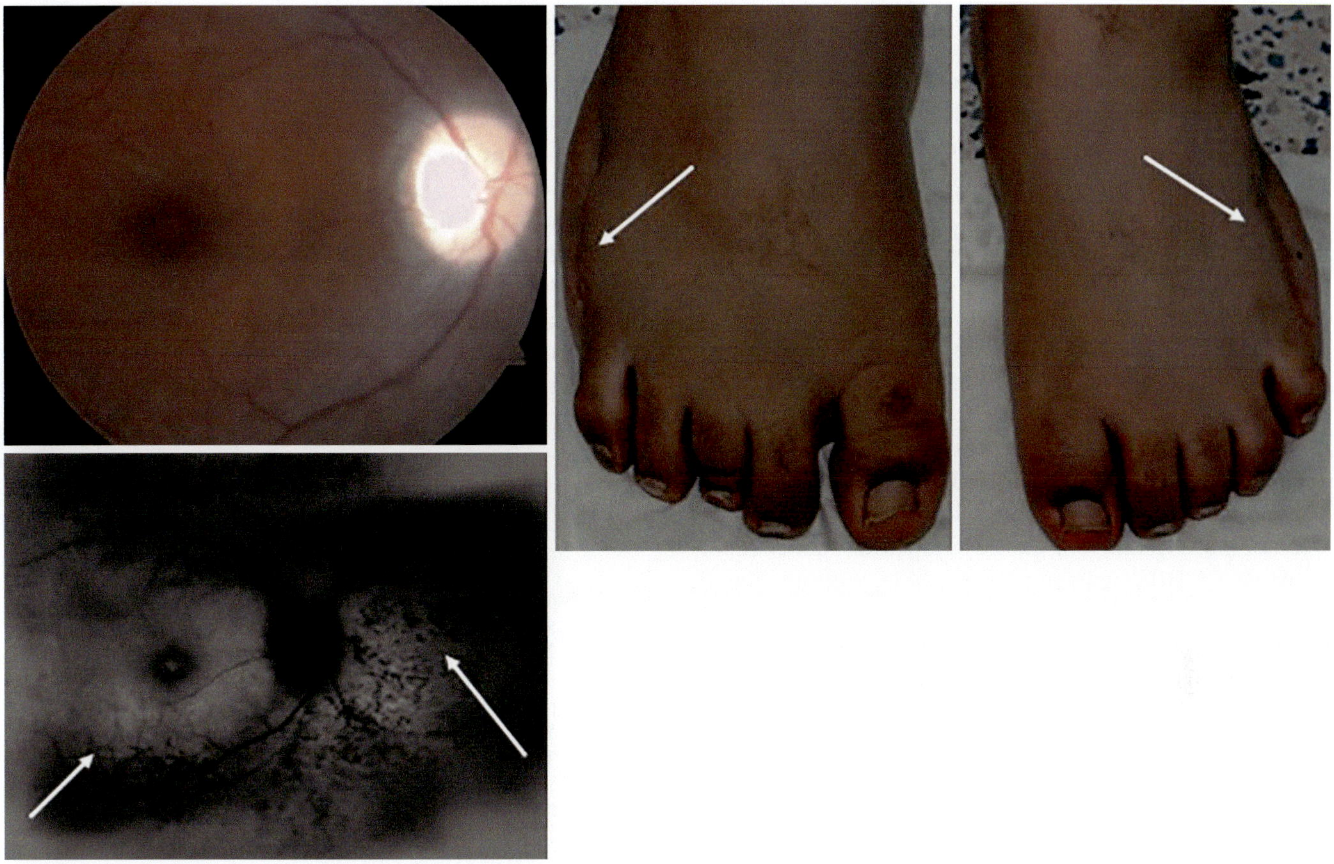

Fig. 33.3 An early case of BBS, with subtle pigment mottling (*arrows*). Scars on the feet (*arrows*) reflect that the patient had undergone surgery for polydactyly

Table 33.1 Criteria for diagnosis of Bardet-Biedl syndrome

Major features	Minor features
• Rod-cone dystrophy • Polydactyly • Central obesity • Hypogonadotropic hypogonadism and/or genitourinary anomalies • Renal anomalies • Cognitive impairment	• Speech delay • Developmental delay • Diabetes mellitus • Dental anomalies • Congenital heart disease • Brachydactyly/syndactyly • Ataxia/poor coordination • Anosmia/hyposmia

- A perimacular hyperfluorescent ring can be seen.
- The clinical diagnosis of BBS relies on the presence of four major features or three major and two minor features (Table 33.1).

Molecular Genetics

- BBS is among the mildest of the multisystem non-motile ciliopathies.
- There are 19 known BBS-related genes (*BBS1–BBS19*) with critical roles in ciliary function. Their proteins are involved in lipid homeostasis, intraflagellar transport, establishing planar cell polarity, and regulation of intracellular trafficking and centrosomal functions. The most common variants are *BBS1* (23%) and *BBS10* (20%).
- Mutations in other seven genes encoding chaperonines and intraflagellar transport complex can also cause BBS.
- In 25% of cases, the molecular cause is unknown.
- Seven BBS proteins (BBS1, BBS2, BBS4, BBS5, BBS7–9) form the BBSome, a protein complex required for ciliogenesis. The BBSome localizes to the nonmembranous centriolar satellites in the cytoplasm with

PCM1, and to the ciliary membrane without PCM1. BBS1 promotes ciliary membrane growth by fusion of exocytic vesicles via RABIN8 and RAB8. The BBSome shares features of the canonical coat complexes (COPI, COPII, and clathrin AP1), and regulates the turnaround from anterograde to retrograde transport at the ciliary tip.

- BBS3 (ARL6) binds to the BBS protein complex; deletion of BBS3 blocks localization of the complex to cilia.
- BBS1 variants often have less severe ophthalmologic involvement, whereas mutations in *BBS2*, *BBS3*, or *BBS4* typically cause early and severe ocular manifestations, with legal blindness by the second decade.
- The phenotypic spectrum of BBS occasionally represents a continuum and occasionally an overlap with other multisystem non-motile ciliopathies, hence recognition of BBS may not be possible without genetic testing.
- Patients with the same genotype and even siblings frequently manifest symptoms differently due to variable expressivity of the genetic error. Hence, it is not possible to make a genotype–phenotype prediction on individual basis.

Management

- The management requires a multidisciplinary approach due to multiple system involvement and heterogeneity in its genetics.
- Impaired cilial signaling in the MC4R pathway underlies the leptin resistance, hyperphagia, and obesity in patients with BBS. Setmelanotide, a melanocortin-4 receptor agonist, has been approved by FDA for treatment for hyperphagia and obesity in BBS.
- No specific treatment exists for the rod-cone dystrophy in BBS.

Further Reading

Ali MU, Rahman MSU, Cao J, Yuan PX. Genetic characterization and disease mechanism of retinitis pigmentosa; current scenario. 3 Biotech. 2017;7:251. https://doi.org/10.1007/s13205-017-0878-3.

Weihbrecht K, Goar WA, Pak T, Garrison JE, DeLuca AP, Stone EM, et al. Keeping an eye on Bardet-Biedl syndrome: a comprehensive review of the role of Bardet-Biedl syndrome genes in the eye. Med Res Arch. 2017;5. https://doi.org/10.18103/mra.v5i9.1526.

CHAPTER 34

Ciliopathy: Senior-Løken Syndrome

Stephen H. Tsang, Alicia R. P. Aycinena, and Tarun Sharma

General Features 190
Molecular Genetics 191
Further Reading 191

S. H. Tsang (✉) · T. Sharma
Department of Ophthalmology, Columbia University,
New York, NY, USA
e-mail: sht2@cumc.columbia.edu

A. R. P. Aycinena
Kaiser Permanente, San Francisco, CA, USA

S. H. Tsang et al. (eds.), *Atlas of Inherited Retinal Diseases*, Advances in Experimental Medicine and Biology 1467, https://doi.org/10.1007/978-3-031-72230-1_34

General Features

- Senior-Løken syndrome is a rare autosomal recessive disease with a prevalence of 1:1,000,000.
- Retinopathy may progress as Leber congenital amaurosis (LCA), retinitis pigmentosa (RP), or sector RP (Figs. 34.1 and 34.2).
- Onset of photophobia, nystagmus, and hyperopia can occur in the first few years of life or later in childhood.

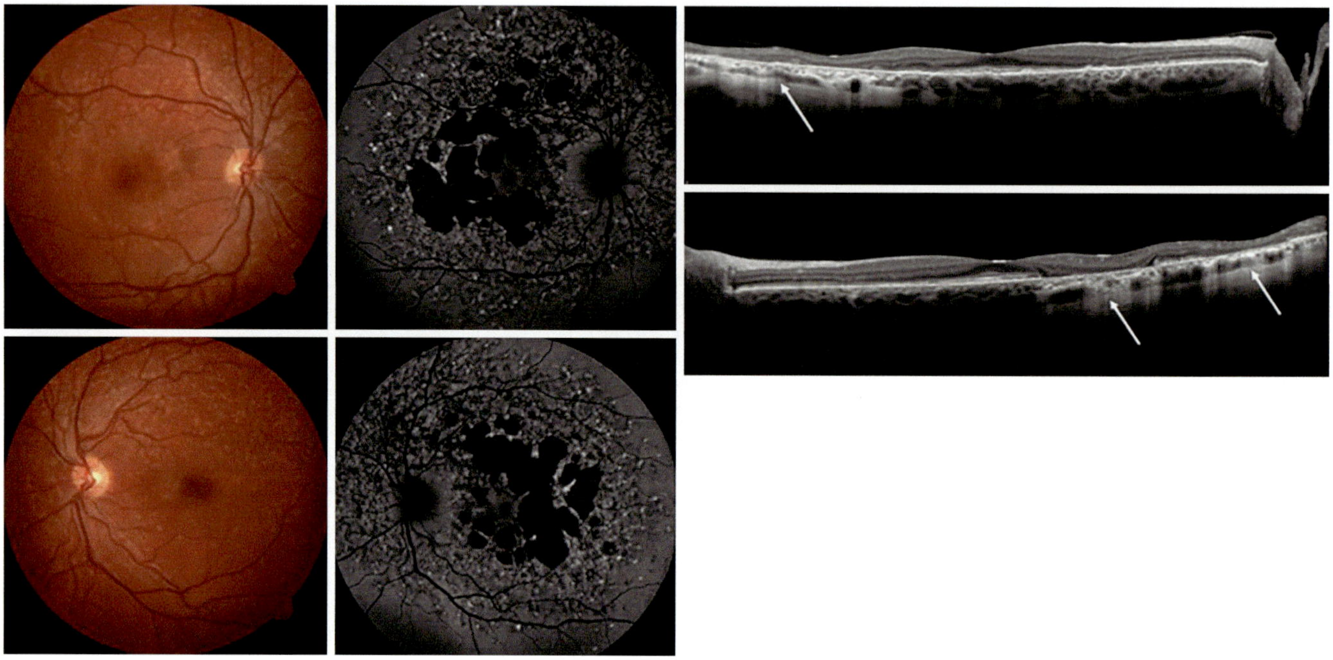

Fig. 34.1 This 37-year-old patient had a history of night blindness for the past 11 years and underwent renal transplantation about 10 years previously. Genetic testing showed a *WDR19(NPHP13)* mutation, characteristic of Senior-Løken syndrome. Areas of retinal pigment epithelium (RPE) changes are better seen on fundus autofluorescence (FAF), though color fundus images show a sort of flecked retina. Optical coherence tomography (OCT) shows disruption of the ellipsoid zone (EZ) line and RPE (*arrows*)

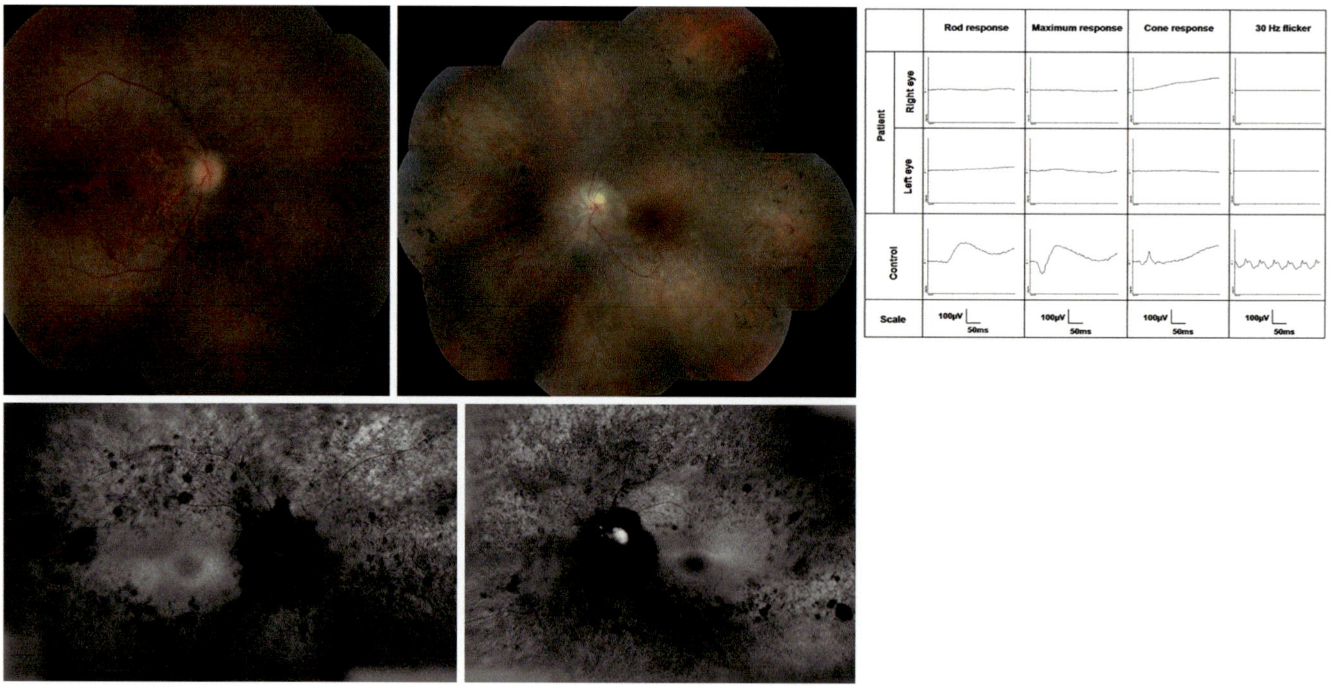

Fig. 34.2 A 20-year-old female with visual impairment (OD 20/200, OS 20/80) since childhood; she had history of renal transplant about 3-years ago. Genetic testing showed an ICQB1/NPHP5 mutation and clinical picture was suggestive of AR-RP

- Patients experience nephronophthisis, characterized by cystic kidney disease (medullary cystic kidney disease), reduced concentrating ability, and chronic tubulointerstitial nephritis, which progresses to end-stage renal disease.
- Hypertension is common.

Molecular Genetics

- Nine genes are known to cause Senior-Løken syndrome: *NPHP1*, *SLSN3*, *NPHP4*, *IQCB1/NPHP5*, *CEP290/NPHP6*, *SDCCAG8*, *WDR19/NPHP13*, *TRAF3IP1*, and *CEP164*. These genes encode for proteins that are likely to be involved in ciliogenesis and regulation of ciliary protein trafficking. They localize to cilia at the ciliary transition zone, inversin compartment, or subunits of the IFT complexes.
- *NPHP5/IQCB1* gene is located on chromosome 3q21.1 and encodes for nephrocystin-5 protein, which is expressed in the connecting cilia of photoreceptors and in primary cilia of renal epithelial cells. The nephrocystin-5 protein co-localized with RP GTPase regulator (RPGR), which is involved in X-linked RP and LCA.
- *NPHP1* and *NPHP4* regulate cell-cell junctions.
- *CEP164*, *SDCCAG8*, and *CEP290* are implicated in DNA damage response.
- *NPHP1* is allelic to LCA type X; *SDCCAG8* is allelic to Bardet-Biedl syndrome 16.
- *NPHP1* and *CEP290* cause tapeto-retinal degeneration. *NPHP1* produces flat ERG.
- With *SLSN3* mutations, patients have LCA, nystagmus, poor pupillary reflexes, and absent ERG.
- *NPHP4* causes RP with rotary nystagmus and diminished amplitude on ERG.
- *IQCB1* presents in infancy to early childhood with LCA; some do not manifest renal disease in the first decade.
- *TRAF3IP1* is associated with onset of RP in early childhood, as well as macular degeneration.

Further Reading

Hemachandar R. Senior-Loken Syndrome – a ciliopathy. J Clin Diagn Res. 2014;8:MD04–05. https://doi.org/10.7860/JCDR/2014/9688.5120.
Kaur A, Dhir SK, Goyal G, Mittal N, Goyal RK. Senior Loken Syndrome. J Clin Diagn Res. 2016;10:SD03–04. https://doi.org/10.7860/JCDR/2016/21832.

CHAPTER 35

Ciliopathy: Alström Syndrome

Stephen H. Tsang, Alicia R. P. Aycinena, and Tarun Sharma

General Features 194
Molecular Genetics 194
Further Reading 194

S. H. Tsang (✉) · T. Sharma
Department of Ophthalmology, Columbia University, New York, NY, USA
e-mail: sht2@cumc.columbia.edu

A. R. P. Aycinena
Kaiser Permanente, San Francisco, CA, USA

S. H. Tsang et al. (eds.), *Atlas of Inherited Retinal Diseases*, Advances in Experimental Medicine and Biology 1467, https://doi.org/10.1007/978-3-031-72230-1_35

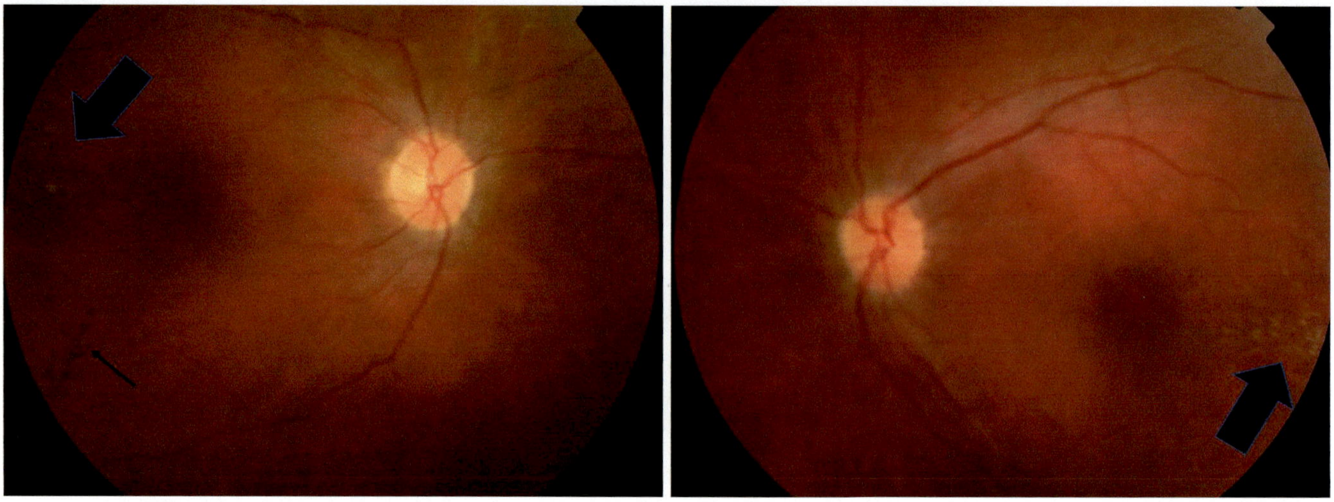

Fig. 35.1 A case of Alström syndrome showing pigmentary abnormality (*arrow*). Note fine dots like drusen located temporal to fovea (*thick arrow*), attenuated arterioles, and pale disc

General Features

- Alström syndrome is an autosomal recessive disease with multisystem involvement, including cone-rod dystrophy, hearing loss, type 2 diabetes, insulin resistance with hyperinsulinemia, dilated cardiomyopathy, and progressive hepatic and renal failure.
- Patients present in childhood with photophobia and nystagmus, and mimic Leber congenital amaurosis (LCA). The fundus shows pigmentary retinopathy with peau d'orange appearance and some fine white dots like drusen around the macula; the disc is pale, with attenuated retinal vessels (Fig. 35.1).
- Patients have short stature; boys have hypogonadotropic hypogonadism and girls have polycystic ovary syndrome (PCOS).
- Obesity is always present, with markedly increased triglyceride and VLDL-C levels; arterial hypertension is diagnosed as early as 2 years of age. There is no polydactyly or syndactyly.
- About half have developmental delay, but intelligence is usually normal.

Molecular Genetics

- The only gene associated with Alström syndrome is *ALMS1*. Pathogenic variants are typically nonsense or frameshift, resulting in a truncated protein. The protein is ubiquitous, but its function is not known.
- The ALMS1 protein localizes to centrosomes and basal bodies of ciliated cells, which suggests a role in intracellular trafficking and ciliary function.
- The ALMS1 protein is involved in PCM1-dependent intracellular transport and is required for localization of NCAPD2 to the proximal ends of centrioles.
- The ALMS1 protein may be involved in endosome recycling through interaction with α-actinin and other components of that pathway.
- One study showed that pathogenic variants in exon 16 cause onset of retinal degeneration before 1 year of age.

Further Reading

Chen JH, Geberhiwot T, Barrett TG, Paisey R, Semple RK. Refining genotype-phenotype correlation in Alstrom syndrome through study of primary human fibroblasts. Mol Genet Genomic Med. 2017;15;390–404.

CHAPTER 36

Ciliopathy: Sjögren-Larsson Syndrome

Stephen H. Tsang, Alicia R. P. Aycinena, and Tarun Sharma

General Features 196
Molecular Genetics 196
Further Reading 196

S. H. Tsang (✉) · T. Sharma
Department of Ophthalmology, Columbia University, New York, NY, USA
e-mail: sht2@cumc.columbia.edu

A. R. P. Aycinena
Kaiser Permanente, San Francisco, CA, USA

S. H. Tsang et al. (eds.), *Atlas of Inherited Retinal Diseases*, Advances in Experimental Medicine and Biology 1467, https://doi.org/10.1007/978-3-031-72230-1_36

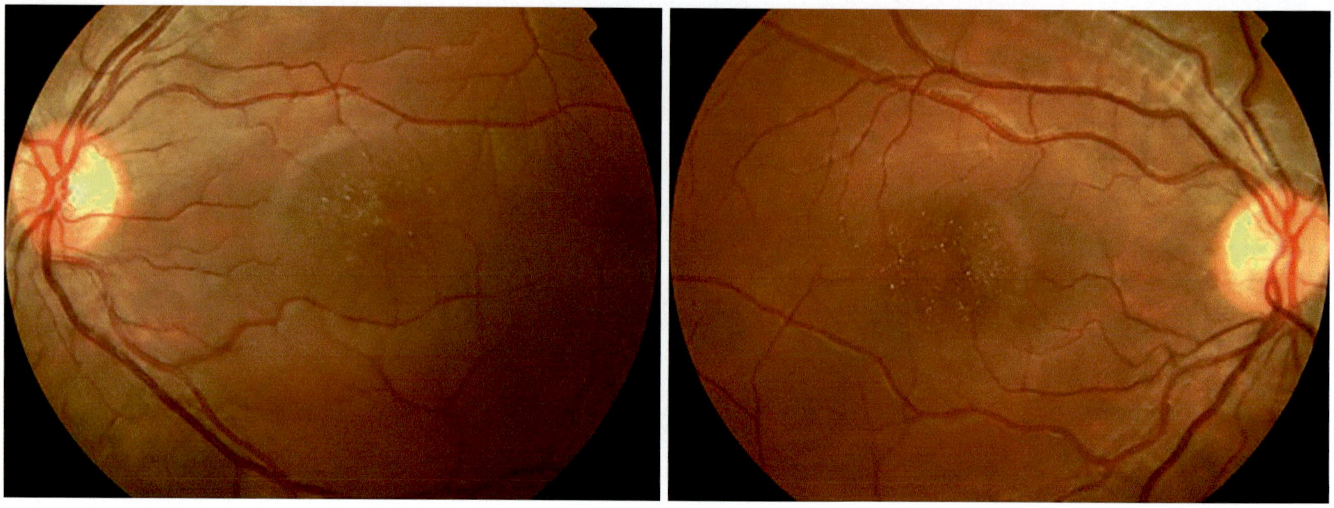

Fig. 36.1 Glistening, yellowish-white crystalline deposits in the central macular area, which are pathognomonic for Sjögren-Larsson syndrome

General Features

- Glistening yellow-white crystalline inclusions in foveal and parafoveal areas are almost pathognomonic (Fig. 36.1). These inclusions are evident at 1–2 years old and increase with age.
- Patients may have corneal stromal opacities, punctate keratitis, myopia, and astigmatism.
- About 50% have pigmentary degeneration of the retina, with decreased visual acuity and marked photophobia.
- Patients have dry, scaly skin (ichthyosis). Affected infants tend to be born prematurely.
- They also have neurological problems due to leukoencephalopathy (affecting the white matter of the brain). Intellectual disability varies from mild to severe, along with dysarthria and delayed speech.

Molecular Genetics

- Sjögren-Larsson syndrome is caused by a mutation in the *ALDH3A2* gene, which produces fatty aldehyde dehydrogenase (FALDH). FALDH is a membrane-bound protein involved in fatty oxidation. Structural disturbance and poor metabolite clearance likely contribute to pathogenesis.
- When FALDH is deficient, excess lipids and aldehyde Schiff adducts accumulate within skin cells, disrupting formation of the plasma membrane, the protective barrier preventing water loss. Myelin formation in the brain is similarly affected.
- It is proposed that impaired degradation of leukotrienes (specifically LTB4) results in severe pruritus and retinal abnormalities.

Further Reading

Cho KH, Shim SH, Kim M. Clinical, biochemical, and genetic aspects of Sjögren-Larsson syndrome. Clin Genet. 2018;93(4):721–30.

Nanda T, Kovach JL. Ophthalmic findings in late stage Sjogren-Larsson syndrome. Retin Cases Brief Rep. 2017 Mar 15. https://doi.org/10.1097/ICB.0000000000000583. [Epub ahead of print].

CHAPTER 37

Inborn Errors of Metabolism: Gyrate Atrophy

Kristina J. Hartung, Stephen H. Tsang, Tarun Sharma, and Vlad Diaconita

General Features 198
Molecular Genetics 200
Further Reading 200

K. J. Hartung
Department of Ophthalmology, University Medical Centre Ljubljana, Ljubljana, Slovenia

Faculty of Medicine, University of Ljubljana, Ljubljana, Slovenia

S. H. Tsang (✉) · T. Sharma · V. Diaconita
Department of Ophthalmology, Columbia University, New York, NY, USA
e-mail: sht2@cumc.columbia.edu; vd2374@cumc.columbia.edu

S. H. Tsang et al. (eds.), *Atlas of Inherited Retinal Diseases*, Advances in Experimental Medicine and Biology 1467, https://doi.org/10.1007/978-3-031-72230-1_37

General Features

Gyrate atrophy is autosomal recessive dystrophy in which night blindness starts in the first decade of life. Peripheral field loss usually begins in the second and third decades as the disease progresses. There is no gender predilection. It is commoner in Finland (1/50,000).

- In the early stage, there are areas of retinal pigment epithelium (RPE) and choriocapillaris (CC) atrophy in the far periphery (lobular shape, Fig. 37.1). (In choroideremia, atrophy appears in the mid periphery.) Later, these areas coalesce and enlarge towards the posterior pole, forming a characteristic scalloped border at the junction of healthy and diseased RPE (Fig 37.2).
- Unlike in choroideremia (which is X-linked), patients with gyrate atrophy show areas of hyperpigmentation of the remaining RPE (Fig. 37.3) and do not show choroidal atrophy (as seen in choroideremia) until the late stages.
- The retinal vessels may be attenuated in later stages, and there may be optic disc pallor.
- Myopia and subcapsular cataracts are common by the end of the second or third decade.
- OCT: CME, outer retinal tubulations, and ERM are possible.
- Full-field ERG: In the early stage, mildly reduced rod and cone amplitudes are present (rods are more affected). The responses deteriorate with disease progression and may become undetectable in the later stages (both rods and cones are equally affected).
- Treatment includes a low-protein, arginine-restricted diet for all patients. In some cases, vitamin B6 (pyridoxine) may help lower plasma ornithine levels. Proline supplementation was shown to slow the disease progression in some patients.

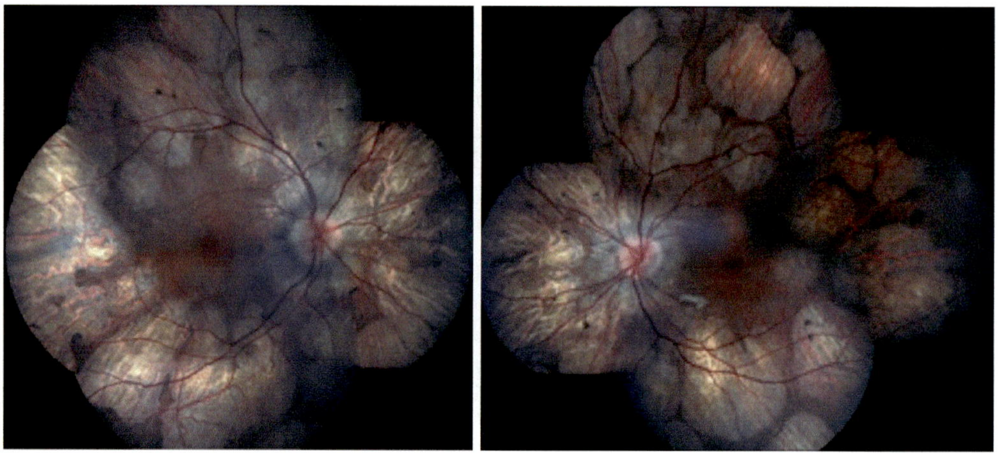

Fig. 37.1 Fundi showing characteristic lobular areas of chorioretinal atrophy with scalloped borders, sparing the macula, in the early stages of gyrate atrophy

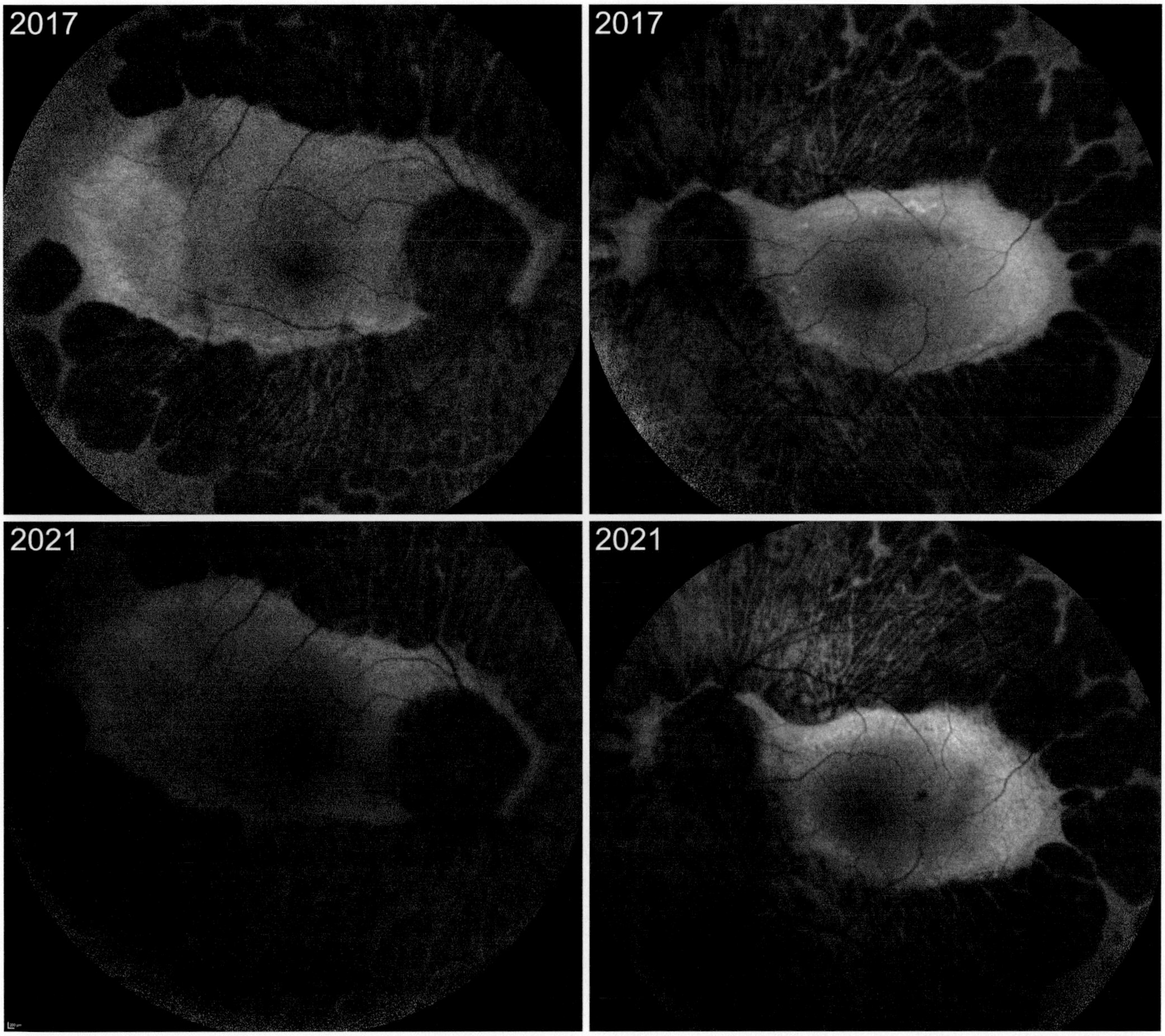

Fig. 37.2 Fundus autofluorescence showing a 4-year progression in a patient with GA and a mutation in OAT (c.473A>C:p.Tyr158Ser)

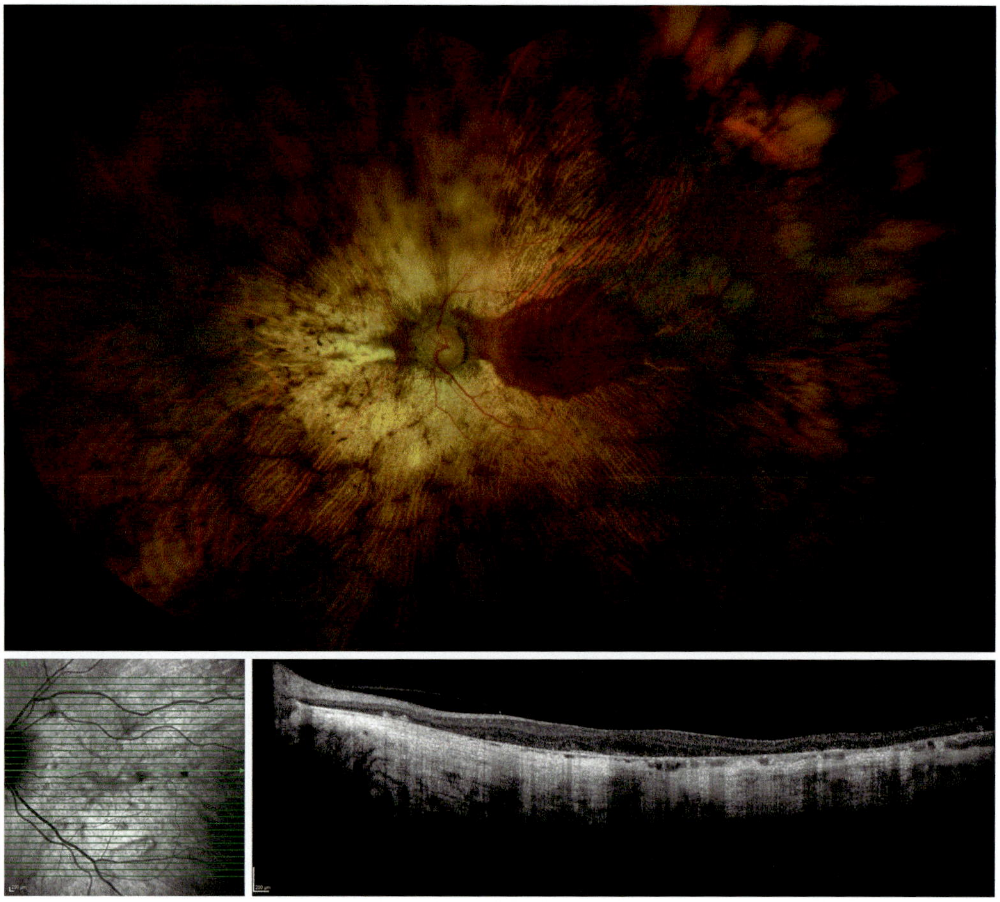

Fig. 37.3 Fundus showing confluent areas of chorioretinal atrophy, sparing a small island at the macula. Note areas of hyperpigmentation. Optical coherence tomography (OCT) shows thinning of the choriocapillaris (CC) and retinal pigment epithelium (RPE)

Molecular Genetics

- Mutation in the *OAT* (ornithine aminotransferase) gene, located on chromosome 10.
- It encodes ornithine aminotransferase, a vitamin B6-dependent mitochondrial matrix enzyme that converts ornithine derived from dietary arginine to glutamate and proline.
- OAT deficiency results in a 10–15× rise in serum plasma ornithine, which is toxic to RPE and choroid. Ornithine is also elevated in urine and CSF.

Further Reading

Kaiser-Kupfer MI, Ludwig IH, de Monasterio FM, Valle D, Krieger I. Gyrate atrophy of the choroid and retina. Early findings. Ophthalmology 1985;92:394–401.

Sergouniotis PI, Davidson AE, Lenassi E, Devery SR, Moore AT, Webster AR. Retinal structure, function, and molecular pathologic features in gyrate atrophy. Ophthalmology 2012;119:596–605.

Takki KK, Milton RC. The natural history of gyrate atrophy of the choroid and retina. Ophthalmology 1981;88:292–301.

CHAPTER 38

Inborn Errors of Metabolism: Pseudoxanthoma elasticum

Justin Cheong, Stephen H. Tsang, and Tarun Sharma

General Features 202
Molecular Genetics 203
Further Reading 203

J. Cheong
Vagelos College of Physicians and Surgeons,
New York, NY, USA

S. H. Tsang (✉) · T. Sharma
Department of Ophthalmology, Columbia University,
New York, NY, USA
e-mail: sht2@cumc.columbia.edu

S. H. Tsang et al. (eds.), *Atlas of Inherited Retinal Diseases*, Advances in Experimental Medicine and Biology 1467, https://doi.org/10.1007/978-3-031-72230-1_38

General Features

- Pseudoxanthoma elasticum (PXE) is an autosomal recessive multisystem disorder that involves the skin, GI tract, and heart, as well as the eye.
- It affects approximately 1 in 50,000 people worldwide and is seen twice as frequently in females as in males.
- Fundus findings include angioid streaks (Fig. 38.1), reticular macular dystrophy, speckled appearance temporal to the macula (peau d'orange, like the dimpled texture of an orange peel), drusen of the optic nerve, and vitelliform-like deposits. Peau d'orange may precede the development of an angioid streak.
- "Comets," with or without a tail, are seen as solitary subretinal, nodular white bodies of retinal pigment epithelium (RPE) atrophy, usually present in the mid periphery (Fig. 38.2). The tail points toward the optic disc.
- Patients sometimes develop choroidal neovascular membrane.
- Skin changes (plucked chicken–like appearance) occur on the flexure areas, including the neck and axilla, as well as increased skin laxity with excessive skin folding.
- Cardiovascular changes include accelerated atherosclerosis with occlusive vascular disease leading to angina, hypertension, restrictive cardiomyopathy, mitral valve prolapse, and others.
- Progressive calcification and fragmentation of elastic fibers in the skin, eye, and cardiovascular system is the underlying pathophysiology.

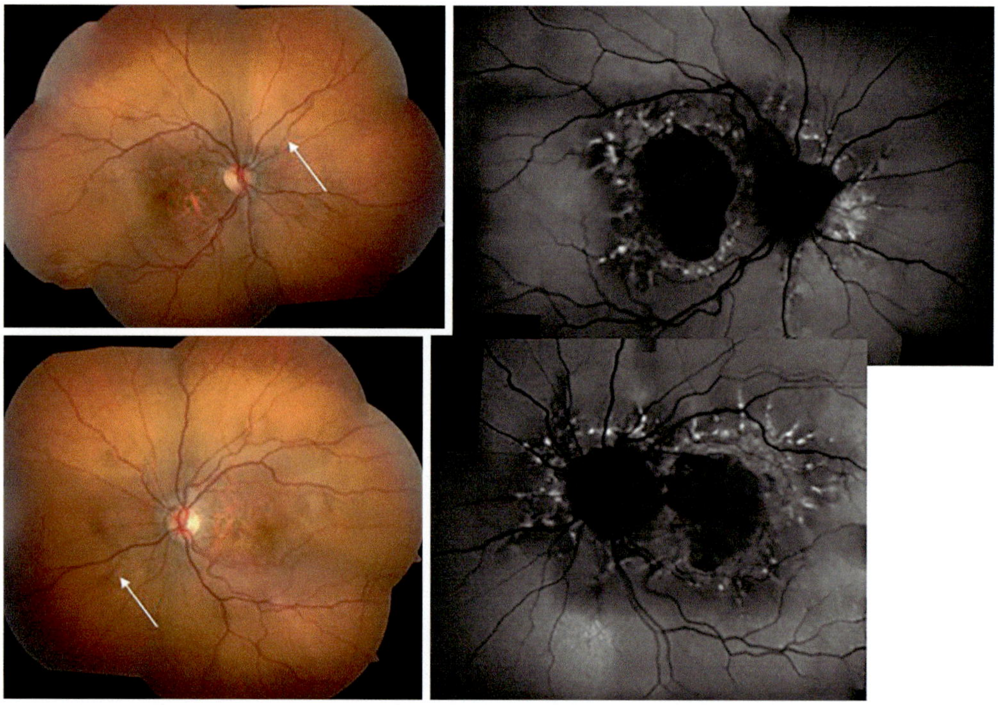

Fig. 38.1 A case of pseudoxanthoma elasticum (PXE) showing angioid streaks (*arrows*) and atrophic RPE changes at the macula, as evident on fundus autofluorescence (FAF)

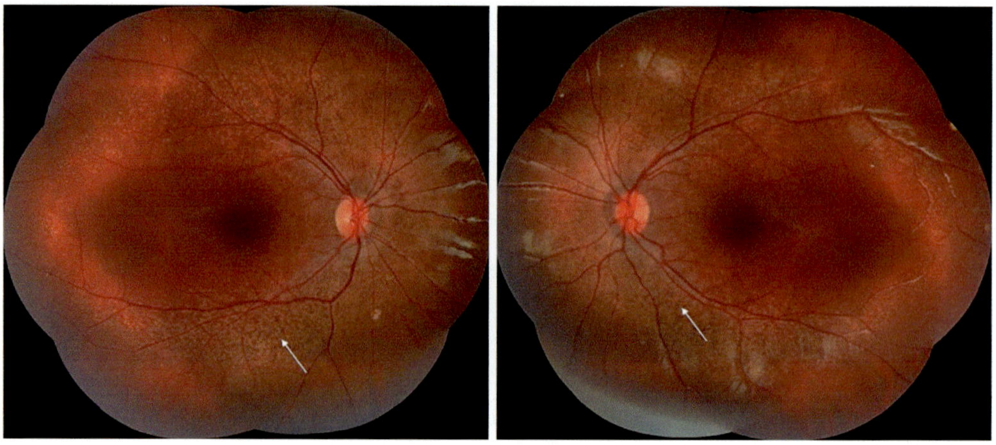

Fig. 38.2 A case of PXE showing peau d'orange (*white arrows*) and comets (*thick arrows*)

Molecular Genetics

- Biallelic mutations occur in the *ABCC6* gene, at chromosome 16.
- The gene provides instructions for making a protein called MRP6 (ABCC6 protein). This protein is found in the liver, kidney, skin, GI tract, blood vessels, and the eye. There is accumulation of deposits of calcium and other minerals in elastic fibers.
- Cytogenetic location: 16p13.11.

Further Reading

Chiu BQ, Tsui E, Hussnain SA, Barbazetto IA, Smith RT. Multimodal imaging of angioid streaks associated with Turner syndrome. Retin Cases Brief Rep 2020;14(4):364-367. https://doi.org/10.1097/ICB.0000000000000713. [Epub ahead of print].

Gliem M, Müller PL, Birtel J, McGuinness MB, Finger RP, Herrmann P, et al. Quantitative fundus autofluorescence in Pseudoxanthoma elasticum. Invest Ophthalmol Vis Sci 2017;58:6159–65.

CHAPTER 39

Inborn Errors of Metabolism: Refsum Disease

Sitara Hirji, Stephen Tsang, Tarun Sharma, and Vlad Diaconita

General Features 206
Molecular Genetics 206
Further Reading 206

S. Hirji · S. Tsang (✉) · T. Sharma · V. Diaconita
Department of Ophthalmology, Columbia University,
New York, NY, USA
e-mail: shh2138@cumc.columbia.edu; sht2@cumc.columbia.edu

S. H. Tsang et al. (eds.), *Atlas of Inherited Retinal Diseases*, Advances in Experimental Medicine and Biology 1467, https://doi.org/10.1007/978-3-031-72230-1_39

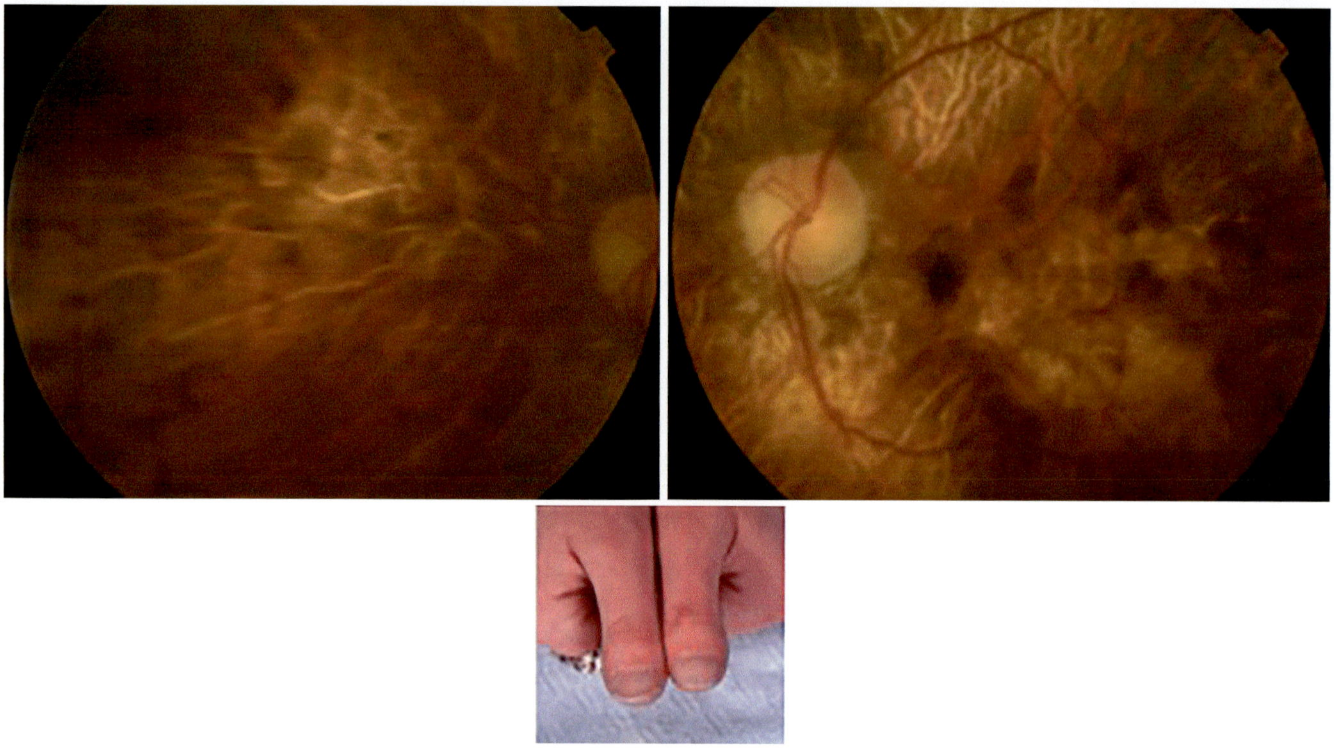

Fig. 39.1 Advanced panretinal degeneration and asymmetrical length of thumbs

General Features

- Autosomal recessive peroxisomal disorder due to a deficiency in phytanic acid oxidation, which results in accumulation of phytanic acid in plasma and tissues.
- Diagnosed by measuring high phytanic acid levels in the blood (0.4–4.0 mg/L).
- Ophthalmologic features of Refsum disease include nyctalopia, cataracts, and progressive panretinal degeneration (Fig. 39.1). Vision gradually decreases, with progressive peripheral constriction. The pupil usually does not dilate well.
- Associated CNS symptoms include ataxia and weakness in the extremities, progressive peripheral neuropathy, and peripheral muscle wasting.
- Other features include sensorineural deafness and epiphyseal dysplasia (fourth toe riding on the dorsum of the foot, short thumbs, or asymmetrical length of toes and fingers).
- Other findings may include cardiac arrhythmia and anosmia.
- Limiting dietary intake of phytanic acid may improve neurologic symptoms. Plasmapheresis has also been used to decrease phytanic acid levels in patients who do not respond to dietary restriction.

Molecular Genetics

- Mutations in the PHYH gene (most common) and PEX7 gene.

Further Reading

Pakzad-Vaezi KL, Maberley DA. Infantile Refsum disease in a young adult: case presentation and brief review. Retin Cases Brief Rep 2014;8:56–9.

Claridge KG, Gibberd FB, Sidey MC. Refsum disease: the presentation and ophthalmic aspects of Refsum disease in a series of 23 patients. Eye (Lond) 1992;6(Pt 4):371-5. doi: https://doi.org/10.1038/eye.1992.76. PMID: 1282471.

CHAPTER 40

Inborn Errors of Metabolism: Bietti Crystalline Dystrophy

Laura A. Jenny, Stephen H. Tsang, and Tarun Sharma

General Features 208
Molecular Genetics 209
Further Reading 210

L. A. Jenny · S. H. Tsang (✉) · T. Sharma
Department of Ophthalmology, Columbia University,
New York, NY, USA
e-mail: lj2539@cumc.columbia.edu; sht2@cumc.columbia.edu

S. H. Tsang et al. (eds.), *Atlas of Inherited Retinal Diseases*, Advances in Experimental Medicine and Biology 1467, https://doi.org/10.1007/978-3-031-72230-1_40

General Features

- Bietti crystalline dystrophy (BCD) has been estimated to occur in 1 in 67,000 people; it is more common in people of East Asian descent.
- This autosomal recessive dystrophy presents with night blindness and paracentral scotomas around the third decade of life; some many have photophobia.
- BCD) is characterized by multiple, small refractile crystalline deposits (fatty/lipid compounds) throughout the fundus, in both the outer and inner retina and choroid (Figs. 40.1 and 40.2). In some cases, these deposits are seen in the mid-stroma of the cornea and lens capsule.
- These crystalline deposits in the retina (some of which are located in front of the blood vessels) tend to disappear with time, leading to atrophic lesions.

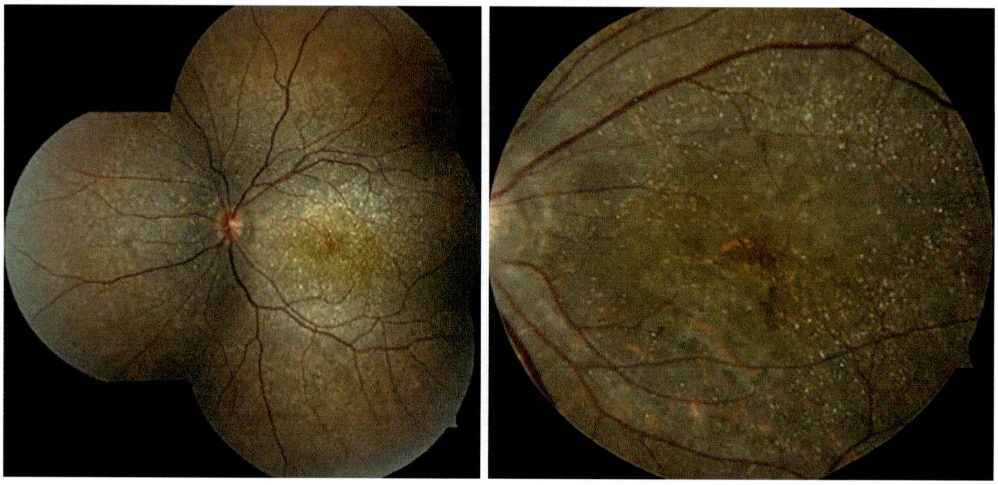

Fig. 40.1 Bietti crystalline dystrophy with mutation in CYP4V2 gene (c.802-8_810del17insGC and c.1062dupA). The whole of the posterior pole shows small, refractile crystalline deposits (*left fundus*). As these crystalline deposits fade, atrophic areas appear (*right fundus*)

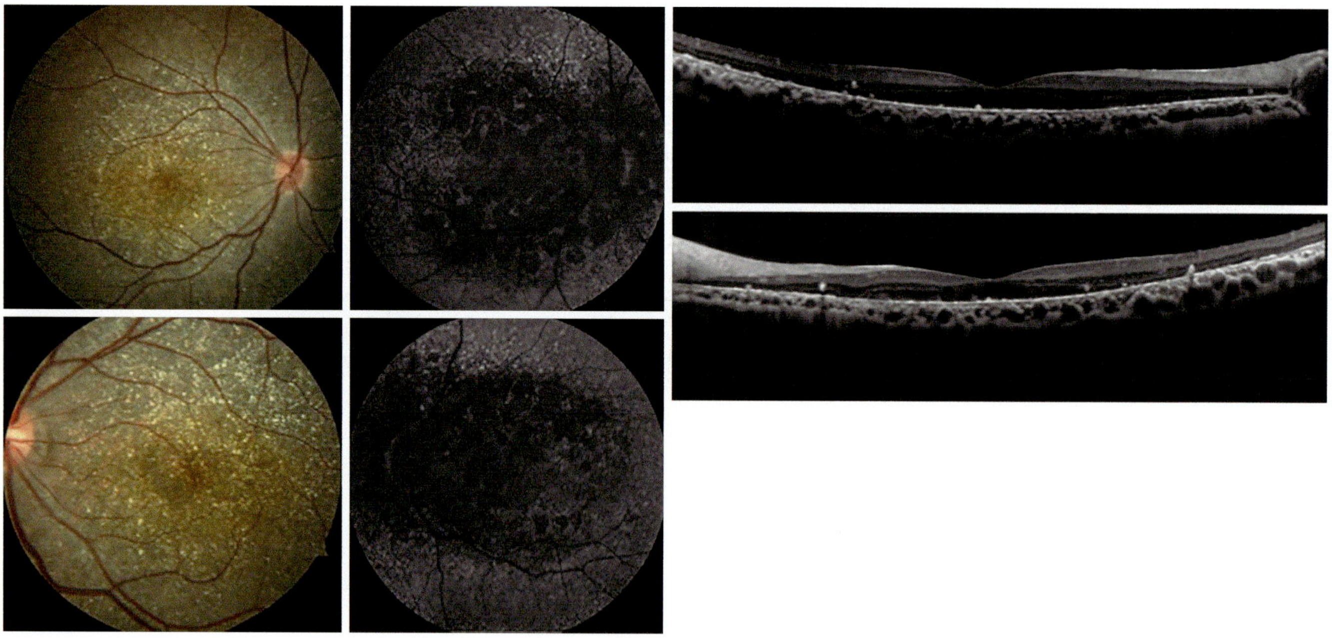

Fig. 40.2 Bietti crystalline dystrophy. Both fundi show crystalline deposits, with associated retinal pigment epithelium (RPE) atrophy on autofluorescence. Optical coherence tomography (OCT) shows the presence of these deposits at different levels of the neurosensory retina

- On fundus autofluorescence (FAF), crystals are not seen, but if there is atrophy, it appears as central hypoautofluorescence surrounded by a ring of hyperautofluorescence.
- Optical coherence tomography (OCT) shows hyperreflective dots in various layers of the retina and choroid.
- Over time, geographic areas of RPE and CC atrophy set in, predominantly at the posterior pole, and as atrophy sets in, crystals seem to disappear. In end-stage disease, the whole of the posterior pole shows areas of chorioretinal atrophy.
- Electroretinography (ERG) shows a pattern like cone-rod dystrophy, or sometimes a rod-cone dystrophy pattern.

Molecular Genetics

- Caused by mutation in the *CYP4V2* gene, located on chromosome 4, involved in lipid metabolism.
- This gene encodes for making a member of the cytochrome P450 family of enzymes. The CYP4V2 enzyme is involved in fatty acid oxidation, a multistep process in which lipids are broken down and converted into energy.
- Cytogenetic location: 4q35/1–4q35.2 (Fig. 40.3).

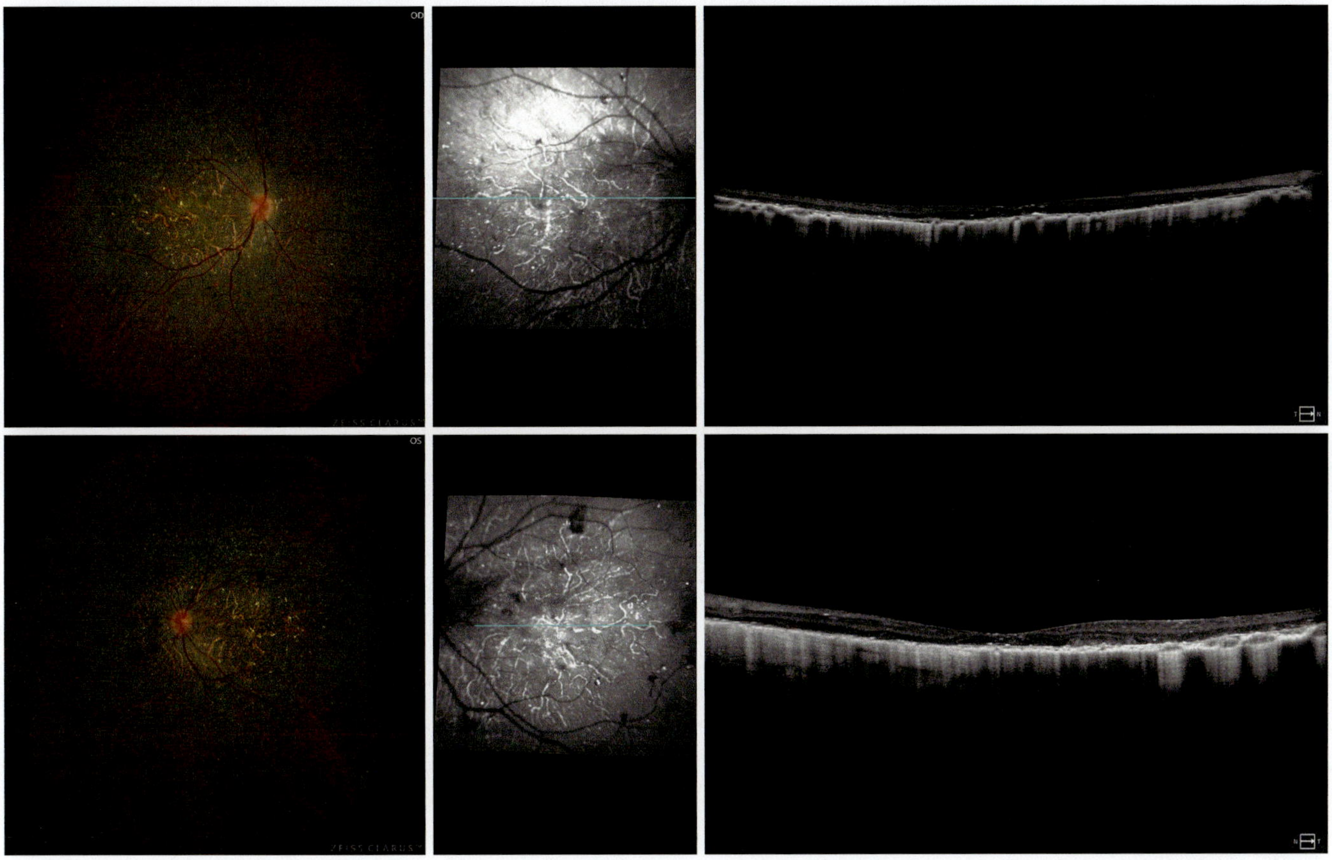

Fig. 40.3 Late-stage Bietti crystalline dystrophy. Extensive retinal pigment epithelium degeneration with diffuse choriocapillaris atrophy and yellow-crystalline deposits

Further Reading

Halford S, Liew G, Mackay DS, Sergouniotis PI, Holt R, Broadgate S, Volpi EV, Ocaka L, Robson AG, Holder GE, Moore AT, Michaelides M, Webster AR. Detailed phenotypic and genotypic characterization of Bietti crystalline dystrophy. Ophthalmology, 2014;121(6): 1174-1184.

Fong AM, Koh A, Lee K, Ang CL. Biettis crystalline dystrophy in Asians: clinical, angiographic and electrophysiological characteristics. Int Ophthalmol 2009;29:459–70.

Ng DS, Lai TY, Ng TK, Pang CP. Genetics of Bietti crystalline dystrophy. Asia Pac J Ophthalmol (Phila) 2016;5:245–52.

Oishi A, Oishi M, Miyata M, Hirashima T, Hasegawa T, Numa S, Tsujikawa A. Multimodal imaging for differential diagnosis of Bietti crystalline dystrophy. Ophthalmol Retina 2018;2(10):1071-1077.

CHAPTER 41

Extracellular Matrix: Alport Syndrome

Jennifer Adeghate, Stephen Tsang, and Tarun Sharma

General Characteristics 212
Molecular Genetics 212
Ocular Features 212
Further Reading 213

J. Adeghate · S. Tsang (✉) · T. Sharma
Department of Ophthalmology, Columbia University,
New York, NY, USA
e-mail: ja3620@cumc.columbia.edu; sht2@cumc.columbia.edu

S. H. Tsang et al. (eds.), *Atlas of Inherited Retinal Diseases*, Advances in Experimental Medicine and Biology 1467, https://doi.org/10.1007/978-3-031-72230-1_41

General Characteristics

- Alport syndrome occurs in approximately 1 in 5000–10,000 newborns.
- Patients may present with nephropathy, neurosensory deafness, and various ocular findings.

Molecular Genetics

- Alport syndrome is caused by mutations in *COL4A3* (chromosome 2), *COL4A4* (chromosome 2), or *COL4A5* (X-chromosome) genes, which encode for the α-chains of Type IV collagen.
- Type IV collagen is an important structural protein found in the basement membranes of glomeruli, inner ear, and the eye. Mutations in type IV collagen result in irregular thickening and progressive splitting of the basal membrane, impairing glomerular filtration (renal failure), auditory processing (neurosensory deafness), and maintenance of ocular structures.
- The disorder is X-linked in about 80% of cases (*COL4A5 mutation)*; autosomal recessive in approximately 15% of cases (*COL4A3 or COL4A4 mutation)*, and autosomal dominant in 5% of cases (*COL4A3* or *COL4A4 mutation)*.

Ocular Features

- Common findings include anterior and posterior lenticonus (about 25%, during the second or third decade), posterior polymorphous corneal dystrophy (about 3%), recurrent corneal erosions, microspherophakia, and anterior/posterior subcapsular cataract. Less commonly, patients may develop keratoconus.
- Fundus examination shows central and peripheral dot and fleck retinopathy with multiple refractile/crystalline deposits in a ring configuration in the macula (about 85%), which is usually diagnosed at an advanced stage of nephropathy (Fig. 41.1). Less common posterior segment findings include a form of vitreomacular traction called "staircase fovea," foveal hypoplasia, macular holes, vitelliform lesions, peripheral retinoschisis, and serous retinal detachment.
- Diagnostic tools include optical coherence tomography and fundus autofluorescence.
- Treatment may include vitrectomy, cataract extraction, and gene therapy, although no evidence has been found to support the efficacy of these management options at this time. Additionally, it is debated that patients with Alport syndrome may have poor postoperative outcomes after vitrectomy due to the abnormalities in type 4 collagen.

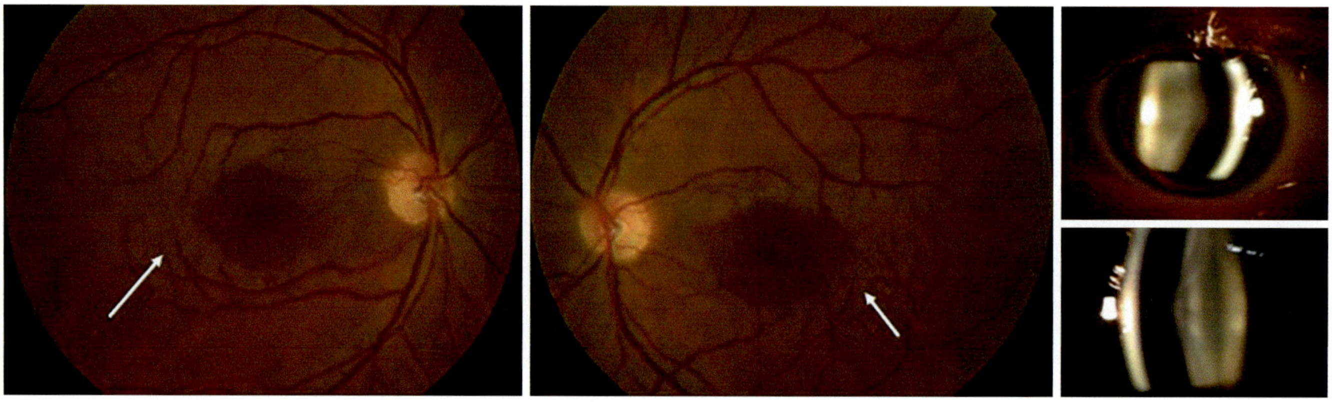

Fig. 41.1 Alport syndrome. Dot and fleck retinopathy (*arrows*) and anterior lenticonus. The patient, who also had renal problems and hearing loss, had a positive family history, with brothers affected

Further Reading

Savige J, Sheth S, Leys A, Nicholson A, Mack HG, Colville D. Ocular features in Alport syndrome: pathogenesis and clinical significance. Clin J Am Soc Nephrol 2015;10:703–9.

Ramakrishnan R, Shenoy A, Meyer D. Ocular manifestations and potential treatments of Alport syndrome: a systematic review. J Ophthalmol 2022;2022:9250367. https://doi.org/10.1155/2022/9250367. PMID: 36119140; PMCID: PMC9477629.

Rare diseases database. https://rarediseases.org/rare-diseases/. Accessed 21 Feb 2023.

CHAPTER 42

Stickler Syndrome

Jennifer Hu, Stephen Tsang, Tarun Sharma, and Vlad Diaconita

General Features 216
Molecular Features of Stickler Syndrome 216
Management of Stickler Syndrome 216
Further Reading 218

J. Hu · S. Tsang (✉) · T. Sharma · V. Diaconita
Department of Ophthalmology, Columbia University,
New York, NY, USA
e-mail: jh4299@cumc.columbia.edu; sht2@cumc.columbia.edu

S. H. Tsang et al. (eds.), *Atlas of Inherited Retinal Diseases*, Advances in Experimental Medicine and Biology 1467, https://doi.org/10.1007/978-3-031-72230-1_42

General Features

- Stickler syndrome represents the most common hereditary hyaloideoretinopathy caused by mutations affecting various types of collagen which make up the vitreous as well as cartilage and other connective tissues.
- Ocular manifestations include myopia, hypoplastic vitreous with an increased risk of giant retinal tears and retinal detachment.
- Systemic Features include (PMC8491173):
  - Oro-facial deformities such as Pierre Robin including cleft palate, micrognathia, and glossoptosis
  - Skeletal deformities with arthritis complications
  - Joint hyper-mobility
- As compared to other types of rhegmatogenous detachments, patients with Stickler's may present with retinal detachments in their second or third decade of life, though even in early childhood depending on the disease severity (PMC4993392).
- Patients with Stickler Syndrome have lower rates of retinal reattachment compared to the general population, ranging from 50 to 60% primary reattachment rate compared to 80–90% in the general population (PMC9318672), in one study requiring an average of 3 surgical interventions (PMC4993392).

Molecular Features of Stickler Syndrome

- The most common, Type I, is characterized in mutations in COL2A1 and represents about 80% of cases.
- The Stickler Highly Specialized Service (HSS) has assessed 1673 patients from 785 families (PMC8491173) and identified several other mutations including COL11A2, COL9A1, COL9A2, COL9A3, BMP4, and LOXL3.
- The structure of collagen of the vitreous is a balance between multiple types of collagen, with a core filament of Type XI surrounded by Type II with Type IX connecting the collagen fibers to the surrounding extracellular matrix (PMC9222743).

	Type I	Type II
Affected gene	COL2A1	COL11A1
Vitreous characteristics	Type 1 Membranous congenital vitreous anomaly	Type 2 Beaded congenital vitreous anomaly
Molecular characteristics	Usually premature stop codons or frameshift mutation, resulting in haploinsufficiency of type II collagen	Exon Skipping and Missense Mutations, resulting in dominant negative type XI collagen
Proposed mechanism	Reduced bulk of vitreous causing a only a portion of membranous vitreous to form	Reduction in the central filament, causing irregular collagen fibers

Management of Stickler Syndrome

- With the wider availability of genetic testing, it may be possible to identify patients prior to presentation with retinal detachments.
- Various retrospective studies have investigated the role of cryotherapy and laser barricade in prevention of retinal detachment, both of which have suggested a beneficial role in those with highest risk of detachment.
- Retrospective analysis of 360 laser prophylaxis in patients with COL2A1 and COL11A1 mutations found the rate of retinal detachment without laser to be 23% compared to 9% in those with (Retina 43(1):88–93, 2023).
- Similarly, in a study of a single family with COL2A1 mutations, 10% of lasered eyes developed retinal detachment, compared to 44% of non-lasered eyes (Eye. 1996;10:701–708).
- Finally, cryotherapy may also provide an option to prevent detachment, with a cohort of COL2A1 patients who were treated with bilateral prophylactic cryotherapy having had 8% failure of prophylaxis whereas those without cryotherapy had a rate of retinal detachment of 73%.
- The exact age at which prophylactic treatment still remains debated and its role in other forms of Stickler Syndrome which may have more mild phenotypes given differences in the involved mutation (Figs. 42.1 and 42.2).

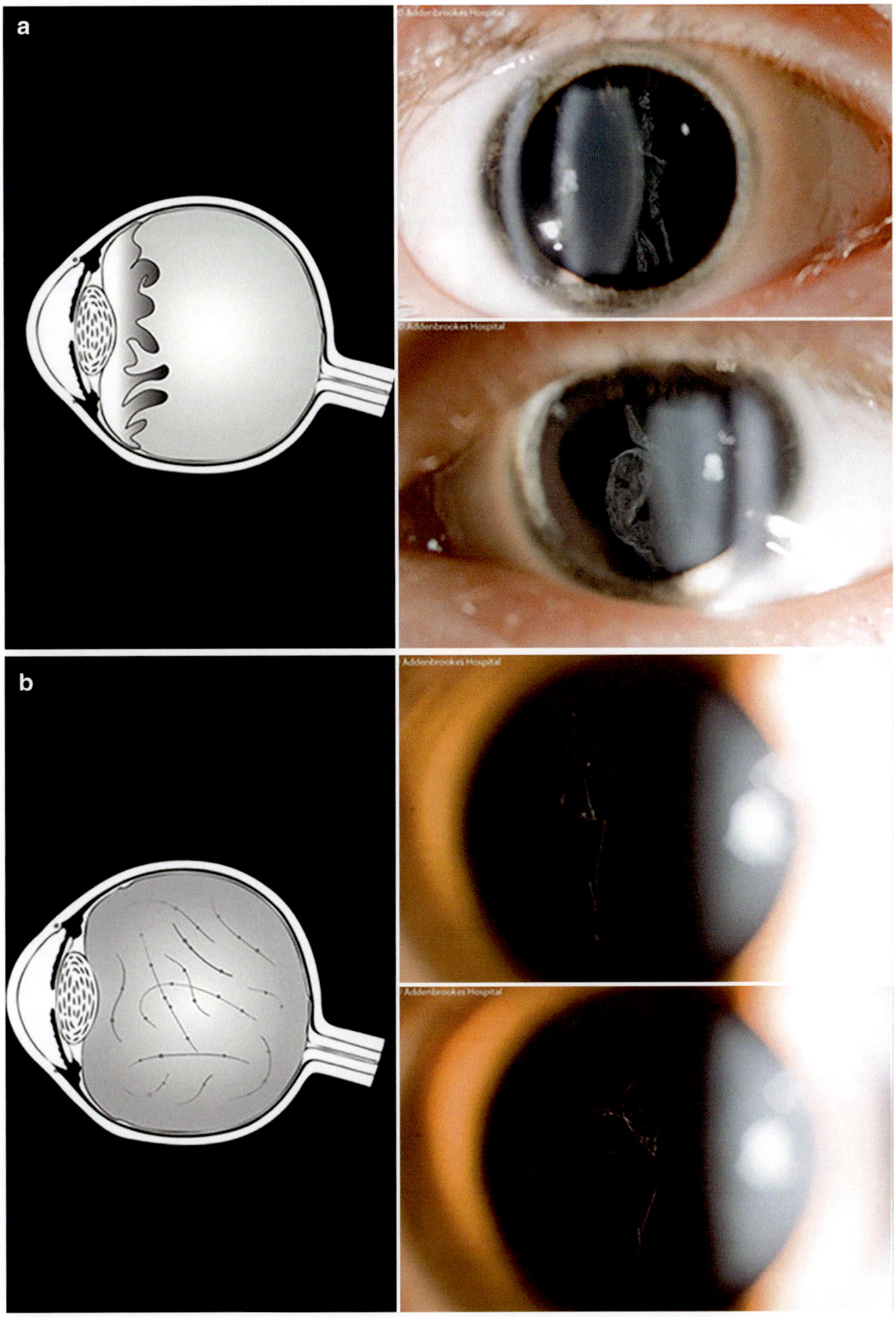

Fig. 42.1 Two vitreous types: Menbranous congenital vitreous anomaly in COL2A1 Haplosufficiency mutations (2a) and Beaded congenital vitreous anomaly in COL11A1 dominant-negative mutations. (Reproduced from Snead MP, Martin H, Bale P, Shenker N, Baguley DM, Alexander P, et al. Therapeutic and diagnostic advances in stickler syndrome. *Ther Adv Rare Dis.* 2020;1:1–13. https://doi.org/10.1177/2633004020978661)

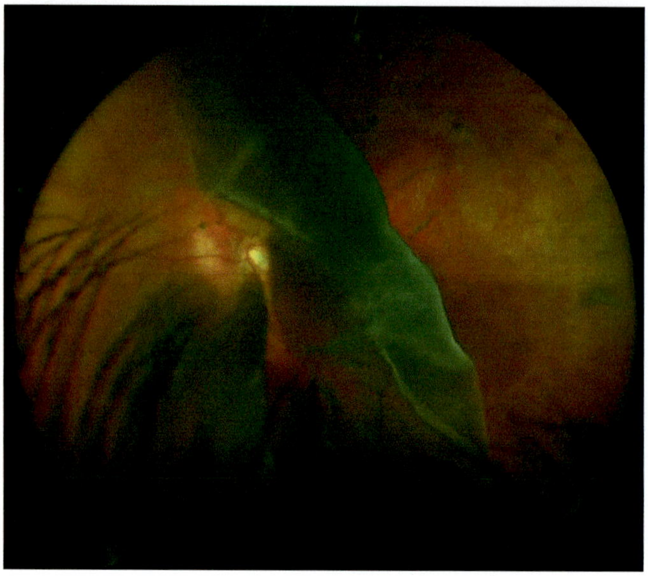

Fig. 42.2 Giant retinal tear in type 1 Stickler syndrome. (Reproduced from Snead MP, Martin H, Bale P, Shenker N, Baguley DM, Alexander P, et al. Therapeutic and diagnostic advances in stickler syndrome. *Ther Adv Rare Dis.* 2020;1:1–13. https://doi.org/10.1177/2633004020978661)

Further Reading

Genes (Basel). 2022; 13(6): 1089.
Eye. 1996;10:701–708.
Eye (Lond). 2022 Oct;36(10): 1966–1972.
Clin Ophthalmol. 2016; 10: 1531–1534.
Retina 43(1):p 88-93, January 2023.
Ophthalmology. 2008 Jan;115(1): 164–8.

CHAPTER 43

OTX2 Syndromes

Chloe Li, Stephen Tsang, Tarun Sharma, and Vlad Diaconita

General Features 220
Molecular Genetics 221
  Specific Genotype and Phenotype Correlations 221
Further Reading 221

C. Li · S. Tsang (✉) · T. Sharma · V. Diaconita
Department of Ophthalmology, Columbia University,
New York, NY, USA
e-mail: cl4128@cumc.columbia.edu; sht2@cumc.columbia.edu

S. H. Tsang et al. (eds.), *Atlas of Inherited Retinal Diseases*, Advances in Experimental Medicine and Biology 1467, https://doi.org/10.1007/978-3-031-72230-1_43

General Features

Patients with *OTX2* mutations have highly variable phenotypes, as the transcription factor is key in the development of retinal, pituitary gland, inner ear, and craniofacial structures. There are two recognized syndromes caused by *OTX2* mutations: syndromic microphthalmia type 5 (MCOPS5), and combined pituitary hormone deficiency type 6 (CPHD6).

In MCOPS5, patients with heterozygous loss-of-function *OTX2* mutations present with unilateral or bilateral anophthalmia or microphthalmia, as well as ocular malformations including coloboma, optic nerve hypoplasia, and/or retinal dystrophy. Some patients with MCOPS5 also have disruption of endocrine function, including agenesis or hypoplasia of the pituitary gland, and deficiency of growth and/or sex hormones.

In MCOPS5 patients with retinal dystrophy, a wide range of phenotypes is possible even among families with the same mutation. Poor distance vision and mildly impaired color vision are the most common symptoms. ERG shows generalized photoreceptor degeneration, with *in vivo* OCT imaging showing abnormal outer photoreceptor segment signal.

CPH6 has been reported less frequently in the literature but is likewise caused by heterozygous loss-of-function mutation in *OTX2*. Patients affected have panhypopituitarism with deficiency of growth, sex, adrenal, and thyroid hormones, and hypoplastic pituitary glands on MRI imaging. They do not have midline or ocular abnormalities.

Beyond these presentations caused by heterozygous mutations in *OTX2*, mutations in this gene have been identified in cases of septo-optic dysplasia, agnathia-otocephaly, and isolated high myopia (Figs. 43.1 and 43.2).

These two images and figure texts are reprinted with the generous permission of Molecular Vision and the authors. Copyright by the Authors, Henderson RH, Williamson KA, Kennedy JS, Webster AR, Holder GE, Robson AG, FitzPatrick DR, van Heyningen V, Moore AT (2009). The work was originally published in *A rare de novo nonsense mutation in OTX2 causes early onset retinal dystrophy and pituitary dysfunction,* Mol Vis, (2009); vol 15:2442–7

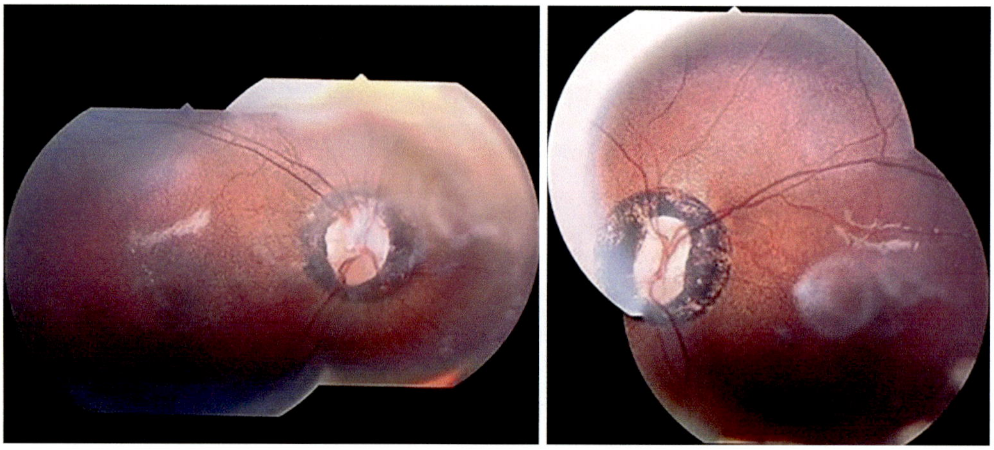

Fig 43.1 Fundus photos from representative patient with MCOPS5. Images are right and left eye color fundus photograph composites, performed using TOPCON retinal camera at 2.45× magnification at 35°. Composite images were created using Adobe Photoshop. Image cropping and reflections are observed secondary to poor pupil dilatation. The images show significant and abnormal peripapillary hyperpigmentation with fine granular pigmentation at the level of the retinal pigment epithelium (RPE)

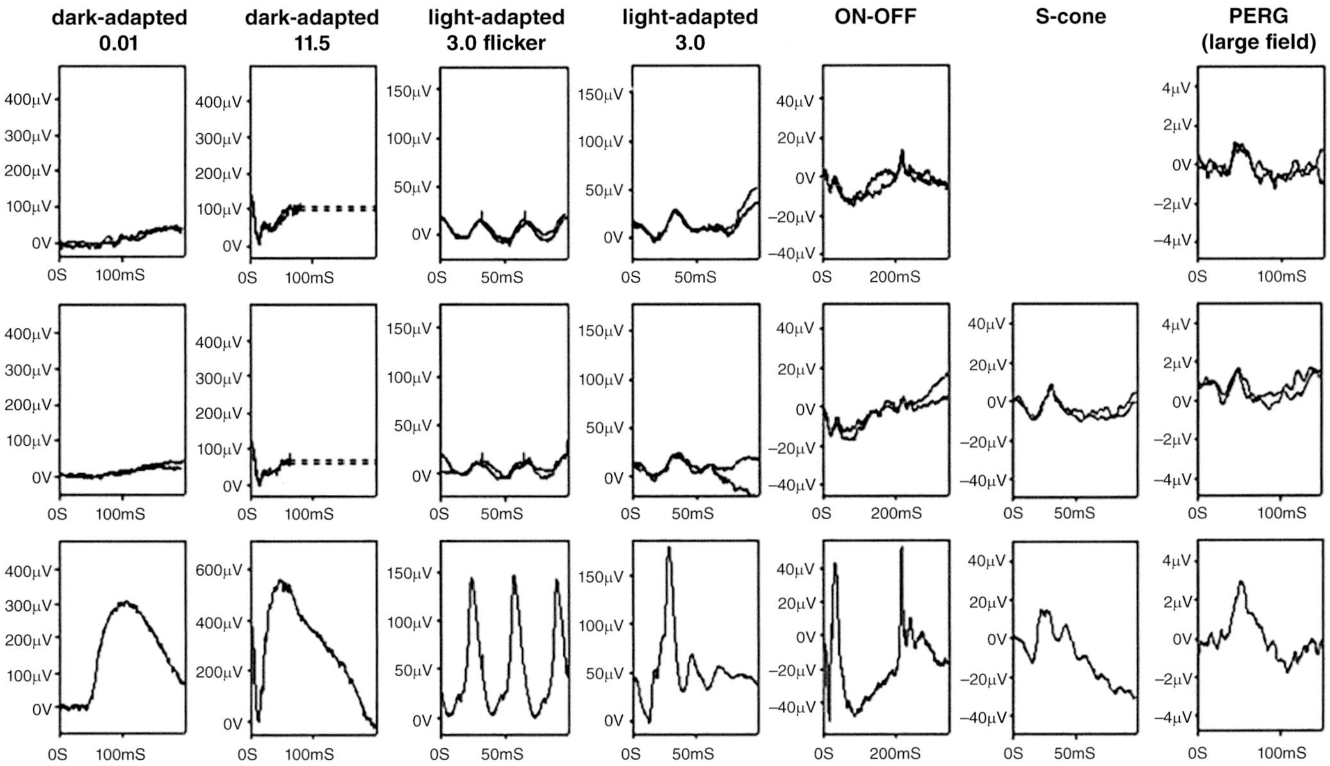

Fig 43.2 Electroretinograms. Full-field ERGs and pattern ERGs from the right (row 1) and left eye (row 2) of the patient and normal examples (row 3) for comparison. Full-field ERGs were performed using corneal electrodes in August 2007; pattern ERGs had been recorded to a 30° field checkerboard stimulus using periorbital surface electrodes two years earlier. Dark-adapted ERGs are shown for flash intensities of 0.01 and 11.5 cd.s.m^{-2}; light-adapted ERGs for a flash intensity of 3.0 cd.s.m^{-2} at 30 Hz (flicker) and at 2 Hz. ON-OFF ERGs used an orange stimulus (560 cd.m^{-2}, duration 200 ms) superimposed on a green background (150 cd.m^{-2}). S-cone ERGs used a blue stimulus (445 nm, 80 cd.m^{-2}) on an orange background (620 nm, 560 cd.m^{-2}). Eye movement artifacts are replaced by broken lines for clarity. S-cone ERGs were unavailable for the right eye

Molecular Genetics

OTX2 is located on chromosome 14q22.3 and is composed of three exons.

Specific Genotype and Phenotype Correlations

Genotype: OTX2 c.708T>A (p.Y179X), heterozygous loss-of-function

Phenotype: MCOPS5 with bilateral microphthalmia and pituitary dysfunction. Even within the same family, two siblings with the same mutation, both with microphthalmia, had very different visual outcomes; one had 20/200 BCVA and one was NLP.

Genotype: c.235G>A (p.E79K), heterozygous loss-of-function

Phenotype: MCOPS5 with pattern dystrophy only

Genotype: OTX2 c.698A>G (p.N233S), heterozygous loss-of-function

Phenotype: CPH6

Further Reading

Ragge NK, Brown AG, Poloschek CM, Lorenz B, Henderson RA, Clarke MP, Russell-Eggitt I, Fielder A, Gerrelli D, Martinez-Barbera JP, Ruddle P, Hurst J, Collin JR, Salt A, Cooper ST, Thompson PJ, Sisodiya SM, Williamson KA, Fitzpatrick DR, van Heyningen V, Hanson IM. Heterozygous mutations of OTX2 cause severe ocular malformations. Am J Hum Genet. 2005 Jun;76(6):1008-22. https://doi.org/10.1086/430721. Epub 2005 Apr 21. Erratum in: Am J Hum Genet. 2005 Aug;77(2):334. PMID: 15846561; PMCID: PMC1196439.

Diaczok D, Romero C, Zunich J, Marshall I, Radovick S. A novel dominant negative mutation of OTX2 associated with combined pituitary hormone deficiency. J Clin Endocrinol Metab. 2008 Nov;93(11):4351-9. https://doi.org/10.1210/jc.2008-1189. Epub 2008 Aug 26. PMID: 18728160; PMCID: PMC2582563.

Vincent A, Forster N, Maynes JT, et al. *OTX2* mutations cause autosomal dominant pattern dystrophy of the retinal pigment epithelium. J Med Genet 2014;**51**:797-805.

Part VI
Phakomatoses

CHAPTER 44

Von Hippel-Lindau Disease

Luanna Chan, Stephen H. Tsang, Tarun Sharma, and Vlad Diaconita

General Features of Phakomatoses 226
General Features of Von Hippel-Lindau Disease 226
Molecular Genetics of Von Hippel-Lindau Disease. 227
Further Reading 227

L. Chan
SUNY Downstate Medical School,
New York, NY, USA

S. H. Tsang (✉) · T. Sharma · V. Diaconita
Department of Ophthalmology, Columbia University,
New York, NY, USA
e-mail: sht2@cumc.columbia.edu

S. H. Tsang et al. (eds.), *Atlas of Inherited Retinal Diseases*, Advances in Experimental Medicine and Biology 1467, https://doi.org/10.1007/978-3-031-72230-1_44

General Features of Phakomatoses

- Phakomatoses (*phakoma* = birthmark) are a group of diseases or syndromes that have hamartomas (tumorous malformations composed of tissues normally present at the location where they develop) of the skin, brain, and eye (oculoneurocutaneous syndromes).
- Phakomatoses include retinal and cerebellar hemangiomatosis (von Hippel-Lindau syndrome [VHL]), tuberous sclerosis (Bourneville syndrome, Chap. 43), neurofibromatosis (NF, von Recklinghausen syndrome, Chap. 44), encephalofacial hemangiomatosis (Sturge-Weber syndrome), and racemose angiomatosis (Wyburn-Mason syndrome).
- Most phakomatoses have an autosomal dominant (AD) inheritance with incomplete penetrance, except for Sturge-Weber and Wyburn-Mason.
- Most of the tumors that develop with phakomatoses are benign and are either stationary or slowly progressive. Occasionally, however, malignant or life-threatening tumors can develop, like cerebellar hemangioblastoma or renal cell carcinoma in VHL, or malignant schwannoma in NF.

General Features of Von Hippel-Lindau Disease

- The incidence of VHL is about 1 in 36,000 live births. Inheritance is AD, with a penetrance of over 90% by 65 years of age.
- These patients have a benign vascular neoplasm occurring either in the retina or optic disc, known as retinal hemangioblastoma (RH), retinal capillary hemangioma or hemangioblastoma, or retinal angioma.
- The mean age of onset of VHL-associated RH is 26 with 5% of these patients presenting under the age of 10.
- Systemic involvement includes hemangioblastoma of the brain (cerebellum) and spinal cord, renal cell carcinoma, pheochromocytoma, pancreatic neuroendocrine tumors, and renal or pancreatic cysts.
- Ocular involvement can be either unilateral (about 40%) or bilateral (about 60%). In about 85% of cases, the tumor involves the mid-peripheral location; about 8% are juxtapapillary, and about 7% affect both locations.
- The number of tumors varies. Usually multiple tumors occur; a new tumor may appear with the passage of time.
- RH is characterized histopathologically by the presence of foamy, vacuolated stromal cells surrounded by vascular structures. Stromal cells have demonstrated complete loss of the VHL gene and increased VEGF expression in RH.
- Small tumors appear as a red dot, like a microaneurysm with no feeding vessels (sessile), but as the tumor grows, characteristic feeding and draining vessels appear, along with retinal edema, exudation, and traction (Fig. 44.1).
- Juxta-papillary tumors are usually sessile with some fine, lacy vascular pattern and fullness of the neural rim or disc margin in one localized sector.
- Fluorescein angiography helps to delineate all of the vascular tumors.
- Optical coherence tomography (OCT) is helpful for juxtapapillary tumors and to characterize macular edema or epiretinal membrane (ERM).
- Ultrasound B-scan shows a solid retinal mass with variable, moderate reflectivity and no orbital or choroidal shadowing.

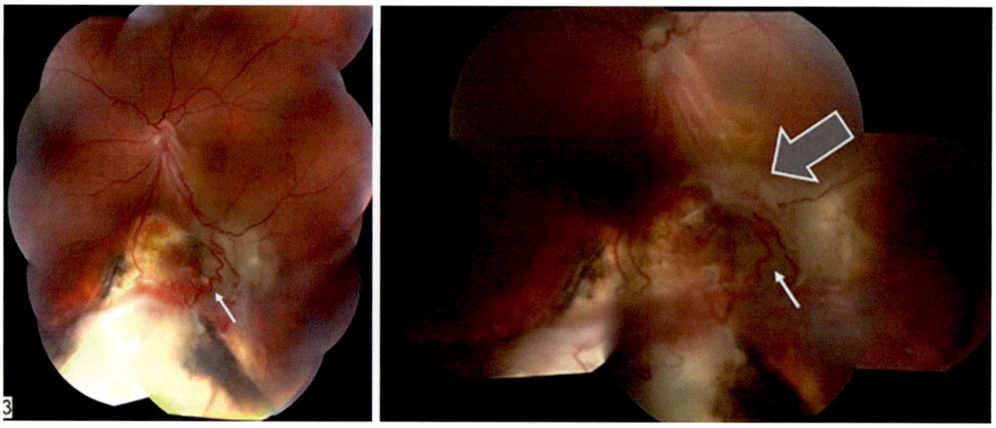

Fig. 44.1 *Left*, partially treated fundus in a case of von Hippel-Lindau disease; vessels are still dilated and tortuous (*small arrow*). *Right*, follow-up at 3 years shows a pigmented scar around a relatively flat angioma, with less tortuous vessels. Also note epiretinal membrane following treatment (*thick arrow*)

- Vision loss is due to exudation involving the macula or traction/exudative retinal detachment. Vision can also be affected if there is an expanding cerebellar hemangioblastoma with intracranial hypertension and papilledema, or pheochromocytoma causing severe hypertension and hypertensive retinopathy.
- Besides an annual eye examination, these patients should be screened for systemic tumors. It is also necessary to rule out pheochromocytoma (with plasma or 24-h urine catecholamines and metanephrines), renal cell carcinoma (ultrasound or CT scan of the abdomen), and CNS hemangioblastoma (MRI of the brain and spine).
- Treatment involves ablative therapy such as thermal laser photocoagulation, cryotherapy, radiation (brachytherapy, external beam), photodynamic therapy, and transpupillary thermotherapy. Patients with multiple lesions or tumors that border the optic nerve may benefit from systemic therapy such as belzutifan.
- Vitreous surgery is needed for traction or rhegmatogenous retinal detachment, with guarded outcome.
- Photodynamic therapy is employed for juxta-papillary lesions.
- Anti–vascular endothelial growth factor (anti-VEGF) agents have been tried as an adjunct to other treatment modalities.

Molecular Genetics of Von Hippel-Lindau Disease

- VHL is caused by a mutation in the *VHL* tumor suppressor gene; the mutation varies, ranging from substitution of a base pair in a single codon to the complete deletion of the gene.
- Cytogenetic location of *VHL* gene: 3p25.3.
- Most patients with VHL inherit a mutant copy of the *VHL* tumor suppressor gene from an affected parent, and a normal copy (wild-type) from the other parent. Neoplasia occurs from the somatic inactivation (second hit) of the normal allele in one or more cells of an individual with a germline mutation (first hit) in the other allele (Knudson's two-hit model).
- The *VHL* gene provides instructions for making a protein that is a part of the complex called VCB-CUL2 complex; this complex helps other proteins to be broken down when they are no longer needed.
- One of the targets of the VCB-CUL2 complex is a protein called hypoxia-inducible factor 2-alpha (HIF-2α), part of the HIF protein complex, which controls several genes involved in cell division, formation of new blood vessels, and production of red blood cells by the regulation of erythropoietin. HIF is particularly important when there is hypoxia. When adequate oxygen is available, the VCB-CUL2 complex prevents inappropriate buildup of HIF in cells.
- With *VHL* gene mutation, HIF builds up and upregulates several growth factors, including VEGF and platelet-derived growth factor (PDGF).

Further Reading

Benjamin C, Yates JRW, Moore AT, et al. Clinical features and natural history of von Hippel-Lindau Disease. Qjm 2012;77(2):1151-1163. doi: https://doi.org/10.1093/qjmed/77.2.1151

Chan CC, Vortmeyer AO, Chew EY, et al. VHL gene deletion and enhanced VEGF gene expression detected in the stromal cells of retinal angioma. Arch Ophthalmol 1999; 117:625-630

Choyke PL, Glenn GM, Walther MM, Patronas NJ, Linehan WM, Zbar B. von Hippel-Lindau disease: genetic, clinical, and imaging features. Radiology 1995;194:629–42.

Jonasch E, Donskov F, Iliopoulos O, et al. Belzutifan for renal cell carcinoma in von Hippel-Lindau disease. N Engl J Med 2021; 385:2036.

Papastefanou VP, Pilli S, Stinghe A, Lotery AJ, Cohen VM. Photodynamic therapy for retinal capillary hemangioma. Eye (Lond) 2013;27:438–42.

Varshney, Neha et al. A review of Von Hippel-Lindau syndrome. J Kidney Cancer VHL 2017;4(3): 20–29.

Wong WT, Agrón E, Coleman HR, Tran T, Reed GF, Csaky K, Chew EY. Clinical characterization of retinal capillary hemangioblastomas in a large population of patients with von Hippel-Lindau disease. Ophthalmology 2008;115:181–8

CHAPTER 45

Tuberous Sclerosis

Vlad Diaconita, Amy Shteyman, Anna Jolia, Stephen H. Tsang, Tarun Sharma, and Brian Marr

General Features 230
Molecular Genetics 233
Further Reading 233

V. Diaconita · A. Jolia · S. H. Tsang · T. Sharma
Department of Ophthalmology, Columbia University, New York, NY, USA

A. Shteyman · B. Marr (✉)
Columbia University, New York, NY, USA
e-mail: bpm2133@cumc.columbia.edu

S. H. Tsang et al. (eds.), *Atlas of Inherited Retinal Diseases*, Advances in Experimental Medicine and Biology 1467, https://doi.org/10.1007/978-3-031-72230-1_45

General Features

- Tuberous sclerosis is characterized by a classic triad: seizures, mental retardation, and cutaneous angiofibromas; this triad occurs in about 30% of cases.
- **Major clinical diagnostic criteria:** Hypomelanotic macules (≥3 in number and at least 5 mm in diameter) or angiofibromas (≥3), ungual fibromas (≥2), shagreen patch, retinal hamartomas, cortical dysplasias, subependymal nodules, subependymal giant cell astrocytoma, cardiac rhabdomyoma, lymphangioleiomyomatosis, angiomyolipomas (>2).
- **Minor clinical diagnostic criteria:** Confetti skin lesions, dental enamel pits (≥3), intraoral fibromas (≥2), retinal achromatic patch, multiple renal cysts, non-renal hamartomas.
- A **definite diagnosis** requires two major criteria or one major and two or more minor criteria, or the presence of *TSC1* or *TSC2* mutation (TSC, tuberous sclerosis complex).
- A **possible diagnosis** requires either one major or two or more minor criteria.
- Astrocytic hamartoma of the retina (Figs. 45.1 and 45.2) usually occurs in close proximity to the optic disc or at the disc; it presents as a large, whitish (calcified) nodular masses or mulberry-like, usually endophytic. They can be associated with neovascularization and the presence of intraretinal and subretinal fluid, which may respond to antiVEGF therapy and photodynamic therapy (PDT) (Figs. 45.3, 45.4, and 45.5).

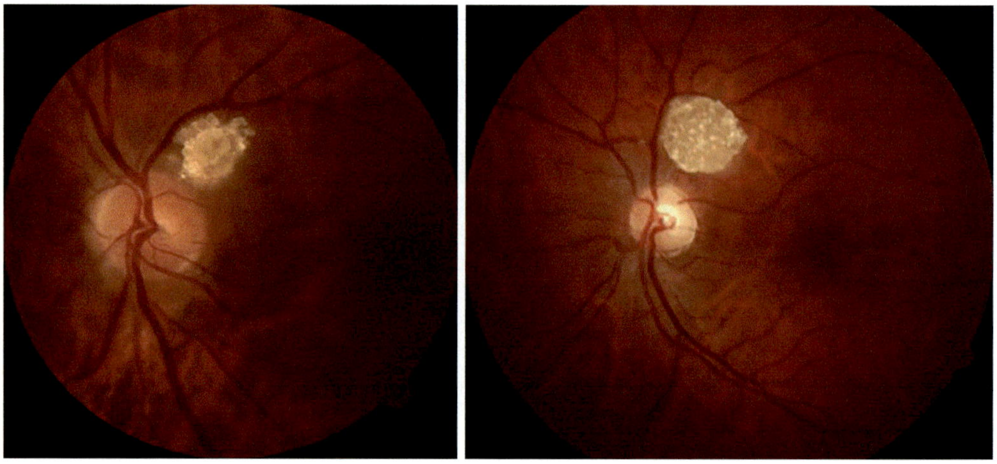

Fig. 45.1 Astrocytic hamartoma, seen as a whitish calcified lesion, mulberry-like, in close proximity to the optic disc

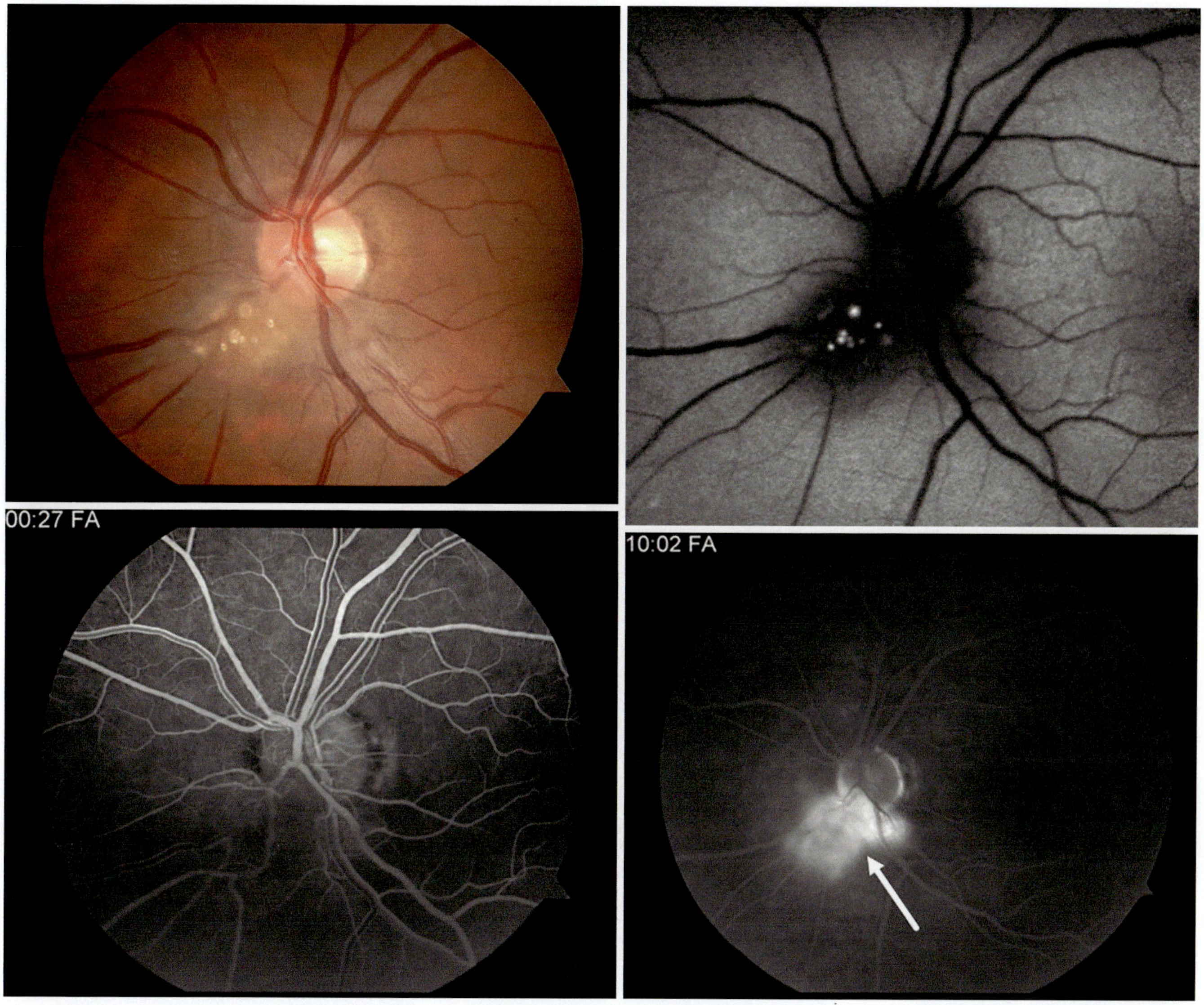

Fig. 45.2 Astrocytic hamartoma, seen as a whitish calcified lesion, located inferonasal to the optic disc. This lesion shows hyperautofluorescent dots, with staining on a fluorescein angiogram (*arrow*)

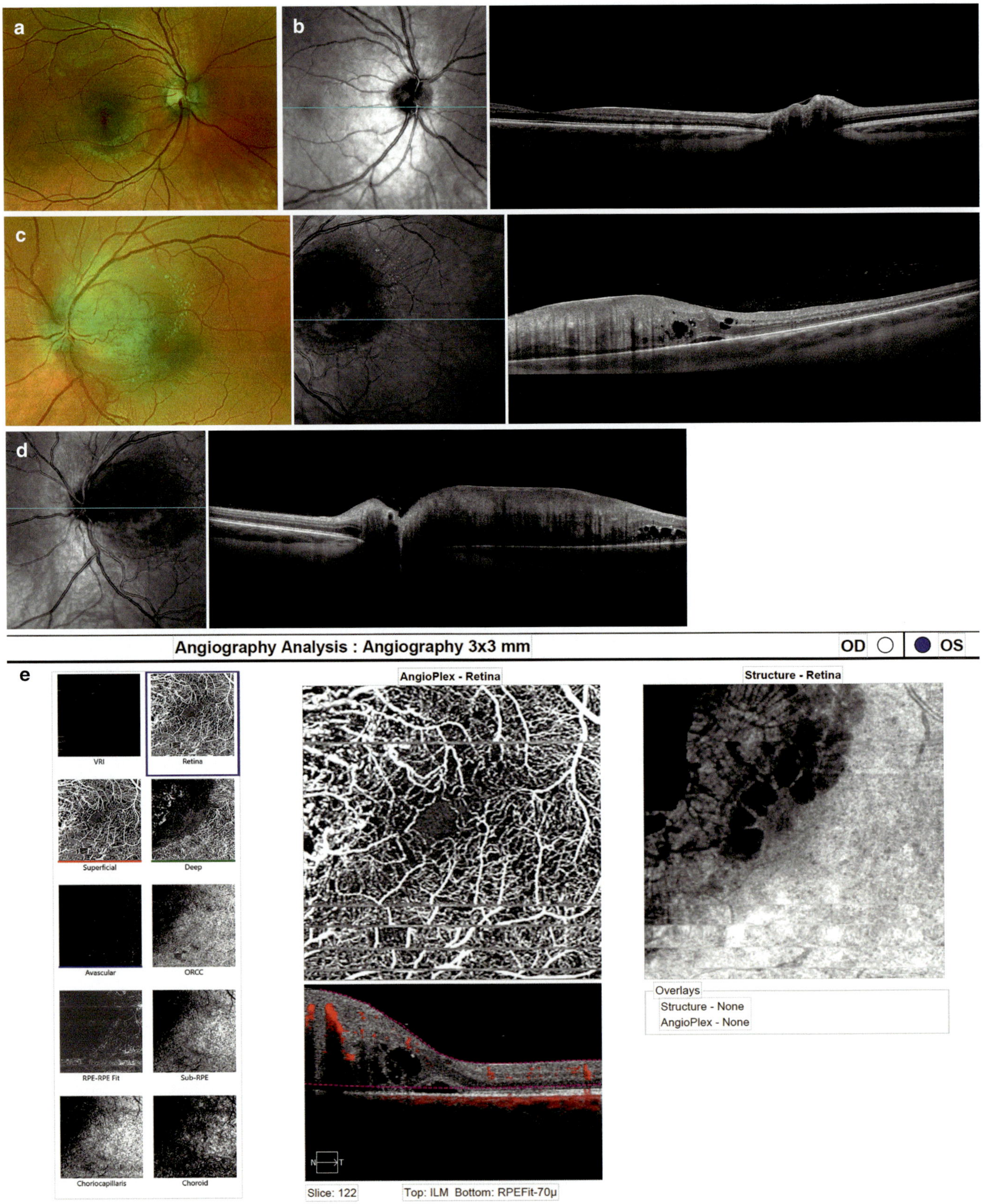

Fig. 45.3 Astrocytic hamartoma seen in the same patient. Panels (**a** and **b**) show the right eye with a lesion nasal to the disc. Panels (**c**–**e**) show the left eye with Optos, OCT and OCT-A imaging modalities. The left eye has visible intraretinal and subretinal fluid and associated vascularity of the lesion

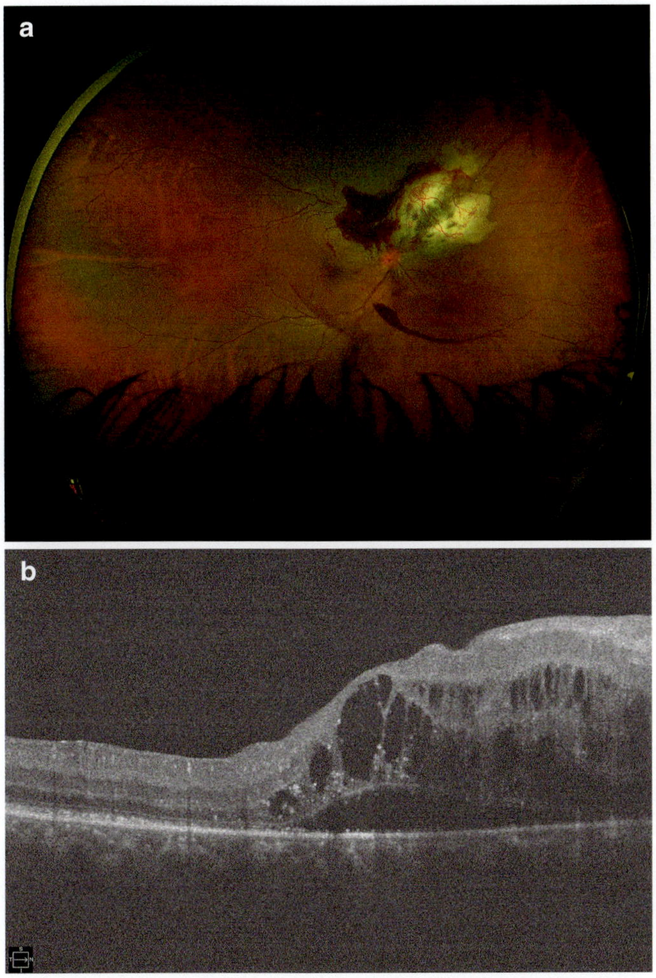

Fig. 45.4 Astrocytyc hamartoma supranasal to the disc with associated neovascularization, intraretinal hemorrhage, preretinal and vitreous hemorrhage (**a**). Panel (**b**) shows associated intraretinal and subretinal fluid on OCT

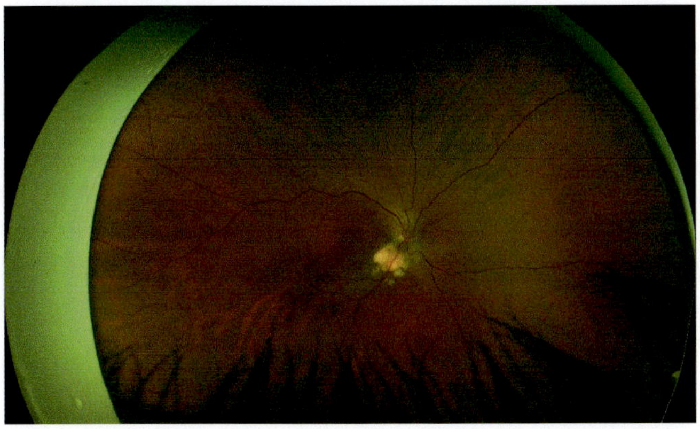

Fig. 45.5 Astrocytic hamartoma seen inferotemporal to disc with associated calcification

Molecular Genetics

- Mutation in *TSC1* or *TSC2* (tuberous sclerosis complex) is an independent diagnostic criterion. About 10–25% of tuberous sclerosis patients have no mutation, so a normal test does not rule out the clinical diagnosis.
- The mutation either prevents or inactivates the function of the TSC1 protein, hamartin, or the TSC2 protein, tuberin.
- Tuberous sclerosis has AD inheritance with incomplete penetrance and variable expressivity.
- The phenotype of *TSC1* mutation is mild; the *TSC2* gene accounts for approximately 90% of clinical cases.
- Cytogenetic location: *TSC1*, 9q34; *TSC2*, 16p13.3.

Further Reading

Northrup H, Krueger DA; International Tuberous Sclerosis Complex Consensus Group. Tuberous sclerosis complex diagnostic criteria update: recommendations of the 2012 International Tuberous Sclerosis Complex Consensus Conference. Pediatr Neurol 2013;49:243–54.

Rowley SA, O'Callaghan FJ, Osborne JP. Ophthalmic manifestations of tuberous sclerosis: a population based study. Br J Ophthalmol 2001;85:420–3.

CHAPTER 46

Neurofibromatosis

Jennifer Adeghate, Stephen Tsang, and Tarun Sharma

General Features 236
Neurofibromatosis Type 1 236
Neurofibromatosis Type 2 236
Molecular Genetics 237
Further Reading 237

J. Adeghate · S. Tsang (✉) · T. Sharma
Department of Ophthalmology, Columbia University,
New York, NY, USA
e-mail: sht2@cumc.columbia.edu

S. H. Tsang et al. (eds.), *Atlas of Inherited Retinal Diseases*, Advances in Experimental Medicine and Biology 1467, https://doi.org/10.1007/978-3-031-72230-1_46

General Features

- There are two types of neurofibromatosis (NF):
  - NF1 (type 1, peripheral NF, or von Recklinghausen syndrome)
  - NF2 (type 2, central or bilateral acoustic NF)
- The mode of inheritance is autosomal dominant (AD), with complete penetrance with 50% expression. About half of all cases have spontaneous mutations with no family history.

Neurofibromatosis Type 1

- NF1 occurs at a rate of approximately 1 in 2500–3000 individuals.
- Diagnosis requires at least two of the following criteria:
  - *Café-au-lait spots:* At least 6 spots larger than 5 mm in diameter if age is <10 years (prepubertal); or at least 15 mm in diameter if age is >10 years (postpubertal).
  - *Freckles:* Axillary or inguinal
  - *Skin neurofibromas:* At least two typical neurofibromas or at least one plexiform neurofibroma
  - *Optic nerve glioma*
  - *Iris Lisch nodules:* At least two lesions
  - *Osseous lesion:* Sphenoid or other skeletal dysplasia
  - *Family history:* First-degree relatives (parent, sibling, or off-spring) with above criteria.
- Ocular signs:
  - Eyelid involvement: diffuse thickening, typical S-shaped curvature of the upper eyelid due to plexiform neurofibroma
  - Multiple iris hamartomas, Lisch nodules
  - Congenital glaucoma, ectropion uveae
  - Optic nerve glioma
  - Multifocal choroidal freckling
  - Vasoproliferative tumor
  - Retinal astrocytic hamartoma
  - Refractive error: Mild to moderate myopia

Neurofibromatosis Type 2

- NF2 occurs with a frequency of 1 in 33,000 people.
- Criteria for diagnosis:
  - Bilateral acoustic neuromas (vestibular schwannomas) *or*
  - Unilateral 8th cranial nerve tumor *plus* NF2 in a first-degree relative *or*
  - Two of the following:
    - Meningioma
    - Glioma
    - Schwannoma
    - Juvenile posterior subcapsular lens opacity plus NF2 in a first-degree relative
- Ocular signs:
  - Posterior subcapsular or cortical cataract in childhood
  - Combined hamartoma of the retina and retinal pigment epithelium (RPE) (Fig. 46.1)
  - Epiretinal membrane
  - Neurotrophic keratitis

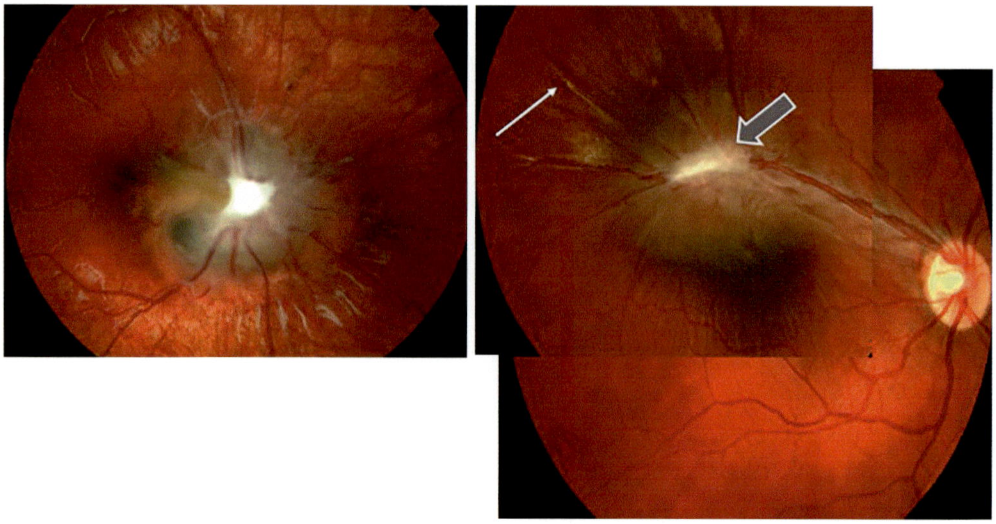

Fig. 46.1 Ocular manifestations of neurofibromatosis type 2, including combined hamartoma of inner retina and retinal pigment epithelium (RPE) and epiretinal membrane (*thick arrow*). Note the radial traction striae (*thin arrow*)

Molecular Genetics

- **Neurofibromatosis Type 1**: is caused by germline mutations in the *NF1* gene (Cytogenetic location: 17q11.2), which codes for a protein called *neurofibromin*, a tumor suppressor that regulates cell growth and division. A variant of NF1 called Segmental NF1 occurs with somatic mutations in the NF1 gene affecting only localized areas of the body. As such, these patients may have isolated Lisch nodules with no other manifestations of NF1.
- **Neurofibromatosis Type 2**: is caused by mutations in the *NF2* gene (Cytogenetic location: 22q12.2). NF2 codes for a protein called *merlin* (schwannomin) that is produced in Schwann cells and acts as a structural protein as well as a tumor suppressor.

Further Reading

Abdolrahimzadeh B, Piraino DC, Albanese G, Cruciani F, Rahimi S. Neurofibromatosis: an update of ophthalmic characteristics and applications of optical coherence tomography. Clin Ophthalmol. 2016;10:851-60. doi: https://doi.org/10.2147/OPTH.S102830.

Evans DG. Neurofibromatosis 2 [bilateral acoustic neurofibromatosis, central neurofibromatosis, NF2, neurofibromatosis type II]. Genet Med. 2009;11:599–610.

Gutmann DH, Aylsworth A, Carey JC, Korf B, Marks J, Pyeritz RE, et al. The diagnostic evaluation and multidisciplinary management of neurofibromatosis 1 and neurofibromatosis 2. JAMA. 1997;278:51–7.

Kinori M, Hodgson N, Zeid JL. Ophthalmic manifestations in neurofibromatosis type 1. Surv Ophthalmol. 2018;63(4):518–533. doi: https://doi.org/10.1016/j.survophthal.2017.10.007.

Rare Diseases Database: https://rarediseases.org/rare-diseases/. Accessed February 21, 2023.

Part VII
Phenocopies

CHAPTER 47

Rubella Retinopathy

Jin Kyun Oh, Stephen H. Tsang, and Tarun Sharma

General Features 242
Further Reading 243

J. K. Oh
Department of Ophthalmology, Edward S. Harkness Eye Institute,
Columbia University Irving Medical Center,
New York, NY, USA

S. H. Tsang (✉) · T. Sharma
Department of Ophthalmology, Columbia University,
New York, NY, USA
e-mail: sht2@cumc.columbia.edu

S. H. Tsang et al. (eds.), *Atlas of Inherited Retinal Diseases*, Advances in Experimental Medicine and Biology 1467, https://doi.org/10.1007/978-3-031-72230-1_47

General Features

- Rubella retinopathy is an ocular manifestation of congenital rubella syndrome characterized by diffuse mottling of the retinal pigment epithelium classically described as a "salt-and-pepper" fundus (Fig. 47.1).
- Rubella retinopathy can be both unilateral or bilateral, center-involving or center-sparing, and the appearance of pigmentation is phenotypically variable, ranging from patchy and stippled to bone spicule patterned.
- Although previously considered a stationary disease, reports of disease with retinal pigment epithelium atrophy and progressive disease have been reported. Electroretinography findings may be normal.
- Complications of rubella retinopathy include subretinal choroidal neovascularization leading to scarring and foveal atrophy.
- Study of optical coherence tomography in rubella retinopathy has demonstrated significantly decreased central foveal thickness and alteration in morphology.
- Rubella retinopathy is the most common ocular disorder associated with congenital rubella with reported prevalence between 23% and 40%. Other common signs and symptoms include cataracts (15%), microphthalmia and microcornea (9%) and glaucoma (5%).

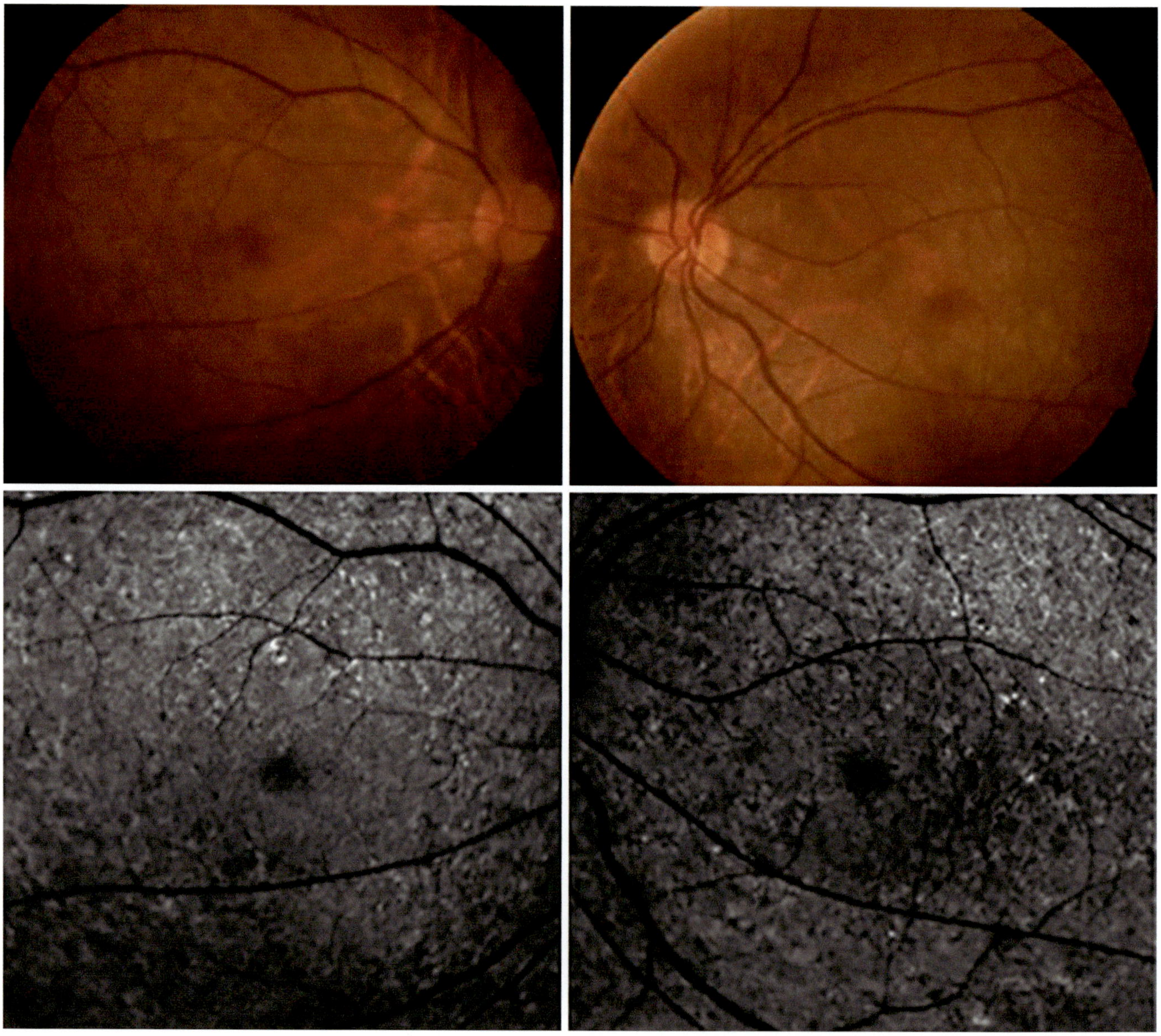

Fig. 47.1 Color fundus photographs and fundus autofluorescence of rubella retinopathy. Color fundus photographs of a patient with congenital rubella retinopathy demonstrate subtle central retinal pigmentary changes greater in the right eye than the left. These changes are better appreciated on fundus autofluorescence which reveals diffuse asymmetric hypoautofluorescent mottling

- The triad of congenital rubella syndrome described by Gregg et al. in 1941 includes cataracts, congenital heart disease (patent ductus arteriosus, interventricular septal defects, pulmonary stenosis), and deafness. Other systemic features of congenital rubella include neurologic anomalies such as microcephaly or seizures.
- Rubella infection is often inconsequential after birth; however, when transmitted transplacentally, rubella causes severe damage to the fetus. Severity of disease is correlated on the virulence and timing of infection. Earlier infection tends to be associated with greater defects.
- Sensitivity of IgM testing is highest before the age of 5 months although it may continue to be synthesized by the fetus up to 18 months after birth.
- The global incidence of congenital rubella syndrome is over 200,000 children annually, with the majority arising from developing countries. However, with the increase in immigration, the prevalence continues to rise across the world.

Further Reading

Frank KE, Purnell EW. Subretinal neovascularization following rubella retinopathy. Am J Ophthalmol 1978;86:462–6.

Goldberg N, Chou J, Moore A, Tsang S. Autofluorescence imaging in rubella retinopathy. Ocul Immunol Inflamm 2009;17(6):400–2.

Gregg NM. Congenital cataract following German measles in the mother. Trans Ophthalmol Soc Aust 1941;335–46.

Matalia J, Vinekar A, Anegondi N, Mangalesh S, Anaspure H, Shetty KB, Sinha Roy A. A prospective OCT study of Rubella retinopathy. Ophthalmol Retina 2018;2(12):1235-1240.

Vijayalakshmi P, Kakkar G, Samprathi A, Banushree R. Ocular manifestations of congenital rubella syndrome in a developing country. Indian J Ophthalmol 2002;50(4):307-11.

Wolff SM. The ocular manifestations of congenital rubella. Trans Am Ophthalmol Soc 1972;70:577-614.

CHAPTER 48

Syphilis

Kristina J. Hartung, Stephen H. Tsang, Tarun Sharma, and Vlad Diaconita

General Features 246
Further Reading 247

K. J. Hartung
Department of Ophthalmology, University Medical Centre Ljubljana, Ljubljana, Slovenia

Faculty of Medicine, University of Ljubljana, Ljubljana, Slovenia

S. H. Tsang (✉) · T. Sharma · V. Diaconita
Department of Ophthalmology, Columbia University, New York, NY, USA
e-mail: sht2@cumc.columbia.edu

S. H. Tsang et al. (eds.), *Atlas of Inherited Retinal Diseases*, Advances in Experimental Medicine and Biology 1467, https://doi.org/10.1007/978-3-031-72230-1_48

General Features

- Syphilis is an infectious disease caused by a spirochete, *Treponema pallidum*. It is mainly acquired by sexual transmission, but transplacental transmission resulting in congenital syphilis is possible.
- **Acquired syphilis** has four stages:
  - *Primary syphilis* (2–6 weeks after infection): painless chancre at the site of inoculation that resolves within 12 weeks; may appear on the eyelid or the conjunctiva.
  - *Secondary syphilis* (6–8 weeks after infection): maculopapular generalized rash involving the palms and soles (presenting complaint in more than 70% of patients); other symptoms include fever/malaise, generalized non-tender lymphadenopathy, headache, nausea, joint pain, mouth ulcer, and hair loss. The hematogenous dissemination can lead to neurologic, ophthalmologic, gastrointestinal, and hepatic disease.
  - *Latent syphilis:* The latent phase between the disappearance of the secondary manifestations and either treatment or the development of tertiary syphilis can range from 1 year to decades after secondary syphilis.
  - *Tertiary syphilis* (years to decades after infection): One-third of untreated patients progress to this stage, characterized by three forms:

    Gummatous syphilis: gumma is a benign, localized granulomatous formation that can affect the skin, bone, and liver but can occur anywhere, including the choroid and iris.

    Neurosyphilis: meningitis, CNS vasculitis, and parenchymatous degeneration (resulting in tabes dorsalis and paresis) can occur.

    Cardiovascular syphilis: aortitis, aortic aneurysms, aortic valve insufficiency, and narrowing of the coronary Ostia may be seen.
- **Congenital syphilis** has two stages:
  - *Early* (≤2 years): Manifests as mucocutaneous rash, periostitis and osteochondritis, hepatosplenomegaly. The most common ocular manifestation is multifocal chorioretinitis (may be accompanied by vasculitis) which results in the bilateral salt and pepper fundus appearance. Glaucoma, congenital cataract, and uveitis may also be seen.
  - *Late* (≥3 years): Hutchinson's triad of congenital syphilis consists of interstitial keratitis, Hutchinson teeth, and sensorineural deafness. Additional signs include saddle nose, frontal bossing, and saber shins.
- **Investigations:**
  - Nontreponemal tests: Venereal Disease Research Laboratory (VDRL) and rapid plasma reagin (RPR) are used for assessing disease activity and therapeutic response. They are also used for screening but have low specificity (a high rate of false positives) and should be confirmed with treponemal tests.
  - *Treponemal* tests: Enzyme immunoassays (EIAs), chemiluminescence immunoassays (CIAs), microhemagglutination assay for T. pallidum antibodies (MHA-TP), and the older fluorescent treponemal antibody absorption (FTA-ABS) assay detect previous or current infection but cannot distinguish the active disease from past infection with syphilis.
  - The *likelihood* of ocular involvement is increased in HIV co-infection.
- **Ocular manifestations: (Fig. 48.1)**
  - *Syphilis* is known as the "great imitator" and can occur in any stage of the disease and involve virtually any ocular structure, causing conjunctivitis, conjunctival chancre, episcleritis, scleritis, interstitial keratitis, uveitis, chorioretinitis, retinitis, vasculitis, neuroretinitis, papillitis, retrobulbar neuritis, papilledema, ocular motor nerve palsies, and Argyll Robertson pupil.
  - *Uveitis* is the most common ocular presentation of syphilis and has various manifestations; it can be unilateral or bilateral, granulomatous or non-granulomatous, and can present as anterior, posterior, or panuveitis. Anterior uveitis is usually unilateral and, more commonly, granulomatous. Atrophy, roseolae, and nodules may be seen, as well as posterior synechiae, lens dislocation, and increased intraocular pressure. Focal or multifocal chorioretinitis with vitritis is the most common posterior segment manifestation. Small, grayish-yellow lesions that may become confluent are seen in the periphery. It may be accompanied by retinal vasculitis, disc edema, and serous retinal detachment.
  - **Retinitis** (necrotizing or non-necrotizing) in the macula, midperiphery, and peripheral retina is a less common presentation. Patches of retinitis may become confluent and are often associated with vasculitis and vascular occlusion, resembling branch retinal vein and/or artery occlusion. It may also resemble acute retinal necrosis or progressive outer retinal necrosis; however, it is slowly progressive, responds well to intravenous penicillin, and heals without disrupting the underlying retinal pigment epithelium.
  - **Acute syphilitic posterior placoid chorioretinopathy (ASPPC)** is a rare *manifestation* of ocular syphilis that is pathognomonic of secondary syphilis. It is defined by the presence of yellow placoid lesions with a faded center and a round and prominent margin. The lesions result from either T. pallidum-induced infection of the RPE or indirect immune-mediated hypersensitivity.

    OCT: disruption of the ellipsoid zone, nodular thickening of the RPE with loss of the linear outer

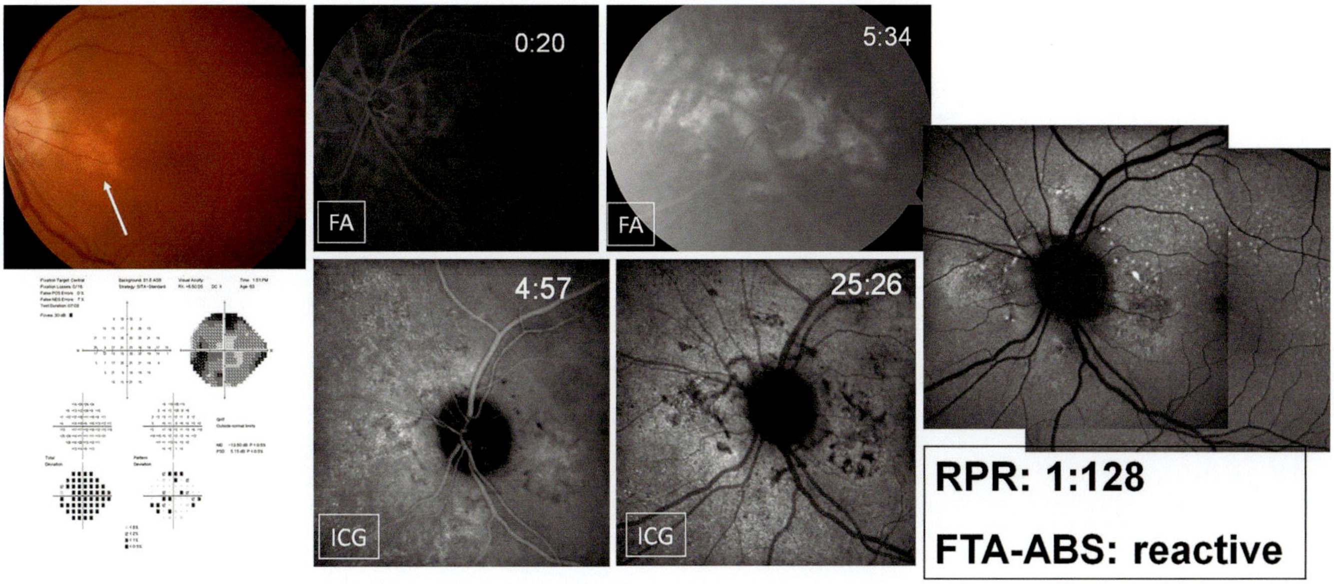

Fig. 48.1 Acute syphilitic posterior placoid chorioretinopathy. A 63-year-old patient presented with enlargement of the blind spot and decreased vision (20/160). Color fundus shows multiple grayish-white lesions at the deeper level (*arrow*) and corresponding scotoma. Fluorescein angiography (FA) shows early hypofluorescence and late staining. Indocyanine green angiography (ICG) shows hypofluorescence, and fundus autofluorescence (FAF) shows characteristic hyperautofluorescent dots in the peripapillary and macular area

segment/RPE junction, loss of the external limiting membrane, serous retinal detachment, and punctate hyperreflectivity in the choroid may be seen. The findings are reversible with treatment; however variable long-term vision loss can result from localized disruption of the ELM, ellipsoid zone, and outer segment/RPE junction.

FA: early central hypofluorescence, followed by progressive hyperfluorescence in the area of the lesion, an active leading edge, identified angiographically as an increased late leakage and staining at the margin may be seen. Perivenous staining and late leakage from the optic disk may be present. Punctate hypofluorescence in a classic leopard spot pattern may be seen after disease resolution. ICG shows hypofluorescent spots corresponding to lesions.

Differential diagnosis: APMPPE, serpiginous choroiditis, and viral retinitis.

- **Pigmentary retinopathy** can occur in both congenital and acquired syphilis and is characterized by attenuated retinal and choroidal vessels, optic disc pallor with sharp margins, and clumpy pigmentary deposits along with chorioretinal scars, mimicking advanced retinitis pigmentosa. Typical bony-spicule pigmentation is uncommon.

- **Treatment:**
  - Ocular syphilis is treated as neurosyphilis, with intravenous aqueous crystalline penicillin G 18–24 million units per day (either as a continuous infusion or 3–4 million units every 4 h) for 10–14 days.
  - Rarely a Jarisch-Herxheimer reaction can occur as a response to the treatment-induced breakdown of the spirochete, which results in the release of toxins and cytokines. In the setting of ocular syphilis, the Jarisch-Herxheimer reaction has been described to cause decreased visual acuity, a swollen optic disc with macular edema, and cotton wool spots. On FA, it may be seen as extensive leakage around the right optic disc and areas of non-perfusion in the inferotemporal quadrant.

Further Reading

Gass JD, Braunstein RA, Chenoweth RG. Acute syphilitic posterior placoid chorioretinitis. Ophthalmology 1990;97(10):1288-97.

Pichi F, Ciardella AP, Cunningham ET Jr, Morara M, Veronese C, Jumper JM, Albini TA, Sarraf D, McCannel C, Voleti V, Choudhry N, Bertelli E, Giuliari GP, Souied E, Amer R, Regine F, Ricci F, Neri P, Nucci P. Spectral domain optical coherence tomography findings in patients with acute syphilitic posterior placoid chorioretinopathy. Retina 2014;34(2):373-84.

Vaze A, De Angelis RE, Simões M, Arantes TE, Moreto R, Oliver GF, Smith JR, Furtado JM. Optical coherence tomography findings in ocular syphilis involving the posterior segment of the eye. Ocul Immunol Inflamm 2021; 15:1-7.

Zhang T, Zhu Y, Xu G. Clinical features and treatments of syphilitic uveitis: a systematic review and meta-analysis. J Ophthalmol 2017; 6594849.

CHAPTER 49

Autoimmune Retinopathy

Fabiha Ahmed, Stephen H. Tsang, and Tarun Sharma

General Features: Autoimmune Retinopathies 250
Paraneoplastic Retinopathy 250
Non-paraneoplastic Autoimmune Retinopathy 250
Further Reading 252

F. Ahmed
Columbia University, New York, NY, USA

S. H. Tsang (✉) · T. Sharma
Department of Ophthalmology, Columbia University,
New York, NY, USA
e-mail: sht2@cumc.columbia.edu

S. H. Tsang et al. (eds.), *Atlas of Inherited Retinal Diseases*, Advances in Experimental Medicine and Biology 1467, https://doi.org/10.1007/978-3-031-72230-1_49

General Features: Autoimmune Retinopathies

- The autoimmune retinopathies (AIRs) are a group of inflammatory-mediated retinopathies that present with unexplained visual loss (both central and peripheral), visual field defects, usually a ring scotoma, photoreceptor dysfunction as evident on electroretinography (ERG), and circulating autoantibodies against retinal antigens. The fundus may be normal or may show vascular attenuation, retinal atrophy with or without pigmentary changes or waxy pallor of the optic disc, and no or minimal inflammatory cells.
- The AIRs can be categorized into two groups:
  - Paraneoplastic retinopathy or paraneoplastic autoimmune retinopathy, which includes cancer-associated retinopathy (CAR), melanoma-associated retinopathy (MAR), paraneoplastic vitelliform maculopathy (PVM), and bilateral diffuse uveal melanocytic proliferation (BDUMP)
  - Non-paraneoplastic autoimmune retinopathy

Paraneoplastic Retinopathy

- Paraneoplastic retinopathies can occur without systemic spread of the cancer tumor and may even precede a clinically evident tumor.
- **Cancer-associated retinopathy (CAR):** Patients typically experience rapidly progressive loss of central and peripheral vision, along with positive visual phenomena (photopsias or sparkles) and a ring scotoma, photosensitivity, impaired color vision, and night blindness. The fundus may show arterial narrowing with no pigment spiculation, though some pigment alteration can occur. ERG shows extinguished response or severely reduced a-waves and b-waves.
  - The most common malignancy associated with CAR is small cell carcinoma of the lung, but other malignancies that may be involved include those of the colon, uterus, cervix, breast, prostate/bladder, and systemic lymphoma.
  - In about 60% of patients with CAR, antibodies are produced against a 23-kDa calcium channel photoreceptor protein called *recoverin*.
  - Prednisone and immunomodulators have been used to stabilize visual function in some patients with CAR.
- **Cutaneous melanoma-associated retinopathy (MAR):** This is different from CAR in several ways: It is nonprogressive, causes only central visual loss, and is associated with vitiligo (20%); ERG shows electronegative waveforms (selective reduction of b-wave), and MAR usually does not manifest before the clinical diagnosis of cutaneous malignant melanoma.
  - Antibodies against transducin, enolase, and aldolase A and C have been reported.
  - Photoreceptor functions are intact; the pathology lies in the bipolar cells.
- **Paraneoplastic vitelliform maculopathy (PVM):** Multiple oval, yellowish lesions are seen at the RPE levels along the arcades and surrounding the disc. Patients have mild visual loss (20/40–20/100) with mild night blindness and photopsias; fields are usually normal and ERG response is variable. An abnormal Arden ratio was noted in less than 50% of cases, suggesting relatively normal RPE function. The clinical picture is similar to acute exudative polymorphous vitelliform maculopathy (AEPVM).
  - PVM is associated with choroidal melanoma, lung carcinoma, and multiple myeloma.
  - Retinal autoantibodies have been noted against bipolar cells, enolase, rod outer segment proteins, and bestrophin.
- **Bilateral diffuse uveal melanocytic proliferation (BDUMP):** This is characterized by multiple, slightly elevated (about 2 mm), pigmented or nonpigmented uveal melanocytic tumors of the choroid, along with diffuse thickening of the uveal tract; multiple, round or oval, subtle red patches at the RPE level; rapidly progressive posterior subcapsular cataract; and exudative retinal detachment. Fluorescein angiography shows multiple areas of early hyperfluorescence.
  - Findings may precede the detection of systemic malignancies (e.g., gastrointestinal, genitourinary, pulmonary, and non-Hodgkin lymphoma).
  - Plasmapheresis to remove the circulating ectopic peptides (presumed) has been used in some patients to stabilize the vision while treatment for the systemic malignancy is underway.

Non-paraneoplastic Autoimmune Retinopathy

- Non-paraneoplastic AIR is almost always bilateral, though asymmetric, with female preponderance (about 65%). The mean age of onset is between 51 and 56 years.

These patients have no prior history of ocular symptoms, but start noticing photopsias, scotomas, acute or subacute visual loss, photoaversion, color vision changes, and night blindness. The fundus is either normal-appearing or shows nonspecific retinal degeneration or RPE atrophy or mottling, attenuated blood vessels, and waxy pallor of the disc, with or without pigmentary changes. ERG shows abnormalities in photoreceptors and bipolar cells.

- Fluorescein angiography may show vascular leakage or staining and/or cystoid macular edema. Spectral domain OCT (SD-OCT) may show disruption of the ellipsoid zone, loss of photoreceptor layer, loss of external limiting membrane, and thinning of the outer nuclear layer (Figs. 49.1 and 49.2). On fundus autofluorescence (FAF), a hyperautofluorescent ring around the fovea is seen in some cases. There is absence of any overt intraocular inflammation.

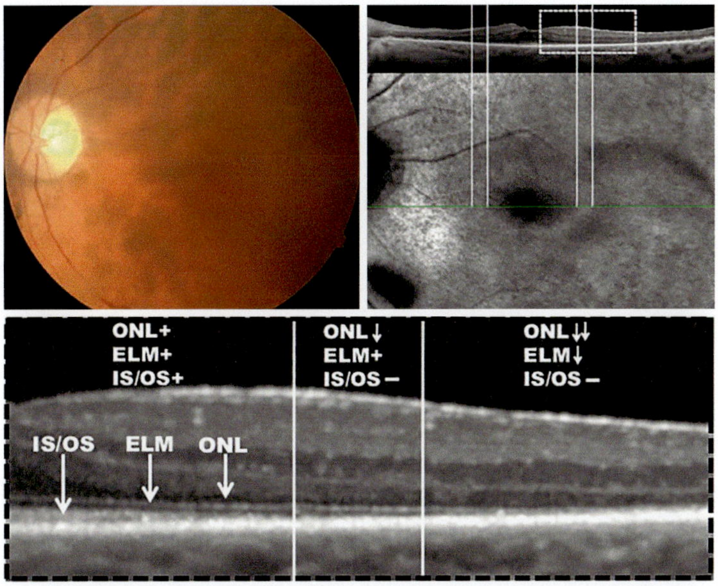

Fig. 49.1 Non-paraneoplastic autoimmune retinopathy. Note attenuation of vessels, thinning of the outer nuclear layer (ONL), disappearance of the external limiting membrane (ELM), and loss of the ellipsoid zone. IS/OS—inner segment/outer segment junction

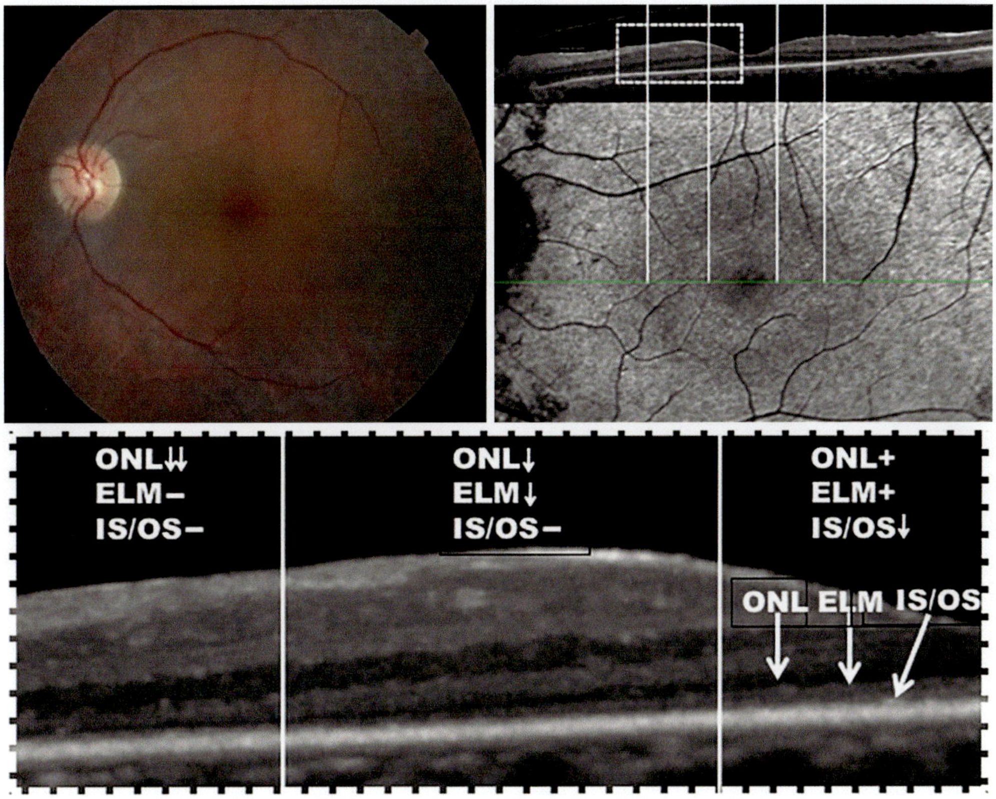

Fig. 49.2 Non-paraneoplastic autoimmune retinopathy. Note mild narrowing of the vessels, thinning of the ONL, disappearance of the ELM, and loss of the ellipsoid zone. IS/OS—inner segment/outer segment junction

- There is a positive history of autoimmune disease in the patient or the patient's family.
- These patients, like those with paraneoplastic retinopathy, demonstrate various antiretinal antibodies, including antirecoverin or antibodies against inner plexiform layer or Müller cells (35 kDa).

Further Reading

Heckenlively JR, Fawzi AA, Oversier J, Jordan BL, Aptsiauri N. Autoimmune retinopathy: patients with antirecoverin immunoreactivity and panretinal degeneration. Arch Ophthalmol 2000;118:1525–33.

Rahimy E, Sarraf D. Paraneoplastic and non-paraneoplastic retinopathy and optic neuropathy: evaluation and management. Surv Ophthalmol 2013;58:430–58.

CHAPTER 50

Drug-Induced Retinal Toxicity

Charu Vyas, Anna Jolia, Stephen H. Tsang, Tarun Sharma, and Vlad Diaconita

General Features . . . 254
Damage to RPE and Photoreceptor Complex . . . 254
Damage to the Vascular Bed or Microvasculopathy/Occlusion . . . 260
Damage to Ganglion Cell Layer or Optic Nerve . . . 260
Drugs Causing Cystoid Macular Edema . . . 260
Drugs Causing Crystalline Retinopathy . . . 260
Drugs Causing Uveitis . . . 262
Further Reading . . . 262

C. Vyas
Vagelos College of Physicians and Surgeons, New York, NY, USA

A. Jolia · S. H. Tsang · T. Sharma · V. Diaconita (✉)
Department of Ophthalmology, Columbia University, New York, NY, USA
e-mail: vd2374@cumc.columbia.edu

S. H. Tsang et al. (eds.), *Atlas of Inherited Retinal Diseases*, Advances in Experimental Medicine and Biology 1467, https://doi.org/10.1007/978-3-031-72230-1_50

General Features

- Drug-induced retinal toxicity can occur from the use of systemic, intravitreal, or topical medications. Despite the presence of the blood-ocular barrier, the retina is vulnerable to toxic effects of systemic medications leading to dysfunction and retinal degeneration.
- These toxicities can be categorized as damage to the retinal pigment epithelium (RPE) and photoreceptor complex, vascular damage, ganglion cell or optic nerve damage, cystoid macular edema, crystalline retinopathy, uveitis, changes in color vision and electroretinography (ERG), and other miscellaneous effects.

Damage to RPE and Photoreceptor Complex

- Chloroquine/hydroxychloroquine
  - *Purpose:* Used for malaria, amoebiasis, rheumatoid arthritis (RA), systemic lupus erythematosus (SLE), and related inflammatory and dermatological conditions.
  - *Mechanism:* Both medications bind to melanin in the RPE and uveal tissue and interfere with metabolic function.
  - *Chloroquine:* The risk of toxicity is low with dosage 3 mg/kg per day, but the risk increases with a cumulative dose of 460 g.
  - *Hydroxychloroquine:* The risk of toxicity is low with dosage 5 mg/kg per day but increases with a cumulative dose of 1000 g. Other risk factors for toxicity are age >60 years, obesity, short stature, age-related macular degeneration (AMD), kidney or liver disease, or concomitant use of tamoxifen.
  - *Onset*: The earliest manifestation of toxicity is bilateral paracentral scotoma or loss of ellipsoid zone (EZ) on SD-OCT (flying saucer sign). Later, subtle RPE stippling, and loss of foveal reflex are seen, followed by classic bull's-eye maculopathy. The end stage shows panretinal degeneration, like retinitis pigmentosa (RP).
  - *Imaging*:
    Fluorescein angiography (FA): RPE window defects or pigmentary changes.
    Fundus autofluorescence (FAF): perifoveal hypoautofluorescence with a border of hyperautofluorescent ring.
    OCT: Atrophy of outer nuclear layer (ONL) and disruption of EZ. Also, perifoveal thinning of the inner plexiform layer (IPL) and ganglion cell layer (GCL) sometimes even precede photoreceptor damage.
    Multifocal ERG (mfERG): reduced paracentral waveforms.
  - *Screening Recommendations*: Baseline screening and follow-up include fundus examination/photo, SD-OCT, and automated threshold field testing (Humphrey, white or red object, 10-2 protocol); FAF and mfERG are also used for assessment. Annual screening (in particular after 5 years of use) should be done with Humphrey visual field 10-2 and SD-OCT.
  - *Practice Recommendations*: Treatment should be stopped at the first sign of toxicity; otherwise toxic effects continue to progress despite discontinuation of the drug.
- **Phenothiazines**
  - *Mechanism*: Both chlorpromazine and thioridazine accumulate in the melanin of RPE and uveal tissue.
  - *Chlorpromazine:* Causes hyperpigmentation in the eyelid, conjunctiva, cornea, or anterior capsule of the lens; retinal toxicity is rare.
  - *Thioridazine:* Toxicity is dependent on the daily dose rather than on the cumulative dose; it is rare at doses of 800 mg/day or lower.
  - *Onset*: Months
  - *Toxicity*: Symptoms include blurred vision, color vision abnormality (reddish/brownish vision), and night blindness. In the early stage, there is a coarse RPE stippling in the posterior pole; at a later stage, nummular areas of RPE/choriocapillaris loss are seen from the posterior pole to mid-periphery. The late-stage mimics choroideremia or Bietti crystalline dystrophy (BCD); vascular attenuation and optic atrophy are seen.

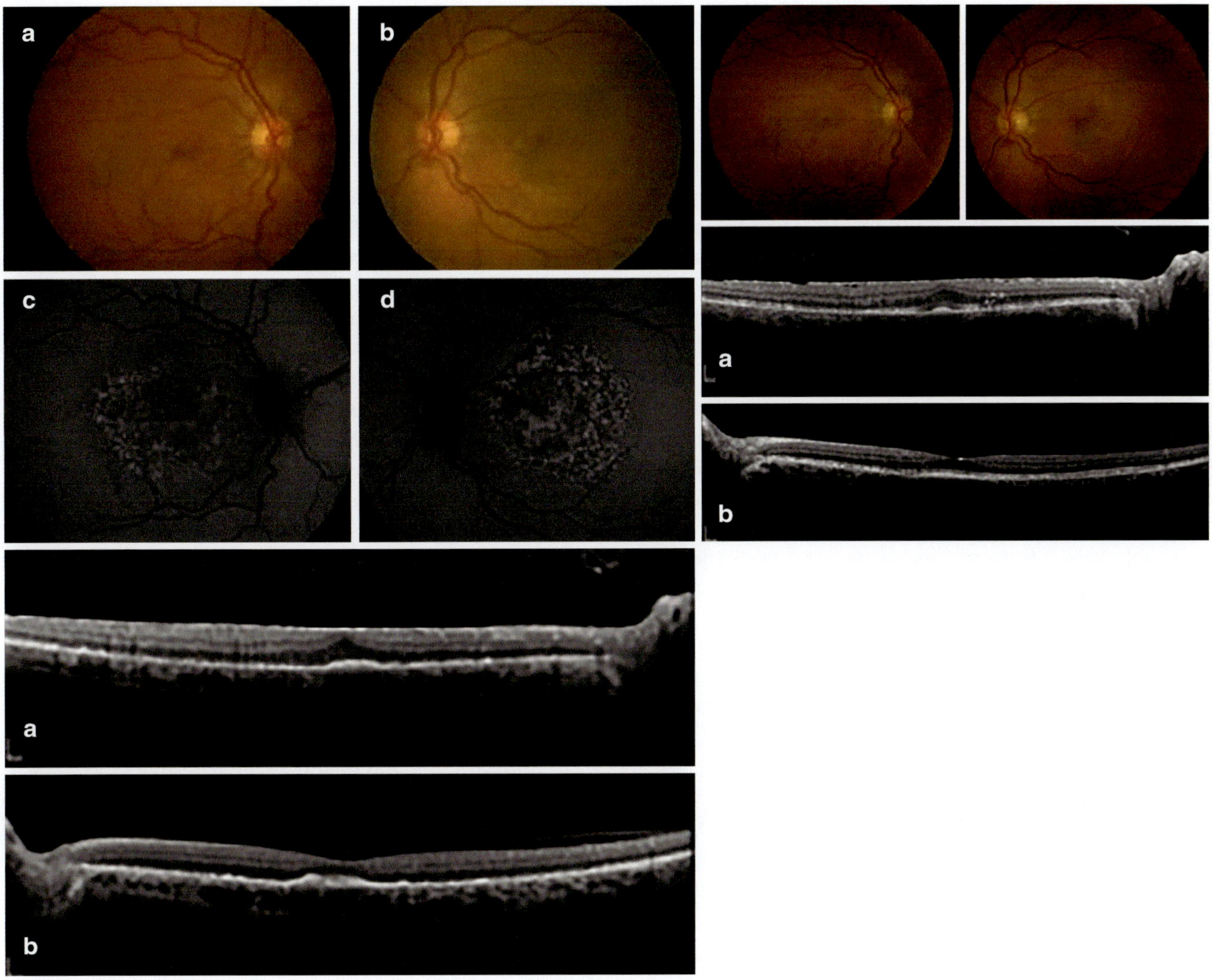

Fig. 50.1 Retinal toxicity due to deferoxamine. **Left,** color photos show pigmentary/RPE changes at the macula; FAF shows a stippled pattern of hypoautofluorescence and hyperautofluorescence; OCT shows hyperreflective material in the subretinal space with disruption of the ellipsoid zone. **Right,** 5 months after discontinuation of the deferoxamine, RPE alterations persist. Note some hyperreflective dots in the outer retina (nasal macula)

- *Screening recommendations*: Baseline evaluation followed by yearly review or sooner for higher doses (>600 mg/day) is recommended, including CV, CFP, OCT, and FAF.
- **Clofazimine**
  - *Purpose*: Used to treat dapsone-resistant leprosy and autoimmune disorders such as psoriasis and lupus.
  - *Toxicity*: bull's-eye maculopathy (BEM) in AIDS patients using dosages between 200 and 300 mg/d.
- **Deferoxamine**
  - *Purpose*: Used to treat iron toxicity/overload, as an iron-chelating agent
  - *Onset*: Months
  - *Toxicity*: Reticular or vitelliform form changes and/or macular edema due to RPE pump failure (Fig. 50.1).
  - *Screening Recommendations*: Baseline evaluation, including color vision, visual field, color fundus imaging, OCT, and FAF, followed by screening at 6-month intervals. Electrophysiology may be useful in determining retinal dysfunction.
- **Dideoxyinosine (DDI)**
  - *Purpose*: Used to treat HIV/AIDS
  - *Toxicity:* Causes mitochondrial toxicity resulting in damage to the optic nerve and RPE; peripheral field loss occurs with concentric loss/mottling of RPE (areas of chorioretinal atrophy), beyond arcades to mid-periphery, bilaterally symmetrical.
- **MEK inhibitors**
  - *Purpose*: Used to treat metastatic cancers
  - *Toxicity:* Multifocal serous retinal detachments (similar to central serous retinopathy [CSCR])
- **Sildenafil**
  - *Purpose*: Used to treat pulmonary arterial hypertension and erectile dysfunction
  - *Toxicity:* Serous macular detachment, possibly due to dilatation of choroidal vasculature. Transient blue tinting of the vision, and temporarily abnormal ERG, a delayed cone b-wave implicit time.
- **Corticosteroid**
  - Purpose: Anti-inflammatory
  - *Toxicity:* Serous macular detachment, complications like cataracts and glaucoma are common with corticosteroids.
- **Poppers/Alkyl nitrates**
  - *Purpose*: Used as recreational drugs for euphoric effect and enhanced sexual arousal
  - *Toxicity:* Disruption of EZ/foveal cones involvement; a yellow spot on fundus with central scotoma/photopsia
- **Cisplatin and carmustine**
  - *Purpose*: Used for metastatic breast cancer and malignant gliomas
  - *Toxicity:* Pigmentary retinopathy at macula, or cotton-wool spots, hemorrhages, exudates, and disc swelling or arterial occlusion, vasculitis, and papillitis
- **Pentosan Polysulfate Maculopathy**
  - *Purpose*: Used to treat interstitial cystitis
  - *Toxicity:* Disruption of the choriocapillaris early in the disease followed by central macular atrophy and CNVM in late stages. Toxicity is related to length of exposure to the drug. Progression is though to occur even after discontinuation of the drug (Figs. 50.2 and 50.3).

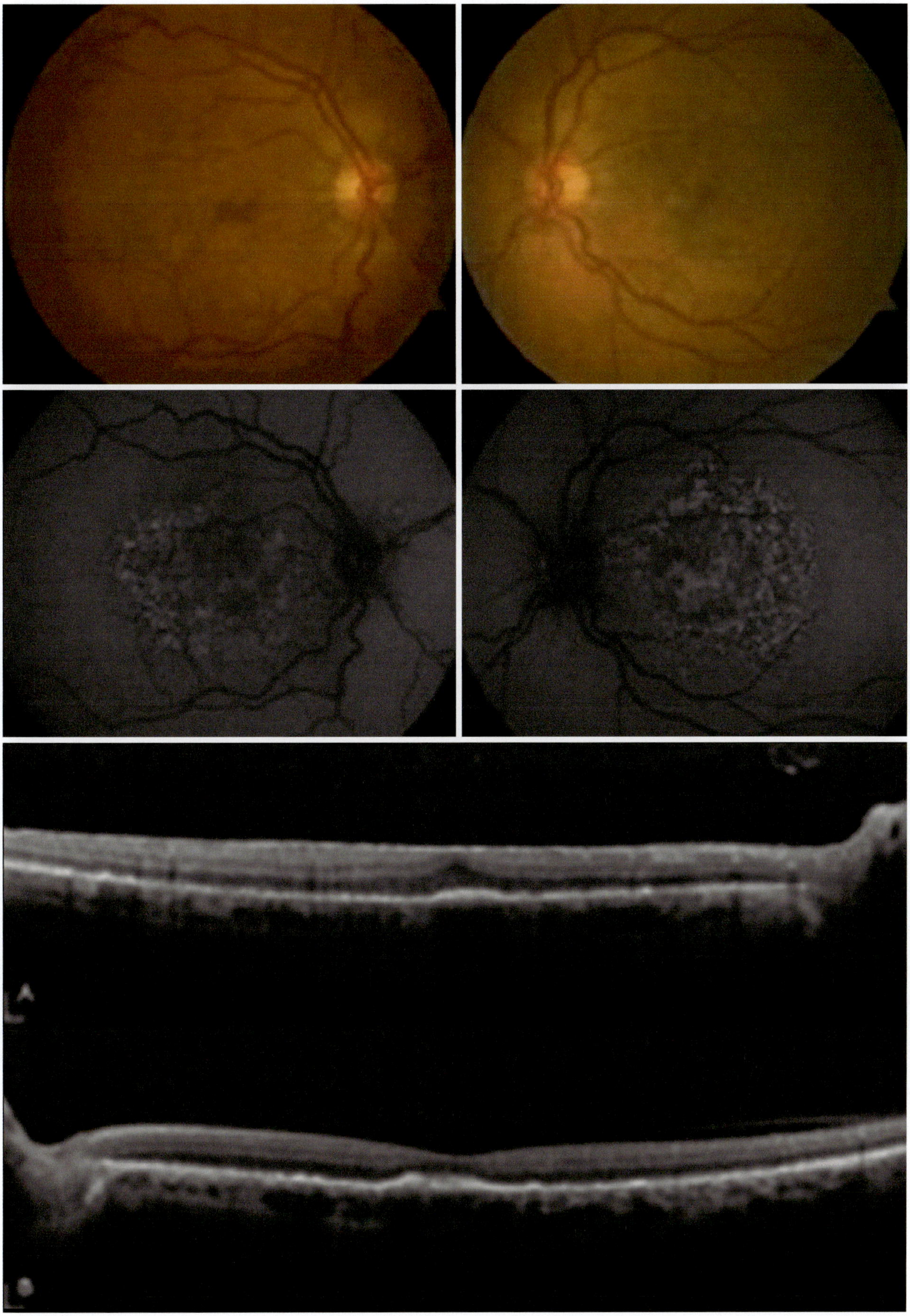

Fig. 50.2 Retinal toxicity due to deferoxamine (see previous publication)

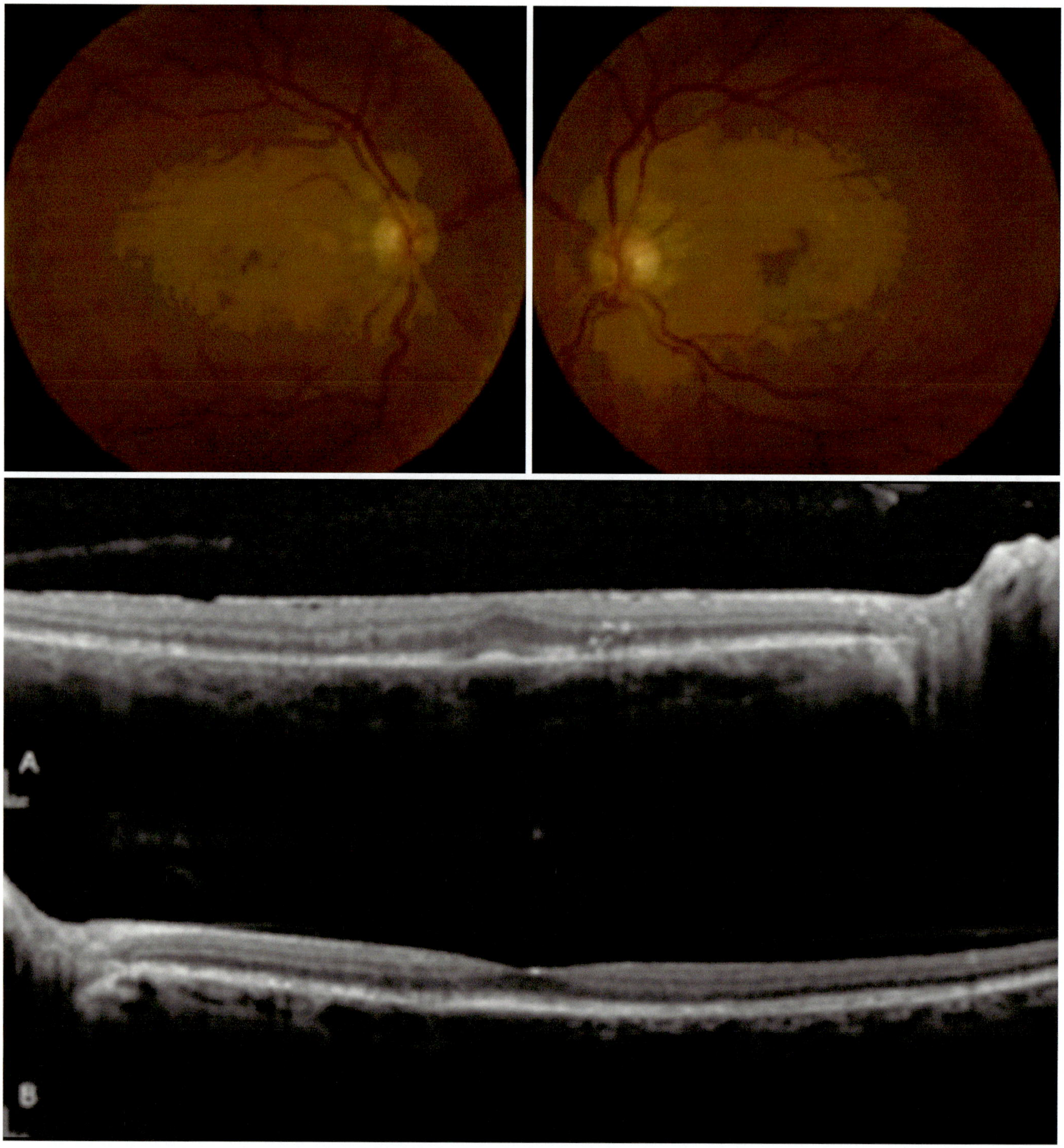

Fig. 50.2 (continued)

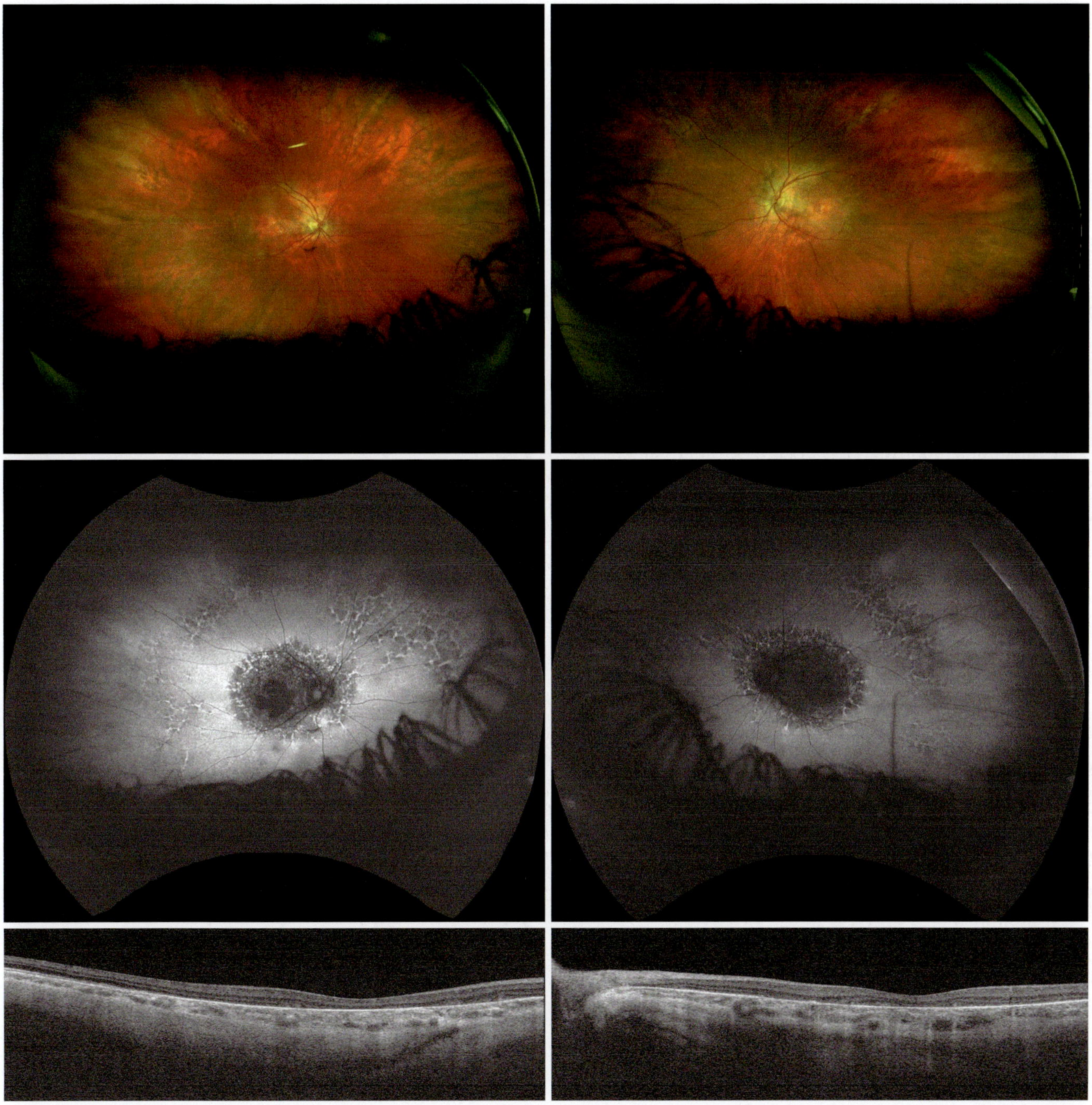

Fig. 50.3 Retinal toxicity due to pentosan polysulfate. Panels show right and left eye optos, FAF and OCT macula of atrophy following pentosan use. Visual acuity decrease below 20/100 in each eye. Extensive genetic workup negative

Damage to the Vascular Bed or Microvasculopathy/ Occlusion

- **Interferon alfa-2a**
  - *Purpose*: Used for viral hepatitis C
  - *Onset*: Months
  - *Toxicity:* Multiple cotton-wool spots and hemorrhages
  - *Screening Recommendations:* Baseline evaluation including color fundus and OCT before treatment initiation, followed by three monthly assessments.
- **Ergot alkaloids**
  - *Purpose*: Used for migraine
  - *Toxicity:* Retinal vasoconstriction
- **Gemcitabine**
  - *Purpose*: Used for non-small-cell lung carcinoma or cancer of breast, ovary, or pancreas
  - *Toxicity:* Purtscher-like retinopathy with non-perfusion
- **Oral contraceptives**
  - *Toxicity:* Vascular occlusions
- **Intraocular gentamicin/tobramycin/amikacin**
  - *Toxicity:* Macular ischemia, infarction
- **Intracameral vancomycin**
  - *Toxicity:* Hemorrhagic occlusive vasculitis, immune-mediated, especially when second eye is operated in proximity
- **Talc**
  - Retinopathy common in intravenous drug users
  - *Toxicity:* Perifoveal yellow-white glistening crystals, later leading to ischemic retinopathy

Damage to Ganglion Cell Layer or Optic Nerve

- **Quinine sulfate**
  - *Purpose*: Used for nocturnal muscle cramps or restless leg syndrome, or as an antimalarial
  - *Toxicity:* Safe if the dose is <2 g; toxic if the dose is >4 g; fatal if >8 g
  - Acute toxicity is like central retinal artery occlusion, cherry red spot, GCL thickening and hyperreflective on OCT, negative waveform on full-field ERG. Blindness is permanent.
- **Methanol**
  - *Purpose*: Solvent, anti-freeze, recreational
  - *Onset*: Within hours of high dose
  - *Toxicity:* Causes acute blindness, optic nerve head and macular edema; later, optic nerve atrophy due to diffuse loss of the GCL.
  - *Screening Recommendation:* Baseline evaluation should include color vision, visual field, color fundus imaging, and OCT. Detailed history should be taken to identify additional factors associated with toxic/nutritional optic neuropathy.

Drugs Causing Cystoid Macular Edema

- Topical epinephrine
- High dose of nicotinic acid/niacin to treat dyslipidemia
- Paclitaxel/docetaxel to treat cancer of breast, lung, or prostate
- Glitazones, rosiglitazone, and pioglitazone, as hypoglycemic agents
- Deferoxamine as iron-chelating agent
- Topical latanoprost

Drugs Causing Crystalline Retinopathy

- **Tamoxifen**
  - *Purpose*: Used as an adjuvant therapy (as antiestrogen agent) in the treatment of estrogen-receptor–positive breast carcinoma
  - *Onset*: >1 year
  - *Toxicity:* Rare if the daily dose is 20 mg/day; may occur with daily dose of 60–100 mg/day or cumulative dose >100 g. There are brilliant yellowish-white crystal deposits in the inner retina, around the fovea, and cystoid macular edema (CME) and loss of EZ on OCT.
  - *Screening Recommendation:* Baseline evaluation within the first year followed by screening every 3–6 months if symptomatic, including color vision, color fundus imaging, and OCT.
- **Canthaxanthin**
  - *Purpose*: Used as an oral tanning agent or for skin pigmentation in vitiligo

– *Toxicity:* With cumulative dose >19 g over 2 years, causes glistening, ring-shaped yellowish-orange crystals in the inner retina, a doughnut pattern around the macula or in the juxtapapillary region. They do not cause visual loss and disappear after the treatment is discontinued.
- **Methoxyflurane**
  – *Purpose*: Used as an inhalational anesthetic agent; no longer used in the United States
  – *Toxicity:* Numerous yellowish-white, punctate deposits in the macula and periarterially, deposited in both RPE and inner retina (more common in those with renal insufficiency)

West African Crystalline Retinopathy

- First described in 2003 in a case series of patients from Nigeria
  – Usually asymptomatic and present in patients from West African regions, where the Kola Nut intake has been postulated to be the causative agent
  – Visual acuity remains good in most cases
  – Genetic predisposition has been hypothesized, but not identified (Fig. 50.4)

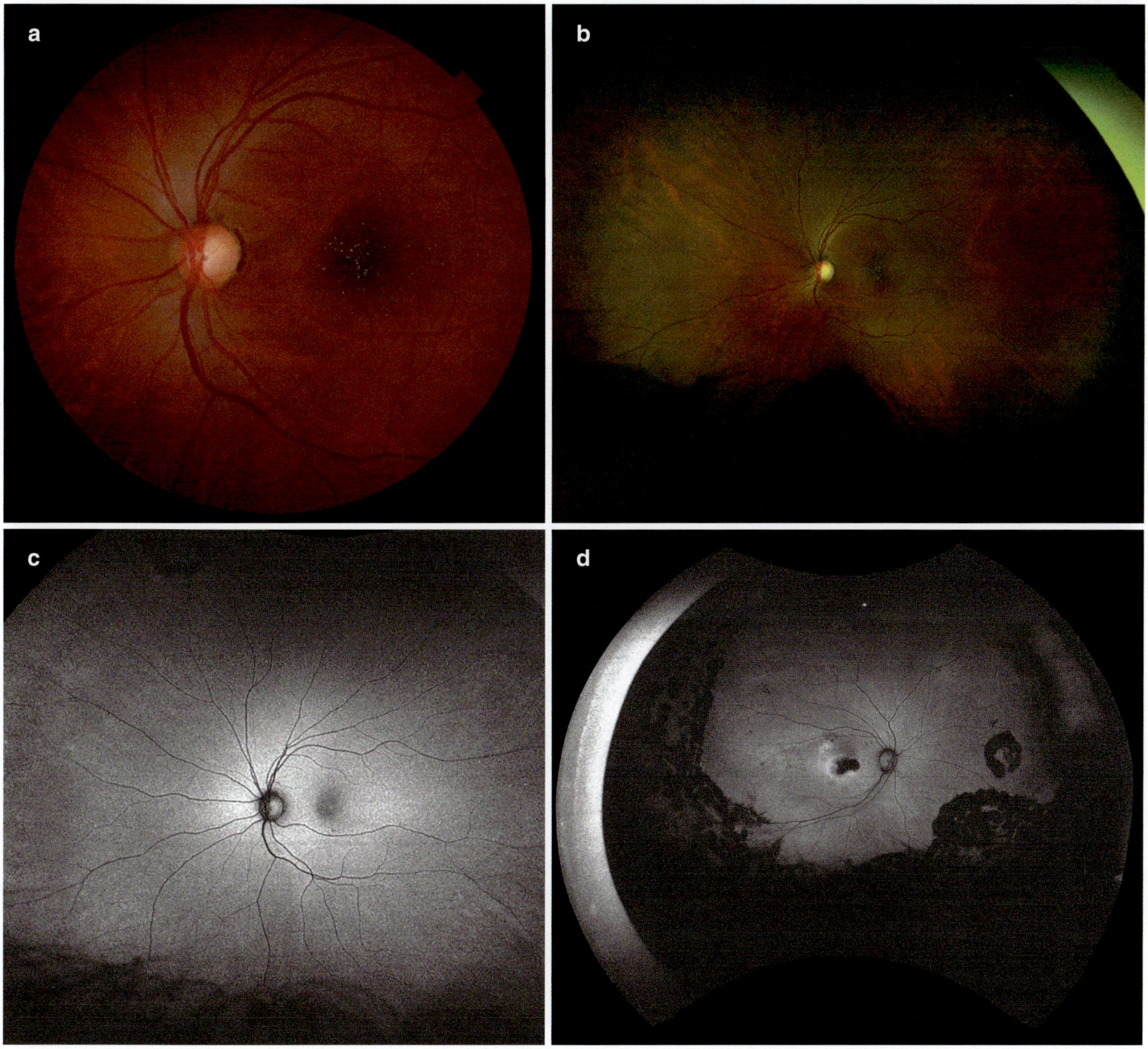

Fig. 50.4 West African crystalline retinopathy: Retinal crystals presumed to be secondary to longstanding Kola Nut intake in a patient from West Africa. Panel (**a**) shows color photo of left eye, (**b**) and (**c**) are Optos pseudocolor and FAF of the left eye. Panels (**d**) and (**e**) are of the contralateral eye, which underwent retinal detachment surgery

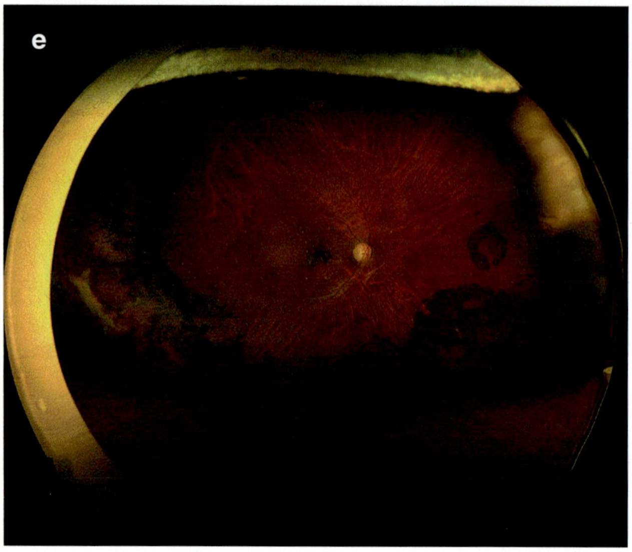

Fig. 50.4 (continued)

Drugs Causing Uveitis

- Alendronate (oral): For the treatment of osteoporosis and Paget's disease of the bone.
- BCG Vaccine (percutaneous, intradermal, intravesical)
- Bevacizumab (intravitreal): Antineoplastic agent used in cancer therapy and for treatment of neovascular age-related macular degeneration.
- Brimonidine (topical): Used topically for the treatment of persistent erythema associated with rosacea.
- Cidofovir (intravenous, intravitreal): For cytomegalovirus (CMV) infection.
- Pamidronate (intravenous): Used for the treatment of moderate to severe hypercalcemia associated with malignant neoplasms, treatment of moderate to severe Paget's disease of bone, and used as an adjunct to antineoplastic therapy for the treatment of osteolytic bone metastases of breast cancer and osteolytic lesions of multiple myeloma.
- Prostaglandin Analogues (topical): Used in treatment of ocular hypertension, glaucoma hypotrichosis of eyelashes.
- Ranibizumab (intravitreal): Used for treatment of neovascular age-related macular degeneration, macular edema following retinal vein occlusion, diabetic macular edema, and diabetic retinopathy.
- Rifabutin (oral): For the prevention and treatment of disseminated *Mycobacterium avium* complex (MAC).
- Sulfonamides (oral): Broad category of drugs with a sulfonamide group used in certain antibiotics and diuretics.

Further Reading

Di Nicola M, Barteselli G, Dell'Arti L, Ratiglia R, Viola F. Functional and structural abnormalities in deferoxamine retinopathy: a review of the literature. Biomed Res Int. 2015;2015:249617.

Hanif, A. M., Shah, R., Yan, J., Varghese, J. S., Patel, S. A., Cribbs, B. E., O'Keefe, G., Hendrick, A. M., Shantha, J. G., Hubbard 3rd, G. B., Patel, P. S., Rao, P., Yeh, S., & Jain, N. (2019). Strength of association between pentosan polysulfate and a novel maculopathy. Ophthalmology, 126(10), 1464–1466. doi: https://doi.org/10.1016/j.ophtha.2019.04.024

Marmor MF. Sildenafil and ophthalmology. Arch Ophthalmol 1999;117:518–9.

Marmor MF, Kellner U, Lai TY, Melles RB, Mieler WF; American Academy of Ophthalmology. Recommendations on screening for chloroquine and hydroxychloroquine retinopathy (2016 revision). Ophthalmology 2016;123:1386–94.

Pearce WA, Chen R, Jain N. (2018). Pigmentary maculopathy associated with chronic exposure to pentosan polysulfate sodium. Ophthalmology, 125(11), 1793–1802.

(2024). Risk Factors for Hydroxychloroquine Retinopathy and Its Subtypes JAMA Network Open. 7(5). e2410677. https://doi.org/10.1001/jamanetworkopen.2024.10677.

Sarraf D, Ceron O, Rasheed K, Drenser KA, Casey R. West African crystalline maculopathy. Arch Ophthalmol 2003;121(3):338-342. doi:https://doi.org/10.1001/archopht.121.3.338

Shah, R., Simonett, J. M., Lyons, R. J., Rao, R. C., Pennesi, M. E., Jain, N. (2020). Disease course in patients with pentosan polysulfate sodium-associated maculopathy after drug cessation. JAMA Ophthalmol 138(8), 894–900. doi: https://doi.org/10.1001/jamaophthalmol.2020.2349

CHAPTER 51

Noninfectious Uveitis Syndromes

Kristina J. Hartung, Omar Moussa, Anna Jolia, Naomi Rivka Goldberg, Royce Chen, Vlad Diaconita, and Jose Pulido

General Features 264
Further Reading 271

K. J. Hartung
Department of Ophthalmology, University Medical Centre Ljubljana, Ljubljana, Slovenia

Faculty of Medicine, University of Ljubljana, Ljubljana, Slovenia

O. Moussa
NYU Langone, New York, NY, USA

A. Jolia
Brooklyn, NY, USA

N. R. Goldberg
Northwell Health, New York, NY, USA

R. Chen
Columbia University, New York, NY, USA

V. Diaconita
Department of Ophthalmology, Edward S. Harkness Eye Institute, NewYork-Presbyterian Hospital Columbia University, New York, NY, USA

J. Pulido (✉)
Department of Ophthalmology, Sidney Kimmel Medical College at Thomas Jefferson University, New York, NY, USA
e-mail: Pulido.Jose@mayo.edu

S. H. Tsang et al. (eds.), *Atlas of Inherited Retinal Diseases*, Advances in Experimental Medicine and Biology 1467, https://doi.org/10.1007/978-3-031-72230-1_51

General Features

Noninfectious uveitis syndromes are a heterogeneous group of disorders affecting the retina and choroid. Their etiology is unknown, but some follow a flu-like illness. Autoimmune pathogenesis has also been postulated. Patients are usually affected from the 2nd to 5th decade, and some syndromes show a female preponderance. The presenting symptoms include photopsias, floaters, decreased visual acuity, and visual field loss. Fundus lesions are bilateral (apart from MEWDS) but may be asymmetric and consist of multiple, discrete, yellow-white, well-circumscribed lesions at the level of the outer retina, RPE, and choroid. Multimodal imaging plays an important role in the diagnosis, as each syndrome exhibits a different pattern. Lesions are often more numerous on imaging than clinically evident. Electrophysiology is within normal limits in some syndromes but can be markedly attenuated in others. Prognosis is widely variable and ranges from a self-limiting course with complete resolution of symptoms to relentless progression of visual loss.

- **AZOOR** (Acute zonal occult outer retinopathy)
  - AZOOR is a presumed inflammatory disorder with outer retinal dysfunction.
  - Acute onset of unilateral photopsia, nasal field loss, and an enlarged blind spot. Later, the other eye is involved in nearly three-fourths of patients. The central vision remains good in most cases.
  - Patients are usually young/middle-aged women with myopia.
  - Fundus: May be normal in the beginning but may show a grayish-white line at the border of the normal and involved retina, usually in the peripapillary area. This line disappears within weeks and is replaced with an orange zone. With time, retinal vessels attenuate, and a large zone of retinal pigment epithelium (RPE) depigmentation appears, resembling a sector retinitis pigmentosa (RP) or unilateral or asymmetric RP (Fig. 51.1).
  - Mild vitritis may be present, relative afferent pupillary defect (RAPD) is present in up to 25% of cases.
  - Fluorescein angiography (FA): In the acute stage, may show evidence of leakage/staining from retinal blood vessels and optic nerve; in the late stage, RPE window defects may be seen.
  - Visual field: Areas of scotomas (usually peripheral and temporal) that enlarge with time and an enlarged blind spot.
  - Indocyanine green angiography (ICG): Area of late hypofluorescence.
  - Optical coherence tomography (OCT): Outer retinal atrophy and ellipsoid zone disruption are present. A trizonal pattern of sequential involvement of the outer retina, RPE and choroid, with zonal progression over time, is characteristic.
  - Fundus autofluorescence (FAF): Shows a characteristic trizonal pattern: hypoautofluorescence (hypo AF) in the involved area, speckled hyper AF at the border, and normal AF beyond the delineating line or border (Figs. 51.2 and 51.3).
  - Electrophysiology: globally affected full-field ERG can be seen, with a delayed 30 Hz response. A reduced light rise may be present on EOG.
  - Prognosis: Usually stabilizes in 6 months but recurs in one-third of patients. The visual acuity of most patients remains in the 20/40 range.
  - Differential diagnosis: RP, cancer-associated retinopathy (CAR), melanoma-associated retinopathy (MAR), autoimmune retinopathy (AIR).
- **MEWDS** (multiple evanescent white dot syndrome)
  - Acute onset of unilateral blurred vision, temporal photopsias, temporal scotoma, and an enlarged blind spot.
  - Predominantly affects young myopic women (75%).
  - Anterior chamber inflammation or mild vitritis is present, along with an afferent pupillary defect. One-third of patients have a flu-like prodrome.
  - *Fundus:* multiple small, grayish-white or yellowish-white dots at the RPE level are present in the posterior pole of one eye. The fovea shows an unusual granularity with tiny yellowish-orange flecks characteristic of MEWDS.
  - *FA:* Punctate hyperfluorescent dots in a wreathlike pattern; disc may show mild leakage/staining.
  - *ICG:* Hypofluorescent spots, which are greater in number than what is observed clinically or on FA.
  - *OCT:* There may be protrusion of hyperreflective material towards the ONL, disruption in the ellipsoid zone, and increased reflectivity in the choroid in the acute phase.
  - *FAF:* Hyper AF spots, usually more numerous than visible clinically.
  - *ERG:* Shows reduced a-wave amplitude (photopic) and/or delayed 30-Hz flicker response suggestive of photoreceptor involvement.
  - *Prognosis:* In most patients, symptoms improve without treatment in 2–6 weeks.
- **APMPPE** (acute posterior multifocal placoid pigment epitheliopathy)
  - Acute sequential vision loss with central/paracentral scotoma; the fellow eye becomes involved after a few days-weeks.

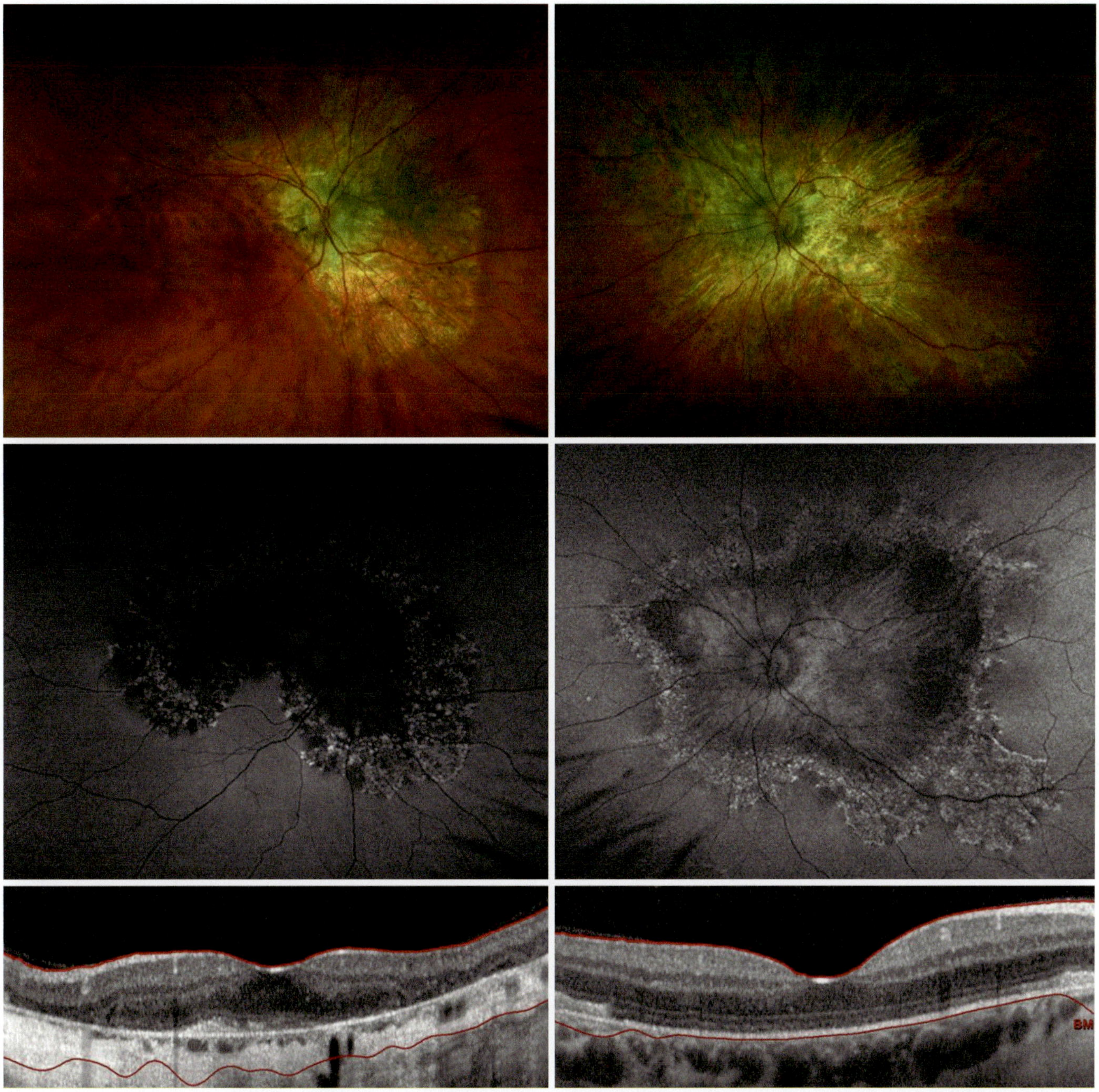

Fig. 51.1 Acute zonal occult outer retinopathy (AZOOR). Funds photographs show the characteristic peripapillary lesion with a demarcating line of progression. The trizonal pattern of hypo AF in the involved area, speckled hyper AF at the border, and normal AF beyond the delineating line is seen on fundus autofluorescence (FAF). Optical coherence tomography (OCT) shows thinning of the outer retinal layers with the loss of the ellipsoid zone and choriocapillaris. (Image courtesy of Naomi Goldberg, MD, PhD)

- Affects otherwise healthy young adults, there is no gender predilection. It may follow a flu-like illness. HLA-B7 and HLA-DR2 association has also been postulated. It can be associated with cerebral vasculitis, which requires urgent neurologic evaluation.
- *Fundus:* multiple yellow-white placoid lesions at the level of RPE, leaving pigmentary changes after the resolution, mild vitritis. Papillitis may be observed, but CME is rare.
- *FA:* early dense hypofluorescence and late hyperfluorescence (block early and stain late).
- *ICG:* hypofluorescence of the corresponding placoid lesions.

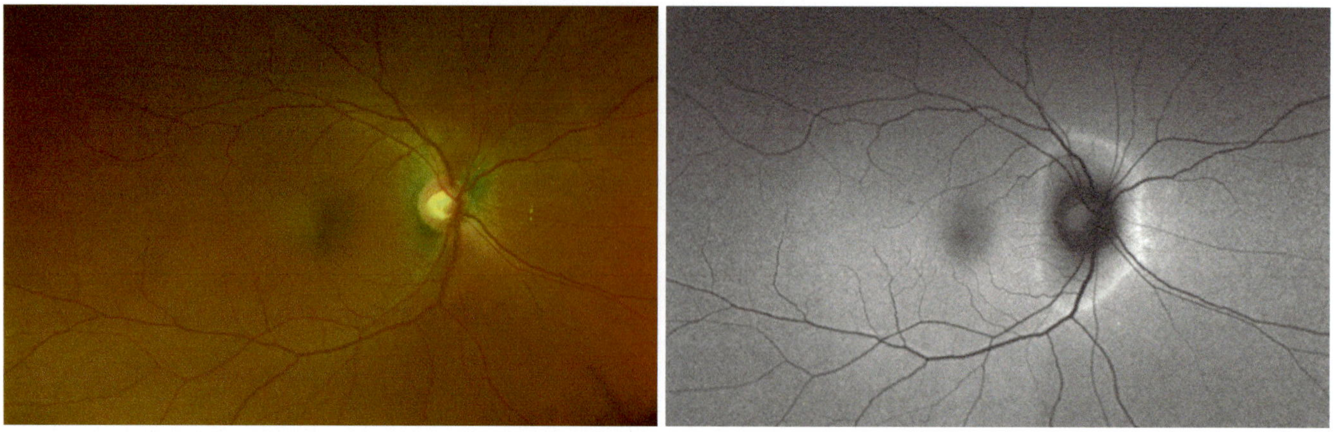

Fig. 51.2 Early AZOOR. Patient with a 3-month history of floaters and photopsias in her right eye. The fundus photograph shows barely visible peripapillary changes, but FAF shows diffuse hyperfluorescence, characteristic of early AZOOR. Note, that the macula is sparred in the early phase

  - *OCT:* hyperreflective outer retinal layers in the acute phase, followed by outer retinal and photoreceptor loss after resolution.
  - *FAF:* hyperAF lesions in the acute phase, followed by hypoAF lesions after resolution.
  - *Prognosis: Most* patients recover spontaneously within 2–3 months.
- **PIC/MCP**(punctate inner choroidopathy/multifocal choroiditis with panuveitis)
  - Acute onset of decreased vision, photopisas, scotomas, and floaters.
  - Patients are usually young, myopic women.
  - Quiet eye without vitritis is seen in PIC, anterior chamber reaction and vitritis are present in MCP.
  - *Fundus:* In the acute phase, bilateral multiple yellowish, round/oval lesions are clustered at the macula and around the optic disc (Fig. 51.4), but they can extend up to mid-periphery or beyond and are frequently arranged in a linear configuration called *Schlaegel lines*. With time, these spots evolve into atrophic chorioretinal scars with pigmentation and subretinal fibrosis or napkin ring around the optic disc, which may appear similar to POHS.
  - *FA:* In the acute phase, early punctate hypo or hyperfluorescent lesions scattered in the posterior pole and late staining are present. Later stages of the disease are characterized by window defects corresponding with RPE atrophy.
  - *ICG:* Hypofluorescent lesions (more than clinically visible), enlarged choroidal vessels, hyperfluorescence of the vessel walls.
  - *OCT:* thickening over the chorioretinal lesions, outer retinal abnormalities, CNV.
  - *FAF*: hyperfluorescent spots in the acute phase, which may enlarge to form an expanding hypofluorescent center with a hyperfluorescent halo.
  - Electrophysiology: within normal limits in PIC, attenuated in MCP.
  - *Prognosis:* Symptoms improve within 1 month; recurrences are common and CNV may develop in one-third of cases (more common in PIC) (Figs. 51.5 and 51.6).
- **Birdshot chorioretinopathy**
  - Patients present with blurred vision, floaters, nyctalopia, and disturbance of color vision.
  - Middle-aged Caucasian females are predominantly affected. Over 95% are HLA-A29 positive.
  - *Fundus:* Bilateral, multifocal, oval, cream-colored lesions radiating from the optic disc, most prominent inferonasally. Mild vitritis is usually present.
  - *FA:* Early hypofluorescnece and late leakage from the retinal vessels and the optic disc, CME may be present.
  - *ICG:* Hypofluorescent spots, more numerous than clinically evident, active lesions show late isofluorescence.
  - *OCT:* CME may be present. Early in the course, patchy or diffuse loss of the ellipsoid zone and choroidal thickening can be seen, while macular and choroidal atrophy can be present in longstanding cases.
  - *FAF:* Hypoautofluorecent spots, more numerous than clinically evident.
  - *Electrophysiology:* ERG shows reduced b-wave amplitude and latency, disease progression can be assessed using a 30-Hz flicker response. A reduced Arden ratio is seen on EOG.

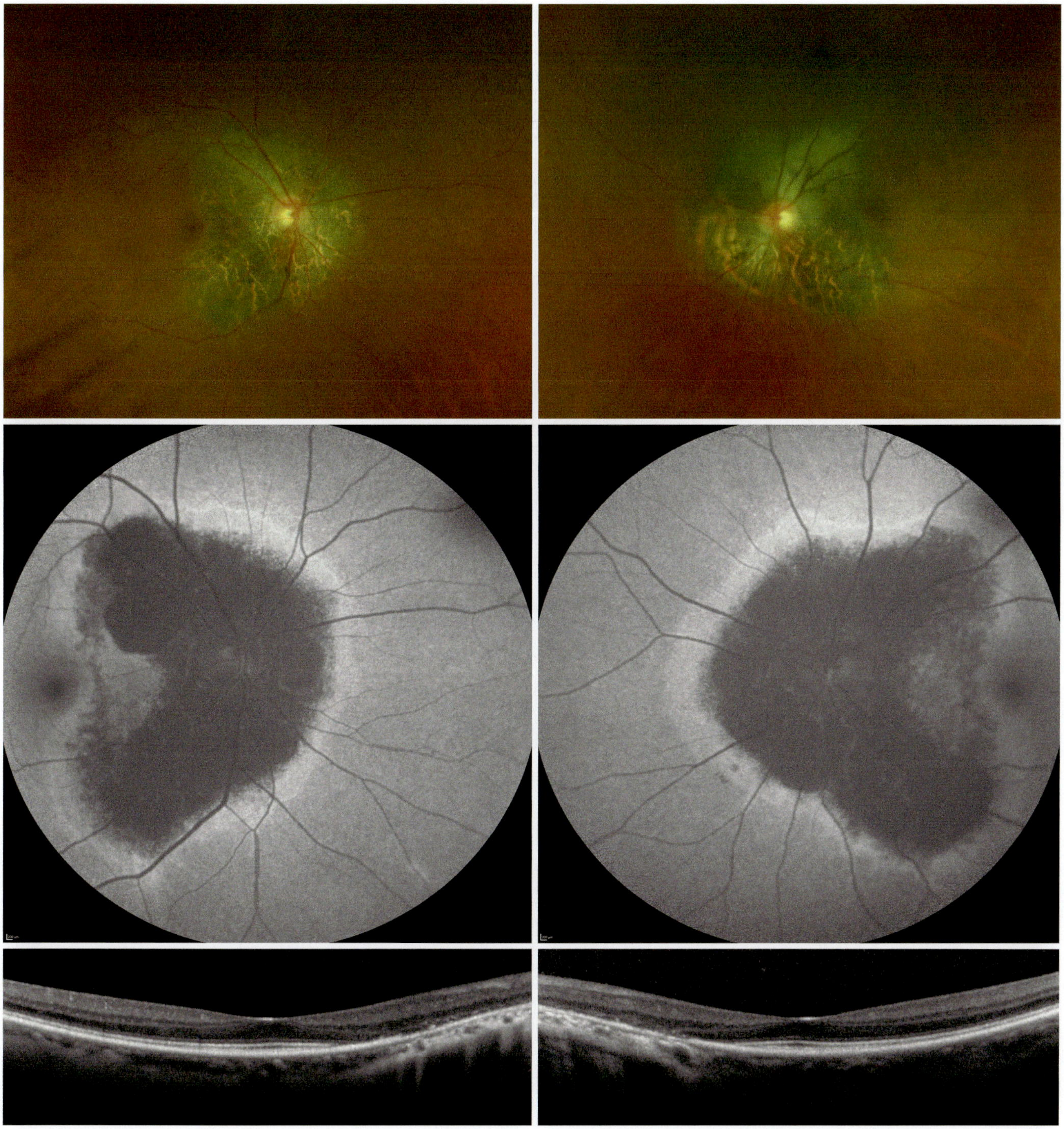

Fig. 51.3 AZOOR. Note the peripapillary extending into the macula in the typical trizonal pattern, but the fovea is still sparred

- *Prognosis:* Progressive worsening of the visual field and ERG responses. Early and aggressive treatment results in better visual outcomes.

- **Serpiginous choroiditis**
  - Patients present with decreased vision and paracentral scotoma but may be asymptomatic until macular involvement.
  - It affects middle-aged patients with no gender predilection. Tuberculosis should be excluded, as it can cause a serpiginous like choroidopathy but is usually unilateral and has more vitritis.
  - *Fundus:* Bilateral but asymmetric serpiginous/ geographic lesions, defined as grey-white lesions at the level of RPE/ inner choroid that spread centrifugally

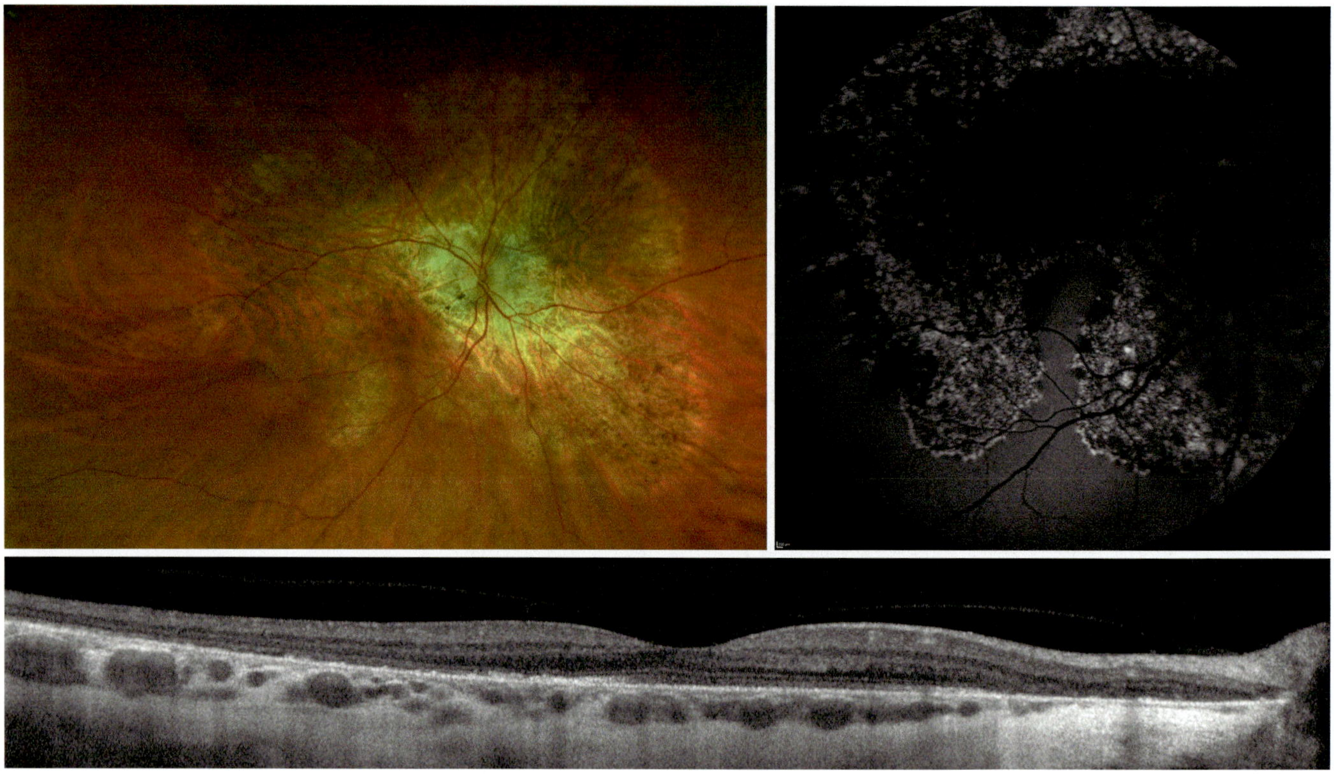

Fig. 51.4 Late AZOOR. Color fundus shows peripapillary atrophy extending into the macula, with a characteristic trizonal pattern. Marked thinning of the outer retinal layers, RPE and choriocapillaris is seen on OCT

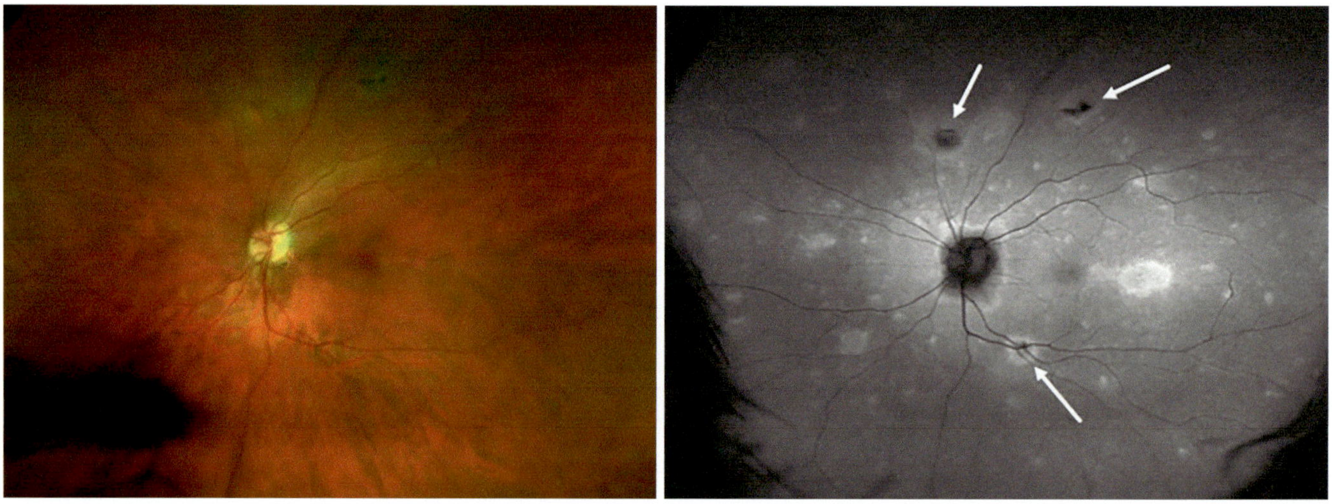

Fig. 51.5 A case of punctate inner choroidopathy (PIC), showing multiple grayish, lightly pigmented lesions in the peripapillary region. They appear more widespread on FAF, with central hypoAF and a hyperAF border (arrows)

from the optic disc. Mild vitritis can be present. Later in the course, atrophy of the RPE and choriocapillaris with subretinal fibrosis ensues. CNV at the border of the scar is present in up to 25%.

- *FA:* early hypofluorescence and late staining of lesions.
- *ICG:* early hypofluorescence and late staining, more than clinically evident.
- *OCT:* hyperreflectivity of the outer retina and RPE in the acute phase, followed by atrophy of the outer retinal layers and RPE in the chronic phase.

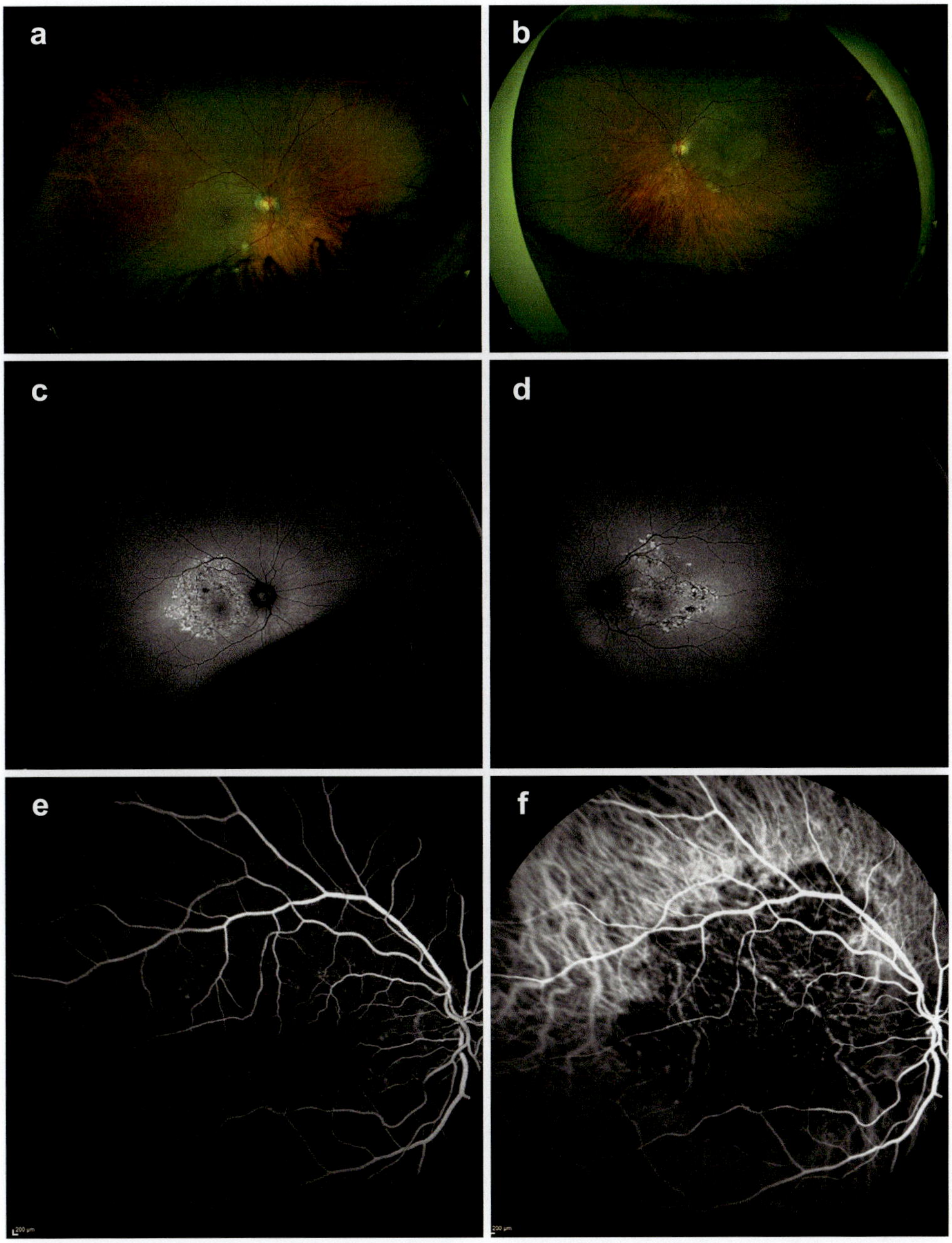

Fig. 51.6 Multimodal imaging of a case of persistent placoid maculopathy (PPM) showing: (**a** and **b**) Optos scans of the right and left eye respectively showing a well demarcated whitish plaque-like lesion involving the macular region and sparing the peripapillary area. (**c** and **d**) Optos FAF of the right and left eye respectively showing areas of hypoautofluorescence and hyperautofluoresence in the macula correlating with the plaque-like lesion. (**e**) FA picture of the right eye showing subtle early hypofluorescence during the arteriovenous phase with small areas of hyperfluoresence within the plaque lesion. (**f** and **h**) ICG of the right eye and left eye respectively showing a well circumscribed area of hypofluorescence involving the macula. (**g**) FA picture of the left eye showing hypofluorescence in the macula during early arteriovenous phase. (**i** and **h**) SD-OCT macular cube scans of the right and left eye respectively showing outer retinal and ellipsoid layer loss, small hyperreflective deposits under the retina, and scattered areas of increased light penetrance through the RPE

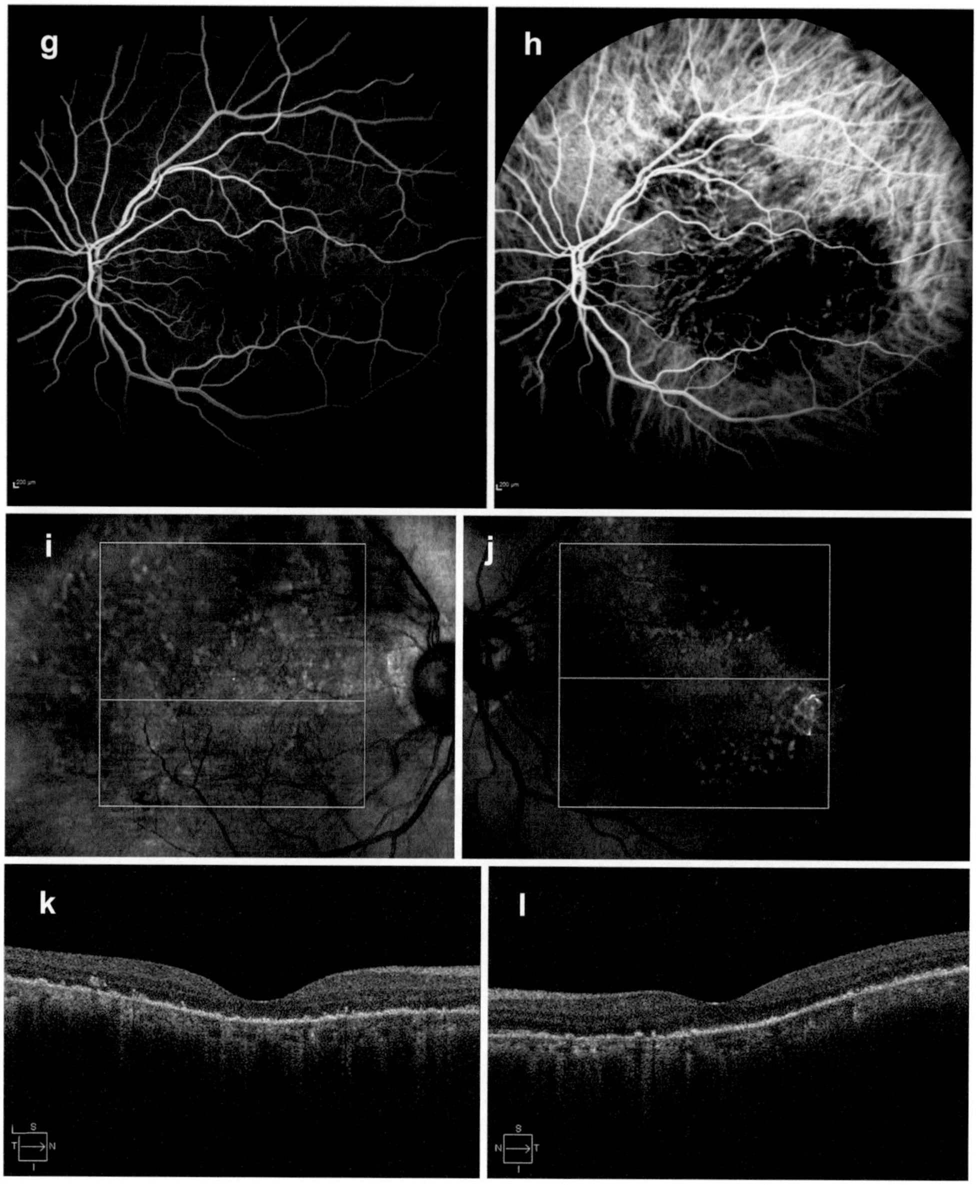

Fig. 51.6 (continued)

 - FAF: hyperfluorescent leading edge and hypofluorescent inactive lesions.
 - Electrophysiology is within normal limits.
 - *Prognosis:* Progressive centrifugal extension, resulting in central vision loss.
- **AIM** (acute idiopathic maculopathy)
 - Patients present with acute, unilateral, or bilateral central or paracentral visual loss (BCVA 20/200 or worse) and may have a history of a flu-like illness or coxsackie virus.
 - Young adults are affected. There is no gender predilection.
 - *Fundus:* Exudative/serous macular detachment, with mild disc swelling, hemorrhages, exudates/subretinal infiltrates, and vasculitis; vitritis, if present is mild. The lesion resolves and gives rise to a bulls-eye pattern.
 - *FA:* Progressive, irregular hyperfluorescence at the RPE level, with late pooling.
 - *ICG:* Early and persistent hypofluorescence.
 - *OCT:* serous macular detachment.
 - *Prognosis:* Spontaneous, almost complete recovery of vision in weeks to months.
- **AMN** (acute macular neuroretinopathy)
 - Either unilateral or bilateral paracentral scotomas, followed by decreased vision.
 - Young (20–40), healthy women are predominantly affected. A history of viral prodrome or the use of

medications (e.g., epinephrine, oral contraceptives) may be present.
  - *Fundus:* one or several triangular or wedge-shaped lesions around the fovea, with the tip pointing centrally. The lesions are red/brown and are best seen on near-infrared imaging.
  - *FA:* usually normal, slight hypoflurescence corresponding with clinically visible lesions may be seen.
  - ICGA: normal.
  - *OCT*: involvement of either the inner nuclear layer (INL) (type 1) or ONL/hyperreflective band at the EZ level (type 2), probably due to ischemia of the deep capillary plexus.
  - *AF:* subtle hypoautofluorescent lesions corresponding with clinically visible lesions.
  - Prognosis: The lesions resolve over a few weeks to months; however, subtle symptoms may persist.
- **PPM** (Persistent Placoid Maculopathy)
  - Rare entity that was first described in 2006.
  - Patients present with acute bilateral central visual loss usually.
  - Reported ages are between 50s and 70s with predominance for Caucasian men in a systematic review of 21 reported cases.
  - *Fundus:* Bilateral symmetric whitish plaque-like lesions at the level of RPE in the macula sparring the peripapillary area. Usually stable in size but few patients showed mild increase in size. The lesion usually fades overtime with follow-up over months to years.
  - *FA:* Early hypofluorescence followed by partial filling in late stages. No significant leakage or staining unless there was an active CNV lesion.
  - *ICG:* Sharply demarcated persistent hypofluorescence throughout the study consistent with the plaque lesion.
  - *OCT:* Hyperreflectivity of the ONL disruption of the ELM and ellipsoid layer. On follow-ups, complete outer retinal and RPE atrophy was seen along with homogenous solid isoreflective pattern of choroidal capillaries.
  - *AF:* Persistent hypoautoflourescence and hypoautofluorescence corresponding to the macular plaque lesion.
  - Prognosis: Visual acuity is usually preserved or mildly affected initially even with the lesion involving the entire macula for months to years. Complications related to CNV or RPE mottling resulted in significant visual loss. In the original case series, 9/10 eyes developed CNV. The white plaque-like lesion fades over months to years but the angiographic findings persisted.

Further Reading

Boudreault KA, Schuerch K, Zhao J, Lee W, Cabral T et al. Quantitative autofluorescence intensities in acute zonal occult outer retinopathy vs healthy eyes. JAMA Ophthalmol 2017;135(12):1330-1338.

Golchet PR, Jampol LM, Wilson D, Yannuzzi LA, Ober M, Stroh E. Persistent placoid maculopathy: a new clinical entity. Trans Am Ophthalmol Soc 2006;104:108-20.

Gendy MG, Fawzi AA, Wendel RT, Pieramici DJ, Miller JA, Jampol LM. Multimodal imaging in persistent placoid maculopathy. JAMA Ophthalmol 2014;132(1):38-49.

Kolomeyer AM, Brucker AJ. Persistent placoid maculopathy: a systematic review. Retina 2018;38(10):1881-1895.

Jung JJ, Khan S, Mrejen S, Gallego-Pinazo R, Cunningham ET Jr, Freund KB, et al. Idiopathic multifocal choroiditis with outer retinal or chorio-retinal atrophy. Retina 2014;34:1439–50.

Monson DM, Smith JR. Acute zonal occult outer retinopathy. Surv Ophthalmol 2011;56:23–35.

Sarraf D, Rahimy E, Fawzi AA, Sohn E, Barbazetto I, Zacks DN, et al. Paracentral acute middle maculopathy: a new variant of acute macular neuroretinopathy associated with retinal capillary ischemia. JAMA Ophthalmol 2013;131:1275–87.

Yannuzzi LA, Jampol LM, Rabb MF, Sorenson JA, Beyrer C, Wilcox LM Jr. Unilateral acute idiopathic maculopathy. Arch Ophthalmol 1991;109:1411–6. Erratum in: Arch Ophthalmol. 1992;110:449.

CHAPTER 52

Diffuse Unilateral Subacute Neuroretinitis (DUSN)

Flavius Beca, Maximilian Kong, Anna Jolia, Vlad Diaconita, and Jose Pulido

General Features 274
Further Reading 275

F. Beca
University of Miami, Miami, FL, USA
e-mail: flaviusbeca@med.miami.edu

M. Kong
Columbia University, New York, NY, USA

A. Jolia · V. Diaconita
Department of Ophthalmology, Columbia University, New York, NY, USA

J. Pulido (✉)
Department of Ophthalmology, Sidney Kimmel Medical College at Thomas Jefferson University, New York, NY, USA
e-mail: Pulido.Jose@mayo.edu

S. H. Tsang et al. (eds.), *Atlas of Inherited Retinal Diseases*, Advances in Experimental Medicine and Biology 1467, https://doi.org/10.1007/978-3-031-72230-1_52

General Features

- A rare helminthic infection, most frequently unilateral, presenting in young healthy patients.
- In the acute phase, patients present with pain and decreased vision, and if a nematode can be identified, laser photocoagulation is the definitive treatment, while systemic anthelminthics have shown to be effective. In the late phase, patients present with poor vision, mild inflammation, and striking fundus changes.
- Differential diagnosis for early phase: toxoplasmosis, histoplasmosis, MCP, MEWDS, APMPPE, Serpiginous chorioretinitis, and optic neuritis; late phase: unilateral RP, posttraumatic chorioretinopathy, occlusive vascular disease, and toxic retinopathy.
- *Fundus*: early, multifocal white-yellowish lesions (Fig. 52.1), rarely a live worm (Fig. 52.2), and late, arterial narrowing, optic atrophy, and diffuse peripheral RPE disruption (Fig. 52.3)
- *FA*: In early phase, early hypofluorescence and late staining in areas of retinitis, with capillary and optic disc leakage. In late phase, prominent window defects are seen in areas of RPE atrophy.
- *ICG*: early hypo and variable hypo to normal)l late fluorescent spots correspond with depth of choroidal infiltration.
- *OCT*: RNLF edema (acute) or loss (late), subretinal tunnels (Garcia's sign), focal and diffuse degenerative RPE changes, and ILM thickening (Oréfice's sign).
- *FAF*: peripapillary and macular granular hyper AF.
- *Electrophysiology:* negative ERG but rarely extinguished distinguishes from RP.

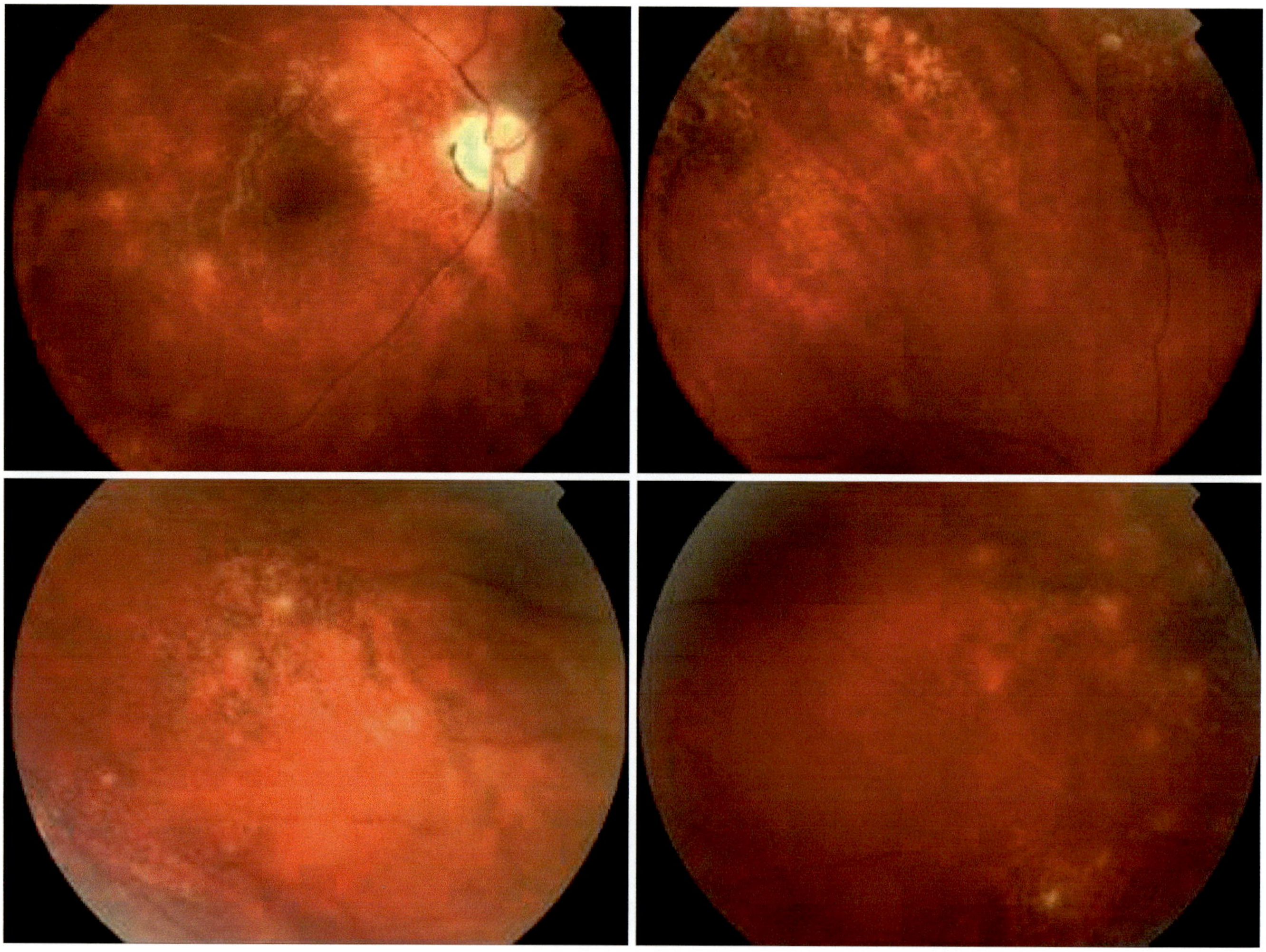

Fig. 52.1 Diffuse unilateral subacute neuroretinitis (DUSN). Pigmentary changes and RPE atrophic patches. Some are oval and some are linear, suggestive of movement tracks of a live nematode

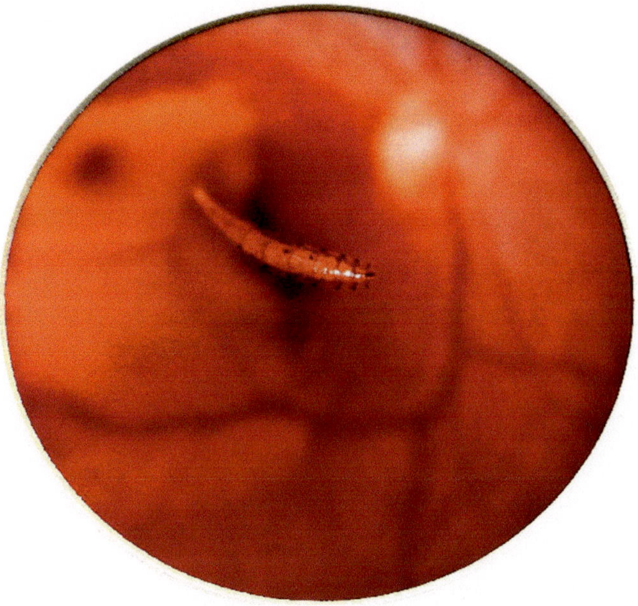

Fig. 52.2 DUSN. A rare living nematode seen free-floating in the vitreous cavity. There is an associated retinal artery occlusion of the eye. (Courtesy of Dr. Don Gass and Dr. Robert Braunstein)

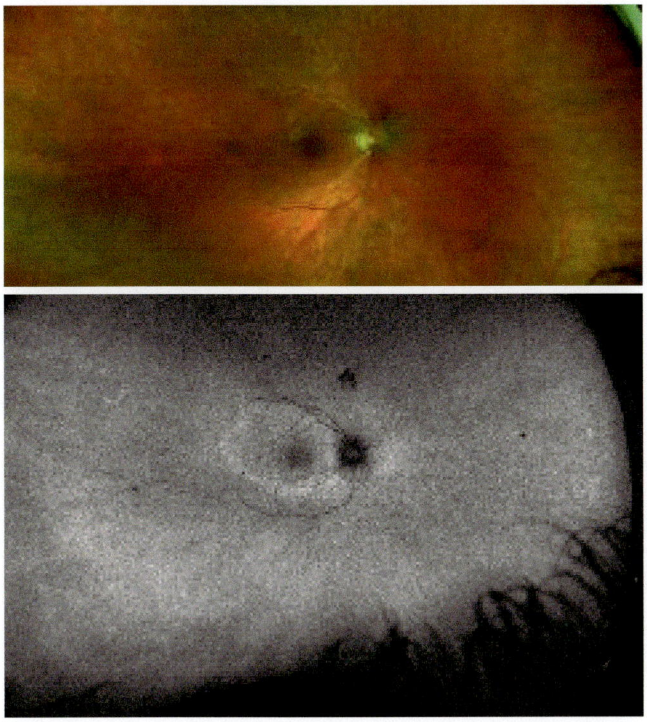

Fig. 52.3 DUSN. In late phase, diffuse vascular attenuation, optic atrophy, and peripheral RPE pigment clumping. On FAF, peripapillary and macular granular hyper AF can be seen both early and late (Courtesy of Dr Jose Pulido)

– *Prognosis*: frequently poor (80% have a final vision of 20/200 or worse) when presenting in late phase. Those with early diagnosis and prompt anti-helminthic or direct photocoagulation of the worm can recover excellent vision. When the worm is in the macula, a bright light can be used to direct it into the periphery where laser can be applied safely.

Further Reading

de Amorim Garcia Filho CA, Gomes AH, Garcia AC, de Amorim Garcia CA. Clinical features of 121 patients with diffuse unilateral subacute neuroretinitis. Am J Ophthalmol 2012;153:743–9.

Garcia CA, et al., Diffuse unilateral subacute neuroretinitis—DUSN. Int Ophthalmol Clin, 2008. 48(3): 119-29.

Mazzeo, T., M.M. Dos Santos Motta, and A.L.L. Curi, Diffuse unilateral subacute neuroretinitis: review article. J Ophthalmic Inflamm Infect, 2019. 9(1): p. 23.

CHAPTER 53

Central Serous Chorioretinopathy

Jin Kyun Oh, Vlad Diaconita, and Tongalp H. Tezel

General Features 278
Further Reading 280

J. K. Oh
Department of Ophthalmology, Edward S. Harkness Eye Institute, Columbia University Irving Medical Center, New York, NY, USA

V. Diaconita (✉)
Department of Ophthalmology, Columbia University, New York, NY, USA
e-mail: vd2374@ccumc.columbia.edu

T. H. Tezel
Columbia University, New York, NY, USA

S. H. Tsang et al. (eds.), *Atlas of Inherited Retinal Diseases*, Advances in Experimental Medicine and Biology 1467, https://doi.org/10.1007/978-3-031-72230-1_53

General Features

- Central serous chorioretinopathy (CSCR) is a disorder characterized broadly by serous detachment of the retina due to retinal pigment epithelium (RPE) dysfunction and leakage of the choroid.
- CSCR is one of the most common non-surgical maculopathies. The highest prevalence of CSCR is seen in men of working age with a reported male to female ratio of nearly 5:1.
- Known risk factors for disease include corticosteroid use, hypercortisolism and hyperaldosteronism, pregnancy, use of vasoactive medications including sildenafil, hyperopia, and "type A personality" as well as other psychosocial comorbidities such as anxiety or depression.
- The pathophysiology of CSCR remains poorly understood but is attributed to primary choroidal dysfunction followed by RPE dysfunction. Thick or pachychoroid vessels are believed to be highly associated with CSCR. Elevated levels of mineralocorticoids and glucocorticoids are also suspected to play a role in disease pathogenesis.
- Clinical symptoms of CSCR include central scotoma, metamorphopsia and distortion, micropsia, dyschromatopsia, and decreased or blurry vision. Symptoms are typically self-limited and resolve within 3–4 months (57%) but can recur (33%) or become permanent.
- CSCR is described as "acute" or "chronic" depending on the duration of subretinal fluid and serous retinal detachment; however, no clear consensus exists on a duration to distinguish between the two.
- Acute CSCR is characterized by subretinal detachment and RPE abnormalities including pigment epithelial detachments and RPE atrophy. On fluorescein angiography, acute CSCR reveals an "ink-blot" or "smoke-stack" pattern of hyperfluorescence (Figs. 53.1 and 53.2).
- Chronic CSCR is defined by significant, multifocal RPE atrophy and photoreceptor degeneration due to longstanding subretinal fluid. Fluorescein angiography demonstrates diffuse leakage due to widespread RPE loss in patches.
- Complications associated with CSCR include subretinal fibrinoid deposition, the development of Type 1 choroidal neovascularization, and subsequent subretinal hemorrhage. In rare cases, bullous serous retinal detachments associated with RPE tears have also been reported.
- Treatment of CSCR is controversial and no preferred treatment has been well established. Proposed treatment modalities have included low dose photodynamic therapy, oral mineralocorticoid antagonists, and anti-VEGF medications.

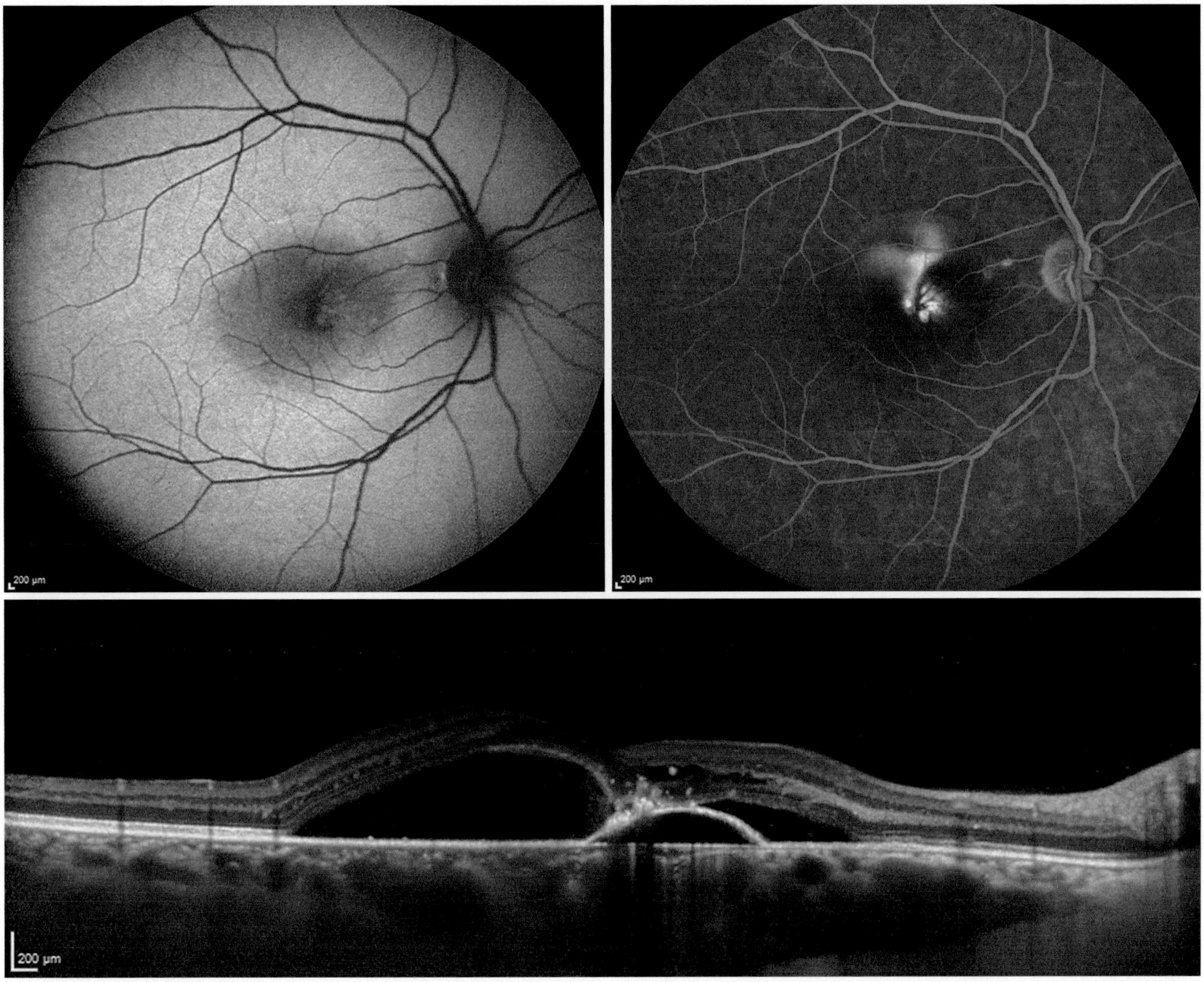

Fig. 53.1 Multimodal imaging of central serous chorioretinopathy. Fundus autofluorescence images of a patient with central serous chorioretinopathy reveals central, well circumscribed hypo-autofluorescence. On fluorescein angiography, a classic "smoke-stack" pattern of leakage is seen. Optical coherence tomography demonstrates a large serous retinal detachment with an additional underlying pigment epithelial detachment. Large pachychoroid vessels are seen

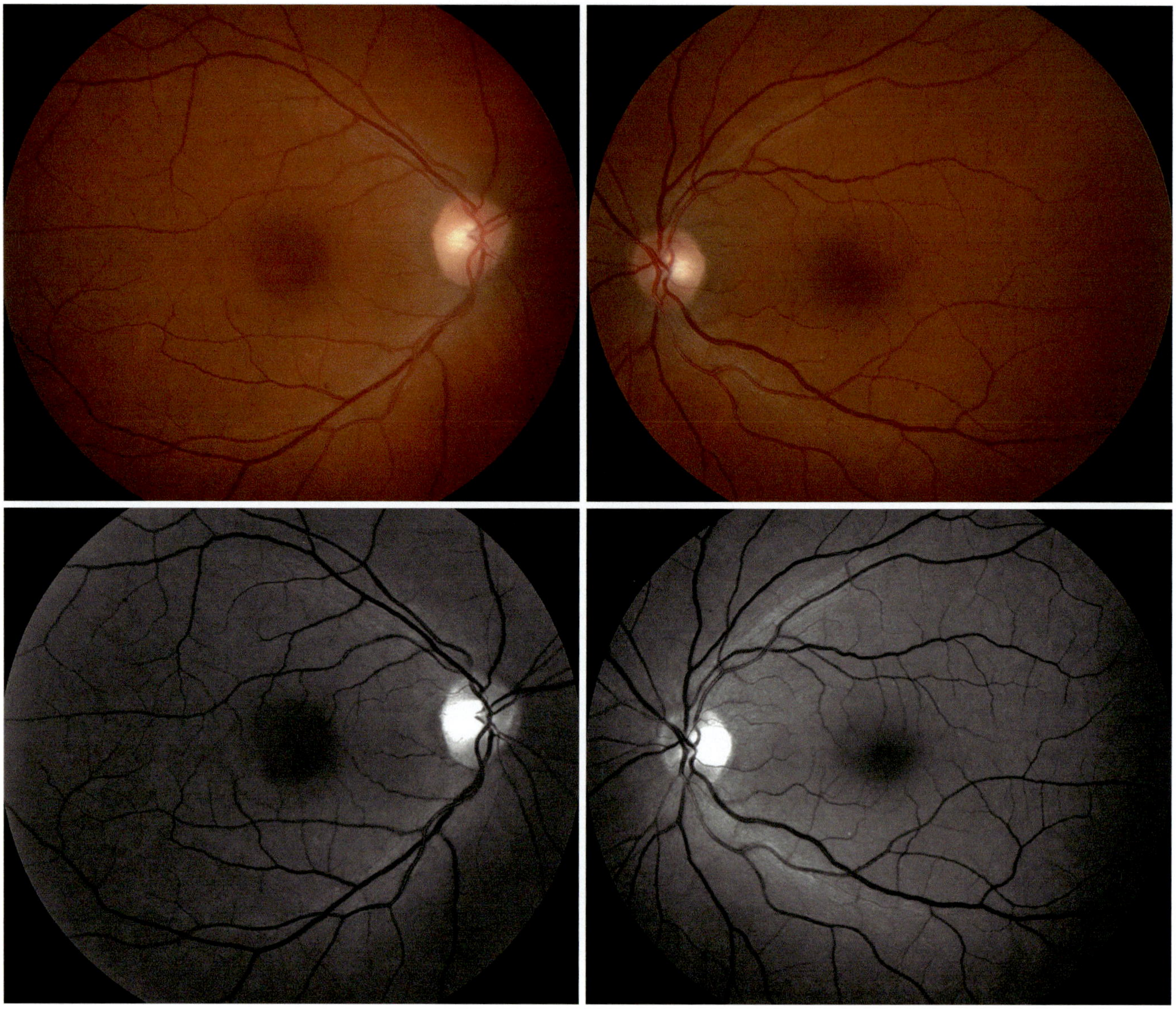

Fig. 53.2 Color fundus photographs and near infrareflectance images of central serous chorioretinopathy. Color fundus photographs of a patient with central serous chorioretinopathy demonstrates blunting of the central foveal reflex in the right eye compared to the unaffected left eye. This area can also be appreciated with relative blurring of the retinal vasculature on near-infrared reflectance imaging

Further Reading

Breukink, M. B., A. J. Dingemans, A. I. den Hollander, J. E. Keunen, R. E. MacLaren, S. Fauser, G. Querques, C. B. Hoyng, S. M. Downes, and C. J. Boon. 2017. Chronic central serous chorioretinopathy: long-term follow-up and vision-related quality of life. Clin Ophthalmol 11:39-46.

Daruich, A., A. Matet, A. Dirani, E. Bousquet, M. Zhao, N. Farman, F. Jaisser, F. Behar-Cohen. Central serous chorioretinopathy: recent findings and new physiopathology hypothesis. Prog Retin Eye Res 2015;48:82-118

Daruich, A., A. Matet, F. Behar-Cohen. Central serous chorioretinopathy. Dev Ophthalmol 2017;58:27-38.

Gass, J. D. Pathogenesis of disciform detachment of the neuroepithelium: II. idiopathic central serous choroidopathy. Am J Ophthalmol 1967;63(3):587/515–615/543.

Kaye R, Chandra S, Sheth J, Boon CJF, Sivaprasad S, Lotery A. Central serous chorioretinopathy: an update on risk factors, pathophysiology and imaging modalities. Prog Retin Eye Res 2020;79:100865.

Kitzmann, A. S., J. S. Pulido, N. N. Diehl, D. O. Hodge, J. P. Burke. The incidence of central serous chorioretinopathy in Olmsted County, Minnesota, 1980–2002. Ophthalmology 2008;115 (1):169-173.

Nicholson, B. P., E. Atchison, A. A. Idris, S. J. Bakri. Central serous chorioretinopathy and glucocorticoids: an update on evidence for association. Surv Ophthalmol 2018;63(1):1-8.

Wang, M. S., B. Sander, and M. Larsen. Retinal atrophy in idiopathic central serous chorioretinopathy. Am J Ophthalmol 2002;133(6):787-793.

Wang, M., I. C. Munch, P. W. Hasler, C. Prunte, M. Larsen. Central serous chorioretinopathy. Acta Ophthalmol 2008;86 (2):126-145.

CHAPTER 54

Pathologic Myopia

Sophie Z. Gu and Stanley Chang

Further Readings 283

S. Z. Gu
Department of Ophthalmology, Columbia University,
New York, NY, USA

S. Chang (✉)
Columbia University, New York, NY, USA
e-mail: zg2172@cumc.columbia.edu

S. H. Tsang et al. (eds.), *Atlas of Inherited Retinal Diseases*, Advances in Experimental Medicine and Biology 1467, https://doi.org/10.1007/978-3-031-72230-1_54

- General Features
  - Pathologic myopia refers to high myopic refractive error (>6.00 diopters) with long axial length (>26.00 m) associated with progressive myopic macular degeneration.
  - Risk factors for pathologic myopia include increased age, increased axial length, and high myopic spherical equivalent.
  - There are many known genetic variants associated with myopia, including variants at the *KCNMA1* and *GJD2* loci associated with high myopia. However, no variant has been specifically associated with the development of myopic macular degeneration.
  - Pathologic myopia generally presents in early childhood and progression may continue past early adulthood. Besides progressive decreased visual acuity at distance, symptoms may include metamorphopsias or scotomas secondary to macular complications.
- Findings
  - Typical fundus findings may include large physiologic cup, tilted optic disc, peripapillary atrophy, posterior staphyloma, foveoschisis, and chorioretinal atrophy. Pathologic myopia also increases the risk of developing lacquer cracks, choroidal neovascularization, Fuchs spots, myopic traction maculopathy, retinal tears, retinal holes, and retinal detachment.

    Tilted optic disc refers to oblique insertion of the optic nerve into the globe (Fig. 54.1).

    Posterior staphyloma refers to posterior outpouching of scleral tissue. Common locations of staphylomas in pathologic myopia are peripapillary, perimacular, and in the inferior or inferotemporal portions of the eye (Fig. 54.2).

    Lacquer cracks refer to yellow-colored breaks in Bruch's membrane and are a risk factor for choroidal neovascularization.

    Fuchs spots refer to RPE hyperplasia in areas of regressed choroidal neovascularization.

    Myopic traction maculopathy (MTM) or myopic foveoschisis refers to separation in various layers of the neurosensory retina at the fovea, thought to be secondary to a combination of anterior traction, such as vitreomacular traction, epiretinal membrane, or retinal vascular traction, and progression of posterior staphyloma (Fig. 54.3.

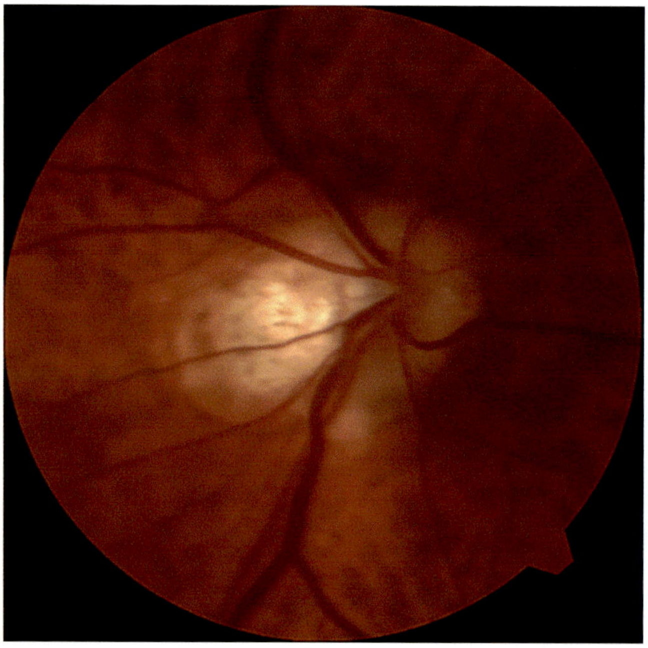

Fig 54.1 Tilted optic disc

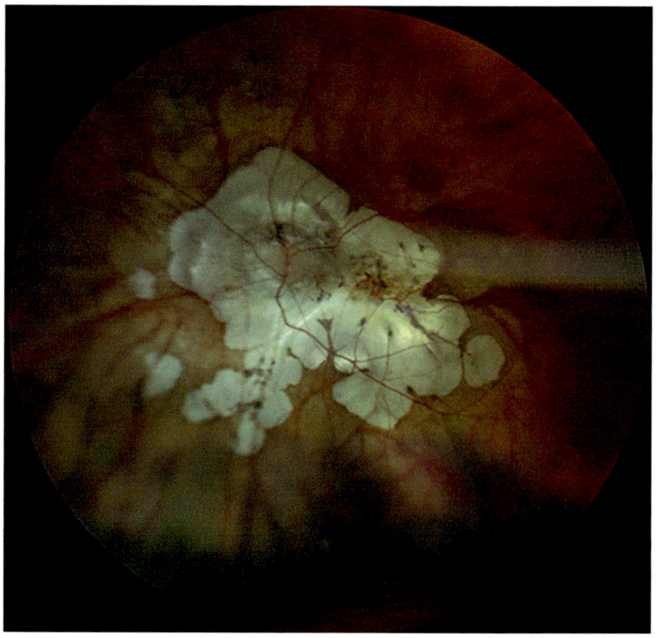

Fig 54.2 Posterior staphyloma with atrophy and a pale fundus. Choroidal vessels and pigment migration are visible

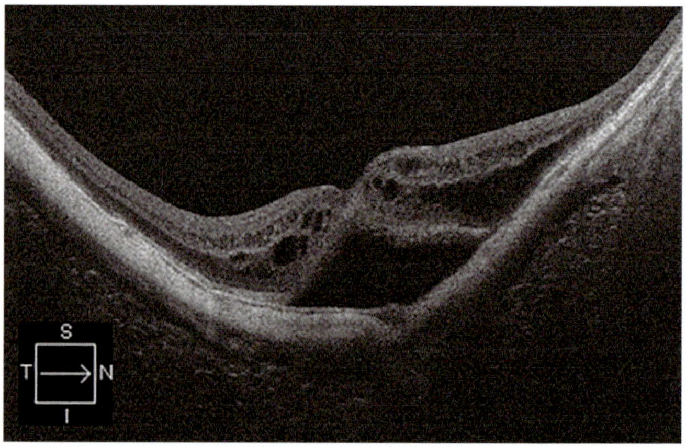

Fig 54.3 Foveoschisis with associated foveal detachment. A thin choroid and mild epiretinal membrane are visible

- Optical coherence tomography (OCT) may help monitor the progression of myopic degeneration and better detect some of the features described above, such as myopic traction maculopathy.
- Fluorescein angiography (FA) may help detect lacquer cracks (signal void in a linear distribution) and choroidal neovascularization (focus of hyperfluorescence with rim of hypofluorescence and late leakage).
- Indocyanine green angiography (ICG) and OCT angiography (OCT-A) may help detect the presence of choroidal neovascularization.

- Classification
  - The International photographic classification of pathologic myopia (META-PM), based on atrophic changes, is as follows:
    - Category 0: no macular degenerative lesions
    - Category 1: tessellated fundus only
    - Category 2: diffuse chorioretinal atrophy
    - Category 3: patchy chorioretinal atrophy
    - Category 4: macular atrophy
  - A newer classification system has also been suggested based on atrophic (A), tractional (T), and neovascular (N) components.

Further Readings

Ip JM, Saw SM, Rose KA, et al. Role of near work in myopia: findings in a sample of Australian school children. Invest Ophthalmol Vis Sci 2008;49:2903-2910

Lam JC, Li FF, Zhang X, Tang SM, et al. Two-year clinical trial of the low-concentration atropine for Myopia Progression (LAMP) study: phase 2 report. Ophthalmology 2020;127(7):910-19.

Malagola R, Pecorella I, Teodori C, Santi G, Mannino G. Peripheral lacquer cracks as an early finding in pathological myopia. Arch Ophthalmol. 2006;124(12):1783–1784. doi: https://doi.org/10.1001/archopht.124.12.1783

Morgan I, Rose K. How genetic is school myopia? Prog Retinal Eye Res 2005;24:1-38.

Ohno-Matsui K. What is the fundamental nature of pathologic myopia? Retina. 2017;37(6):1043–8.

Ohno-Matsui K, Kawasaki R, Jonas JB, et al. International photographic classification and grading system for myopic maculopathy. Am J Ophthalmol. 2015;159(5):877-83. Epub 2015 Jan 26.

Pineles SL, Kraker RT, VanderVeen DK, et al. Atropine for the prevention of myopia progression in children: a report by the American Academy of Ophthalmology. Ophthalmology 2017;124(12):1857-66.

Ruiz-Medrano J, Montero JA, Flores-Moreno I, et al. Myopic maculopathy: current status and proposal for a new classification and grading system (ATN). Prog Retin Eye Res 2019;69:80-115

VanderBeek BL, Johnson MW. The diversity of traction mechanisms in myopic traction maculopathy. Am J Ophthal. 2012;153(1):93-102.

Willams, A, et al. The tilted disc syndrome. Practical Neurology, 2005a;5:54-55

Willams A, et al. The tilted disc syndrome. Pract Neurol 2005b;5:54-55

Wong YL, Hysi P, Cheung G, Tedja M, Hoang QV, Tompson SWJ, et al. (2019) Genetic variants linked to myopic macular degeneration in persons with high myopia: CREAM Consortium. PLoS ONE 14(8): e0220143.

Part VIII
Managing IRDs in Clinics

CHAPTER 55

A Practical Approach to Retinal Dystrophies

Ahmet Hondur, Irena Tsui, Brian J. Song, Chyuan-Sheng Lin, Stephen Tsang, Tarun Sharma, and Vlad Diaconita

Examination and Testing 289
Stationary Retinal Dystrophies 290
  Stationary Rod Dystrophies 290
  Stationary Cone Dysfunction Syndromes 291
Progressive Retinal Dysfunction 291
Central Dystrophies 293
  Stargardt Disease 293
  Other Central Dystrophies 294

A. Hondur · S. Tsang (✉) · T. Sharma · V. Diaconita
Department of Ophthalmology, Columbia University, New York, NY, USA
e-mail: sht2@cumc.columbia.edu

I. Tsui
Department of Ophthalmology, University of California—Los Angeles, Los Angeles, CA, USA

B. J. Song
Massachusetts Eye and Ear Infirmary, Department of Ophthalmology, Harvard Medical School, Boston, MA, USA

C.-S. Lin
Department of Pathology and Cell Biology, Herbert Irving Comprehensive Cancer Center, Columbia University Medical Center, New York, NY, USA

S. H. Tsang et al. (eds.), *Atlas of Inherited Retinal Diseases*, Advances in Experimental Medicine and Biology 1467, https://doi.org/10.1007/978-3-031-72230-1_55

Treatments 299
Dietary Treatment and Vitamins 299
Other Treatments 300
Treatment of Complications 300
Gene Therapy 300
Other Emerging Therapies 301
Summary 302
References 302

Genomic approaches to developing new diagnostic and therapeutic strategies in retinal dystrophies are among the most advanced applications of genetics (Tsang and Gouras 1996). The notion that "nothing can be done" for patients with retinal dystrophies is no longer true. Electrophysiological testing and autofluorescence imaging help to diagnose and predict the patient's course of disease. Better phenotyping can contribute to better-directed, cost-efficient genotyping. Combining fundoscopy, autofluorescent imaging, and electrophysiological testing is essential in approaching patients with retinal dystrophies. The first FDA-approved gene therapy, Voretigene Neparvovec (VN, Luxturna• , Spark Therapeutics, Philadelphia, PA), is now available for eyes with AR RP and LCA with biallelic RPE65 mutation. Emerging are more new gene-based treatments for these devastating retinal dystrophies.

Classification of retinal dystrophies can be confusing because they are both clinically and genetically heterogeneous (Bird 1995; Taylor and Hoyt 2005). There are several disease classification schemas; we present them as (1) stationary *versus* progressive and (2) central (macular) dystrophies *versus* generalized (Table 55.1). Important patient history includes specific symptoms such as central scotomas or night blindness, age of onset to determine severity of disease, and family history to establish the inheritance pattern. Examining other family members is often helpful. A complete past medical history and focused physical exam are important to relate systemic associations.

Table 55.1 Some genes implicated in inherited retinal degeneration

Stationary retinal dystrophies			***Central retinal dystrophies***		
Disease	**Gene**	**Protein**	**Disease**	**Gene**	**Protein**
Achromatopsia	*CNGA3*	Alpha subunit of cone cGMP-gated channel	Best disease	*VMD2*	Bestrophin
	CNGB3	Beta subunit of cone cGMP-gated channel	Doyne honeycomb retinal dystrophy	*EFEMP1*	EGF-containing fibrillin-like extracellular matrix protein 1
	GNAT2	Alpha subunit of cone transducin	North Carolina macular dystrophy	*MCDR1*	Unknown gene on chromosome 6
CSNB (dominant)	*GNAT1*	Alpha subunit of rod transducin	Sorsby fundus dystrophy	*TIMP3*	Tissue inhibitor of metalloproteinase-3
	PDE6B	Beta subunit of rod cGMP-PDE	Stargardt disease	*ABCA4*	ATP-binding cassette transporter
	RHO	Rhodopsin	Stargardt-like dominant macular dystrophy	*ELOVL4*	Elongation of very long chain fatty acid protein
CSNB (XL complete)	*NYX*	Nyctalopin	***Progressive generalized dystrophies (functional class)***		
CSNB (XL incomplete)	*CACNA1F*	Alpha1 subunit of Cav1.4 calcium channel	Juvenile retinoschisis	*XLRS1*	Retinoschisin

(continued)

Table 55.1 (continued)

Stationary retinal dystrophies			***Central retinal dystrophies***		
Disease	**Gene**	**Protein**	**Disease**	**Gene**	**Protein**
Fundus albipunctatus	*RDH5*	11-cis retinol dehydrogenase	Autosomal dominant RP	*RP1*	RP1
Oguchi disease	*GRK1*	G-protein coupled receptor kinase 1		*PRPF31, PRPF3, and PRPF8*	Pre-mRNA processing factor
	SAG	Arrestin		*RHO*	Rhodopsin
Early onset progressive retinal dystrophies			Autosomal recessive RP	*CNGA1*	Alpha subunit of rod cGMP-gated channel
Leber congenital amaurosis	*AIPL1*	Aryl hydrocarbon receptor interacting protein like-1		*CNGB1*	Beta subunit of rod cGMP-gated channel
	CRB1	Crumbs homolog 1		*PDE6A*	Alpha subunit of PDE
	CRX	Cone-rod homeobox		*PDE6B*	Beta subunit of PDE
	GUCY2D	Retinal guanylate cyclase		*RHO*	Rhodopsin
	RDH12	Retinal dehydrogenase 12		*SAG*	Arrestin
	RPE65	RPE-specific protein, 65 kD		*USH2A*	Usherin
	RPGRIP1	RP GTPase interacting protein 1			
	CEP290	Centrosomal protein			

Mutations in structural proteins and transcription factors tend to cause dominant disease; abnormalities in metabolic enzymes tend to cause recessive diseases

Examination and Testing

After a family history and clinical exam (Yannuzzi et al. 1995), the electroretinogram (ERG) is usually the first ancillary test for classifying retinal dystrophies (Berson et al. 1968). Full-field ERG is a noninvasive test of retinal function that measures mass response electrical activity after light stimulation and helps to exclude generalized retinal dysfunction. Full-field ERGs are performed on patients with dilated pupils using standards published by the International Society for Clinical Electrophysiology of Vision (ISCEV) (International Society for Clinical Electrophysiology of Vision 2007). The minimum protocol incorporates the rod-specific (Fig. 55.1, Column 1) and standard bright-flash ERGs (Fig. 55.1, Column 2), both recorded after a minimum of 20 min dark adaptation, and the photopic 30 Hz flicker (Fig. 55.1, Column 3) and transient photopic ERGs (Fig. 55.1, Column 4), both recorded after a standard period and intensity of light adaptation (Heckenlively and Arden 2006).

Though full-field ERG evaluates mass cone and rod system function, pattern ERG (checkerboard stimulus) is useful to distinguish decreased vision secondary to optic nerve disease *versus* macular disease (Holder 2001). An electrical response in a 30-Hz flicker ERG is an important prognostic tool because its presence confirms residual cone function (Berson et al. 1993; Berson 1974).

Scanning laser ophthalmoscopy (SLO) autofluorescence is a noninvasive imaging technique that detects the lipophilic cation N-retinyl-N-retinylidene ethanolamine (A2E) (a by-product of the visual cycle) in lipofuscin fluorophores, which accumulates in the retinal pigment epithelium (RPE) (Delori et al. 1995; Sparrow and Boulton 2005; von Ruckmann et al. 1995). Hypofluorescent or hyperfluorescent areas in SLO images are associated with abnormal accumulation or depletion of A2E fluorophore in RPE. Besides diagnosing malfunctioning RPE cells, using autofluorescence to follow patients with pathology can frequently predict their clinical course (Robson et al. 2003, 2004).

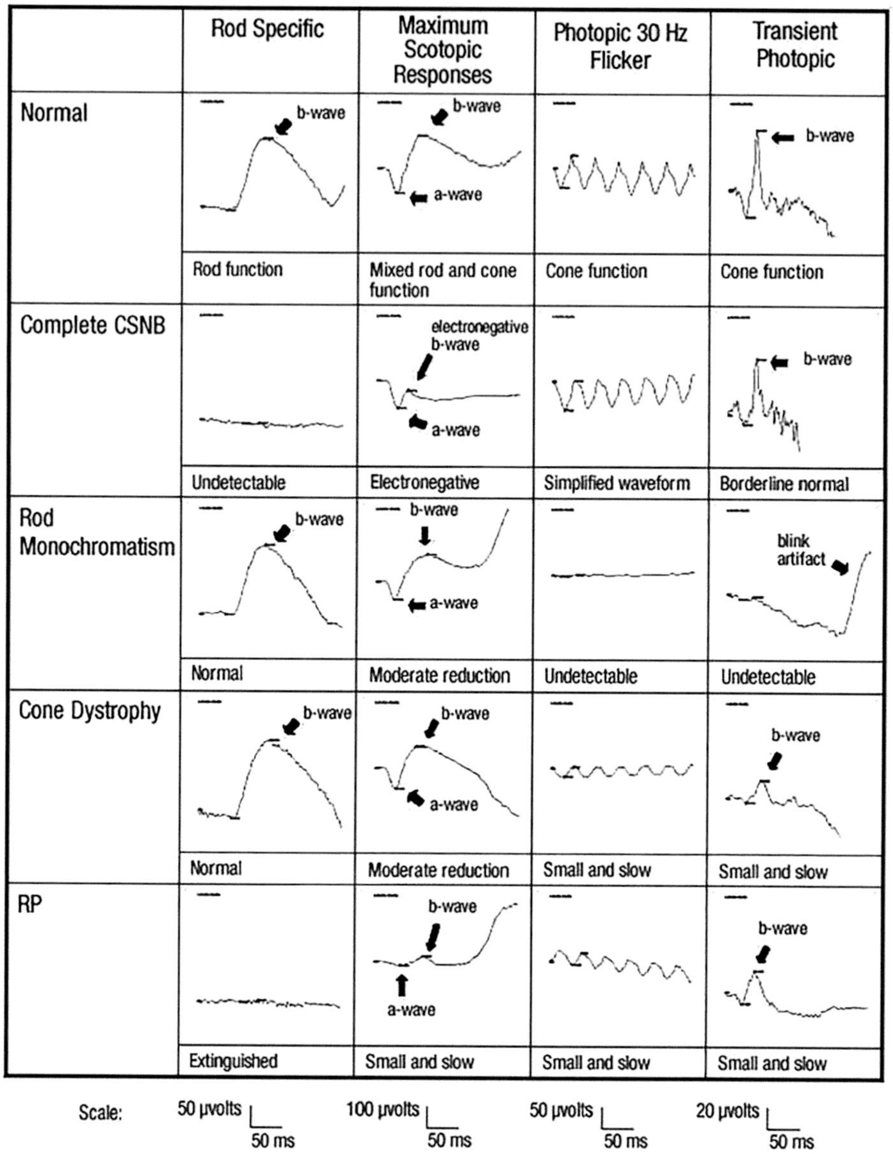

Fig. 55.1 Electroretinogram (ERG) testing. Full-field ERG measures mass response of the retina: photoreceptor (a-wave) and inner nuclear layers (b-wave). Full field ERG was performed using Dawson, Trick, and Litzkow recording electrodes and Ganzfeld stimulation according to International Society for Clinical Electrophysiology of Vision (ISCEV) standards. ERG tracings from a normal 40-year-old individual are shown in the top row. Typical tracings from different patients at 40–50 years of age are shown in other rows. *Hz* Hertz, *CSNB* congenital stationary night blindness, *RP* retinitis pigmentosa, *ms* milliseconds

Stationary Retinal Dystrophies

It is important to distinguish stationary *versus* progressive disease because the patient's primary concern is whether sight will deteriorate. There are two common forms of stationary night blindness and several forms of stationary cone dysfunction syndrome.

Stationary Rod Dystrophies

Congenital stationary night blindness (CSNB), although most commonly inherited as an X-linked recessive (XLR) disorder, can also demonstrate autosomal dominant (AD) or autosomal recessive (AR) inheritance. All modes of inheritance result in normal-appearing fundi. The AR XLR forms of CSNB present in infancy with nystagmus, strabismus, and reduced vision. In contrast, the AD form of CSNB typically presents in teenage years with normal visual acuity and symptomatic night blindness (Salchow et al. 1999; Tsang et al. 1998, 2006a, b).

Because fundoscopy is normal in CSNB, ERG is important in making the diagnosis (Tsang et al. 2006b), as it detects severe dysfunction and an electronegative waveform (see Fig. 55.1 and Table 55.2). The defect in X-linked complete CSNB has been suggested to lie downstream from the photoreceptor neurons, possibly in the ON bipolar cells or their interconnections (Hood and Birch 1996). Visual acuity (VA) is diminished in X-linked CSNB patients because their ON

Table 55.2 Inherited causes of electronegative (b-wave negative) ERG

Table 55.2 Inherited causes of electronegative (b-wave negative) ERG
Congenital stationary night blindness
X-linked retinoschisis
Inner retinal dystrophy
Batten disease
Duchenne muscular dystrophy

bipolar cells fail to have proper interpretation of photoreceptor signal to noise.

Distinctive radial flecks in the retina characterize another form of night blindness, fundus albipunctatus, caused by delayed rhodopsin regeneration. Although usually stationary, there is a subtype of fundus albipunctatus associated with progressive cone dysfunction (Nakamura et al. 2000). Patients have discrete white flecks at the level of the RPE, most numerous in the mid-periphery. Either patients present with night blindness, or the flecks are found on routine ophthalmoscopy. Fundus albipunctatus can be confused with retinitis punctata albescens, a form of retinitis pigmentosa (RP) with white spots on fundus exam. ERG recovers after prolonged dark adaptation in fundus albipunctatus because it allows the delayed rhodopsin regeneration to recover.

Stationary Cone Dysfunction Syndromes

There are three categories of individuals with congenital achromatopsia (Michaelides et al. 2004):

- Typical rod monochromats (complete achromats) lack all sensitivity mediated by cone pigments.
- Atypical rod monochromats (incomplete achromats) have some residual cone function and thus better VA and residual color vision.
- S-cone monochromats have rod and blue-cone function; these patients have better VA than either type of rod monochromat.

The two types of achromatopsia, complete and incomplete, are both autosomal recessive and have normal rod function.

Patients usually present in infancy with nystagmus, poor vision, and a preference for dim illumination. Because of the sensory nystagmus, CSNB, foveal hypoplasia, optic nerve hypoplasia, and cone dystrophy are in the differential. Extinguished-cone ERGs in achromatopsia help to establish this non-progressive dystrophy.

Clinically, S-cone monochromatism presents like the other achromatopsias, with nystagmus and photophobia during infancy, but the disease is less severe. Family history can help distinguish this entity, because S-cone monochromatism is X-linked recessive, whereas the other achromatopsias are autosomal recessive. Special color testing or an ERG at a specified wavelength (440 nm) are useful diagnostic tools when interpreted in the correct clinical context (Michaelides et al. 2003, 2005).

Progressive Retinal Dysfunction

Retinitis pigmentosa (RP), also known as *rod-cone dystrophy*, has both genetic and allelic heterogeneity (Figs. 55.2 and 55.3) (Daiger et al. 2007). RP inheritance can be autosomal dominant, autosomal recessive, or X-linked recessive, and there is genetic heterogeneity even within each group. Retinal degeneration slow (RDS) protein and retinal-specific adenosine triphosphate (ATP)-binding cassette transporter (ABCA4) are examples of allelic heterogeneity in that they can cause macular dystrophy in some patients and cone-rod dystrophy in others.

Approximately 50% of patients have no family history of RP (termed *simplex RP*) or evidence of parental consanguinity; many of these cases are likely to be autosomal recessive, but some may be males who have X-linked disease from female carriers. Others may be new AD mutations or manifestations of AD disease with reduced penetrance. Accurate genetic counseling depends on identifying the causative mutation and mode of inheritance.

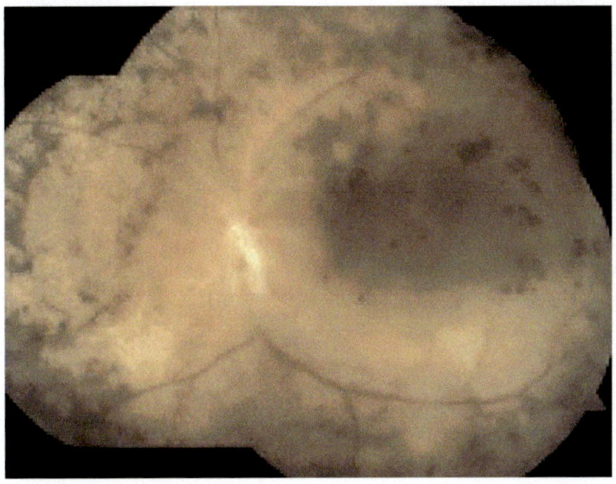

Fig. 55.2 Leber congenital amaurosis at its end stage in a 37-year-old woman with retinal pigment epithelium (RPE) migration, near-total loss of RPE cells, severe retinal atrophy, and peripapillary ghost vessels in her left eye. She retains light perception vision and has an extinguished full-field electroretinogram response

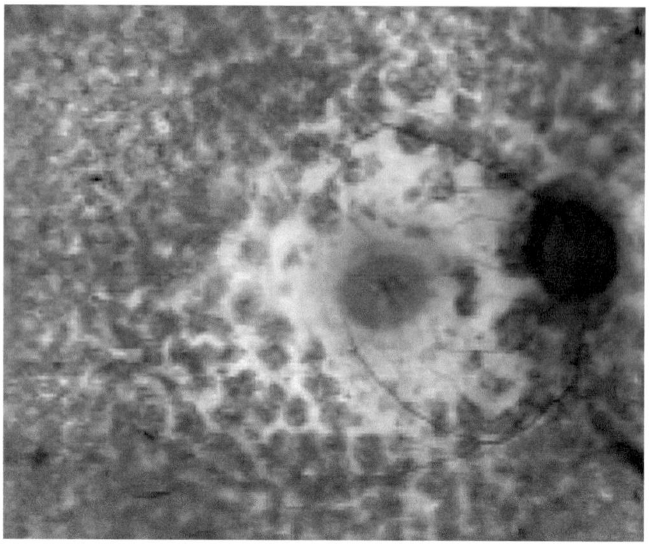

Fig. 55.3 Usher syndrome in a 34-year-old woman. Autofluorescence of the right eye shows retinal pigment epithelium atrophy under degenerated rods. Note the cystoid macular edema, a common complication of rod-cone dystrophies

The age of onset of RP is variable, and patients do not always present with the classic triad of intraretinal pigment migration, optic nerve pallor, and attenuated vessels (Fig. 55.3). In general, patients who develop symptoms at younger ages have worse prognoses, and AD disease has a less severe natural history with later onset than XLR and AR variants. In RP, normal VA and smaller, higher-density autofluorescent rings correlate with worse pattern ERGs, representing a shrinking area of photoreceptor function as the disease progresses (Robson et al. 2003, 2004). In time, VA declines, visual fields coalesce to give the classic peripheral ring scotoma, and posterior subcapsular cataracts or cystoid macular edema (CME) may develop (Fig. 55.3).

Enhanced S-cone, also known as Goldmann-Favre syndrome, is a subtype of AR RP featuring larger photopic single-flash ERG a-wave than 30-Hz b-wave amplitudes. Patients also have supernormal S-cone function due to an increased number of S-cones, low numbers of L-cones and M-cones, and a lack of rods (Greenstein et al. 1996; Milam et al. 2002; Sharon et al. 2003). Goldmann-Favre is one of the most frequently described hereditary vitreoretinal disorders (Table 55.3). Although the fundus appearance is variable, patients may have deep, nummular-shaped RPE clumping in the mid-periphery, with central or peripheral retinoschisis.

Choroideremia is an XLR rod-cone dystrophy that has marked atrophy of the choroid and RPE but does not feature intraretinal pigment migration (MacDonald et al. 2004). It is important to distinguish carriers of this trait from carriers of X-linked RP (XLRP) because they have similar fundi, but the prognosis for affected children is different (Fig. 55.4).

Table 55.3 Hereditary vitreoretinal disorders
X-linked juvenile retinoschisis
Goldmann-Favre syndrome
Wagner disease
Stickler syndrome
Familial exudative vitreoretinopathy
Autosomal dominant neovascular inflammatory vitreoretinopathy (large Iowan pedigree)
Autosomal dominant vitreoretinochoroidopathy

Boys with XLRP are more severely impaired than those with choroideremia. The distinction can be made with ERG of their mothers at childbearing age: carriers of choroideremia are normal, whereas carriers of XLRP are often abnormal and sometimes asymmetrical (Berson et al. 1979).

Gyrate atrophy, an AR dystrophy, is part of the differential diagnosis of choroideremia. The characteristic fundus appearance of gyrate atrophy is hyperpigmented fundi with lobular loss of the RPE and choroid starting in the periphery. A defect in ornithine aminotransferase, which results in a tenfold increase in plasma ornithine, causes the disease. Although a diet restricted in arginine limits the disease (Kaiser-Kupfer et al. 2004), the diet is difficult to maintain and requires careful monitoring. Vitamin B6 lowers plasma ornithine levels in some patients.

Other well-known types of syndromic RP include Alström syndrome, Kearns-Sayre syndrome, Refsum syndrome, Werner syndrome, Cockayne syndrome, Flynn-Aird syndrome, Bardet-Biedl syndrome, Usher syndrome, Joubert syndrome, Batten disease, Zellweger syndrome, and spinocerebellar ataxia.

There are also progressive cone dystrophies, which, in contrast to the stationary cone dysfunction syndromes discussed above, usually present in older childhood or adulthood (Kurz-Levin et al. 2002). Patients' presenting symptoms are decreasing VA and color vision. In the late stages, fundus exam may show a typical bull's eye maculopathy, but subtle cases can be picked up earlier with ERG and occasionally autofluorescence imaging (Fig. 55.1) (Simunovic and Moore 1998).

Progressive cone-rod dystrophies (Simunovic and Moore 1998; Gouras et al. 1983) show the clinical features of cone dystrophy early on; eventually, rod involvement with associated night blindness becomes evident (Holopigian et al. 2004; Michaelides et al. 2006). There are reports of AD, AR, and XLR inheritance. The fundus exam progresses from macular atrophy in the early stages to peripheral retinal pigmentation, arteriolar attenuation, and optic nerve pallor in the later stages.

A number of acquired disorders with an abnormal ERG can phenocopy retinal dystrophies. These include cancer-associated retinopathy (CAR), melanoma-associated reti-

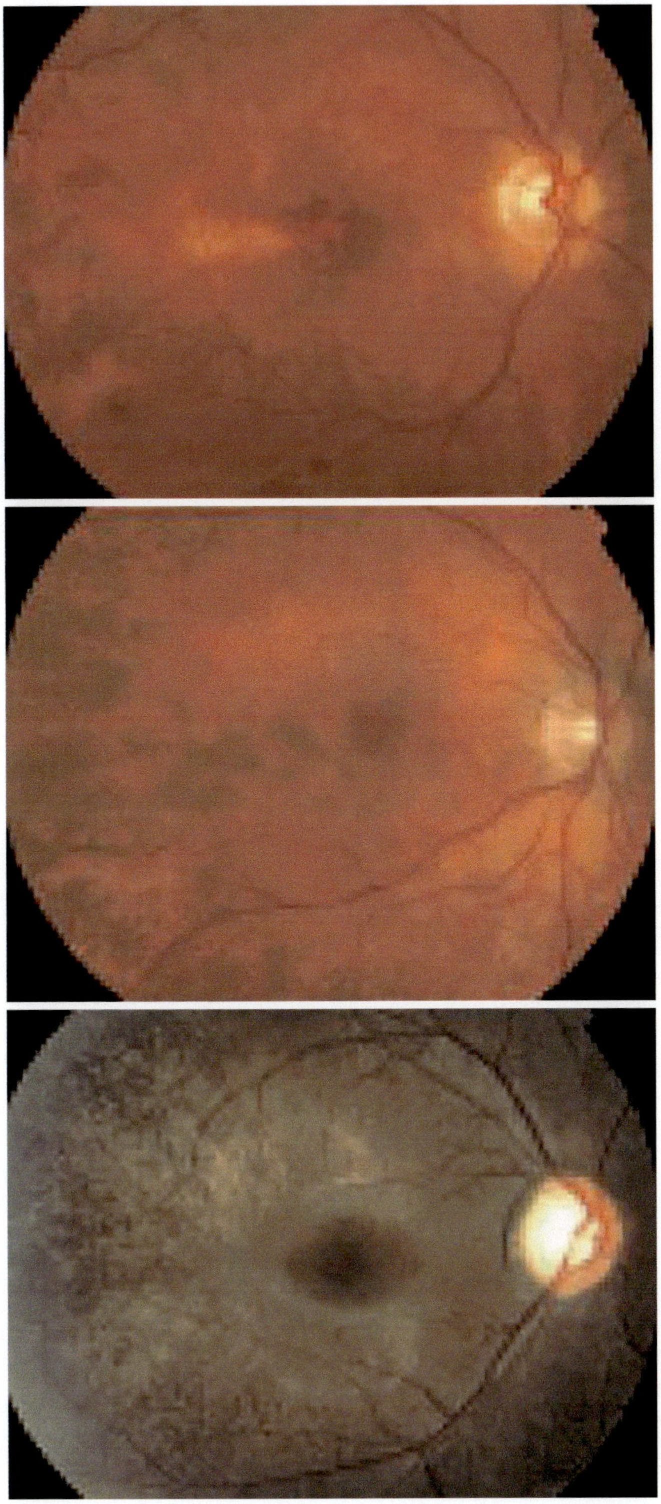

Fig. 55.4 Fundus photographs. (*Top*) Choroideremia carrier with retinal pigment epithelium (RPE) mottling. (*Middle*) Ocular albinism carrier with radially oriented mosaic RPE hypopigmentation. (*Bottom*) X-linked recessive retinitis pigmentosa carrier with a tapeto-like retinal sheen. Correct diagnosis in X-linked heterozygous carriers is important, but fundus distinctions may be subtle

nopathy, acute zonal occult outer retinopathy, diffuse unilateral subacute neuroretinitis, syphilis, foreign body, drug toxicity, and trauma. Clinical history and/or unilateral presentation distinguish these entities from inherited retinal dystrophies.

Central Dystrophies

In general, only the ERG can establish the functional phenotype in patients with maculopathy. In contrast to global retinal diseases and cone-predominant degenerations, central or macular dystrophies have normal full-field ERG implicit time responses and normal peripheral visual fields. Macular dystrophies cause a reduction in central vision, but they usually progress slowly and patients retain some central vision. Thus, most individuals with central dystrophies can enjoy normal activities of daily living (unlike those with progressive, generalized, degenerative dystrophies).

A review of patients with bull's eye lesions showed normal full-field ERGs in most cases. Others showed cone dystrophy, cone-rod dystrophy, or rod-cone dystrophy, but the functional phenotype could not be established by fundus appearance (Kurz-Levin et al. 2002). Similarly, the functional phenotype in Stargardt disease could not be predicted from the fundus phenotype; only the ERG was concordant within affected families, and only the ERG gave prognostic information regarding retention of peripheral vision.

Central dystrophies appear in childhood or early adulthood. Discussed here are Stargardt disease, pattern dystrophy, Best disease, Sorsby pseudoinflammatory disease, and Doyne honeycomb retinal dystrophy.

Stargardt Disease

The most common form of inherited juvenile macular degeneration is Stargardt disease. Classically, Stargardt patients present in childhood with decreased central vision, foveal atrophy, and yellow pisciform flecks at the level of the RPE in the macula. Stargardt and fundus flavimaculatus are manifestations of the same genetic disorder, due to a recessive defect in the *ABCA4* gene (Allikmets et al. 1997). Stargardt patients should avoid a diet high in vitamin A, because the defective gene (*ABCA4*) encodes for a transmembrane transporter of A2E intermediates, a toxic by-product of vitamin A (Koenekoop 2003).

At least 80% of Stargardt patients have a "silent choroid," which is a dark choroid on fluorescein angiogram. A2E accumulation in the RPE causes this phenomenon (Bui et al.

2006). Stargardt patients' functional phenotypes may not be predictable from fundus exam, but ERG can help prognose peripheral vision of patients by dividing them into three groups:

- Type 1 patients (Figs. 55.5 and 55.6) have a normal full-field ERG.
- Type 2 patients have more loss of photopic function.
- Type 3 patients (Fig. 55.7) have both abnormal scotopic and photopic ERG, with a worse prognosis (Lois et al. 2001).

It is important to subtype Stargardt patients for better counseling regarding their prognoses. Fundus flavimaculatus frequently manifests as type 1 Stargardt disease. Macular autofluorescence is usually abnormally high in Stargardt patients, but normal levels of lipofuscin could also indicate late disease in which the RPE cells have burnt out (Lois et al. 2004).

Other Central Dystrophies

Pattern dystrophies are a group of heterogeneous diseases that are caused by a defect in the *RDS* gene (Francis et al. 2005) and other genes. There are four recognized types, based on fundoscopic appearance: adult-onset vitelliform, butterfly, reticular, and fundus pulverulentus (Fig. 55.8). They present in adulthood with decreased vision and can be mistaken for age-related macular degeneration (AMD) (Daniele et al. 1996).

Best disease, or vitelliform dystrophy, is an AD maculopathy caused by a defect in the bestrophin gene. The yolk-like yellow lesion is hyperautofluorescent and usually bilateral (Fig. 55.9). Vision is good during this initial stage but decreases during the scrambled-egg stage or secondary to geographic atrophy in the later stages. In addition, about 20% of patients develop a choroidal neovascular membrane, which, although self-limited, causes poorer vision. Electrooculogram (EOG) shows a reduced light rise in Arden ratio in affected Best disease as well as AD vitreoretinochoroidopathy (ADVIRC). EOG is mildly subnormal or normal in adult vitelliform macular dystrophy. Best disease and ADVIRC (both caused by bestrophin gene defects) are examples of allelic heterogeneity, whereas Best disease and adult vitelliform macular dystrophy are examples of genetic heterogeneity.

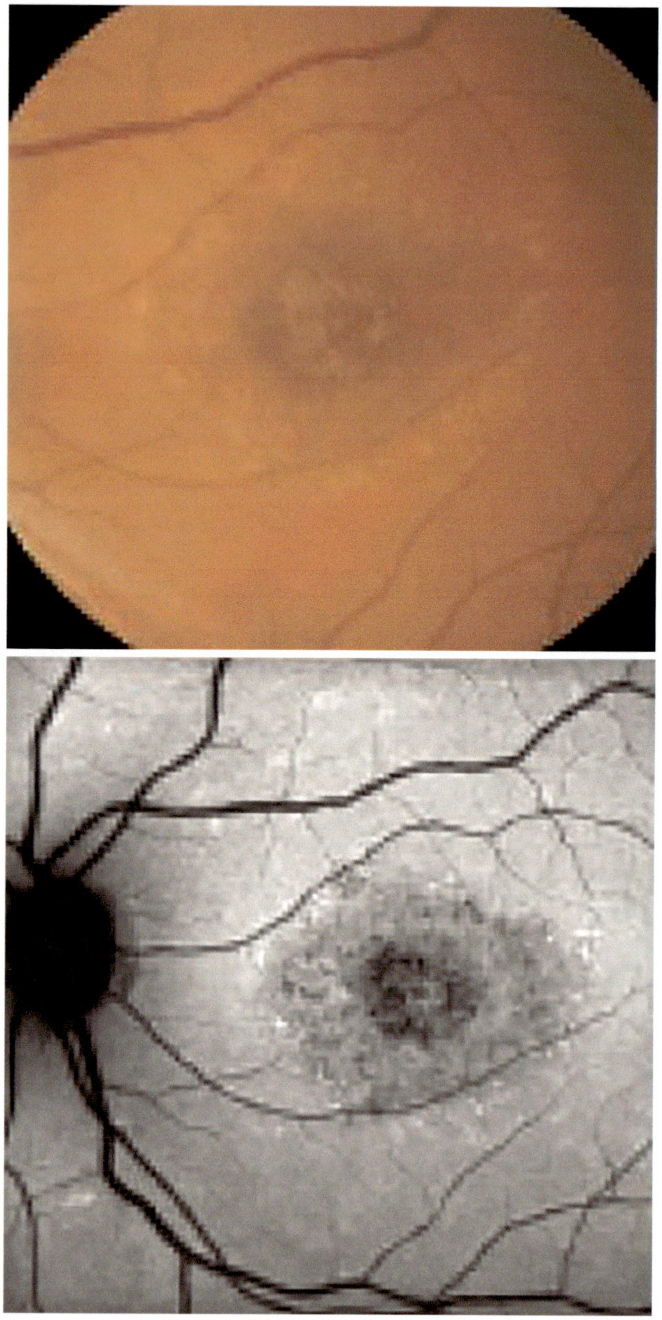

Fig. 55.5 Group 1 Stargardt disease in a 36-year-old woman. (*Top*) Color fundus photograph of the left eye shows macular flecks with a classic beaten bronze appearance. (*Bottom*) Autofluorescence of the same eye shows hyperfluorescent flecks and hypofluoresent, diseased retinal pigment epithelium. This patient has extinguished P50 responses on pattern electroretinogram (ERG), but normal full-field ERG responses

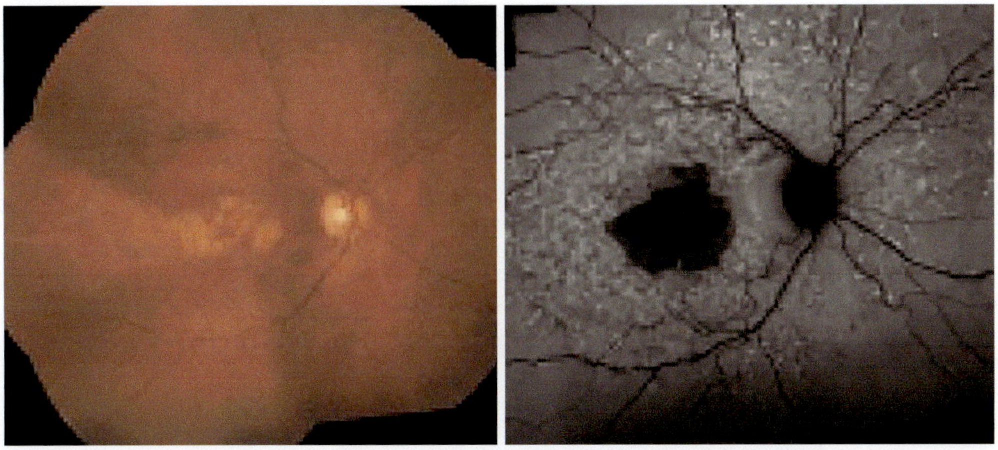

Fig. 55.6 Group 1 Stargardt disease in a 36-year-old man. Color fundus photograph of the right eye with macular atrophy and generalized flecks (*left*) and corresponding autofluorescent imaging (*right*) shows retinal pigment epithelium atrophy (hypofluorescence) centrally and hyperfluorescent flecks. This patient has extinguished P50 responses on pattern electroretinogram (ERG), but normal full-field ERG responses

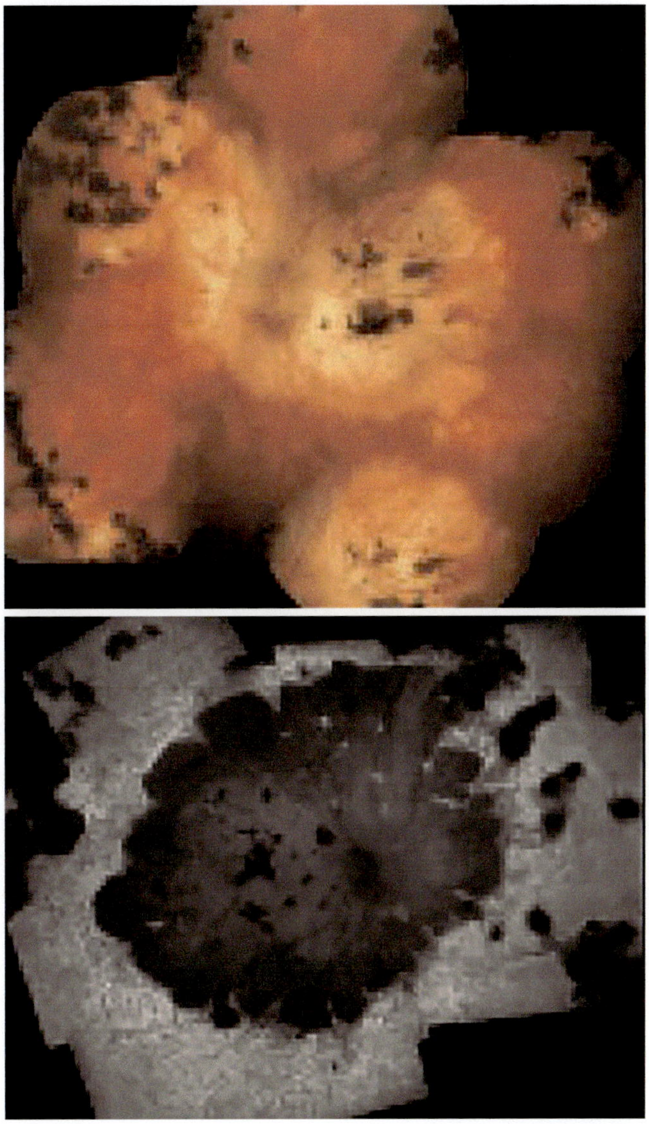

Fig. 55.7 Group 3 Stargardt disease in a 57-year-old woman with bilaterally symmetric disease. (*Top*) Color fundus photo of the left eye, with large areas of atrophic retina and pigment clumping. (*Bottom*) Autofluorescence of the right eye with a large area of missing retinal pigment epithelium and generalized hyperfluorescent flecks peripherally. Both her scotopic and photopic electroretinogram responses are diminished

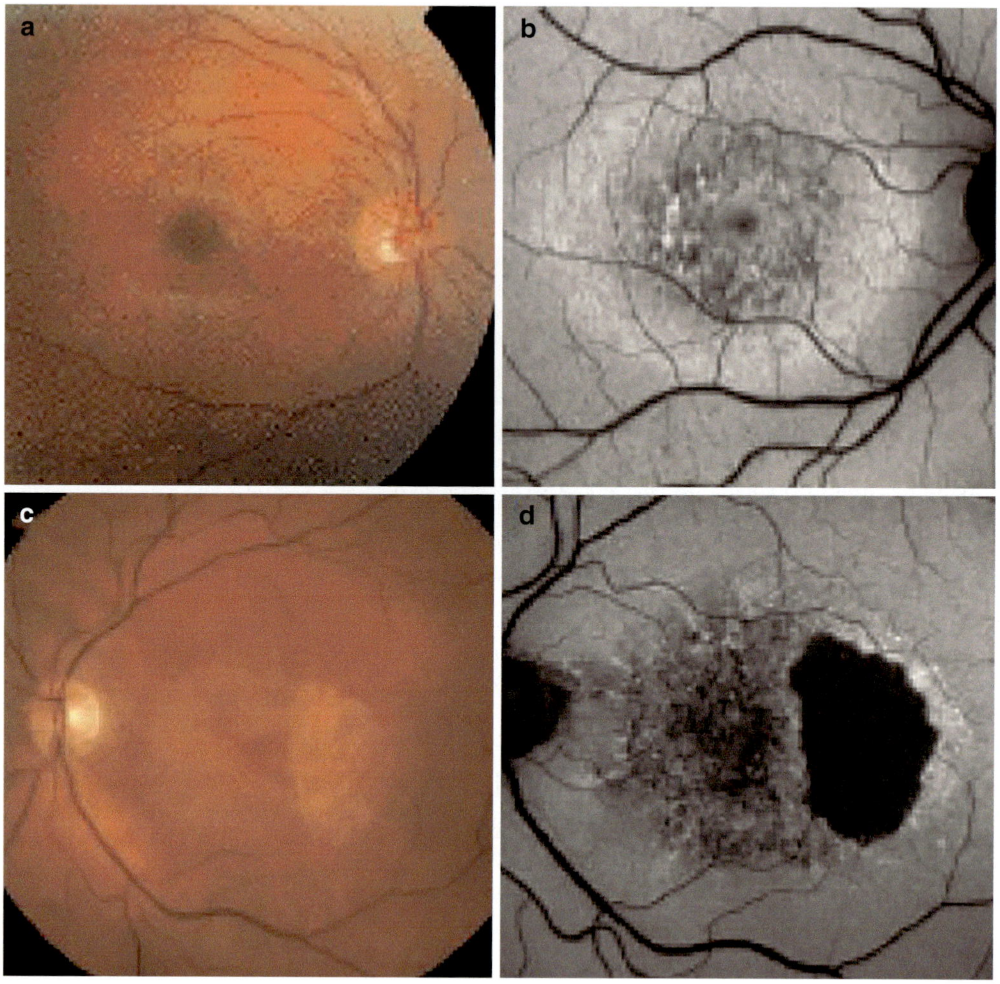

Fig. 55.8 Pattern dystrophy in a 26-year-old woman with 20/15 visual acuity in each eye. (**a**), Color fundus photograph of the right eye with subtle retinal pigment epithelium (RPE) irregularities in the macula. (**b**), Corresponding autofluorescence with speckled hyperfluorescence. (**c**, **d**) Her father had a larger area of RPE atrophy seen on a color fundus photograph of the left eye and corresponding autofluorescence. He had 20/40 vision in each eye

Sorsby macular dystrophy, a dominantly inherited disease with a defect in the tissue inhibitor of the metalloproteinase-3 gene, leads to eventual bilateral subfoveal choroidal neovascularization, although the disease may progress asymmetrically (Fig. 55.10). Beginning around the fifth decade of life, patients present with complaints of problems transitioning between light and dark, followed by central-vision abnormalities. Early on, fine drusen accumulate in the macula, and they later develop a pseudoinflammatory appearance, which may be misdiagnosed as punctate inner choroidopathy (Atan et al. 2004). Individuals with Sorsby have normal pattern ERGs until neovascularization develops.

Doyne honeycomb retinal dystrophy (malattia Leventinese) is a dominantly inherited dystrophy with drusen-like deposits in the macula and peripapillary retina. It can be mistaken for AMD but is distinguished by its characteristic finding of nasal drusenoid material and autofluorescence (Gregory et al. 1996). Individuals with Doyne honeycomb retinal dystrophy have normal full-field ERG responses.

North Carolina macular dystrophy is AD with complete penetrance. Initially, researchers hypothesized that the genetic defect originated from a single family, but the disease is actually present in many distinct genealogies. The clinical appearance is bilateral drusen and macular changes in a young child that may resemble a macular coloboma or staphyloma (Fig. 55.11) (Traboulsi 1998). Despite the severe-appearing fundus, the visual prognosis is good, and the disease is non-progressive. Individuals with North Carolina macular dystrophy have normal full-field ERG recordings. CHAPTER

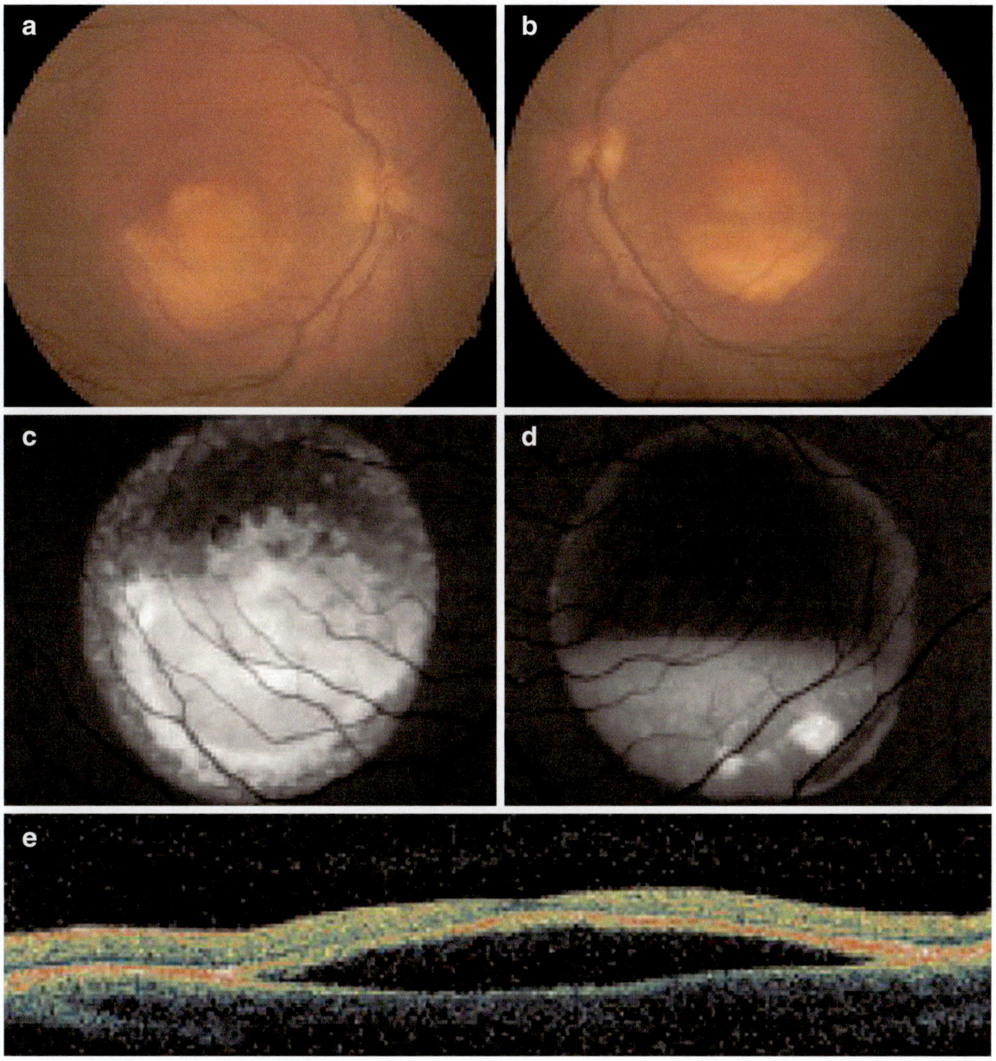

Fig. 55.9 Best vitelliform dystrophy in an 18-year-old man. Color fundus photo of the left eye (**a**) and right eye (**b**) with resolving egg-yolk lesion. Autofluorescence of the left eye (**c**) and right eye (**d**) with corresponding hyperfluorescence. Optical coherence tomography (**e**) shows splitting of the retinal pigment epithelium with elevation. Visual acuity is 20/200 OD and 20/20 OS

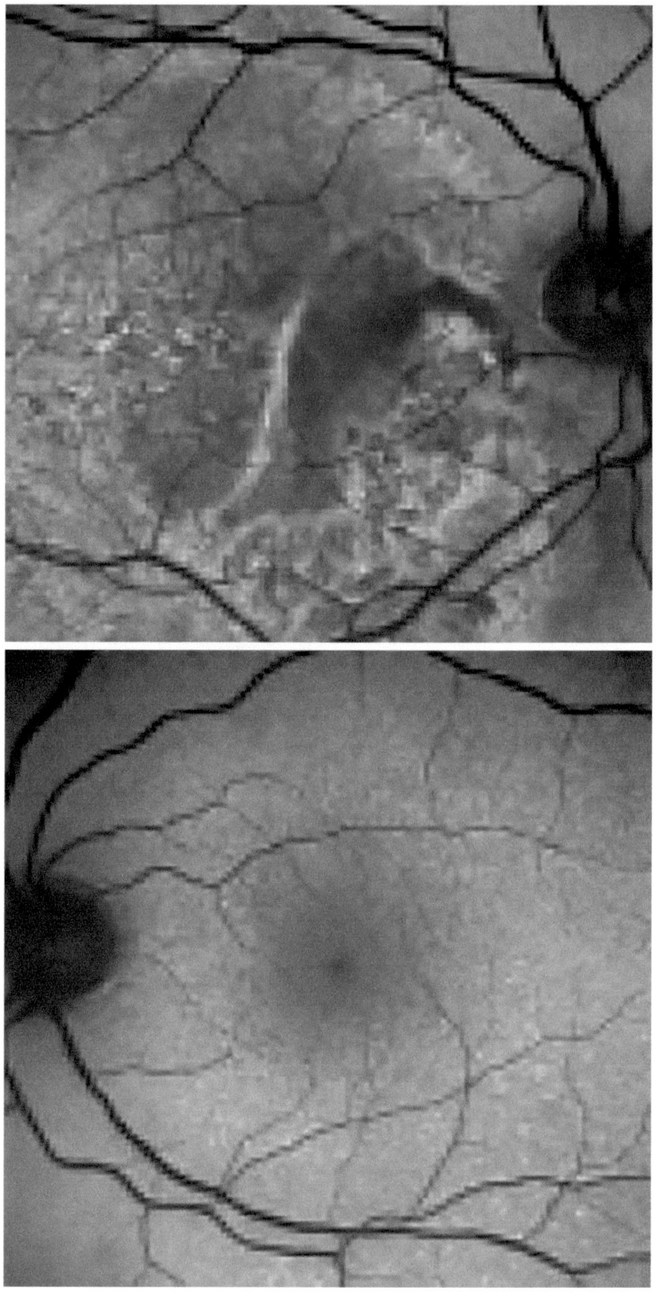

Fig. 55.10 Sorsby pseudoinflammatory dystrophy in a 53-year-old man with asymmetric disease who has an retinal pigment epithelium tear in the right eye (count-finger vision) (*Top*) secondary to choroidal neovascularization, and fine drusen-like deposits in the left eye (20/20 visual acuity) on autofluorescence (*Bottom*)

Fig. 55.11 North Carolina macular dystrophy in a 12-year-old girl. Color fundus photography shows coloboma-like defects with scalloped edges worse in her right eye (**a**) than her left eye (**b**). Her father (**c**) has a milder presentation (grade 1) of North Carolina macular dystrophy, with fine, drusen-like deposits that appear hyperfluorescent on autofluorescence (**d**)

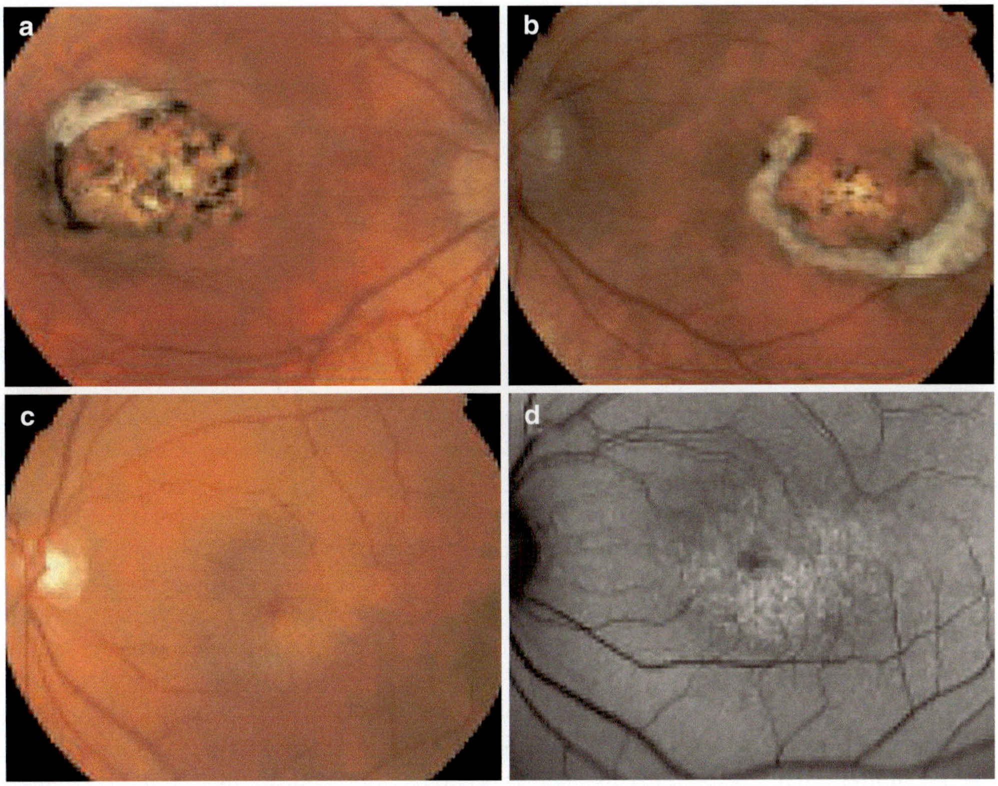

Treatments

Dietary Treatment and Vitamins

It is important to recognize gyrate atrophy and the three secondary causes of retinitis pigmentosa, which are Refsum syndrome, alpha tocopherol transfer protein deficiency, and abetalipoproteinemia. For these retinal dystrophies, dietary treatment can help significantly. The main clinical manifestations of Refsum syndrome are RP, cataracts, chronic polyneuropathy, cerebellar ataxia, and cardiac arrhythmias. Elevated serum phytanic acid levels are diagnostic. Avoiding foods high in phytanic acid (e.g., fat and butter) and plasmapheresis help to improve all neurologic signs (Claridge et al. 1992; Leroy et al. 2003).

Bassen-Kornzweig syndrome (abetalipoproteinemia) is due to malabsorption of cholesterol, fats, and fat-soluble vitamins from the small intestine. Deficiencies of vitamin A and vitamin E cause failure to thrive, peripheral neuropathy with muscle weakness, spinocerebellar ataxia, and RP. Vitamin A (300 IU/kg/ day) and vitamin E (100 IU/kg/ day) restore function and slow the progression of retinal degeneration (Grant and Berson 2001).

Alpha tocopherol transfer protein (α-TTP) deficiency is associated with defective intracellular transport of α-tocopherol, the most biologically active form of vitamin E. Disease progression may be slowed with vitamin E supplementation.

There has also been interest in whether vitamin A supplementation can help other forms of RP. A study in two mouse RP models with different alleles showed that vitamin A supplementation decreased the rate of photoreceptor degeneration caused by a class II rhodopsin mutation (*T17 M*, defective in thermal stability/ folding and cannot easily reconstitute with 11-cis-retinal), but it did not help the mice with a class I rhodopsin mutation (*P347S*, defective in outer-segment localization). It was hypothesized that vitamin A supplementation may work by stabilizing mutant opsins through increased availability of the chromophore (Li et al. 1998). Therefore, vitamin A may benefit a subset of patients with RP.

A randomized controlled study sponsored by the National Institutes of Health (NIH) looking at vitamin therapy for RP in adults showed that vitamin A (15,000 IU) delays the progression of cone ERG loss. The same study showed that vitamin E (400 IU) may have a deleterious effect on RP patients (Berson et al. 1993). These results have not been universally accepted in ophthalmic centers (Norton 1993). Subgroup analysis revealed that patients who started taking docosahexaenoic acid (1200 mg/day) at the same time as vitamin A may have a modest additional benefit of slowing RP (Berson et al. 2004). A safety study did not find substantial side effects of high-dose vitamin A therapy (Sibulesky et al. 1999). Currently, the NIH recommends that adult RP patients take a supplement of 15,000 IU of vitamin A daily under the supervision of an ophthalmologist and avoid use of high-dose vitamin E supplements (National Eye Institute 2007a, b).

Other Treatments

In RP, increased physical activity has been found to be associated with greater self-reported visual function and quality of life (Levinson et al. 2017).

A phase 3 trial of oral N-acetylcysteine for retinitis pigmentosa is underway. Earlier studies demonstrated that some patients may benefit in terms of visual acuity and macular microperimetric sensitivity, but no change in any structural parameter such as the ellipsoid zone width.

Trials of visual cycle modulators are underway for Stargardt disease. These molecules target to prevent retinal degeneration and subsequent vision loss by slowing the accumulation of retinal toxins. Direct visual cycle modulators target the retinal pigment epithelium-specific 65 kDa protein (RPE65), reducing the production of visual chromophore (11-*cis*-retinal) and thus all-trans-retinal, the substrates necessary for the production of toxic bis-retinoids such as A2E. "Safety and Efficacy of Emixustat in Stargardt Disease (SeaSTAR)" is a phase 3 study involving the visual cycle modulator Emixustat (Kubota et al. 2022).

Other molecules inhibit the visual cycle indirectly by targeting vitamin A delivery to the RPE. The RBP4 antagonists reduce blood concentrations of the protein RBP4, which delivers vitamin A to the retina. By reducing the delivery of vitamin A to the retina, these molecules may potentially reduce the accumulation of retinal toxins and prevent retinal degeneration and subsequent vision loss. Trials with two RBP4 antagonists for Stargardt disease, specifically Tinlarebant (Belite Bio, San Diego, CA) and STG-001 (Stargazer Pharmaceuticals, Boston, MA), are in phase 3 and phase 2, respectively. Another similar molecule being tested for Stargardt disease in phase 2 is ALK-001 (Tolerability and Effects of ALK-001 on Stargardt Disease (TEASE), Alkeus Pharmaceuticals, Cambridge, MA). This molecule is a deuterated vitamin A and acts by slowing vitamin A dimer formation 4-5 folds without inhibiting the visual cycle (Scholl et al. 2016).

An NIH sponsored phase 1–2 trial for Stargardt disease has been "Oral metformin for treatment of ABCA4 retinopathy." The mechanism of action of metformin seems to be not limited to AMP dependent kinase activation. It helps scavenging reactive oxygen species generated by mitochondria, enhances autophagy by lysosomal activation, reduces levels of inflammatory cytokines, and inhibits apoptosis.

Treatment of Complications

Physicians should follow RP patients at infrequent but regular intervals to detect treatable complications such as posterior subcapsular cataract, macular edema, and the rare autoimmune reaction. Cataract extraction can improve visual perception and brightness.

Cystoid macular edema (CME) is an important and common cause of central vision loss in later stages of RP. As RP patients already have a limited reserve of photoreceptors and severe peripheral field loss, the impact of CME in RP is tremendous compared to that in other retinal pathologies (Ruff et al. 2022). In most cases, oral or topical carbonic anhydrase inhibitors are first-line treatment options. Subtenon or intravitreal steroids have also been successfully used for CME associated with RP (Schaal et al. 2016).

RP patients with a rapid progression of visual symptoms and ERG progression should have a Western blot analysis for anti-retinal antibody. If this serum test is positive, patients presumably have developed autoimmunoretinopathy or a CAR-like syndrome and should undergo evaluation for an occult malignancy (Heckenlively et al. 2000).

The Coats'-type variant of RP is thought to occur in as many as 3% of all RP patients. These fundi have typical bone spicules of RP with Coats-type changes in the inferotemporal quadrants (Fig. 55.12). Retinal telangiectasia and exudative retinal detachments are treated with photocoagulation or cryosurgery as indicated (Khan et al. 1998).

Gene Therapy

The key to gene-based treatments is efficient and accurate genotyping (Weleber 2005; Sieving and Collins 2007; Stone 2007). If genetic testing is not available on-site, blood can be drawn in a tube with ethylenediaminetetraacetic acid (EDTA) and sent at room temperature to a Clinical Laboratory Improvement Amendment (CLIA)-certified laboratory for DNA testing (National Eye Institute 2007a, b; GeneTests Home Page 2007).

One commercially available method is a genotyping microarray (gene chip) for the *ABCA4* gene (Asper Ophthalmics 2007). Allelic heterogeneity in the *ABCA4* gene has been associated with five distinct phenotypes, including Stargardt disease/fundus flavimaculatus, cone-rod dystrophy, and AMD. This gene chip screens for more than 400 different *ABCA4* variants with >98% efficacy of finding them (Jaakson et al. 2003).

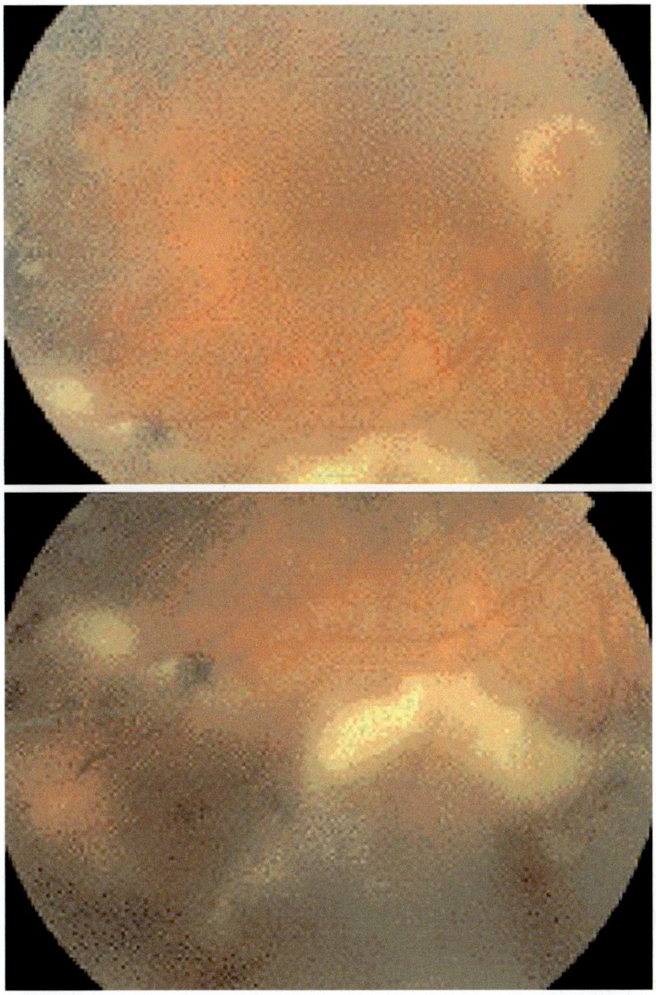

Fig. 55.12 Coats'-type autosomal recessive retinitis pigmentosa in a 55-year-old man with vitreous haze centrally (*Top*) and subretinal exudates peripherally (*Bottom*) in his right eye

Gene therapy is a promising treatment for retinal degeneration (Bennett et al. 1996; Dunaief et al. 1995; Stone 2005). It is applicable when there is a loss of function due to genetic defect. The eye is an ideal organ for gene therapy because it is easily accessible, highly compartmentalized, and is an immune-privileged site. In 2001, vision was restored in *RPE65* deficient dogs, an animal model for one type of early-onset retinal dystrophy, known as Leber congenital amaurosis (LCA), by injecting subretinal adeno-associated virus (AAV) carrying *RPE65* (Acland et al. 2001; Bainbridge et al. 2006). Regain of function was demonstrated with ERG, pupillometry, and behavioral testing and was maintained for over 3 years (Acland et al. 2005).

This proof of principle demonstrating functional improvement following gene replacement of *RPE65* led to the first ever FDA-approved gene therapy. Voretigene Neparvovec (Luxturna, Spark Therapeutics, Philadelphia, PA) is a recombinant AAV-mediated gene therapy for patients with LCA and AR-RP with biallelic RPE65 mutation.

A phase 1-2 study of subretinal injection of AAV-mediated gene therapy for X-Linked Retinitis Pigmentosa (XLRP) due to mutations in the gene encoding retinitis pigmentosa GTPase regulator (RPGR) (4D-125 in Patients With XLRP, 4D Molecular Therapeutics, Emeryville, CA) is underway. Emerging on the horizon are more new gene-based treatments for many other of these devastating retinal dystrophies.

IRD mostly involve a loss of RPE and/or photoreceptor cells. However, the remaining visual pathway including the bipolar and retinal ganglion cells (RGC) remains mostly intact. Optogenetics for IRD involves the transfer of opsin or other genes to the retinal ganglion cells or the bipolar cells to gain visual function within these cells and hence achieve an alternate light-sensitive system in the retina. This treatment may be applied to any genotype and in late-stage degeneration, when the photoreceptors are almost completely lost. However, the gained neural response does not follow the regular and well-organized neural networks of retina; hence, even if successful, optogenetic treatment may produce incoherent responses upstream in the visual pathway. Targeting the bipolar cells, instead of the retinal ganglion cells, may be theoretically more advantageous in an optogenetic approach for a number of reasons. First, bipolar cells are closer to photoreceptors, and hence visual processing circuitry may be utilized to a greater extent theoretically. The greater number of the bipolar cells compared to the RGCs may potentially yield a higher spatial resolution. And unlike the RGCs, the absence of lateral extending processes in bipolar cells may provide the possibility of a focal activation. Two optogenetic trials in phase 2 involve the study of intravitreal injection of multi-characteristic opsins (MCOs) for retinitis pigmentosa (Efficacy and Safety of of vMCO-010 Optogenetic Therapy in Adults With Retinitis Pigmentosa [RESTORE study], Nanoscope Therapeutics Inc., Dallas TX) and Stargardt disease (Safety and Effects of a Single Intravitreal Injection of vMCO-010 Optogenetic Therapy in Subjects With Stargardt Disease [STARLIGHT study], Nanoscope Therapeutics Inc., Dallas TX).

Other Emerging Therapies

Another promising idea in treating retinal degenerations has come from using ciliary neurotrophic factor (CNTF) to rescue photoreceptors. At least 13 animal models of RP show improvement with CNTF. Recently, a phase 1 trial evaluated surgically implanted capsules loaded with human RPE cells transfected with the CNTF gene. The treatment was well tolerated with minimal direct side effects, and the visual results were promising (Sieving et al. 2006).

Retinal-cell transplantation is another way to replace damaged photoreceptors, which has been worked on for many years (Gouras et al. 2002). Studies in mouse models (Tsang et al. 1996) show that timing of transplantation is crucial (MacLaren et al. 2006). There is a specific window period, coincident with the peak of rod genesis, during which the donor cells can be harvested that will enable the generation of new photoreceptors in diseased mice (MacLaren et al. 2006).

Clinical use of stem-cell transplantation is on the horizon (Columbia University Medical Center 2023).

Summary

Retinal dystrophies are a heterogeneous group of disorders whose classifications are evolving as retinal physicians better understand their phenotypes, genotypes, and pathophysiology. Genetic counseling is an important part of taking care of the patient. ERG continues to be the mainstay of diagnosis, and autofluorescence imaging has become essential for better phenotyping and following of disease progression. Electrophysiological studies are most helpful to phenotype different retinal conditions that ophthalmoscopy may show to be equivalent or similar.

Eyecare professionals have an active role in caring for patients with retinal dystrophies (Fishman and Apushkin 2007; Sieving 2005). We manage complications that arise, such as cataracts, leaking vessels, and macular edema, and we also give genetic counseling and low-vision evaluations. Looking to the horizon, retinal physicians should help enroll patients in genetic registries so that they can contact patients when a specific genetic cure becomes available.

Many trials of different treatment approaches for IRDs at different stages are underway. Providing relevant information to each patient is becoming part of the eyecare.

Acknowledgment The 1st edition of this chapter previously appeared in the April 2007 issue of *Retinal Physician*, published by Pentavision LLC; *reprinted with permission*.

References

Acland GM, Aguirre GD, Ray J, Zhang Q, Aleman TS, Cideciyan AV, et al. Gene therapy restores vision in a canine model of childhood blindness. Nat Genet. 2001;28:92–5.

Acland GM, Aguirre GD, Bennett J, Aleman TS, Cideciyan AV, Bennicelli J, et al. Long-term restoration of rod and cone vision by single dose rAAV- mediated gene transfer to the retina in a canine model of childhood blindness. Mol Ther. 2005;12:1072–82.

Allikmets R, Singh N, Sun H, Shroyer NF, Hutchinson A, Chidambaram A, et al. A photoreceptor cell-specific ATP-binding transporter gene (ABCR) is mutated in recessive Stargardt macular dystrophy. Nat Genet. 1997;15:236–46.

Asper Ophthalmics. ABCR genetic testing. Available at: http://www.asperoph-thalmics.com/ABCRgenetest.htm; 2007. Accessed 20 Mar 2007.

Atan D, Gregory Evans CY, Louis D, Downes SM. Sorsby fundus dystrophy presenting with choroidal neovascularisation showing good response to steroid treatment. Br J Ophthalmol. 2004;88:440–1.

Bainbridge JW, Tan MH, Ali RR. Gene therapy progress and prospects: the eye. Gene Ther. 2006;13:1191–7.

Bennett J, Tanabe T, Sun D, Zeng Y, Kjeldbye H, Gouras P, Maguire AM. Photoreceptor cell rescue in retinal degeneration (rd) mice by in vivo gene therapy. Nat Med. 1996;2:649–54.

Berson E. Electroretinographic testing as an aid in determining visual prognosis in families with hereditary retinal degenerations. New York: Retina Congress Appleton-Century-Crofts; 1974.

Berson EL, Gouras P, Gunkel RD. Rod responses in retinitis pigmentosa, dominantly inherited. Arch Ophthalmol. 1968;80:58–67.

Berson EL, Rosen JB, Simonoff EA. Electroretinographic testing as an aid in detection of carriers of X-chromosome-linked retinitis pigmentosa. Am J Ophthalmol. 1979;87:460–8.

Berson EL, Rosner B, Sandberg MA, Hayes KC, Nicholson BW, Weigel-DiFranco C, Willett W. A randomized trial of vitamin A and vitamin E supplementation for retinitis pigmentosa. Arch Ophthalmol. 1993;111:761–72.

Berson EL, Rosner B, Sandberg MA, Weigel-DiFranco C, Moser A, Brockhurst RJ, et al. Further evaluation of docosahexaenoic acid in patients with retinitis pigmentosa receiving vitamin A treatment: subgroup analyses. Arch Ophthalmol. 2004;122:1306–14.

Bird AC. Retinal photoreceptor dystrophies LI. Edward Jackson Memorial Lecture. Am J Ophthalmol. 1995;119:543–62.

Bui TV, Han Y, Radu RA, Travis GH, Mata NL. Characterization of native retinal fluorophores involved in biosynthesis of A2E and lipofuscin-associated retinopathies. J Biol Chem. 2006;281:18112–9.

Claridge KG, Gibberd FB, Sidey MC. Refsum disease: the presentation and ophthalmic aspects of Refsum disease in a series of 23 patients. Eye. 1992;6:371–5.

Columbia University Medical Center. Stem Cell Consortium. Available at: http://www.cumc.columbia.edu/stemcell; 2023. Accessed 14 Jan 2023.

Daiger SP, Bowne SJ, Sullivan LS. Perspective on genes and mutations causing retinitis pigmentosa. Arch Ophthalmol. 2007;125:151–8.

Daniele S, Carbonara A, Daniele C, Restagno G, Orcidi F. Pattern dystrophies of the retinal pigment epithelium. Acta Ophthalmol Scand. 1996;74:51–5.

Delori FC, Staurenghi G, Arend O, Dorey CK, Goger DG, Weiter JJ. In vivo measurement of lipofuscin in Stargardt's disease—fundus flavimaculatus. Invest Ophthalmol Vis Sci. 1995;36:2327–31.

Dunaief JL, Kwun RC, Bhardwaj N, Lopez R, Gouras P, Goff SP. Retroviral gene transfer into retinal pigment epithelial cells followed by transplantation into rat retina. Hum Gene Ther. 1995;6:1225–9.

Fishman GA, Apushkin MA. Continued use of dorzolamide for the treatment of cystoid macular oedema in patients with retinitis pigmentosa. Br J Ophthalmol. 2007;91:743–5.

Francis PJ, Schultz DW, Gregory AM, Schain MB, Barra R, Majewski J, et al. Genetic and phenotypic heterogeneity in pattern dystrophy. Br J Ophthalmol. 2005;89:1115–9.

GeneTests Home Page. Available at: http://www.genetests.org; 2007. Accessed 20 Mar 2007.

Gouras P, Eggers HM, MacKay CJ. Cone dystrophy, nyctalopia, and supernormal rod responses. A new retinal degeneration. Arch Ophthalmol 1983;101:718–24.

Gouras P, Kong J, Tsang SH. Retinal degeneration and RPE transplantation in Rpe65(−/−) mice. Invest Ophthalmol Vis Sci. 2002;43:3307–11.

Grant CA, Berson EL. Treatable forms of retinitis pigmentosa associated with systemic neurological disorders. Int Ophthalmol Clin. 2001;41:103–10.

Greenstein VC, Zaidi Q, Hood DC, Spehar B, Cideciyan AV, Jacobson SG. The enhanced S cone syndrome: an analysis of receptoral and post-receptoral changes. Vis Res. 1996;36:3711–22.

Gregory CY, Evans K, Wijesuriya SD, Kermani S, Jay MR, Plant C, et al. The gene responsible for autosomal dominant Doyne's honeycomb retinal dystrophy (DHRD) maps to chromosome 2p16. Hum Mol Genet. 1996;5:1055–9.

Heckenlively JR, Arden GB. Principles and practice of clinical electrophysiology of vision. 2nd ed. Cambridge, MA: MIT Press; 2006.

Heckenlively JR, Fawzi AA, Oversier J, Jordan BL, Aptsiauri N. Autoimmune retinopathy: patients with antirecoverin immunoreactivity and panretinal degeneration. Arch Ophthalmol. 2000;118:1525–33.

Holder GE. Pattern electroretinography (PERG) and an integrated approach to visual pathway diagnosis. Prog Retin Eye Res. 2001;20:531–61.

Holopigian K, Greenstein VC, Seiple W, Hood DC, Carr RE. Rod and cone photoreceptor function in patients with cone dystrophy. Invest Ophthalmol Vis Sci. 2004;45:275–81.

Hood DC, Birch DG. Beta wave of the scotopic (rod) electroretinogram as a measure of the activity of human on-bipolar cells. J Opt Soc Am A Opt Image Sci Vis. 1996;13:623–33.

International Society for Clinical Electrophysiology of Vision. Standards, recommendations and guidelines. Available at: http://www.iscev.org/standards/index.html; 2007. Accessed 20 Mar 2007.

Jaakson K, Zernant J, Külm M, Hutchinson A, Tonisson N, Glavac D, et al. Genotyping microarray (gene chip) for the ABCR (ABCA4) gene. Hum Mutat. 2003;22:395–403.

Kaiser-Kupfer MI, Caruso RC, Valle D, Reed FG. Use of an arginine-restricted diet to slow progression of visual loss in patients with gyrate atrophy. Arch Ophthalmol. 2004;122:982–4.

Khan JA, Ide CH, Strickland MP. Coats'-type retinitis pigmentosa. Surv Ophthalmol. 1998;32:317–32.

Koenekoop RK. The gene for Stargardt disease, ABCA4, is a major retinal gene: a mini-review. Ophthalmic Genet. 2003;24:75–80.

Kubota R, Birch DG, Gregory JK, Koester JM. Randomised study evaluating the pharmacodynamics of emixustat hydrochloride in subjects with macular atrophy secondary to Stargardt disease. Br J Ophthalmol. 2022;106:403-408.

Kurz-Levin MM, Halfyard AS, Bunce C, Bird AC, Holder GE. Clinical variations in assessment of bull's-eye maculopathy. Arch Ophthalmol. 2002;120:567–75.

Leroy BP, Hogg CR, Rath PR, McBain V, Kestelyn P, Bird AC, Holder GE. Clinical features & retinal function in patients with adult Refsum syndrome. Adv Exp Med Biol. 2003;544:57–8.

Levinson JD, Joseph E, Ward LA, Nocera JR, Pardue MT, Bruce BB, Yan J. Physical activity and quality of life in retinitis pigmentosa. J Ophthalmol. 2017;2017:6950642.

Li T, Sandberg MA, Pawlyk BS, Rosner B, Hayes KC, Dryja TP, Berson EL. Effect of vitamin a supplementation on rhodopsin mutants threonine-17 --> methionine and proline-347 --> serine in transgenic mice and in cell cultures. Proc Natl Acad Sci U S A. 1998;95:11933–8.

Lois N, Holder GE, Bunce C, Fitzke FW, Bird AC. Phenotypic subtypes of Stargardt macular dystrophy-fundus flavimaculatus. Arch Ophthalmol. 2001;119:359–69.

Lois N, Halfyard AS, Bird AC, Holder GE, Fitzke FW. Fundus autofluorescence in Stargardt macular dystrophy-fundus flavimaculatus. Am J Ophthalmol. 2004;138:55–63.

MacDonald IM, Sereda C, McTaggart K, Mah D. Choroideremia gene testing. Expert Rev Mol Diagn. 2004;4:478–84.

MacLaren RE, Pearson RA, MacNeil A, Douglas RH, Salt TE, Akimoto M, et al. Retinal repair by transplantation of photoreceptor precursors. Nature. 2006;444:203–7.

Michaelides M, Aligianis IA, Holder GE, Simunovic M, Mollon JD, Maher ER, et al. Cone dystrophy phenotype associated with a frameshift mutation (M280fsX291) in the α-subunit of cone specific transducin (GNAT2). Br J Ophthalmol. 2003;87:1317–20.

Michaelides M, Hunt DM, Moore AT. The cone dysfunction syndromes. Br J Ophthalmol. 2004;88:291–7.

Michaelides M, Johnson S, Simunovic MP, Bradshaw K, Holder G, Mollon JD, et al. Blue cone monochromatism: a phenotype and genotype assessment with evidence of progressive loss of cone function in older individuals. Eye. 2005;19:2–10.

Michaelides M, Hardcastle AJ, Hunt DM, Moore AT. Progressive cone and cone-rod dystrophies: phenotypes and underlying molecular genetic basis. Surv Ophthalmol. 2006;51:232–58.

Milam AH, Rose L, Cideciyan AV, Barakat MR, Tang WX, Gupta N, et al. The nuclear receptor NR2E3 plays a role in human retinal photoreceptor differentiation and degeneration. Proc Natl Acad Sci U S A. 2002;99:473–8.

Nakamura M, Hotta Y, Tanikawa A, Terasaki H, Miyake Y. A high association with cone dystrophy in fundus albipunctatus caused by mutations of the RDH5 gene. Invest Ophthalmol Vis Sci. 2000;41:3925–32.

National Eye Institute. Information for doctors who follow patients with retinitis pigmentosa. Available at: http://www.nei.nih.gov/news/clinicalalerts/alert-rp.asp; 2007a. Accessed 20 Mar 2007.

National Eye Institute. National Ophthalmic Disease Genotyping Network (eyeGENE). Available at: http://www.nei.nih.gov/resources/eyegene.asp; 2007b. Accessed 20 Mar 2007.

Norton EW. A randomized trial of vitamin a and vitamin E supplementation for retinitis pigmentosa. Arch Ophthalmol. 1993;111:1460. author reply 1463–5

Robson AG, El-Amir A, Bailey C, Egan CA, Fitzke FW, Webster AR, et al. Pattern ERG correlates of abnormal fundus autofluorescence in patients with retinitis pigmentosa and normal visual acuity. Invest Ophthalmol Vis Sci. 2003;44:3544–50.

Robson AG, Egan CA, Luong VA, Bird AC, Holder GE, Fitzke FW. Comparison of fundus autofluorescence with photopic and scotopic fine-matrix mapping in patients with retinitis pigmentosa and normal visual acuity. Invest Ophthalmol Vis Sci 2004;45:4119–25.

Ruff A, Tezel A, Tezel TH. Anatomical and functional correlates of cystic macular edema in retinitis pigmentosa. PLoS One. 2022;17:e0276629.

Salchow DJ, Gouras P, Doi K, Goff SP, Schwinger E, Tsang SH. A point mutation (W70A) in the rod PDE-gamma gene desensitizing and delaying murine rod photoreceptors. Invest Ophthalmol Vis Sci. 1999;40:3262–7.

Schaal Y, Hondur AM, Tezel TH. Subtenon triamcinolone for cystoid macular edema due to retinitis pigmentosa unresponsive to oral acetazolamide. Can J Ophthalmol. 2016;51:e113–e115.

Scholl HP, Shah SM, Kay CN, Tsang SH, Stepien KE, Bernstein PS, Lam BL, Gorin MB, Washington I, Saad L. TEASE: a phase 2 clinical trial assessing the tolerability and effects of oral once-a-day ALK-001 on Stargardt disease. Invest Ophthalmol Visual Sci 2016;57:2685

Sharon D, Sandberg MA, Caruso RC, Berson EL, Dryja TP. Shared mutations inNR2E3 in enhanced S-cone syndrome, Goldmann-Favre syndrome, and many cases of clumped pigmentary retinal degeneration. Arch Ophthalmol. 2003;121:1316–23.

Sibulesky L, Hayes KC, Pronczuk A, Weigel-DiFranco C, Rosner B, Berson EL. Safety of <7500 RE (<25000 IU) vitamin A daily in adults with retinitis pigmentosa. Am J Clin Nutr. 1999;69:656–63.

Sieving PA. The National Eye Institute: translational clinical research initiatives on inherited and orphan retinal diseases: personal observations. Retina. 2005;25:S8–9.

Sieving PA, Collins FS. Genetic ophthalmology and the era of clinical care. JAMA. 2007;297:733–6.

Sieving PA, Caruso RC, Tao W, Coleman HR, Thompson DJ, Fullmer KR, Bush RA. Ciliary neurotrophic factor (CNTF) for human retinal degeneration: phase I trial of CNTF delivered by encapsulated cell intraocular implants. Proc Natl Acad Sci U S A 2006;103:3896–901.

Simunovic MP, Moore AT. The cone dystrophies. Eye 1998;12:553–65.

Sparrow JR, Boulton M. RPE lipofuscin and its role in retinal pathobiology. Exp Eye Res 2005;80:595–606

Stone EM. Challenges in genetic testing for clinical trials of inherited and orphan retinal diseases. Retina. 2005;25:S72–3.

Stone EM. Genetic testing for inherited eye disease. Arch Ophthalmol. 2007;125:205–12.

Taylor D, Hoyt CS. Pediatric ophthalmology and strabismus. 3rd ed. Elsevier Saunders: New York, NY; 2005.

Traboulsi EI. Genetic diseases of the eye. New York: Oxford University Press; 1998.

Tsang SH, Gouras P. Molecular physiology and pathology of the retina. In: Duane TD, Tasman W, Jaeger AE, editors. Duane's clinical opthalmology. Philadelphia: Lippincott-Raven; 1996.

Tsang SH, Gouras P, Yamashita CK, Kjeldbye H, Fisher J, Farber DB, Goff SP. Retinal degeneration in mice lacking the gamma subunit of rod cGMP phosphodiesterase. Science. 1996;272:1026–9.

Tsang SH, Burns ME, Calvert PD, Gouras P, Baylor DA, Goff SP, Arshavsky VY. Role for the target enzyme in deactivation of photoreceptor G protein in vivo. Science. 1998;282:117–21.

Tsang SH, Woodruff ML, Chen CK, Yamashita CY, Cilluffo MC, Rao AL, et al. GAP-independent termination of photoreceptor light response by excess gamma subunit of the cGMP-phosphodiesterase. J Neurosci. 2006a;26:4472–80.

Tsang SH, Woodruff ML, Jun L, Mahajan V, Yamashita CK, Pedersen R, et al. Transgenic mice carrying the H258N mutation in the gene encoding the beta-subunit of phosphodiesterase-6 (PDE6B) provide a model for human congenital stationary night blindness. Hum Mutat. 2006b;28:243–54.

von Ruckmann A, Fitzke FW, Bird AC. Distribution of fundus autofluorescence with a scanning laser ophthalmoscope. Br J Ophthalmol. 1995;79:407–12.

Weleber RG. Inherited and orphan retinal diseases: phenotypes, genotypes, and probable treatment groups. Retina. 2005;25:S4–7.

Yannuzzi LA, Guyer DR, Green WR. The retina atlas. St. Louis: Mosby; 1995.

CHAPTER 56

Genetic Testing for Inherited Retinal Dystrophies: Basic Understanding

Megan E. Soucy

What Are the Structure and Functions of the Genome? 306
What Is Genomic or Genetic Variation?. 306
What Is Genetic Code? 307
What Is a Mutation?. 307
What Is a Polymorphism? 307
What Is Genetic Testing?. 307
Who Are Genetics Professionals?. 308
Which Retinal Disease Would Benefit from Genetic Testing? 308
What Is the Role of the Ophthalmologist in Genetic Testing?. 308
What Are the Benefits of Genetic Testing?. 309
What Are the Risks of Genetic Testing? 310
Does Everyone with the Same Diagnosis Have the Same Genetic Mutation? 310
What Information Is Given by Genetic Testing?. 310
What Are the Types of Genetic Testing? 311
Does Every Variant Result in Disease?. 312
References 312

M. E. Soucy (✉)
Department of Ophthalmology, Columbia University, New York, NY, USA
e-mail: mes2350@cumc.columbia.edu

S. H. Tsang et al. (eds.), *Atlas of Inherited Retinal Diseases*, Advances in Experimental Medicine and Biology 1467, https://doi.org/10.1007/978-3-031-72230-1_56

What Are the Structure and Functions of the Genome?

The genome is all of the genetic information needed to build and maintain an organism. The instructions in the genome are encoded in deoxyribonucleic acid (DNA). DNA can differ between species and individuals. Within the human genome, 99.9% of the DNA sequence is the same. That 0.1% of change is what makes each individual unique and makes DNA one of the best identifiers of an individual. All DNA is made up of four chemical bases: adenine (A), guanine (G), cytosine (C), and thymine (T). DNA is made up of nucleotides. A nucleotide is a chemical base pair—the A pairs with the T, and the C pairs with the G—that is bound to a sugar and a phosphate. The nucleotides are linked together in a strand. Two pairs of DNA strands are twisted together like a spiral staircase with the two chemical base pairs binding together in the middle, with each strand running in the opposite direction, known as a double helix.

A gene is made of DNA. A gene is a sequence of DNA that has a specific function and is considered a hereditary unit. In the human genome, there is estimated to be approximately 20,000 genes. About 1–2% of DNA encode protein sequences (known as exons). The rest of the non-coding DNA is utilized for a variety of other functions. The genes are packaged into condensed structures called chromosomes.

Humans typically have 46 chromosomes, split in 23 pairs in which one half of each pair is inherited from each parent. The first 22 pairs are the same between sexes, called autosomes. The 23rd pair is known as the sex chromosome, in which biological females typically have two X chromosomes and a biological male typically has an X and a Y chromosome. Each cell in the body has a complete set of all chromosomes. The chromosomes are located in the nucleus of each cell. The specific location of a gene on a chromosome is known as a locus (or loci). Different variants of the DNA sequence, at a loci, are called an "allele."

Additionally, a small number of genes are located in the DNA that is present in the mitochondria of cells. The mitochondria are only inherited from the mother, and thus, mitochondrial DNA is only inherited maternally.

When a gene encodes for a protein, it is read so that every codon (three bases) encodes for a specific amino acid. As the DNA is not able to leave the nucleus, the DNA is translated into messenger ribonucleic acid (mRNA) which transports the message out of the nucleus to the ribosomes. The mRNA is typically edited through a process called splicing prior to being sent to the ribosome. The parts of the gene that will be cut out of the mRNA product are known as introns, and the parts that will be made into the final mRNA product are called exons. The process of going from DNA to mRNA is called transcription. Once the final mRNA product arrives at the ribosome, the amino acids are then assembled together, through the help of the ribosome, to synthesize a protein. This process, known as translation, uses the mRNA coding to produce the corresponding amino acid sequence that will make up that protein.

A common analogy utilized to describe the genome is that of a library in a town. The bases of DNA are letters that are arranged into words (codons). The words are used to make sentences (genes). The sentences are arranged into chapters (chromosomes). The chapters are arranged into a book (known as the nuclear DNA). The book is not able to leave the library in which it housed (the nucleus). So, to utilize the knowledge in the book, a copy must be made (RNA), which can be transported out of the library. Once out of the library, the copy can be brought to a factory (ribosomes) where it can be utilized to create something (protein). Another common analogy is that a gene is a recipe for a cake and the cake is the protein. The recipe analogy can be useful for describing to patients that not all variation to the recipe may significantly alter the cake.

The function of the genome is to store the DNA, copy the DNA for future generations, and express the DNA in the body. Only 1–2% of the DNA's function is to make proteins, which are the key to the cells structure and function. The remaining non-coding DNA's function is not fully known/understood. Some of the function is known to be regulatory in nature, which helps the DNA to be turned on or off in certain cells.

What Is Genomic or Genetic Variation?

Genetic or genomic variation is a change (or changes) in the DNA from one organism to the next. As humans, we share approximately 99.9% of our DNA, allowing for 0.1% of variation from one individual to the next. This variation, while small in percentage, accounts for millions of letters of DNA code.

Genetic variation occurs typically when the DNA is replicated and divided. There are two points in the body where this can happen, mitosis and meiosis. When a germ cell (egg or sperm) is created, the DNA must duplicate and go through the meiosis process. Typically, in meiosis, the cell starts with

46 chromosomes and ends with 4 germ cells with 23 chromosomes. In this process, crossing over may occur, which will allow for alleles (a specific version of a gene) to switch chromosomes. This naturally causes variation between offspring. Additionally, a copy error may occur, or the chromosomes may not evenly split, which may cause problems in a person. These changes are known as germline variation or germ cell variation. Variation in the germ cells result in genetic changes to occur in every cell of the body and can be passed on to the next generation. In mitosis, one somatic cell (all cells other than germ cells) duplicates and divides to become two cells as part of the cell cycle. During this process, changes can sporadically occur in the DNA or can be caused by environmental damage, subsequently causing somatic variation. Somatic variation, such as what one would see in a cancer cell, is not passed on from parent to child. When multiple cell lines are present in an individual, they are considered mosaic. Depending on when and where a mitotic error occurs in embryogenesis, sometimes it can also be present in the germline. Additionally, an individual may only have genetic variation present in the germline and not in any of the other cells in the body, which would be considered germline mosaicism.

What Is Genetic Code?

The genetic code is made up of the bases of the DNA or mRNA that are read in groups of three, known as codons. The codons are read during mRNA translation to signal which amino acid should be synthesized next in the protein. There is redundancy built into the codon signal to help adjust for genetic variants, as there are 64 codons and only 20 amino acids. Not all codons will code for an amino acid, as some are used to identify when the process should begin and when it should stop.

What Is a Mutation?

The term mutation means a change in the DNA. Historically it has been used when the genetic variant is thought to be damaging. Today this term has mostly been replaced by the by American College of Medical Genetics (ACMG) classification criteria terms of pathogenic or likely pathogenic variant in the literature and on clinical reports. This is change to the DNA that is thought to potentially cause disease, depending on how it is inherited. These are typically rare in the general population.

What Is a Polymorphism?

This term is used to describe normal variants in the population. Typically, these genetic variants would be considered benign or likely benign by ACMG classification criteria. The most common type of polymorphism is a single-nucleotide polymorphism (SNP), which is often utilized for ancestry testing analysis, along with comparison of DNA for microarray analysis.

What Is Genetic Testing?

Genetic testing is a type of laboratory testing that identifies abnormal changes to DNA. Genetic testing may be used to identify somatic mutations, such as characterizing tumors for potential targets of therapy, or to identify germline mutations. For the purposes of this chapter, we will be dissecting the use of germline genetic testing as a means to diagnose potential diseases.

Germline genetic testing is utilized to identify the genetic makeup (DNA sequencing) with which a person is born, and can potentially pass on to their offspring. This testing may be used to confirm a diagnosis, predict the risk of an inherited disease associated with a genetic mutation, or may be used to identify individuals who are carriers of that disease. Carriers, though they may not manifest symptoms of the disease, are at risk of passing that genetic mutation on to their child—when both parents pass the same mutated gene to a child, their child would typically be expected to have the resulting disease. Through genetic testing, however, we are able to screen our patients who carry the gene mutations, to provide them with this information. This empowers them to consider health implications for both themselves along with potential offspring.

Genetic testing can be performed in both research and clinical settings. Research genetic testing is typically less expensive and should be available at no personal cost to the patient. Testing can be used to identify genes that have tenuous or perhaps no known association with disease in order to strengthen/discover new disease association. However, research genetic testing results should not be shared with patients unless they have been confirmed by a clinical test. Many research studies have added this into their intuitional review board (IRB) protocol, allowing for seamless confirmation of genetic variants to return results to participants. Clinical laboratories are approved under the Clinical Laboratory Improvement Amendments (CLIA) and may need additional licenses and certifications. When clinical testing is ordered, a report will be generated that should be entered into the medical record for the patient. Clinical laboratories will also provide an interpretation of testing results, which would include information about the pathogenicity of a variant based on a set of classification guidelines (described below).

There are many different types of germline genetic testing, each with their own sets of benefits, risks, and limitations. Multiple genetic tests may be necessary depending on the indication. Genetic counseling either with a genetics professional or ordering physician is recommended.

Who Are Genetics Professionals?

Genetic counselors are members of the healthcare team that receive specialized training in genetics and communication. Terminal degree for this profession is currently a master's degree in either genetics or genetic counseling. They are accredited through the American Board of Genetic Counseling (ABGC) and may need to be licensed depending on the state of practice. Genetic counselors can work independently or with a physician to help identify which genetic testing to order, interpret genetic test results and genetic risk for families, and to communicate effectively with patients. Many genetic counselors are now working within ophthalmology departments or with diagnostic laboratories to assist ophthalmologists directly with the increasing demands of genetic testing.

Medical geneticists are physicians who specialize in genetics in relation to health and receive special training in hereditary disorders. They are certified by the American Board of Medical Genetics and Genomics (ABMGG) after two to three years of training (American Board of Medical Genetics and Genomics: An ABMS member board 2023). Ophthalmologists may refer their patients to a medical geneticist to do a genetics evaluation. Ophthalmologists who primarily work with hereditary forms of eye diseases may wish to become dual board certified in genetics as well as ophthalmology.

Genetic nurses are nurses with special training in medical genetics. This typically means a registered nurse (RN) has on the job genetics training, though some may also have a Master of Science in nursing with a concentration in genetics (*How to become a genetic nurse—salary* 2022).

Which Retinal Disease Would Benefit from Genetic Testing?

Many retinal conditions may have a genetic/hereditary component but not all are a good candidate for clinical genetic testing. The American Academy of Ophthalmology Task Force on Genetic testing recommends clinical genetic testing be considered for all patients where clinical findings are suggestive of a Mendelian disorder (diseases caused by a single defective gene that can follow various inheritance patterns) where one or more causative genes have been identified (Stone et al. 2014). Examples of retinal diseases that would be a good candidate for genetic testing include, but are not limited to, Leber congenital amaurosis (LCA), retinitis pigmentosa (RP), and Stargardt disease. More complex or multifactorial genetic conditions have been advised against routine genetic analysis, such as age-related macular degeneration and late-onset primary open-angle glaucoma (Stone et al. 2014). Most clinical genetic testing laboratories do not offer assays for these multifactorial conditions at this time. Additionally, predictive genetic testing can be considered for patients who are interested in better understanding their hereditary risk for managing disease. Genetic testing in asymptomatic minors for non-treatable adult-onset conditions is also not recommended.

What Is the Role of the Ophthalmologist in Genetic Testing?

All patients with a suspected or presumed diagnosis of an inherited retinal dystrophy (IRD) should consult a provider about genetic testing. It is also essential that ophthalmologists characterize the IRD in their initial diagnosis to provide clear phenotypic information regarding the patient's condition—this is a critical step for clinical genetic testing interpretation. The ophthalmologist should begin the conversation about genetic testing with the patient, and if knowledgeable and comfortable, can complete the genetic counseling and testing process. Many ophthalmologists will make a referral to a retinal specialist and/or a genetic counselor, professionals who possess greater genetics knowledge to complete the genetic counseling/testing process.

Before testing is ordered, patients and/or family should be counseled on the genetic testing procedure, risks, limitations, and benefits in addition to the types of genetic testing results that may arise. Many types of genetic testing may offer information not related to the primary indication of testing, and patients should be given ample time to consider if they want this information (i.e., hereditary cancer risk). Additional questions about privacy, insurance coverage, and cost should be addressed as well. Many laboratories have the option to utilize leftover DNA samples for genetic research and to share or use the data to help with variant curation efforts. Most consent forms will be purposefully vague in order to allow for a wide possibility of use. Patients and/or their family members should review these options with the providers to ensure they feel comfortable with how this information/their sample may be used. Some patients and/or family members may experience levels of anxiety, depression, guilt, and shame around the genetic testing process and results. Additionally there are religious and cultural differences/considerations that may come up, such as stigma around marriage, encouragement to marry within the family, not speaking openly about medical ailments with the family or even trying to keep diagnosis a secret from the person who is sick. Ample time and space should be given to these individuals to address and help them begin to navigate those feelings and concerns. Patients should have a basic enough understanding of genetics to grasp all this information in order for the provider to obtain *informed* consent. The written informed consent should be kept in the patients' medical records. Minors should give informed assent to the genetic testing, when age appropriate.

The genetic testing results will have an interpretation of pathogenicity for any variants identified. Knowledge of the sensitivity and specificity of the test is important and should be considered in conjunction with the clinical history to either confirm or rule out a diagnosis. For many IRDs, patients may have a diagnosis with negative genetic testing results. However, this does not rule out the initial diagnosis, as we are continuing to identify additional novel genes that cause known IRDs. Alternatively, a positive genetic test in the absence of pertinent clinical findings might be suggestive of a reduced penetrance, variable expressivity, multiple genetic diagnosis, or a classification error. Often, variants will come back uncertain, in which further clinical information may be valuable to the laboratory in order to assist with classification evidence. To move forward with these uncertain results, ophthalmologists should consider the ocular and family history, retinotoxic exposures, family member evaluations, clinical examinations, and ancillary testing. Recommended ancillary testing would include color fundus photos, fundus autofluorescence, spectral domain ocular coherence tomography, as well as functional testing such as visual fields (including microperimetry) and electroretinography (full-field or multifocal).

The ophthalmologist should provide a copy of the genetic testing results to the patient. The ophthalmologist should explain results of the genetic testing or should ensure proper referral to a genetic counselor or medical geneticist to ensure the patient received a thorough explanation of the results. Further consultation on the familial testing options and reproductive risk should be explained. Age-appropriate information should be given about partner testing (for autosomal recessive diseases), in vitro fertilization (IVF) with pre-implantation genetic testing for monogenic diseases (PGT-M) along with prenatal testing options. Recommendations for cascade testing should be given, along with recommendations to follow up with non-ocular specialists related to any syndromic manifestations (i.e., follow-up with nephrology consult for Bardet-Biedl syndrome).

What Are the Benefits of Genetic Testing?

Genetic testing can improve the accuracy of a diagnosis. Depending on sensitivity and specificity of the testing, a negative test result may be able to rule out a diagnosis on a differential. Positive genetic testing may be able to confirm a clinical diagnosis and/or give a more specific diagnosis. To date, there are over 341 genes and mapped loci known to cause retinal disease (Daiger, S.P. & University of Texas-Houston Health Science Center 2023). It's thought that a genetic diagnosis can be identified in approximately 60–80% of patients with retinal diseases (Eisenberger et al. 2013). Identifying the gene may be helpful when giving information on prognosis. As the databases on genetic mutations are growing, some genes have shown genotype to phenotype correlation—through ongoing research and clinical findings, these correlations only continue to grow. This has allowed for specific counseling on prognosis based on the individual mutation(s) that a person possesses. The genetic testing may help physicians with treatment plans and can give helpful risk information for the family and reproductive risk. Some couples might find out the risk of passing on the condition is very low and thus feel more empowered to start families.

In December of 2017, the FDA approved the first gene therapy to treat LCA caused by mutations in the *RPE65* gene. Additional gene therapies are currently being studied in clinical trials for a plethora of IRDs. Clinical trials are constantly updating, and as such, providers should check online to stay up to date on the latest available therapies in order to identify those available for patient enrollment (clincialtrials.gov).

What Are the Risks of Genetic Testing?

Genetic testing can have risks associated with undue anxiety, guilt, shame, and depression. It can disrupt familial relationships and a parent may feel blame or guilt for having "given" their child a condition. Incidents regarding familial relations may also be brought to light, sometimes known by the family and kept secret, but they are also sometimes a surprise to those involved. Examples of this may be adoption, use of egg/sperm donors, and non-paternity, as well as incidents of consanguinity, incest, and/or rape. It is important that patients are explicitly told these tests are not meant to test paternity and so cannot be used in a court of law for such purpose, but that paternity relativeness may be tested to determine if a variant was inherited or *de novo*. Parents are often in a state of heightened anxiety/stress during the time of consent and may not realize that these incidental findings are a possibility of the genetic testing unless they are explicitly told.

Depending on if the genetic testing was done presymptomatically, there may be insurance discrimination concerns for patients. The Genetic Information Nondiscrimination Act (GINA) helps mitigate some of the discrimination concerns but does not apply to long-term care insurance, private disability insurance, and life insurance (2009).

Lastly, even though genetic testing is available, not all mutations are identifiable. Some mutations may only be identifiable on specific assays and may need to be added on after a test is negative. False positives and false negatives may occur, though chances are low if using a CLIA-certified lab.

Does Everyone with the Same Diagnosis Have the Same Genetic Mutation?

Not necessarily. Genetic testing can be challenging due to heterogeneity. There are three type of heterogeneity that should be considered with genetic testing:

- *Genetic heterogeneity*: Mutations in different genes may cause the same disease. For example, there are over 100 different genes associated with RP.
- *Phenotypic heterogeneity*: Different mutations in the same gene may cause different diseases or phenotypes. For example, the gene *PRPH2* is associated with LCA, RP, retinitis punctate albescens, pattern macular dystrophy, vitelliform macular dystrophy, and central areolar choroidal dystrophy [MIM: 179605].
- *Clinical heterogeneity*: The same mutation in different individuals, even within the same family, may have differences in clinical presentation. For example, those homozygous with the 2314delG mutation in the *USH2A* gene, both in the same family and those in different families, gave rise to usher syndrome type 2 and to atypical usher syndrome (Liu et al. 1999).

What Information Is Given by Genetic Testing?

Genetic testing reports can be put into the following categories:

- *Positive result*: A disease causing variant(s) was identified in a gene. Depending on the inheritance of the gene(s), this could mean that the cause of the patient's disease may be identified, or that the patient is a carrier of disease. It is possible for multiple genes to have a positive result on a test.
- *Negative result*: All the genes analyzed on the genetic test yielded no disease causing variants or variants of interest. This alone does not rule out the presence of an IRD. The disease causing gene may not have been analyzed, the gene function might currently be unknown (and therefore was not analyzed), or the mutation may not be detectable by a certain technology.
- *Inconclusive result*: This is when a genetic test provides a candidate or potential diagnosis. A variant of uncertain significance (VUS or VOUS) in a gene that is a good clinical fit may be identified. A single disease causing variant in an autosomal recessive disease was identified that might be a good clinical fit and so further genetic analysis on that gene might be recommended. Perhaps a gene of uncertain significance was purposed with a variant that might be causative. Usually further genetic analysis on family members or on the patient themself is recom-

mended to help gain clarity. Sometimes further testing in the population must be completed in order to see the variant in other probands and conclusively resolve a VUS.

Genetic testing results will typically only give information back on the disease for which it was ordered. For example, an IRD panel will not provide information on someone's risk to develop diabetes. Reporting on secondary findings from the testing has more caveats in those secondary results, compared to a diagnostic test for the listed indication. Thus, screening for that condition specifically, even in the presence of a negative genetic testing results, should be done based on personal and family history. For example, an individual with a strong family history of colon cancer that does not have a known genetic etiology of the condition (they were never tested for genes associated with colon cancer), who receives a negative secondary findings report, should still be screened at increased risk for colon cancer due to the family history.

What Are the Types of Genetic Testing?

There are many different technologies that can be utilized with genetic testing.

- *Karyotype:* This test looks at the number and arrangement of the chromosomes. Cells are typically cultured and analyzed at metaphase. The number of the chromosomes are counted. Large deletions/duplications and balanced rearrangements are all identifiable on chromosome analysis. Most retinal diseases will not be caused by an aneuploidy and so this test is rarely utilized for the purpose of IRD screening.
- *Chromosomal Microarray:* Microarrays are used to detect imbalances in the amount of DNA present. Microarray testing allows for testing of deletions smaller than what can be picked up on karyotype and larger than what would be able to be picked up on next-generation sequencing (NGS). It typically allows for exact deletion breakpoints to be identified, which can be important for preimplantation genetic testing. Microarrays can also identify aneuploidy. Limitations of the size of mutations detectable depend by laboratory as it is typically dependent on the number and spacing of the probes. Microarrays cannot identify balanced translocations. Additionally, there are two types of microarrays: comparative genomic hybridization (CGH) array and single nucleotide polymorphism (SNP) array. The different array technologies also have slightly different limitations. A CGH array compares the proband's DNA to a control DNA to detect areas that may be missing or repeated. With a SNP array, the proband's DNA is tagged with fluorescence SNP probes and then measured against a library of control samples also tagged with SNP probes. SNP arrays alone can also identify large areas of homozygosity, which can indicate uniparental disomy (UPD) or consanguinity as well as triploidy. Chromosomal microarray should be considered in instances where there is a negative or single hit on a whole exome sequencing (WES) analysis, or to further define breakpoints in a deletion/duplication identified on a panel. Approximately 10% of IRD diagnoses could have a copy number identifiable on this testing.
- *Variant Testing*: Targeted testing is performed on a single variant. The result will typically only be negative or positive for the variant in question. This is often completed through Sanger sequencing. It may be used to confirm a finding on a research test or on a clinical test to reduce chances of a false positive. Providers may also order this testing for family member analysis to reduce cost and turnaround time.
- *Single Gene Sequencing*: Only one gene is sequenced. This is a useful test when there is only one gene known to cause the suspected clinical diagnosis. Examples of these diagnosis would be retinoschisis (*RS1* gene) or choroideremia (*CHM* gene).
- *Multi-gene (panel) testing*: Multiple genes are analyzed together as they all relate to a similar phenotype. There are many differences in the sensitivity of different panels. Older panels used genotyping analysis where only specific mutations were tested for in each gene. Modern panels typically utilize NGS to analyze the gene more completely. Most panels will also include deletion/duplication analysis and may also include intronic mutation analysis for certain genes. A multi-gene panel would be recommended in the case of an individual who is diagnosed with an IRD for which multiple genes are known to be a cause; for example RP, Usher syndrome, and Bardet-Biedl syndrome.
- *Whole exome sequencing (WES)*: The protein coding regions (exons) of all the genes in the patient's DNA are sequenced. This is approximately 2% of the total DNA. The sequencing data is then filtered to look for variants that might cause a specific phenotype. A known limitation of WES is the inability to detect large deletion/duplications. Many clinical WES analyses will include additional copy number variant (CNV) analysis without additional specification for their WES tests. Often WES will not be covered by insurance for patients with non-syndromic IRDs.
- *Whole genome sequencing (WGS)*: The entire genome (every gene—both coding and non-coding—in the DNA)

is sequenced. The coverage of the exons in the DNA is increased through this technology. Large areas of deletion/duplication can be identified on this test as well. Additionally, WGS would allow for complete sequencing of the intronic areas of the DNA, which are not typically analyzed. This can be helpful in resolving cases of IRDs where only one mutation was previously identified (in the case of autosomal recessive diseases, two mutations *in trans* would need to be identified in order to have genetic confirmation of the IRD). Since most of the genome does not have a history of clinical sequencing, WGS may yield increased VUS identified. Until very recently, WGS was considered experimental by all insurances in the United States. Recently, some insurance companies have begun to reimburse the cost of WGS.

Does Every Variant Result in Disease?

Simply no. There are many changes to the DNA that do not cause any problems and simply contribute the unique diversity of humanity. It is not always easy to determine if a variant is disease causing, as sometimes a single base pair change can cause disease, and other times, a large area of the DNA may be missing or repeated, and there will be no resulting disease. Guidelines were created to help guide geneticists in determine the pathogenicity of genetic variation in order to try and standardize this across the field.

References

American Board of Medical Specialties. (2023, February 10). *American Board of Medical Genetics and Genomics: An ABMS member board*. American Board of Medical Specialties. Retrieved February 13, 2023, from https://www.abms.org/board/american-board-of-medical-genetics-and-genomics/

Daiger, S. P., & University of Texas-Houston Health Science Center. (2023, January 10). Genes and mapped loci causing retinal diseases. RetNet: Disease Table. Retrieved February 13, 2023, from https://sph.uth.edu/RetNet/

Eisenberger, T., Neuhaus, C., Khan, A. O., Decker, C., Preising, M. N., Friedburg, C., Bieg, A., Gliem, M., Issa, P. C., Holz, F. G., Baig, S. M., Hellenbroich, Y., Galvez, A., Platzer, K., Wollnik, B., Laddach, N., Ghaffari, S. R., Rafati, M., Botzenhart, E., ... Bolz, H. J. (2013). Increasing the yield in targeted next-generation sequencing by implicating CNV analysis, non-coding exons and the overall variant load: The example of retinal dystrophies. *PLoS ONE*, *8*(11). https://doi.org/10.1371/journal.pone.0078496

International Organization for Standardization (ISO), ISO 15189:2022 Medical laboratories – Requirements for quality and competence. 2002, December. ISO: Geneva.

The Genetic Information Nondiscrimination Act of 2008 (Gina) (2009). bill.

Liu, X.-Z., Hope, C., Liang, C. Y., Zou, J. M., Xu, L. R., Cole, T., Mueller, R. F., Bundey, S., Nance, W., Steel, K. P., & Brown, S. D. M. (1999). A mutation (2314DELG) in the Usher Syndrome Type IIA gene: High prevalence and phenotypic variation. *The American Journal of Human Genetics*, *64*(4), 1221–1225. https://doi.org/10.1086/302332

RegisteredNursing.org staff. (2022, October 29). *How to become a genetic nurse - salary*. How to Become a Genetic Nurse. Retrieved February 13, 2023, from https://www.registerednursing.org/specialty/genetic-nurse/

Stone, E. M., Aldave, A. J., Drack, A. V., MacCumber, M. W., Sheffield, V. C., Traboulsi, E., & Weleber, R. G. (2014, February 25). *Recommendations for Genetic Testing of Inherited Eye Diseases - 2014*. American Academy of Ophthalmology - Clinical Statements. Retrieved February 13, 2023, from https://www.aao.org/clinical-statement/recommendations-genetic-testing-of-inherited-eye-d

Further Reading

https://findageneticcounselor.nsgc.org/
https://clinics.acmg.net/
https://www.isong.org/
https://www.ncbi.nlm.nih.gov/books/NBK1116/
https://medlineplus.gov/genetics/
https://clinicaltrials.gov/
https://www.clinicaltrialsregister.eu/
http://clinvar.com/
https://www.omim.org/
https://www.hgvs.org/content/guidelines
https://www.clinicalgenome.org/

Richards, S., Aziz, N., Bale, S., Bick, D., Das, S., Gastier-Foster, J., Grody, W. W., Hegde, M., Lyon, E., Spector, E., Voelkerding, K., Rehm, H. L., & ACMG Laboratory Quality Assurance Committee (2015). Standards and guidelines for the interpretation of sequence variants: a joint consensus recommendation of the American College of Medical Genetics and Genomics and the Association for Molecular Pathology. Genetics in medicine: official journal of the American College of Medical Genetics, 17(5), 405–424. https://doi.org/10.1038/gim.2015.30

Uhlmann, W. R., Schuette, J. L., & Yashar, B. M. (2009). A Guide to Genetic Counseling. Wiley.

CHAPTER 57

Genetic Testing for Inherited Retinal Dystrophies: Genetic Reports

Megan E. Soucy

How Are Genomic Variants Classified in a Genetic Report? 314
How Do I Read and Understand the Genetic Testing Report?. 314
How Are Variants Reported or Described?. 315
Are Further Investigations Needed After the Genetic Testing? 316

M. E. Soucy (✉)
Department of Ophthalmology, Columbia University,
New York, NY, USA
e-mail: mes2350@cumc.columbia.edu

S. H. Tsang et al. (eds.), *Atlas of Inherited Retinal Diseases*, Advances in Experimental Medicine and Biology 1467, https://doi.org/10.1007/978-3-031-72230-1_57

How Are Genomic Variants Classified in a Genetic Report?

Genetic variants are put into five classifications based on the likelihood that they would cause disease. This is determined by a variety of evidence, which is assigned a point value. The classifications are:

- *Benign*: This is the highest confidence level that a variant does not cause disease.
- *Likely benign*: The variant is considered unlikely to cause disease.
- *Uncertain significance*: VUS either do not have enough evidence to determine pathogenicity or the evidence available is conflicted. Additional information is needed to better understand the pathogenicity of the variant.
- *Likely pathogenic*: The variant is likely to be disease causing.
- *Pathogenic*: This is the highest level of confidence that a variant causes disease.

Even though the confidence of the disease assertion is different between pathogenic and likely pathogenic variants, both are considered clinically actionable. VUS are typically not considered clinically actionable and are often treated as a benign variant as a large number are reclassified to benign over time. In addition, the presence of one or more likely pathogenic or pathogenic variant(s) may not always equate to a diagnosis, as a person may just be a carrier of a genetic disease or there could be non-penetrance (when a disease causing variant does not always manifest in disease).

Guidance for gene variation classification is constantly being updated in general and for specific indications. Further reading is available in the suggested reading section to explore variant classification further, along with resources on how to stay up to date with current guidelines.

How Do I Read and Understand the Genetic Testing Report?

Over the years, organizations have worked to standardize genetic testing reports, not just in the USA but abroad as well [The International Organization for Standardization (ISO), last revised in 2022]. While labs will have similar information on the reports, they are often arranged differently and it may be necessary to contact the testing laboratory directly in order to obtain any missing information (i.e., the reference sequence). Additionally, older genetic testing reports may be missing variant classifications—what is now considered standard information to be reported in genetic testing reports.

As genetic testing reports are considered part of the medical record, both the patient's name and date of birth are typically included. The medical record number (MRN) may also be listed. Typically, the person's gender and/or sex assigned at birth is listed. The date of collection, the date of receipt to the laboratory, and the date the report was issued should all be included. The type of sample provided to the lab for DNA analysis (i.e., EDTA blood, saliva, buccal, etc.) will be listed along with the assay name. Lastly, the name of the ordering provider and any other provider to whom the report was issued will be listed on the report as well. This is all typically found at the top of the testing report.

For larger genomic assays, such as microarray, WES and WGS, the indication for testing is typically included (ex: rod-cone dystrophy or possible connective tissue disorder). A summary of the genetic findings is typically found on the first page. Sometimes the variants will come with a descriptor such as negative, positive, possibly positive, or inconclusive. Sometimes laboratories will clarify that a person is positive to be a *carrier* when it is a diagnostic test.

For all variants identified, it is recommended that the gene and the gene reference sequence be included along with the coding in Human Genome Variant Society (HGVS) nomenclature. Sometimes the reference sequence number may be found later in the report and not at the top where the initial result summary is located. The zygosity of the variant identified should be included along with the ACMG variant classification category that the laboratory has issued for the variant. If multiple variants in the same gene are identified, the lab will make a comment on phase of the variants (*in cis*, *in trans*, or unknown).

A summary of the evidence the laboratory used for the variant interpretation must be included. This includes the use of any databases, primary literature, functional studies, in silico predictors, etc. A note about risks for family members should be present and an indication if further testing is indicated is typically listed.

Additionally, information on the gene is typically included. This would usually include the OMIM nomenclature (MIM number), along with information on the disease with which the gene is associated, along with its inheritance. Many labs will provide a brief clinical summary on symptoms the disease may present with, but this is not universal.

Much of the report will be taken up with information on the testing methodologies and limitations of the testing. It should also include a complete list of all genes analyzed and any additional limitations to the analysis that was present for each gene.

How Are Variants Reported or Described?

Reference sequences are updated over time and so it is important to have the reference sequence clearly marked in order to insurance accuracy of comparing mutations in different reports over time. If there are questions about what certain nomenclature means, highly recommend visiting the Human Genome Variation Society (HGVS) website's glossary of nomenclature terms. The reference sequence used should be written first with a colon followed by the variant description (Reference:Description). All the reference sequences should be public and clearly described. Most commonly, the reference utilized is through the National Center for Biotechnology Information (NCBI) known as RefSeq. However, it may also be through the European Bioinformatics Institute (EBI) known Ensembl. Locus Reference Genomic (LRG) is still acceptable but it not usually used due to the reference no longer being updated.

Both RefSeq and Ensembl have different nomenclature before the reference number to specify the type of sequence that is being reviewed. For RefSeq, a report will commonly use NC_#, NG_#, NM_#, and NP_#. This is not an exhaustive list of possible nomenclature. NC refers to chromosomal, NG refers to gene/genomic region, NC is the coding transcript, and NP is the protein transcript. In Ensembl, you will typically see ENSG, ENST, or ENSP (not a conclusive list), corresponding to the gene/genomic region, the coding transcript, and the protein, respectively.

Variant descriptions will have a letter prefix to describe the type of residue that is changed. Some commonly utilized ones are "c.", "g."," m.", and "p." The "c." nomenclature stands for coding DNA, "g." for linear genomic, "m." for mitochondrial DNA, and "p." for protein.

After the prefix there will be the location of the specific variant, along with nomenclature to denote the type of genetic change and the nucleotide or protein that is affected. Nucleotides are numbered from the first nucleotide in the reference sequence to the last in the reference sequencing, starting with the start codon and ending with the stop codon. Nucleotides outside of the coding sequence are denoted with a "−" (minus) or "+" (plus). If a nucleotide is located at the 5′ end of the intron, it will be denoted with the last nucleotide directly upstream followed by the "+" (plus) and the number of nucleotides upstream the variant is located at. Conversely, if the variant is located on the 3′ end of the codon, the last nucleotide directly downstream is noted with the "−" (minus) and the number of nucleotides outside of the intron. A range of nucleotides, often used for deletions/duplications, is notated with an underscore (_). Angled bracts [and] are used to denote alleles and when multiple alleles are noted they can often use a semicolon (;) to separate.

Common types of variants noted are substitutions, deletions, duplications, insertions, deletions/insertions (indel), and frameshifts. Each has a specific nomenclature marking to quickly identify the type of mutation presented.

- *Substitutions* will denote the reference nucleotide/protein a greater than sign ">" and the nucleotide/protein that was denoted in the patient. For example, "c.4587A > C" shows that at coding DNA position 4587, the reference showed the DNA nucleotide C was changed to A.
- *Deletions* are when one or more nucleotides from the DNA code are missing. This is noted by the use of "del". An example of this would be "c.5879_5881del" where nucleotides 5879 through 5881 are deleted from the DNA sequence code.
- *Duplications* are when one or more nucleotides are repeated. This is noted by the use of "dup". For example, "c.5879_5881dup" indicated that nucleotides 5879 through 5881 are duplicated or repeated in the DNA sequence.

- *Insertion* is when one or more nucleotides are newly added (inserted) into the DNA sequence. This is indicated by the term "ins". For example, "c.5879_5881insACCT" indicates that the new DNA sequence of "ACCT" is inserted between nucleotides 5879 and 5881.
- *Deletion/Insertion (indel)* is when one or more nucleotides of the DNA code are missing and are replaced by a new sequence. This is noted by the term "delins". For example, "c.5879_5881delinsACCT" would indicate that nucleotides 5879 through 5881 are deleted, and in that location, the sequence "ACCT" was inserted.
- *Frame shifts (fs)* refer to the fact that the reading frame (the alignment of the codons) shifted from the reference reading frame. This typically occurs due to a change in the DNA that creates a number of nucleotides that is not divisible by three. An example of a frameshift nomenclature is p.Arg456GlyfsTer17 or p.Arg456Glyfs*17; the nomenclature here will denote that in the protein, at position 456, there should be an arginine but instead there is a glycine. The glycine is the start of the frameshift which ends with a termination of the protein at termination position 17 (counting from the first amino acid change to the stop codon).

Are Further Investigations Needed After the Genetic Testing?

Often further investigation is needed. For example, segregation analysis of two variants should be completed when two disease causing variants are identified in an autosomal recessive disease. This confirms that the variants are on different alleles (rather than inherited on the same allele, which would not be disease causing), thus ensuring the correct orientation for disease. It can also provide information if the variant is de novo, which can impact classification as well as familial risks. Often, familial testing can be considered when a VUS is identified to gain further information and to help with classification.

Part IX
Novel Therapies for Inherited Retinal Disease

Inherited retinal dystrophies (IRDs) are a heterogenous group of genetic conditions, many of which have a pathophysiologic endpoint of photoreceptor dysfunction and death. In the past, management of these conditions has focused on supportive care, with emphasis on refractive correction, low vision rehabilitation, and occupational and educational supports. However, with the 2017 approval of a gene therapy, voretigene neparvovec-rzyl (Luxturna), for patients with *RPE65*-associated retinal dystrophy, we have entered an exciting era in which treatments are emerging to halt, and even, reverse vision loss from IRDs.

The chapters in the section review the genetic basis of inherited retinal diseases, before discussing voretigene neparvovec-rzyl and other gene therapies currently in clinical trials. Subsequent chapters address a variety of other novel therapeutic strategies under investigation for IRDs. For ease of presentation and discussion, therapies are classified into gene specific treatment strategies, which target a specific gene or mutation, and global treatment strategies, which may be useful in treating a variety of conditions, irrespective of the underlying genetic defect.

CHAPTER 58

The Genetic Basis of IRDS and the Role of Genetic Testing

Gareth D. Mercer, Brian G. Ballios, and Peter J. Kertes

References 321

G. D. Mercer · B. G. Ballios · P. J. Kertes (✉)
Department of Ophthalmology & Vision Sciences, Temerty Faculty of Medicine, University of Toronto, Toronto, ON, Canada
e-mail: peter.kertes@sunnybrook.ca

S. H. Tsang et al. (eds.), *Atlas of Inherited Retinal Diseases*, Advances in Experimental Medicine and Biology 1467, https://doi.org/10.1007/978-3-031-72230-1_58

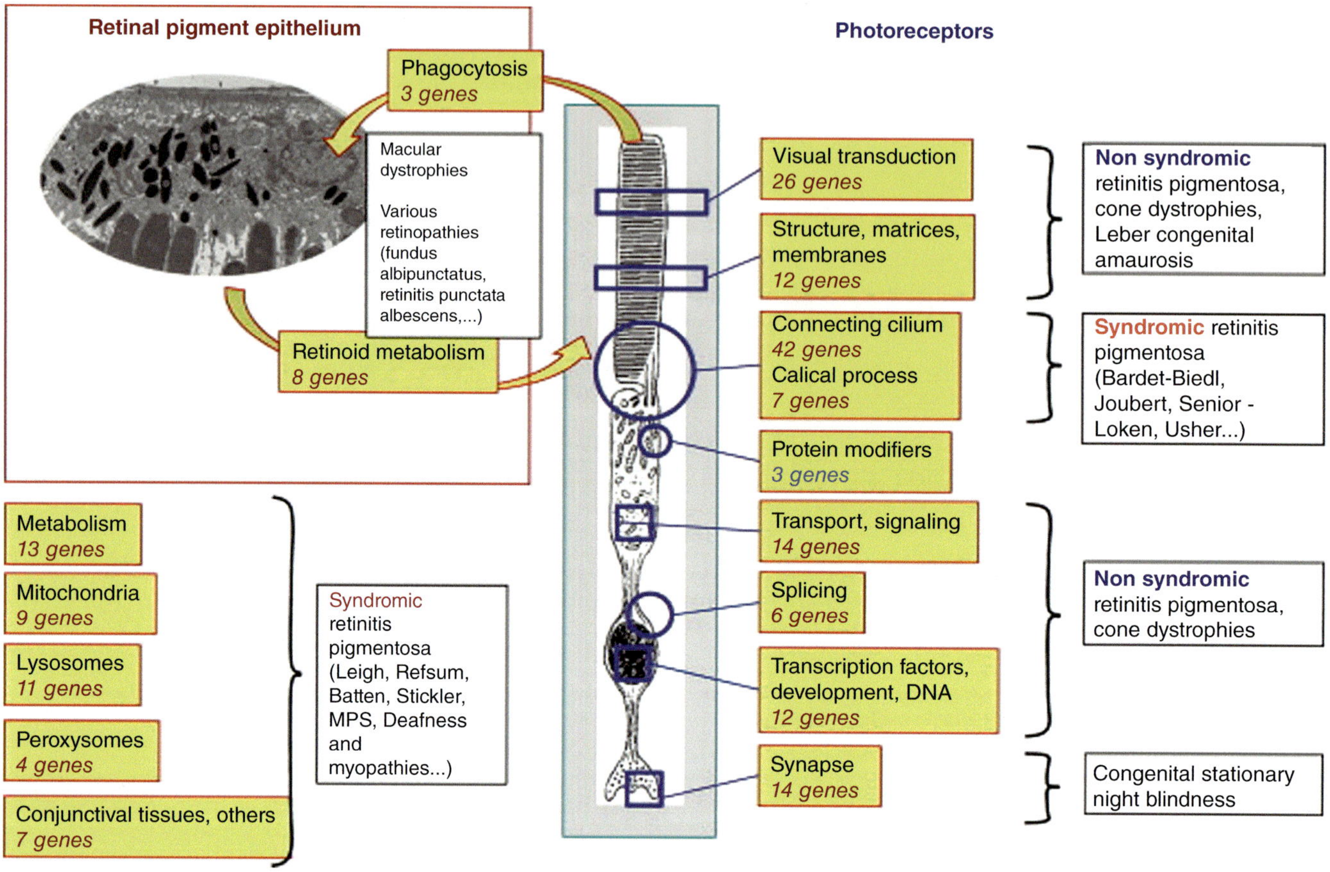

Fig. 58.1 Schematic depiction of the subtypes of IRD-associated genes and their sites of activity within photoreceptor and retinal pigment epithelium cells. (Reprinted from Hamel 2014, with permission from Elsevier Masson SAS)

Most of the major clinical IRD phenotypes are monogenic conditions and demonstrate classic Mendelian inheritance patterns. More than 300 genes have been implicated in IRDs, many of which are involved in phototransduction, the retinoid cycle, photoreceptor cell structure, structure and function of the connecting cilium, transcription factors, and RNA splicing (Fig. 58.1; Hamel 2014; Dias et al. 2018). Studies involving relatively large patient cohorts have documented 4798 discrete variants in 194 genes (Schneider et al. 2022). The majority of variants either alter one or more amino acid in sequence (missense mutations) or prematurely truncate proteins (nonsense mutations). Smaller percentages of mutations affect RNA splicing, start and stop codons, or control regions (Schneider et al. 2022).

Among IRDs, the mode of inheritance is often predictive of the underlying molecular disease mechanism, which has implications in the design of effective genetic treatments (Fig. 58.2). Recessive mutations (autosomal and X-linked) tend to cause disease through a loss of function mechanism and are, therefore, amenable to treatments aimed at restoring function (e.g., gene replacement/augmentation). Autosomal Dominant (AD) mutations may cause loss of function through haploinsuffiency (i.e., one normal gene copy is insufficient to produce the wildtype phenotype). However, AD mutations may also cause disease through the dominant-negative mechanism, in which the mutant protein interferes with the function of the wildtype, or through a gain of function mechanism, in which the mutated protein acquires novel functionality. While gene augmentation may be effective in situations of haploinsufficiency, a treatment strategy aimed at correcting or silencing the underlying gene mutation, such as gene editing, may be necessary for dominant-negative and gain of function mutations (Daich Varela et al. 2022; Diakatou et al. 2019).

Classically, IRDs have been categorized and diagnosed based on their clinical and electrophysiologic features, with more than 50 major subtypes described. However, it is well-established that pathogenic variants in different genes can produce almost indistinguishable clinical IRD phenotypes. This genetic heterogeneity is exemplified in non-syndromic retinitis pigmentosa, which can be caused by mutations in 69 different genes (https://web.sph.uth.edu/RetNet/sum-dis.htm). In addition, the same causative familial variant can present with significant intrafamilial heterogeneity in the clinical phenotype (Cremers et al. 2020; Dias et al. 2018; Rivolta et al. 2002). Heterogenous genetic diseases, like

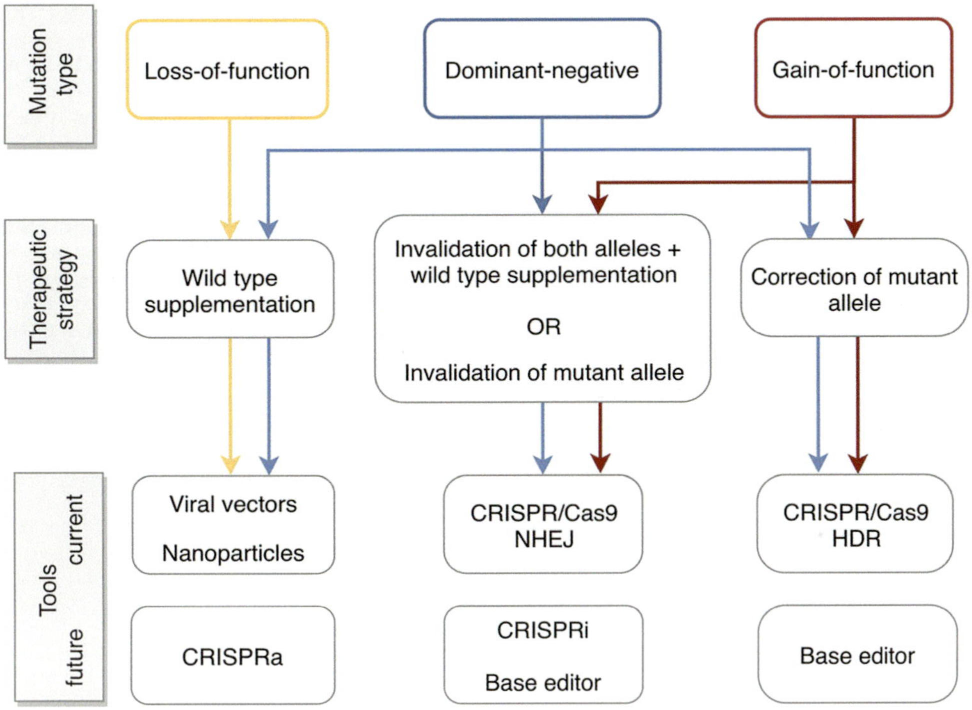

Fig. 58.2 Flow chart showing the appropriate gene therapy strategy based on the type of mutation. Loss-of-function mutations (in yellow) could be treated by gene supplementation using, for example, viral vectors or non-viral nanoparticles to vehicle the transgene into cells. A future alternative would be to use CRISPR activation (CRISPRa) to enhance gene expression. Dominant-negative (in blue) and gain-of-function (in red) mutations can be treated by mutation-independent or -dependent gene invalidation using the CRISPR/Cas system and the error-prone non-homologous end-joining (NHEJ) repair pathway. In the case of mutation-independent invalidation, this would systemically need to be coupled to gene supplementation. The more recent CRISPR interference (CRISPRi) or base-editor technology, which can be used to suppress gene expression, also hold potential. Alternatively, these mutations can be treated by mutation-dependent correction using the CRISPR/Cas system and homology-directed repair (HDR). The recent base-editor technology also holds promise as a future gene correction strategy. (Reprinted from Diakatou et al. 2019, under Creative Commons Attribution (CC BY) 4.0 licence)

IRDs, are characterized by a high prevalence of heterozygous recessive mutations in unaffected individuals. As many as one in three people in the general population are asymptomatic carriers of pathologic variants associated with IRDs (Hanany et al. 2020). Similarly, it is not uncommon for IRD patients to carry heterozygous recessive variants that are not the cause of their disease, but may modify it (Schneider et al. 2022). This is one of the factors contributing to the genetic complexity and phenotypic variability of these conditions, and these potential modifier genes and their role in variable expressivity is the subject of much current study (https://www.biosino.org/PhenoModifier/).

Though clinical and electrophysiologic testing helps to narrow the list of potential genetic causes in IRDs, clinical assessment in isolation has inherent diagnostic, prognostic, and therapeutic limitations. It is, therefore, recommended that genetic testing be offered to all patients suspected of having an IRD with a known causative gene (Stone et al. 2012). Identification of a molecular etiology enables appropriate genetic counselling of patients and their family members, and positions them to potentially benefit from approved, or to participate in a variety of investigational therapies. Nevertheless, even after comprehensive genetic testing, the genetic cause of a clinically diagnosed IRD remains elusive in 30–40% of individuals (Daiger et al. 2013).

References

Cremers FPM, Lee W, Collin RWJ, Allikmets R. Clinical spectrum, genetic complexity and therapeutic approaches for retinal disease caused by ABCA4 mutations. Prog Retin Eye Res. 2020;79:100861. https://doi.org/10.1016/j.preteyeres.2020.100861. Epub 2020 Apr 9.

Daich Varela M, Georgiadis A, Michaelides M. Genetic treatment for autosomal dominant inherited retinal dystrophies: approaches, challenges and targeted genotypes. Br J Ophthalmol. 2022:bjophthalmol-2022-321903. https://doi.org/10.1136/bjo-2022-321903

Daiger SP, Sullivan LS, Bowne SJ. Genes and mutations causing retinitis pigmentosa. Clin Genet. 2013;84(2):132-41. https://doi.org/10.1111/cge.12203. Epub 2013 Jun 19.

Diakatou M, Manes G, Bocquet B, Meunier I, Kalatzis V. Genome Editing as a Treatment for the Most Prevalent Causative Genes of Autosomal

Dominant Retinitis Pigmentosa. Int J Mol Sci. 2019;20(10):2542. https://doi.org/10.3390/ijms20102542.

Dias MF, Joo K, Kemp JA, Fialho SL, da Silva Cunha A Jr, Woo SJ, Kwon YJ. Molecular genetics and emerging therapies for retinitis pigmentosa: Basic research and clinical perspectives. Prog Retin Eye Res. 2018;63:107-131. https://doi.org/10.1016/j.preteyeres.2017.10.004. Epub 2017 Oct 31. Erratum in: Prog Retin Eye Res. 2018 Sep;66:220-221.

Hamel CP. Gene discovery and prevalence in inherited retinal dystrophies. C R Biol. 2014;337(3):160-6. https://doi.org/10.1016/j.crvi.2013.12.001.

Hanany M, Rivolta C, Sharon D. Worldwide carrier frequency and genetic prevalence of autosomal recessive inherited retinal diseases. Proc Natl Acad Sci U S A. 2020;117(5):2710-2716. https://doi.org/10.1073/pnas.1913179117. Epub 2020 Jan 21.

Rivolta C, Sharon D, DeAngelis MM, Dryja TP. Retinitis pigmentosa and allied diseases: numerous diseases, genes, and inheritance patterns. Hum Mol Genet. 2002;11(10):1219-27. https://doi.org/10.1093/hmg/11.10.1219. Erratum in: Hum Mol Genet. 2003 Mar 1;12(5):583-4.

Schneider N, Sundaresan Y, Gopalakrishnan P, Beryozkin A, Hanany M, Levanon EY, Banin E, Ben-Aroya S, Sharon D. Inherited retinal diseases: Linking genes, disease-causing variants, and relevant therapeutic modalities. Prog Retin Eye Res. 2022;89:101029. https://doi.org/10.1016/j.preteyeres.2021.101029.

Stone EM, Aldave AJ, Drack AV, Maccumber MW, Sheffield VC, Traboulsi E, Weleber RG. Recommendations for genetic testing of inherited eye diseases: report of the American Academy of Ophthalmology task force on genetic testing. Ophthalmology. 2012;119(11):2408-10. https://doi.org/10.1016/j.ophtha.2012.05.047.

CHAPTER 59

"Seeing Stars": The Gene Therapy Revolution

Gareth D. Mercer, Brian G. Ballios, and Peter J. Kertes

Vectors for DNA Delivery. 326
Case Vignette: Gene Replacement Therapy for RPE65-Associated Retinal Dystrophy 328
Voretigene Neparvovec-rzyl 331
Other Gene Therapies for IRDs 334
References 334

G. D. Mercer · B. G. Ballios · P. J. Kertes (✉)
Department of Ophthalmology & Vision Sciences, Temerty Faculty of Medicine, University of Toronto, Toronto, ON, Canada
e-mail: peter.kertes@sunnybrook.ca

S. H. Tsang et al. (eds.), *Atlas of Inherited Retinal Diseases*, Advances in Experimental Medicine and Biology 1467, https://doi.org/10.1007/978-3-031-72230-1_59

Gene therapy is the delivery of exogenous DNA for the treatment of human disease (Friedmann and Roblin 1972). The advantage of gene therapy over other treatments for IRDs is the potentially curative effect of replacing, repairing, or silencing the responsible gene defect. The European Medicines Agency and the American Food and Drug Administration (FDA) approved the first gene therapy products in 2012 and 2017, respectively. Among them was voretigene neparvovec-rzyl (Luxturna), a revolutionary step in the treatment of IRDs.

Several features of inherited retinal diseases have made them promising candidates for gene therapy: knowledge of the molecular pathophysiology for many of the single-gene Mendelian IRD disorders; the relative ease of surgical access to the retina and RPE via intravitreal and subretinal injection; the eye's compartmental structure and immune privilege, which theoretically reduces the risk of systemic dissemination and significant inflammatory reaction; and the availability of high-resolution, non-invasive imaging modalities and functional tests for interrogating the effects of therapy (Schneider et al. 2022). At the time of writing, around twenty gene therapy products for IRDs were in human clinical trials, including one phase III study (Table 59.1).

Table 59.1 Gene therapies for inherited retinal diseases currently in clinical trials

Intervention type	Disease	Gene (mutation)	Trial identifier	Phase	Status	Sponsor	Intervention description
Gene augmentation	Achromatopsia	*CNGB3*	NCT02599922	1/2	Active, not recruiting	Beacon Therapeutics	Unilateral subretinal injection of AGTC-401, a recombinant mutant AAV2 carrying the human *CNGB3* gene (rAAV2tYF-PR1.7-hCNGB3)
Gene augmentation	Achromatopsia	*CNGA3*	NCT02935517	1/2	Active, not recruiting	Beacon Therapeutics	Unilateral subretinal injection of AGTC-402, a recombinant AAV2 carrying the human *CNGA3* gene
Gene augmentation	Achromatopsia	*CNGA3*	NCT02610582	1/2	Recruiting	Tubingen Hospital	Bilateral single subretinal injection of rAAV.hCNGA3, a recombinant AAV vector expressing the human *CNGA3* gene
Gene augmentation	Achromatopsia	*CNGA3*, *CNGB3*	NCT03278873	1/2	Active, not recruiting	MeiraGTx/ Janssen	Recombinant AAV vector (AAV2/8-hCARp.hCNGB3 and AAV2/8-hG1.7p. coCNGA3) expressing human *CNGB3* or *CNGA3* genes, long-term follow-up study
Gene augmentation	Bietti's Crystalline Dystrophy	*CYP4V2*	NCT04722107 & NCT05714904	1	Recruiting	Beijing Tongren Hospital and Peking University Third Hospital, China	Subretinal injection of ZVS101e, a recombinant AAV8 vector carrying the human *CYP4V2* gene
Gene augmentation	Bietti's Crystalline Dystrophy	*CYP4V2*	NCT05399069	1	Recruiting	Beijing Tongren Hospital, China	Subretinal injection of VGR-R01, a recombinant AAV vector carrying the human *CYP4V2* gene

(continued)

Table 59.1 (continued)

Intervention type	Disease	Gene (mutation)	Trial identifier	Phase	Status	Sponsor	Intervention description
Gene augmentation	Choroideremia	*CHM*	NCT04483440	1/2	Active, not recruiting	4D Molecular Therapeutics	Single uniocular intravitreal injection of 4D-110, an AAV capsid variant (4D-R100) carrying a transgene encoding a codon-optimized human *REP1* gene.
Gene augmentation	Leber's Congenital Amaurosis	*LCA5*	NCT05616793	1/2	Not yet recruiting	Opus Genetics	Subretinal OPGx-001 a recombinant AAV8 vector carrying human *LCA5* gene
Gene augmentation	Leber's Congenital Amaurosis	*GUCY2D*	NCT03920007	1/2	Active, not recruiting	Atsena Therapeutics	Subretinal SAR439483, an AAV5 vector carrying the human *GUCY2D* gene (AAV8-hGRK1-hGUYC2D)
Gene augmentation	Leber's Congenital Amaurosis	*RPE65*	NCT02781480	1/2	Completed	MeiraGTx/ Janssen	Subretinal injection of AAV-RPE65, a recombinant AAV expressing a codon optimized human *RPE65* gene
Gene augmentation	Neuronal Ceroid Lipofuscinosis (Batten) disease	*CLN2*	NCT05791864	1/2	Recruiting	REGENXBIO	Subretinal injection of RGX-381, an AAV9 vector carrying the human *CLN2* gene (AAV9.CB7.hCLN2)
Gene augmentation	Neuronal Ceroid Lipofuscinosis (Batten) disease	*CLN5*	NCT05228145	1/2	Recruiting	Neurogene	Intracerebroventricular and intravitreal injection of AAV9 vector carrying *CLN5* gene
Gene augmentation	Retinitis Pigmentosa	*PDE6A*	NCT04611503	1/2	Active, not recruiting	STZ eyetrial	Subretinal injection rAAV.hPDE6A, a recombinant AAV vector carrying the human *PDE6A* gene
Gene augmentation	Retinitis Pigmentosa	*PDE6B*	NCT03328130	1/2	Recruiting	Coave Therapeutics	Subretinal injection AAV2/5-hPDE6B, an AAV vector carrying the human *PDE6B* gene
Gene augmentation	Retinitis Pigmentosa	*RLBP1*	NCT03374657	1/2	Active, not recruiting	Novartis	Subretinal injection of CPK850, a self-complementary AAV8 vector carrying the gene for human *RLBP1*
Gene augmentation	X-linked RP	*RPGR*	NCT05874310	1	Recruiting	Frontera Therapeutics	Intraocular FT-002, an AAV carrying human *RPGR* gene
Gene augmentation	X-linked RP	*RPGR*	NCT04671433 & NCT04794101	3	Recruiting	MeiraGTx/ Janssen	Subretinal injection of RPGR2e11, a recombinant AAV3 vector carrying the human *RPGR* gene

(continued)

Table 59.1 (continued)

Intervention type	Disease	Gene (mutation)	Trial identifier	Phase	Status	Sponsor	Intervention description
Gene augmentation	X-linked RP	*RPGR*	NCT04517149	1/2	Active, not recruiting	4D Molecular Therapeutics	Single intravitreal injection of 4D-125, an AAV capsid variant carrying a codon-optimized human *RPGR* transgene
Gene augmentation	X-linked RP	*RPGR*	NCT03316560	1/2	Active, not recruiting	Beacon Therapeutics	Subretinal injection of AGTC-501, a recombinant AAV2 expressing human *RPGR* (rAAV2tYF-GRK1-RPGR)
Gene augmentation	X-linked retinoschisis	*RS1*	NCT05878860	1/2	Not yet recruiting	Atsena Therapeutics	Single subretinal injection of ATSN-201, a recombinant AAV vector expressing human *RS1* gene (AAV.SPR-hGRK1-hRS1syn)
Gene augmentation	X-linked retinoschisis	*RS1*	NCT02317887	1/2	Active, not recruiting	National Eye Institute, USA	Unilateral intravitreal injection of a recombinant AAV vector expressing human *RS1* gene (AAV8-scRS/IRBPhRS)

Vectors for DNA Delivery

In general, gene therapy can be delivered in vitro or in vivo. In the former, cells are isolated from the patient, treated and reintroduced into the target organ. This strategy is used in an approved treatment for large B cell lymphoma using chimeric antigen receptor T cells (i.e., CAR T-cell therapy) (Locke et al. 2022). The in vivo approach requires exogenous DNA to be directly introduced into patient tissues. Viruses are attractive vectors in this regard because of their ability to be internalized into host cells and induce the host transcriptional and translational machinery to produce proteins encoded by the viral genome (Fig. 59.1). The engineering of adeno-associated viral (AAV) vectors has been a major milestone in bring in vivo gene therapies to clinical trials, particularly in the retina (Li and Samulski 2020). Voretigene neparvovec-rzyl was the first approved in vivo gene therapy to use an AAV vector, and similar vectors are used in the majority of IRD gene therapeutics currently in clinical trials.

AAVs are single-stranded DNA parvoviruses (Ong et al. 2019). Thirteen wildtype serovars have been discovered (AAV 1–13), which differ in their host cell specificity (termed tissue tropism). Most trials for retinal disease have used AAV serotypes 2, 4, 5, and 8, as these have all been demonstrated to efficiently transduce photoreceptors and RPE cells (Li and Samulski 2020; Trapani et al. 2014). Benefits of AAV as a gene therapy vector include the fact that it is non-pathogenic in humans and has low immunogenicity. Because photoreceptors are terminally differentiated and non-dividing, retinal gene therapy requires a vector capable of transfecting non-dividing cells. AAVs are able to do this and achieve long-term expression of transgenes without integration into host chromosomes, avoiding the risks of mutagenesis and promotion of malignancy (Li and Samulski 2020; Ong et al. 2019).

The wildtype AAV genome is composed of two genes (*rep* and *cap*) flanked by two inverted terminal repeats (ITRs). Investigators have determined that only the ITRs are essential for transgene expression and persistent transfection, meaning 96% of the viral genome can be re-engineered for gene therapy (Li and Samulski 2020). The essential components of the expression cassette in recombinant AAV vectors are a promoter, the therapeutic transgene and a poly-A tail (Fig. 59.2). As reviewed by Li and Samulski 2020, modifications can be made to these components to decrease immunogenicity and improve transduction efficiency (Fig. 59.2).

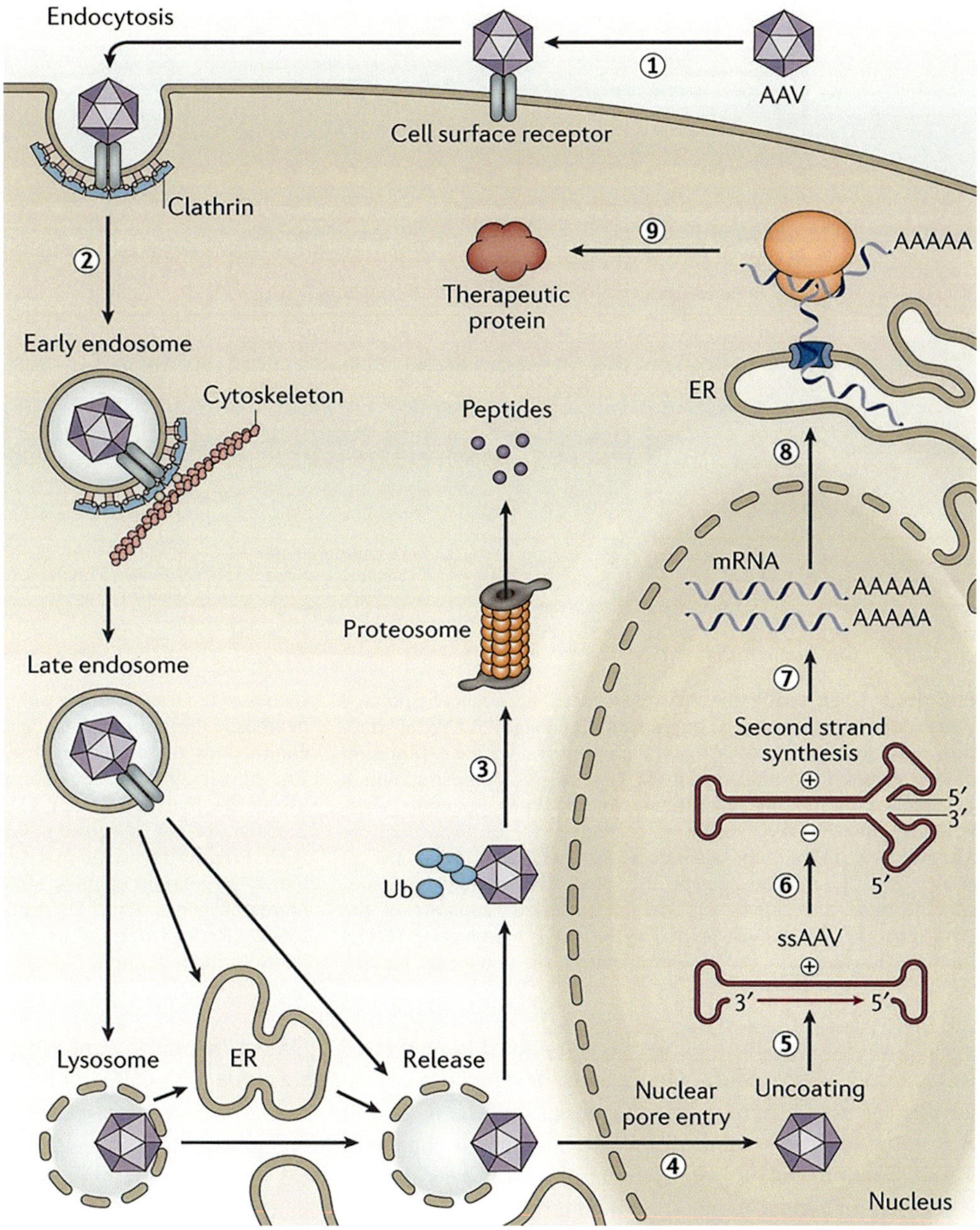

Fig. 59.1 AAV vector transduction pathway: Adeno-associated virus (AAV) vector virions bind to receptors and co-receptors on the surface of target cells (step 1) and are taken into endosomes within these cells through endocytosis (step 2). Following their release from endosomes, AAV virions are either ubiquitylated and targeted for proteasome-mediated degradation (step 3) or intracellularly trafficked to the nucleus (step 4). Once in the nucleus, AAV virions are uncoated and the AAV genome is released (step 5). The AAV single-stranded DNA genome then is converted into double-stranded DNA (step 6), followed by transcription (step 7) and the nuclear export of mRNA (step 8) for translation and expression of the therapeutic transgene (step 9). Engineering AAV vectors to affect any step of their transduction pathway impacts their transduction efficiency. *ER* endoplasmic reticulum, *ssAAV* single-stranded AAV, *Ub* ubiquitin. (Reprinted from Li and Samulski 2020, with permission from Springer Nature)

A limitation of AAV vectors is that the largest transgene that can be packaged into the capsid (4.7-kilobases, kb) is smaller than genes associated with some important inherited retinal diseases, including *ABCA4* associated with Stargardt disease and *USH2A* associated with autosomal recessive retinitis pigmentosa and Usher syndrome. Strategies to overcome this limitation are being investigated in pre-clinical studies. These include dual AAV vector "split-gene" approaches, where large transgenes are split across two or more vectors, with trans-splicing or homologous recombination of the fragments occurring inside transfected cells (Trapani et al. 2014). Viral vectors with larger packaging capacity, such as recombinant lentiviral vectors, have been explored in animal models of retinal degeneration (Bemelmans et al. 2005). Lentiviral vectors can package transgenes up to 8 kb in length but, compared to AAV, they have poor efficiency in transfecting adult photoreceptors (Bemelmans et al. 2005) and have the disadvantage of integrating into the host genome, potentially inducing mutations in host cells (Ong et al. 2019). Two phase 1/2 clinical trials of lentiviral gene therapy have been completed, one involving patients with Stargardt disease (NCT01367444) and another in patients with Usher syndrome type 1B (NCT01505062). Both therapies were based on a recombinant equine infectious anemia virus (EIAV) vector delivered via subretinal injection. Three-year safety results have been published for the trial involving patients with Stargardt disease (Parker et al. 2022). All 22 study subjects experienced

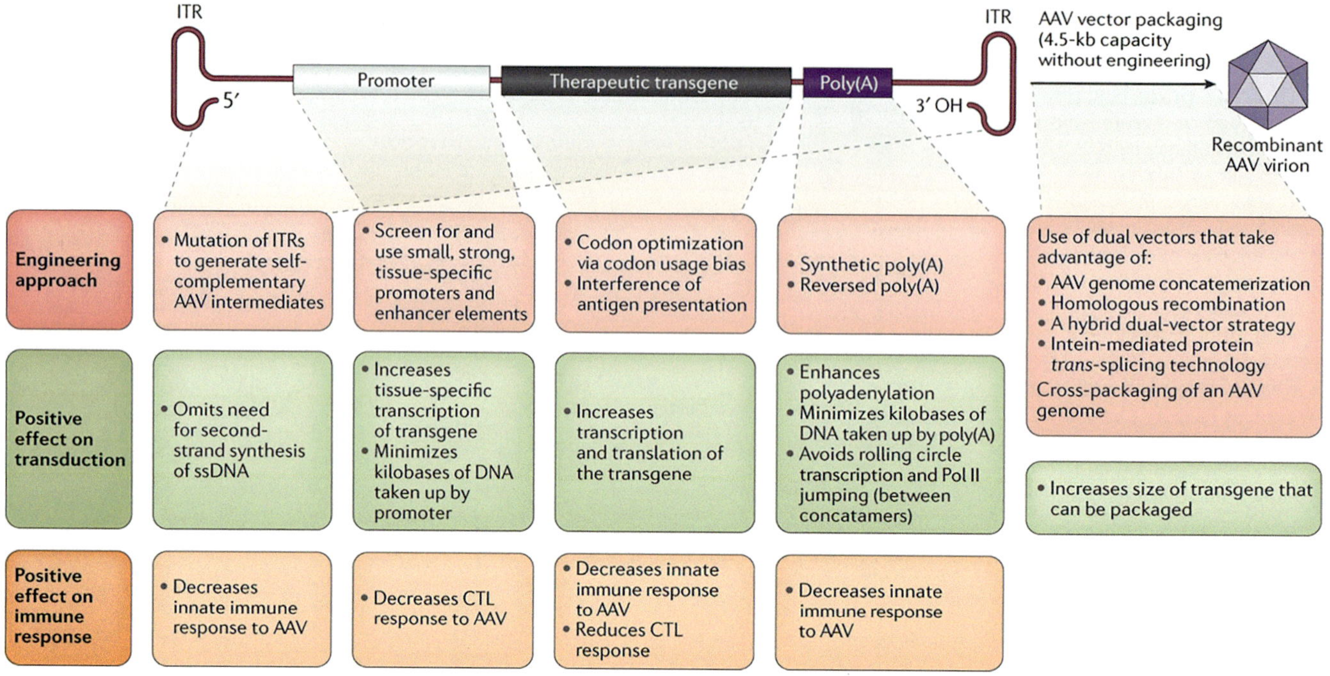

Fig. 59.2 Engineering the AAV cassette. The adeno-associated virus (AAV) cassette can be engineered to enhance AAV transduction and also to enable AAV to escape immune responses. Mutation of one inverted terminal repeat (ITR) on the AAV vector, which prevents the nicking of Rep protein, can generate self-complementary AAV vector to enhance vector transduction. Mutation of ITRs may also decrease the innate response to AAV. Use of small tissue-specific promoters increases tissue-specific transgene expression and the packaging capacity of the AAV genome and minimizes the cytotoxic T lymphocyte (CTL) immune response to AAV. Optimization of transgene codons increases the transcription and translation of the AAV transgene and decreases the immune response to AAV. Using synthetic poly(A) can increase the nuclear export, translation, and stability of mRNA (by enhancing polyadenylation) and using reversed poly(A) can avoid the transcription of ITR; both approaches enhance AAV transduction, and reversed poly(A) decreases the innate response to AAV. Finally, the use of dual AAV vectors or the cross-packaging of the AAV genome enables effective and functional expression of large transgenes. *Pol II* DNA polymerase II, *ssDNA* single-stranded DNA. (Reprinted from Li and Samulski 2020, with permission from Springer Nature)

adverse events, but most were mild and judged to be related to the surgical procedure. One patient developed a severe adverse event of chronic ocular hypertension. The treatment was not associated with any clinically significant change in visual function, and, in fact, a quarter of treated eyes showed worsening of retinal pigment epithelium atrophy compared to untreated eyes. The sponsor of both trials has reportedly halted further development of these products (Taylor 2019).

Case Vignette: Gene Replacement Therapy for *RPE65*-Associated Retinal Dystrophy

A 6-year-old boy was referred to our paediatric ophthalmology department for evaluation of a suspected inherited retinal disorder. His parents had noticed he was having difficulty navigating in the dark from an early age but did not seem to have any difficulty during the day. At his initial evaluation, his best corrected visual acuities were 20/40 OD and 20/50 OS, color plates were 1/6 and 3/6, and contrast sensitivity was reduced to 6% and 8%. On fundus examination, he was found to have mid-peripheral retinal pigment epithelium depigmentation, corresponding to outer retinal loss on optical coherence tomography, with relative preservation of the maculae (Fig. 59.3a, b). His visual fields were relatively full to the III4e stimulus, with midperipheral scotomata (Fig. 59.3c). Electroretinography findings were consistent with severe generalized rod and cone dysfunction, and full field stimulus testing (FST) showed moderately-reduced sensitivity to blue and red stimuli in both eyes. We arranged for genetic testing using a retinal dystrophy panel, which revealed he had biallelic pathogenic missense mutations in the *RPE65* gene (p.Leu341Ser/p.Gly187Glu) consistent with *RPE65*-associated Early-childhood onset retinal dystrophy (ECORD). In 2020, when he was 13 years old, Health Canada approved voretigene neparvovec-rzyl (Luxturna). He became one of the first patients at our hospital to receive this therapy (Fig. 59.3d). Following treatment, he has noticed an improvement in his night vision, saying that, for the first time, he is able to see stars and aeroplanes. At his most recent

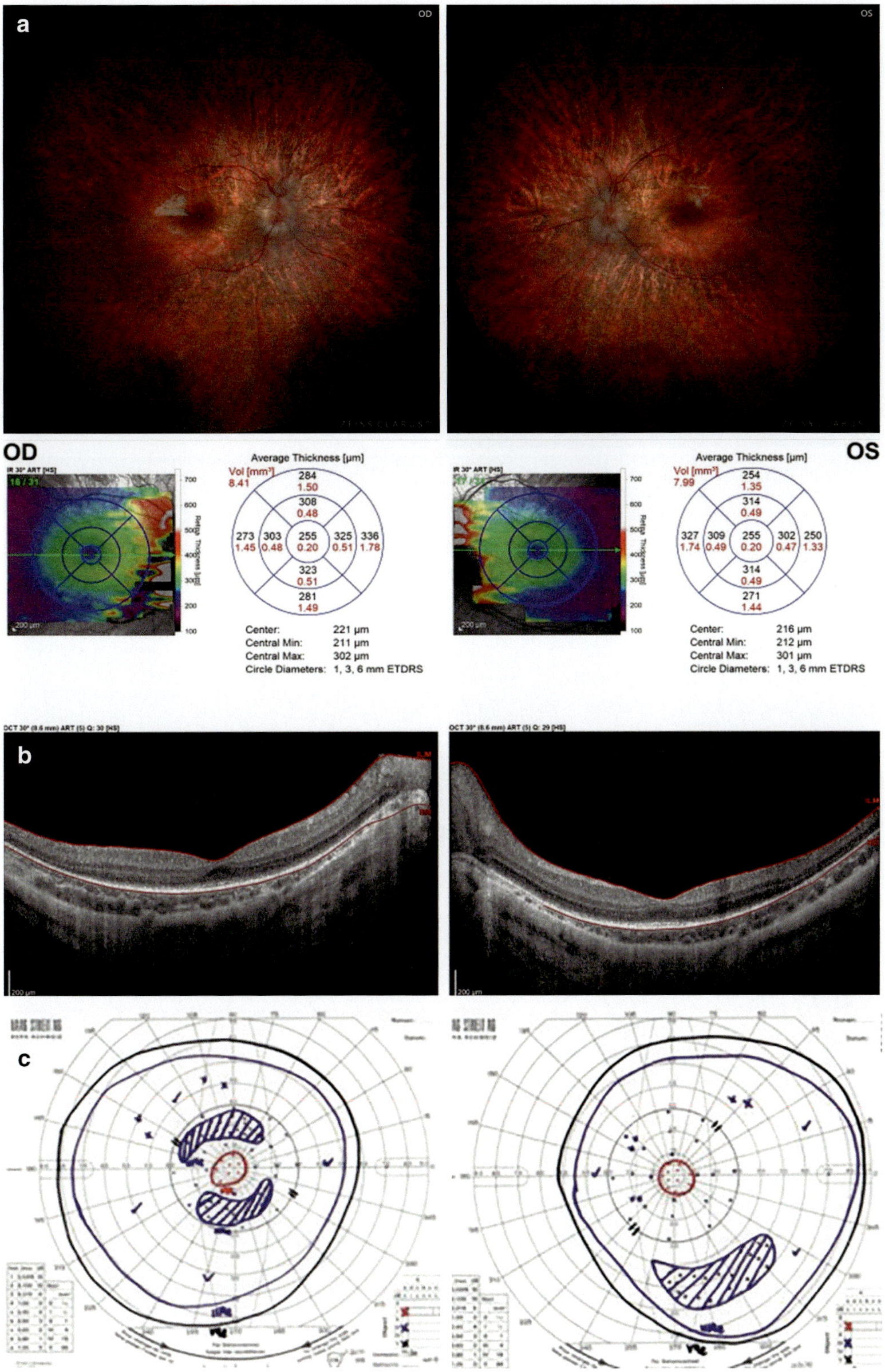

Fig. 59.3 Pre-treatment: color fundus photographs showing mid-peripheral retinal pigment epithelium atrophy (**a**), optical coherence tomograms showing extensive outer retinal loss with sparing of the subfoveal maculae (**b**), and Goldmann visual field testing demonstrating preservation of the III4e isopter and paracentral scotomata (**c**). Intra-operative photograph showing subretinal injection of voretigene neparvovec-rzyl in the right eye (**d**). Post-treatment: color fundus photograph showing circumscribed patches of chorioretinal atrophy (**e**) and Goldmann visual fields demonstrating progression of the paracentral scotomata (**f**) with stable macular sensitivity

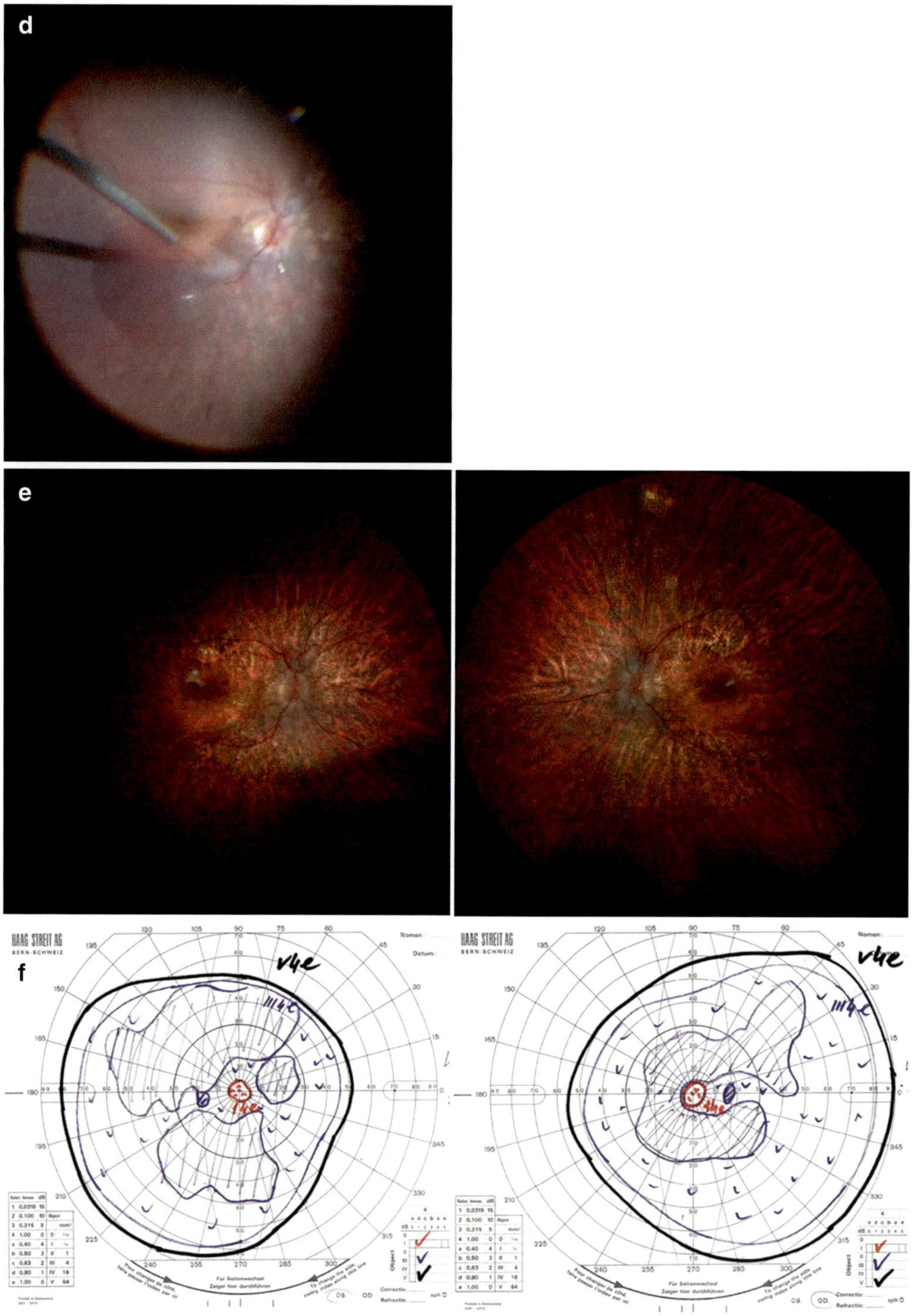

Fig. 59.3 (continued)

visit, 9 months following treatment, his visual acuity has remained stable and FST sensitivities have normalized. He has developed circumscribed patches of chorioretinal atrophy around the vascular arcades (Fig. 59.3e). His visual fields are stable peripherally, although there has been expansion of the paracentral scotomata (Fig. 59.3f).

Voretigene Neparvovec-rzyl

Voretigene neparvovec-rzyl is a gene replacement therapy for patients with autosomal recessive *RPE65*-associated retinal dystrophy. The *RPE65* gene encodes for retinoid isomerohydrolase, an enzyme expressed in the RPE, the function of which is to replenish the 11-*cis* retinal chromophore required for phototransduction (Cai et al. 2009). Biallelic *RPE65* mutations result in an absence of functional enzyme, which most commonly presents with a phenotype of severe, early-onset visual impairment and nystagmus. Patients tend to present between birth and 5 years of age, with further visual decline generally occurring during adolescence (Chao et al. 2019). Most affected individuals are legally blind by the end of their second decade of life (Chao et al. 2019). There is, however, clinical heterogeneity in the onset and course of degeneration associated with the disorder. Characteristically, retinal structure remains relatively preserved until late in the disorder, despite profoundly depressed ERG response amplitudes. This function-structure dissociation makes *RPE65*-associated retinal dystrophy an ideal candidate for gene replacement therapy (Maguire et al. 2021).

Following successful proof-of-concept studies for AAV2-mediated gene replacement in a canine model of *RPE65*-associated retinal degeneration (Acland et al. 2005), several phase I human trials were completed (Maguire et al. 2008; Hauswirth et al. 2008; Bainbridge et al. 2008; Banin et al. 2010; Le Meur et al. 2018). Delivered via subretinal injection, these therapies are designed to restore expression of the normal *RPE65* gene in transfected RPE cells, thereby improving photoreceptor function. The phase I trials differed in terms of exact gene therapy product used, varying, for example, in the AAV vector capsid, gene promoter sequence and codon optimization of the *RPE65* cDNA and in the dose and volume delivered (Maguire et al. 2021). All demonstrated acceptable safety, without harmful immune responses (Maguire et al. 2021).

Voretigene neparvovec-rzyl was the only one of these therapies to proceed to a phase III trial (NCT00999609). The trial enrolled subjects aged 3 years and older with a molecular diagnosis of *RPE65*-associated LCA (Russell et al. 2017). To be eligible, subjects had to have sufficient viable photoreceptors as determined by ophthalmoscopy and retinal thickness >100 µm on optical coherence tomography (OCT). Intervention subjects received bilateral subretinal injections of voretigene neparvovec-rzyl 6–18 days apart and were followed for 1 year.

In patients with severe visual impairment and nystagmus, traditional psychometric and electrophysiologic testing is challenging to perform and insensitive to therapeutic effects. To improve sensitivity and clinical relevance, the voretigene neparvovec-rzyl phase III trial employed two novel outcome measures as primary and secondary endpoints. The primary outcome measure was the change in performance on the Multi Luminance Mobility Test (MLMT), a visual navigation test specially developed to test functional vision in people with severe visual impairment and nyctalopia (Russell et al. 2017). The test employs a 5 × 10-foot standardized obstacle course, which subjects complete under varying levels of environmental illumination (Fig. 59.4). The score is

Light Levels	Examples
1 lux	Moonless summer night; Indoor nightlight
4 lux	Cloudless night with half moon; Parking lot at night
10 lux	1 hour after sunset in city; Bus stop at night
50 lux	Outdoor train station at night; Inside of lighted stairwell
125 lux	30 minutes before sunrise; Interior of train / bus at night
250 lux	Interior of elevator or office hallway
400 lux	Office environment or food court

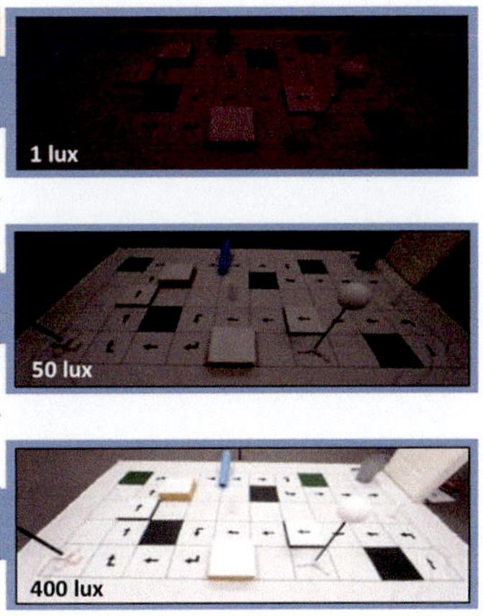

Fig. 59.4 Multi luminance mobility test. One possible iteration of the mobility course layout, shown under varying levels of environmental illumination, with examples of corresponding real-world light levels. (Reprinted with permission from Spark Therapeutics 2017)

the lowest light intensity at which the subject can complete the course within 3 min while making fewer than four errors. The second novel outcome measure was the change in full-field sensitivity threshold testing (FST; Roman et al. 2007; Klein and Birch 2009). FST is based on presenting light stimuli of differing intensities to the full visual field and is performed in the dark-adapted state (Roman et al. 2022). Similar to automated perimetry, subjects click a button indicating that the stimulus is perceived, and a full threshold algorithm is used to identify the intensity at which the stimulus is detectable 50% of the time. It has been shown that FST captures the responses of the best-functioning retinal regions, without the need for fixation or interrogation of specific retinal loci (Roman et al. 2022).

In the phase III trial, 29 subjects received bilateral treatments with voretigene neparvovec-rzyl and were followed to study completion. The trial demonstrated significant improvements in MLMT performance and FST sensitivity from baseline to 1 year, significantly improved visual field testing (Goldmann and static automated perimetry), and significant improvement in patient/parent-reported visual functioning on a modified version of the VFQ-25 (Russell et al. 2017). Best corrected visual acuity tended to improve more in treated versus control subjects, but the difference was not statistically significant (Russell et al. 2017). Treatment effects were found to be nearly maximal 30 days after treatment (Russell et al. 2017), and a 4-year follow-up a study found persistence of benefit up to this time (Maguire et al. 2019). It should be noted that studies have found that retinal degeneration may progress after treatment if a certain threshold (taken to be 30%) of photoreceptor loss has been surpassed (Gardiner et al. 2020). Consequently, the FDA drug label for the treatment recommends that prospective patients must undergo spectral domain OCT testing to demonstrate the presence of sufficient treatable photoreceptors.

In clinical trials, subretinal injection of voretigene neparvovec-rzyl was found to be safe and well tolerated. Although ocular adverse events were relatively common, most were mild and generally transient, including elevated intraocular pressure (20%), cataract (15%), retinal tear (10%) and mild intraocular inflammation (10%) (Russell et al. 2017). There were no major systemic adverse events (Russell et al. 2017). A post-marketing retrospective case review of patients treated at four institutions in the USA identified 10 patients (18 eyes) who developed stereotypic patterns of progressive chorioretinal atrophy following treatment (Gange et al. 2022; Fig. 59.5). The atrophy was first evident an average of 4.7 months after treatment and was progressive, in all cases, for up to a mean of 11.3 months of follow-up (Gange et al. 2022). Most often, the affected area included but was not limited to the site of the subretinal bleb. Notably, all patients continued to show post-treatment improvements in FST and Goldmann visual fields despite the appearance of atrophy. However, three patients developed paracentral scotomata corresponding to the areas of atrophy (Gange et al. 2022). Currently, the mechanism and incidence of this chorioretinal atrophy is unknown.

Fig. 59.5 Bilateral progressive chorioretinal atrophy following subretinal voretigene neparvovec-rzyl injection. Preoperative baseline ultra-widefield pseudocolor images of the right (**a**) and left (**b**) eyes. Intraoperative photographs demonstrating the extent of the subretinal bleb (**c, d**). Note the red asterisks indicating the bleb initiation site. Fundus photos from post-operative month 1, with early signs of chorioretinal atrophy in both eyes (**e, f**). Fundus photos from postoperative year 1.5, with progressive, more confluent areas of chorioretinal atrophy (**g, h**). Optical coherence tomography at post-op year 1.5, with areas of outer retinal loss and increased signal transmission from the choroid noted temporally in both eyes, correlating with areas of atrophy noted on the previously described fundus photos (**i, j**). (Reprinted from Gange et al. 2022, with permission from the American Academy of Ophthalmology)

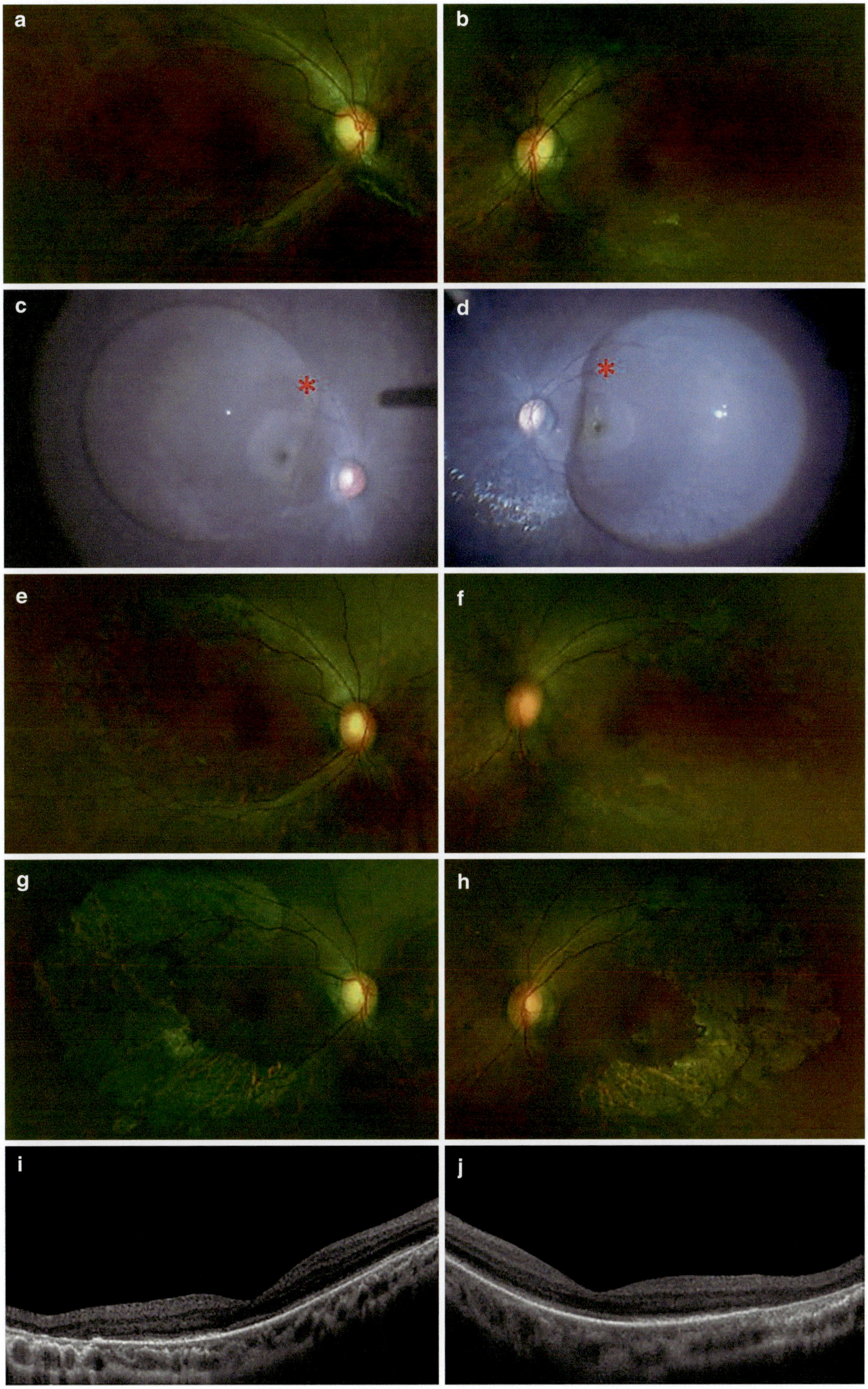
a
b
c
d
e
f
g
h
i
j

Other Gene Therapies for IRDs

Since the approval of voretigene neparvovec-rzyl, clinical trials of AAV-mediated gene replacement therapy have been conducted for several other IRDs. Most advanced in the development pipeline is a subretinal-delivered AAV-2 vector expressing Rab Escort Protein 1 (REP1), the gene defective in choroideremia. A phase III trial was completed in 2020 (NCT03496012), which failed to reach its primary endpoint of ≥15 letter improvement in BCVA at 12 months (MacLaren et al. 2023). A follow-up long-term safety and efficacy study is ongoing (NCT03584165). A meta-analysis of five phase I/II trials of *REP1* gene replacement therapy similarly found no overall improvement in BCVA in treated subjects, although there was variation across the studies (Abbouda et al. 2021). Despite these disappointing results, the authors of the phase III study found that greater percentages of treated compared to control patients experienced ≥10-letter improvements in visual acuity (MacLaren et al. 2023). They argue that future trials ought to enrol patients with greater area of preserved retina and consider alternative study endpoints (MacLaren et al. 2023).

Phase I/II trial results have been reported for X-linked RP (AAV8-RPGR, cotoretigene toliparvovec; von Krusenstiern 2023), Achromatopsia (AAV8-CNGA3; Fischer et al. 2020), and X-linked retinoschisis (AAV8-RS1; Cukras et al. 2018, and rAAV2tYF-CB-hRS1; Pennesi et al. 2022). The AAV8-RPGR and AAV8-CNGA3 vectors were both delivered via subretinal injection, whereas AAV8-RS1 and rAAV2tYF-CB-hRS1 were the first gene therapy product tested in humans as an intravitreal injection. Intravitreal delivery has several advantages over subretinal delivery, including less invasive surgery and potential to reach a wider area of the retina, and is the preferred method for targeting cells of the inner retina (Trapani et al. 2014). However, this route of delivery has much lower transduction efficiency for photoreceptors and RPE cells (Dalkara et al. 2009). An important finding of these trials was the generation of dose-dependent intraocular inflammatory responses, which may have been more significant for the intravitreal therapeutic. In the AAV8-RS1 trial, all three subjects who received the highest vector dosage demonstrated clinically significant intraocular inflammation during the first 2 months after therapy. The inflammation was responsive to topical and systemic corticosteroids, but one patient with severe vitritis developed a retinal tear and vitreous hemorrhage and required vitrectomy (Cukras et al. 2018). Similarly, all three subjects who received the highest vector dose of rAAV2tYF-CB-hRS1 developed chronic uveitis, and two experienced retinal detachments (Pennesi et al. 2022). The AAV8-RPGR trial also reported significant rates of ocular inflammation and non-infectious retinitis among participants in the four highest dosage groups (4/12 and 8/12 subjects, respectively), with two subjects experiencing permanent reductions in visual acuity (von Krusenstiern 2023). Although both retinoschisis trials failed to find clear evidence of clinical benefit, encouragingly, both the X-linked RP and achromatopsia studies have reported statistically significant and sustained improvements in their secondary endpoints, i.e., macular integrity assessment (MAIA) microperimetry and low light VA (XLRP); and BCVA, contrast sensitivity and chromatic pupillography (ACHM).

Several additional trials are underway (Table 59.1). Among these is a phase III trial of AAV-mediated *RPGR* gene therapy for X-linked RP, results of which are expected in 2024.

References

Abbouda A, Avogaro F, Moosajee M, Vingolo EM. Update on Gene Therapy Clinical Trials for Choroideremia and Potential Experimental Therapies. Medicina (Kaunas). 2021;57(1):64. https://doi.org/10.3390/medicina57010064.

Acland GM, Aguirre GD, Bennett J, Aleman TS, Cideciyan AV, Bennicelli J, Dejneka NS, Pearce-Kelling SE, Maguire AM, Palczewski K, Hauswirth WW, Jacobson SG. Long-term restoration of rod and cone vision by single dose rAAV-mediated gene transfer to the retina in a canine model of childhood blindness. Mol Ther. 2005;12(6):1072-82. https://doi.org/10.1016/j.ymthe.2005.08.008.

Bainbridge JW, Smith AJ, Barker SS, Robbie S, Henderson R, Balaggan K, Viswanathan A, Holder GE, Stockman A, Tyler N, Petersen-Jones S, Bhattacharya SS, Thrasher AJ, Fitzke FW, Carter BJ, Rubin GS, Moore AT, Ali RR. Effect of gene therapy on visual function in Leber's congenital amaurosis. N Engl J Med. 2008;358(21):2231-9. https://doi.org/10.1056/NEJMoa0802268. Epub 2008 Apr 27.

Banin E, Bandah-Rozenfeld D, Obolensky A, Cideciyan AV, Aleman TS, Marks-Ohana D, Sela M, Boye S, Sumaroka A, Roman AJ, Schwartz SB, Hauswirth WW, Jacobson SG, Hemo I, Sharon D. Molecular anthropology meets genetic medicine to treat blindness in the North African Jewish population: human gene therapy initiated in Israel. Hum Gene Ther. 2010;21(12):1749-57. https://doi.org/10.1089/hum.2010.047. Epub 2010 Nov 3.

Bemelmans AP, Bonnel S, Houhou L, Dufour N, Nandrot E, Helmlinger D, Sarkis C, Abitbol M, Mallet J. Retinal cell type expression specificity of HIV-1-derived gene transfer vectors upon subretinal injection in the adult rat: influence of pseudotyping and promoter. J Gene Med. 2005;7(10):1367-74. https://doi.org/10.1002/jgm.788.

Cai X, Conley SM, Naash MI. RPE65: role in the visual cycle, human retinal disease, and gene therapy. Ophthalmic Genet. 2009;30(2):57-62. https://doi.org/10.1080/13816810802626399.

Chao DL, Burr A, Pennesi M. RPE65-Related Leber Congenital Amaurosis/Early-Onset Severe Retinal Dystrophy. 2019 Nov 14. In: Adam MP, Mirzaa GM, Pagon RA, et al., editors. GeneReviews® [Internet]. Seattle (WA): University of Washington, Seattle; 1993–2023. Available from: https://www.ncbi.nlm.nih.gov/books/NBK549574/

Cukras C, Wiley HE, Jeffrey BG, Sen HN, Turriff A, Zeng Y, Vijayasarathy C, Marangoni D, Ziccardi L, Kjellstrom S, Park TK, Hiriyanna S, Wright JF, Colosi P, Wu Z, Bush RA, Wei LL, Sieving PA. Retinal AAV8-RS1 Gene Therapy for X-Linked Retinoschisis: Initial Findings from a Phase I/IIa Trial by Intravitreal Delivery. Mol Ther. 2018;26(9):2282-2294. https://doi.org/10.1016/j.ymthe.2018.05.025. Epub 2018 Jul 7.

Dalkara D, Kolstad KD, Caporale N, Visel M, Klimczak RR, Schaffer DV, Flannery JG. Inner limiting membrane barriers to AAV-mediated retinal transduction from the vitreous. Mol Ther. 2009;17(12):2096-102. https://doi.org/10.1038/mt.2009.181.

Fischer MD, Michalakis S, Wilhelm B, Zobor D, Muehlfriedel R, Kohl S, Weisschuh N, Ochakovski GA, Klein R, Schoen C, Sothilingam V, Garcia-Garrido M, Kuehlewein L, Kahle N, Werner A, Dauletbekov D, Paquet-Durand F, Tsang S, Martus P, Peters T, Seeliger M, Bartz-Schmidt KU, Ueffing M, Zrenner E, Biel M, Wissinger B. Safety and Vision Outcomes of Subretinal Gene Therapy Targeting Cone Photoreceptors in Achromatopsia: A Nonrandomized Controlled Trial. JAMA Ophthalmol. 2020;138(6):643-651. https://doi.org/10.1001/jamaophthalmol.2020.1032.

Friedmann T, Roblin R. Gene therapy for human genetic disease? Science. 1972;175(4025):949-55. https://doi.org/10.1126/science.175.4025.949. PMID: 5061866.

Gange WS, Sisk RA, Besirli CG, Lee TC, Havunjian M, Schwartz H, Borchert M, Sengillo JD, Mendoza C, Berrocal AM, Nagiel A. Perifoveal Chorioretinal Atrophy after Subretinal Voretigene Neparvovec-rzyl for RPE65-Mediated Leber Congenital Amaurosis. Ophthalmol Retina. 2022;6(1):58-64. https://doi.org/10.1016/j.oret.2021.03.016.

Gardiner KL, Cideciyan AV, Swider M, Dufour VL, Sumaroka A, Komáromy AM, Hauswirth WW, Iwabe S, Jacobson SG, Beltran WA, Aguirre GD. Long-Term Structural Outcomes of Late-Stage RPE65 Gene Therapy. Mol Ther. 2020;28(1):266-278. https://doi.org/10.1016/j.ymthe.2019.08.013. Epub 2019 Sep 3.

Hauswirth WW, Aleman TS, Kaushal S, Cideciyan AV, Schwartz SB, Wang L, Conlon TJ, Boye SL, Flotte TR, Byrne BJ, Jacobson SG. Treatment of leber congenital amaurosis due to RPE65 mutations by ocular subretinal injection of adeno-associated virus gene vector: short-term results of a phase I trial. Hum Gene Ther. 2008;19(10):979-90. https://doi.org/10.1089/hum.2008.107.

Klein M, Birch DG. Psychophysical assessment of low visual function in patients with retinal degenerative diseases (RDDs) with the Diagnosys full-field stimulus threshold (D-FST). Doc Ophthalmol. 2009;119(3):217-24. https://doi.org/10.1007/s10633-009-9204-7. Epub 2009 Nov 3.

Le Meur G, Lebranchu P, Billaud F, Adjali O, Schmitt S, Bézieau S, Péréon Y, Valabregue R, Ivan C, Darmon C, Moullier P, Rolling F, Weber M. Safety and Long-Term Efficacy of AAV4 Gene Therapy in Patients with RPE65 Leber Congenital Amaurosis. Mol Ther. 2018;26(1):256-268. https://doi.org/10.1016/j.ymthe.2017.09.014. Epub 2017 Sep 19.

Li C, Samulski RJ. Engineering adeno-associated virus vectors for gene therapy. Nat Rev Genet. 2020;21(4):255-272. https://doi.org/10.1038/s41576-019-0205-4.

Locke FL, Miklos DB, Jacobson CA, Perales MA, Kersten MJ, Oluwole OO, Ghobadi A, Rapoport AP, McGuirk J, Pagel JM, Muñoz J, Farooq U, van Meerten T, Reagan PM, Sureda A, Flinn IW, Vandenberghe P, Song KW, Dickinson M, Minnema MC, Riedell PA, Leslie LA, Chaganti S, Yang Y, Filosto S, Shah J, Schupp M, To C, Cheng P, Gordon LI, Westin JR; All ZUMA-7 Investigators and Contributing Kite Members. Axicabtagene Ciloleucel as Second-Line Therapy for Large B-Cell Lymphoma. N Engl J Med. 2022;386(7):640–654. https://doi.org/10.1056/NEJMoa2116133.

MacLaren RE, Fischer MD, Gow JA, et al. Subretinal timrepigene emparvovec in adult men with choroideremia: a randomized phase 3 trial [published online ahead of print, 2023 Oct 9]. Nat Med. 2023; https://doi.org/10.1038/s41591-023-02520-3

Maguire AM, Bennett J, Aleman EM, Leroy BP, Aleman TS. Clinical Perspective: Treating RPE65-Associated Retinal Dystrophy. Mol Ther. 2021;29(2):442-463. https://doi.org/10.1016/j.ymthe.2020.11.029.

Maguire AM, Russell S, Wellman JA, Chung DC, Yu ZF, Tillman A, Wittes J, Pappas J, Elci O, Marshall KA, McCague S, Reichert H, Davis M, Simonelli F, Leroy BP, Wright JF, High KA, Bennett J. Efficacy, Safety, and Durability of Voretigene Neparvovec-rzyl in RPE65 Mutation-Associated Inherited Retinal Dystrophy: Results of Phase 1 and 3 Trials. Ophthalmology. 2019;126(9):1273-1285. https://doi.org/10.1016/j.ophtha.2019.06.017

Maguire AM, Simonelli F, Pierce EA, Pugh EN Jr, Mingozzi F, Bennicelli J, Banfi S, Marshall KA, Testa F, Surace EM, Rossi S, Lyubarsky A, Arruda VR, Konkle B, Stone E, Sun J, Jacobs J, Dell'Osso L, Hertle R, Ma JX, Redmond TM, Zhu X, Hauck B, Zelenaia O, Shindler KS, Maguire MG, Wright JF, Volpe NJ, McDonnell JW, Auricchio A, High KA, Bennett J. Safety and efficacy of gene transfer for Leber's congenital amaurosis. N Engl J Med. 2008;358(21):2240-8. https://doi.org/10.1056/NEJMoa0802315.

Ong T, Pennesi ME, Birch DG, Lam BL, Tsang SH. Adeno-Associated Viral Gene Therapy for Inherited Retinal Disease. Pharm Res. 2019;36(2):34. https://doi.org/10.1007/s11095-018-2564-5.

Parker MA, Erker LR, Audo I, Choi D, Mohand-Said S, Sestakauskas K, Benoit P, Appelqvist T, Krahmer M, Ségaut-Prévost C, Lujan BJ, Faridi A, Chegarnov EN, Steinkamp PN, Ku C, da Palma MM, Barale PO, Ayelo-Scheer S, Lauer A, Stout T, Wilson DJ, Weleber RG, Pennesi ME, Sahel JA, Yang P. Three-Year Safety Results of SAR422459 (EIAV-ABCA4) Gene Therapy in Patients With ABCA4-Associated Stargardt Disease: An Open-Label Dose-Escalation Phase I/IIa Clinical Trial, Cohorts 1-5. Am J Ophthalmol. 2022;240:285-301. https://doi.org/10.1016/j.ajo.2022.02.013.

Pennesi ME, Yang P, Birch DG, et al. Intravitreal Delivery of rAAV2tYF-CB-hRS1 Vector for Gene Augmentation Therapy in Patients with X-Linked Retinoschisis: 1-Year Clinical Results. Ophthalmol Retina. 2022;6(12):1130–1144. https://doi.org/10.1016/j.oret.2022.06.013

Roman AJ, Cideciyan AV, Aleman TS, Jacobson SG. Full-field stimulus testing (FST) to quantify visual perception in severely blind candidates for treatment trials. Physiol Meas. 2007;28(8):N51–6. https://doi.org/10.1088/0967-3334/28/8/N02.

Roman AJ, Cideciyan AV, Wu V, Garafalo AV, Jacobson SG. Full-field stimulus testing: Role in the clinic and as an outcome measure in clinical trials of severe childhood retinal disease. Prog Retin Eye Res. 2022;87:101000. https://doi.org/10.1016/j.preteyeres.2021.101000.

Russell S, Bennett J, Wellman JA, Chung DC, Yu ZF, Tillman A, Wittes J, Pappas J, Elci O, McCague S, Cross D, Marshall KA, Walshire J, Kehoe TL, Reichert H, Davis M, Raffini L, George LA, Hudson FP, Dingfield L, Zhu X, Haller JA, Sohn EH, Mahajan VB, Pfeifer W, Weckmann M, Johnson C, Gewaily D, Drack A, Stone E, Wachtel K, Simonelli F, Leroy BP, Wright JF, High KA, Maguire AM. Efficacy and safety of voretigene neparvovec (AAV2-hRPE65v2) in patients with RPE65-mediated inherited retinal dystrophy: a randomised, controlled, open-label, phase 3 trial. Lancet. 2017;390(10097):849–860. https://doi.org/10.1016/S0140-6736(17)31868-8. Epub 2017 Jul 14. Erratum in: Lancet. 2017 Aug 26;390(10097):848.

Schneider N, Sundaresan Y, Gopalakrishnan P, Beryozkin A, Hanany M, Levanon EY, Banin E, Ben-Aroya S, Sharon D. Inherited retinal diseases: Linking genes, disease-causing variants, and relevant therapeutic modalities. Prog Retin Eye Res. 2022;89:101029. https://doi.org/10.1016/j.preteyeres.2021.101029.

Spark Therapeutics, Inc. Luxturna FDA Advisory Committee Briefing Document. Oct 12, 2017. Accessed on Aug 10, 2023 from <https://pink.pharmaintelligence.informa.com/-/media/supporting-documents/pink-sheet/2017/10/voretigene_ac_spark_brfg.pdf>

Taylor, NP *Sanofi punts 38 R&D projects to narrow pipeline focus.* Fierce Biotech, Feb. 7, 2019. Retrieved Aug 10, 2023 <https://www.fiercebiotech.com/biotech/sanofi-punts-38-r-d-projects-to-narrow-pipeline-focus>

Trapani I, Colella P, Sommella A, Iodice C, Cesi G, de Simone S, Marrocco E, Rossi S, Giunti M, Palfi A, Farrar GJ, Polishchuk R, Auricchio A. Effective delivery of large genes to the retina by dual AAV vectors. EMBO Mol Med. 2014;6(2):194–211. https://doi.org/10.1002/emmm.201302948.

von Krusenstiern L, Liu J, Liao E, Gow JA, Chen G, Ong T, Lotery AJ, Jalil A, Lam BL, MacLaren RE; XIRIUS Part 1 Study Group XOLARIS Study Group. Changes in Retinal Sensitivity Associated With Cotoretigene Toliparvovec in X-Linked Retinitis Pigmentosa With RPGR Gene Variations. JAMA Ophthalmol. 2023;141(3):275–283. https://doi.org/10.1001/jamaophthalmol.2022.6254. Erratum in: JAMA Ophthalmol. 2023 Mar 1;141(3):293.

Mutation Specific Treatments for Inherited Retinal Diseases

Gareth D. Mercer, Brian G. Ballios, and Peter J. Kertes

Gene Editing 338
Post-transcriptional Gene Silencing (Antisense Oligonucleotides). 339
References 340

G. D. Mercer · B. G. Ballios · P. J. Kertes (✉)
Department of Ophthalmology & Vision Sciences,
Temerty Faculty of Medicine, University of Toronto,
Toronto, ON, Canada
e-mail: peter.kertes@sunnybrook.ca

S. H. Tsang et al. (eds.), *Atlas of Inherited Retinal Diseases*, Advances in Experimental Medicine and Biology 1467, https://doi.org/10.1007/978-3-031-72230-1_60

Table 60.1 Mutation specific treatments for inherited retinal diseases currently in clinical trials

Intervention type	Disease	Gene (mutation)	Trial identifier	Phase	Status	Sponsor	Intervention description
Gene editing	Retinitis Pigmentosa	*RHO*	NCT05805007	1	Not yet recruiting	Peking University Third Hospital	Subretinal injection of ZVS203e, a recombinant AAV-mediated gene editing drug that silences RHO mutant protein expression by CRISPR/Cas9 editing
Gene editing	LCA10	*CEP290* (c.2991 + 1655A > G)	NCT03872479	1/2	Active, not recruiting	Editas Medicine, Inc.	Subretinal injection of EDIT-101, a CRISPR/Cas9 based gene editor to edit the mutation c.2991 + 1655A > G in intron 26 (IVS26) of the CEP290 gene
Post-transcriptional variant silencing	*PRPF31*-associated RP (RP11)	*CNOT3*	NCT05902962	1	Recruiting	PYC therapeutics	Intravitreal injection of VP-001, antisense oligomer that causes down expression of *CNOT3*, a negative regulator of *PRPF31*

The next frontier in genetic therapy for IRDs is the correction or silencing of specific mutations. This is of relevance to conditions with dominant negative and gain-of-function disease mechanisms, or with causative genes that are too large for gene replacement using AAV vectors. We discuss two approaches that have reached the stage of human clinical trials: CRISPR-Cas9 based gene editing and post-transcriptional gene silencing using antisense oligonucleotides (Table 60.1). Other mutation-specific treatment approaches in pre-clinical development include adenosine deaminases acting on RNA (ADAR)-based RNA editing, RNA interference, and translational read through inducing drugs (TRADs). These have been reviewed elsewhere (e.g., Martinez Velazquez and Ballios 2021).

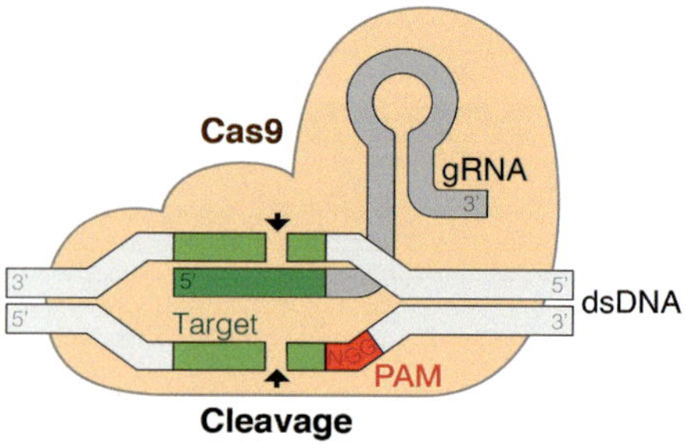

Fig. 60.1 Depiction of the CRISPR-Cas9 enzyme (peach color), with the guide RNA (in dark grey), double stranded break in DNA (light green); PAM—Protospacer Adjacent Motif. (Reprinted from wikimedia commons <https://commons.wikimedia.org/wiki/File:GRNA-Cas9.png> under Creative Commons Attribution (CC BY) 4.0 license)

Gene Editing

The clustered regularly interspaced short palindromic repeats (CRISPR)-Cas9 system occurs naturally in bacteria and archaea as the basis for immune responses against bacteriophage (Barrangou et al. 2007). Cas9 is an endonuclease that cleaves double stranded DNA (dsDNA). Specificity of the cleavage is determined by a short DNA sequence termed the Protospacer Adjacent Motif (PAM), present in the phage but not the host genome. The CRISPR/Cas9 system has been modified for the purposes of human gene editing through incorporation of a short sequence of single-stranded guide RNA (sgRNA) (Bansal et al. 2021). The sgRNA binds through complementary base pairing to a section of DNA carrying the target mutation, thereby directing the Cas9 dsDNA cleavage (Fig. 60.1a). dsDNA breaks are naturally repaired by one of two mechanisms:

non-homologous end joining, or homology directed repair (Saleh-Gohari and Helleday 2004). Homology directed repair (HDR) is directed by a DNA template, which can be engineered to incorporate desired edits into the genome. However, HDR is primarily active during cell division, therefore is not available in terminally differentiated cells like photoreceptors (Saleh-Gohari and Helleday 2004). In contrast, non-homologous end-joining, which is active during all stages of the cell cycle, directly ligates cleaved DNA without using a template, potentially introducing undesirable insertions/deletions. The DNA encoding CRISPR-Cas9 therapies can be delivered using AAV vectors, similarly to gene replacement therapies, or using non-viral vectors (Salman et al. 2022).

The EDIT-101 trial is the first human trial of CRISPR-Cas9-based gene editing (NCT03872479). This therapy targets the aberrant splice mutation, c.2991 + 1655A > G, in intron 26 of the *CEP290* gene, responsible for LCA10. The study sponsor recently reported acceptable interim safety data, and an encouraging efficacy signal among the subset of patients who were homozygous for the c.2991 + 1655A > G mutation (Editas 2022). Recruitment for the study is currently on hold pending identification of collaborators to help recruit this subset of patients.

Further refinements to the CRISPR-Cas9 system, particularly base editing and prime editing, have the potential to be safer and more versatile for genome editing in IRDs, but their clinical efficacy has yet to be evaluated. These are reviewed in detail elsewhere (Ballios et al. 2021; Bansal et al. 2021; da Costa et al. 2021).

Post-transcriptional Gene Silencing (Antisense Oligonucleotides)

Antisense oligonucleotides (AONs) are single- or double-stranded RNA or DNA molecules designed to bind, through complementary base pairing, to specific target RNA sequences (Altay et al. 2022). Binding of the target reverses the effects of pathogenic gene mutations through various mechanisms, including modulation of aberrant pre-mRNA splicing, and degradation of toxic gain-of-function mRNAs (Fig. 60.2a). AONs offer a theoretic safety advantage over CRISPR-Cas9 genome editing because their effects are transient and reversible.

Two RNA antisense oligonucleotide therapies for IRDs have been studied in phase II/III clinical trials (NCT05158296, NCT04855045 and NCT03913143). Sepofarsen targets the c.2991 + 1655A > G (p.Cys998) mutation in *CEP290,* which is responsible for LCA10, thereby restoring wild type pre-mRNA splicing (Collin et al. 2012; Russell et al. 2022). Ultevursen targets exon 13 of the *USH2a* gene, causing skipping of the exon and any mutations within it, during pre-mRNA splicing, resulting in a shortened but functional gene product (Dulla et al. 2021). In 2022, it was announced that

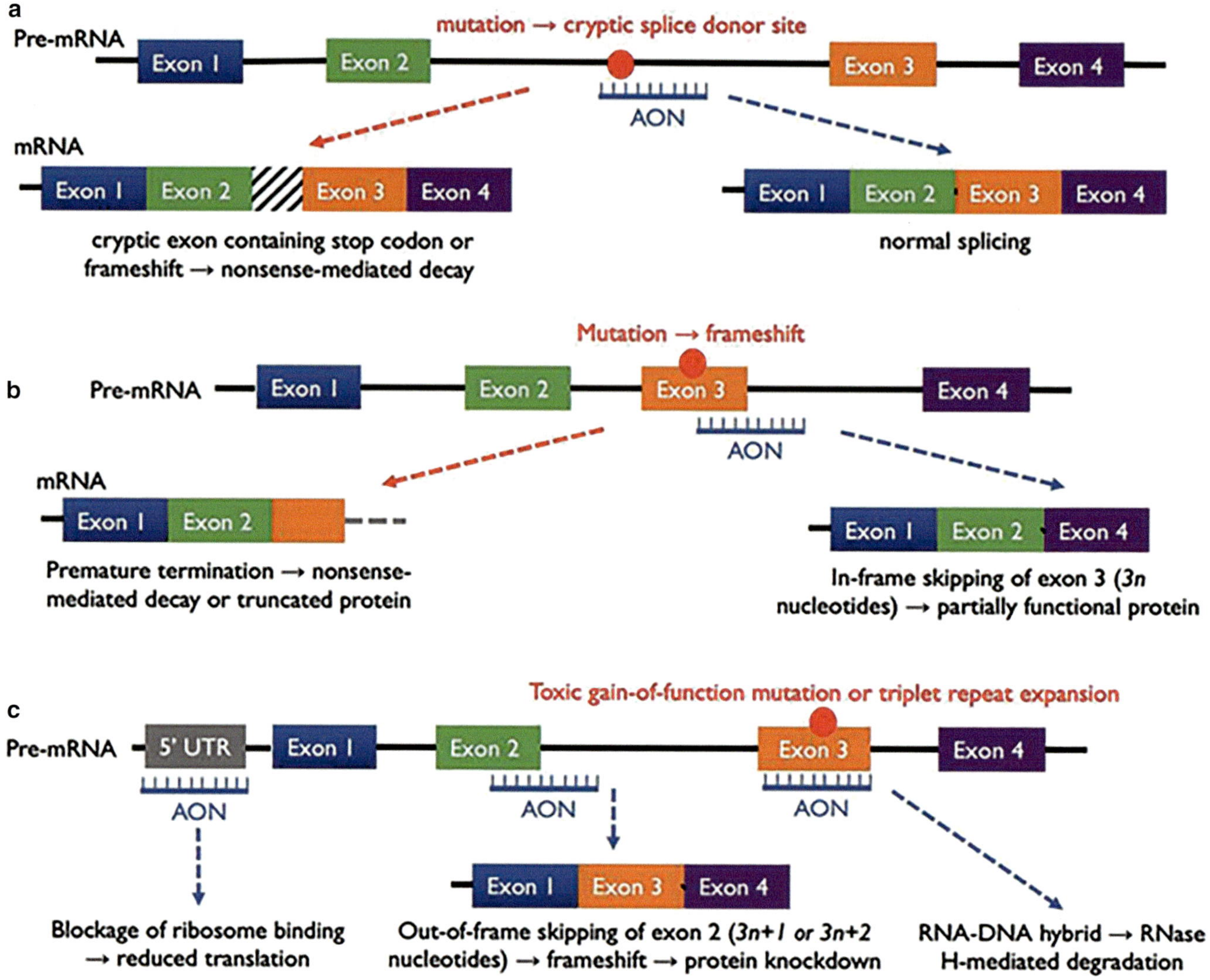

Fig. 60.2 Mechanisms of action of antisense oligonucleotide (ASO) therapies. (**a**) An ASO hybridizing with a deep-intronic mutation could prevent aberrant splicing which would otherwise lead to the incorporation of a cryptic exon or frameshift. (**b**) An ASO could be designed to block a splice donor site in order to cause skipping of a mutation-containing exon, so that a shortened but partially functional protein may be produced. (**c**) ASOs could be designed to knockdown a dominant negative allele by direct binding to the mutant RNA sequence to generate a RNA-DNA hybrid, which would be recognized and degraded by RNase H. Alternatively, knockdown of both alleles may be achieved through ASO binding to the 5′-untranslated region of a target mRNA or ASO-mediated exon skipping with frameshift. (Reprinted from Xue and MacLaren 2020, under Creative Commons Attribution (CC BY) 4.0 licence)

the *Illuminate* trial of Sepofarsen in patients with *CEP290*-mediated LCA10 failed to meet its primary and secondary endpoints (ProQR 2022). In this trial, outcomes from Sepofarsen-treated eyes were compared to those from sham-treated eyes. However, post-hoc analyses comparing Sepofarsen treated eyes to contralateral untreated eyes in the same patient showed more encouraging improvements in visual acuity and full field stimulus testing thresholds, with similar improvements not observed in sham-treated patients (ProQR 2022). Phase II/III trial results have yet to be reported for Ultevursen. While the results of the Sepofarsen trial are disappointing, there is still considerable enthusiasm about the potential benefit of ASO-based therapies for IRDs.

References

Altay HY, Ozdemir F, Afghah F, Kilinc Z, Ahmadian M, Tschopp M, Agca C. Gene regulatory and gene editing tools and their applications for retinal diseases and neuroprotection: From proof-of-concept to clinical trial. Front Neurosci. 2022;16:924917. https://doi.org/10.3389/fnins.2022.924917.

Ballios BG, Pierce EA, Huckfeldt RM. Gene editing technology: Towards precision medicine in inherited retinal diseases. Semin Ophthalmol. 2021;36(4):176-184. https://doi.org/10.1080/08820538.2021.1887903.

Bansal M, Acharya S, Sharma S, Phutela R, Rauthan R, Maiti S, Chakraborty D. CRISPR Cas9 based genome editing in inherited retinal dystrophies. Ophthalmic Genet. 2021;42(4):365-374. https://doi.org/10.1080/13816810.2021.1904421. Epub 2021 Apr 6.

Barrangou R, Fremaux C, Deveau H, Richards M, Boyaval P, Moineau S, Romero DA, Horvath P. CRISPR provides acquired resistance against viruses in prokaryotes. Science. 2007;315(5819):1709-12. https://doi.org/10.1126/science.1138140.

Collin RW, den Hollander AI, van der Velde-Visser SD, Bennicelli J, Bennett J, Cremers FP. Antisense Oligonucleotide (AON)-based Therapy for Leber Congenital Amaurosis Caused by a Frequent Mutation in CEP290. Mol Ther Nucleic Acids. 2012;1(3):e14. https://doi.org/10.1038/mtna.2012.3.

da Costa BL, Levi SR, Eulau E, Tsai YT, Quinn PMJ. Prime Editing for Inherited Retinal Diseases. Front Genome Ed. 2021;3:775330. https://doi.org/10.3389/fgeed.2021.775330.

Dulla K, Slijkerman R, van Diepen HC, Albert S, Dona M, Beumer W, Turunen JJ, Chan HL, Schulkens IA, Vorthoren L, den Besten C, Buil L, Schmidt I, Miao J, Venselaar H, Zang J, Neuhauss SCF, Peters T, Broekman S, Pennings R, Kremer H, Platenburg G, Adamson P, de Vrieze E, van Wijk E. Antisense oligonucleotide-based treatment of retinitis pigmentosa caused by USH2A exon 13 mutations. Mol Ther. 2021;29(8):2441-2455. https://doi.org/10.1016/j.ymthe.2021.04.024.

Editas Medicine Announces Clinical Data Demonstrating Proof Of Concept Of EDIT-101 From Phase 1/2 BRILLIANCE Trial. Editas Medicine, Inc, Cambridge, Mass., November 17, 2022. Webpage retrieved June 18, 2023. <https://ir.editasmedicine.com/news-releases/news-release-details/editas-medicine-announces-clinical-data-demonstrating-proof>

Martinez Velazquez LA, Ballios BG. The Next Generation of Molecular and Cellular Therapeutics for Inherited Retinal Disease. Int J Mol Sci. 2021;22(21):11542. https://doi.org/10.3390/ijms222111542.

ProQR Announces Top-Line Results from Phase 2/3 Illuminate Trial of Sepofarsen in CEP290-mediated LCA10. ProQR Therapeutics, Leiden, Netherlands, February 11, 2022. Webpage retrieved June 18, 2023. <https://www.proqr.com/press-releases/proqr-announces-top-line-results-from-phase-23-illuminate-trial-of-sepofarsen-in-cep290-mediated-lca10>

Russell SR, Drack AV, Cideciyan AV, Jacobson SG, Leroy BP, Van Cauwenbergh C, Ho AC, Dumitrescu AV, Han IC, Martin M, Pfeifer WL, Sohn EH, Walshire J, Garafalo AV, Krishnan AK, Powers CA, Sumaroka A, Roman AJ, Vanhonsebrouck E, Jones E, Nerinckx F, De Zaeytijd J, Collin RWJ, Hoyng C, Adamson P, Cheetham ME, Schwartz MR, den Hollander W, Asmus F, Platenburg G, Rodman D, Girach A. Intravitreal antisense oligonucleotide sepofarsen in Leber congenital amaurosis type 10: a phase 1b/2 trial. Nat Med. 2022;28(5):1014-1021. https://doi.org/10.1038/s41591-022-01755-w.

Saleh-Gohari N, Helleday T. Conservative homologous recombination preferentially repairs DNA double-strand breaks in the S phase of the cell cycle in human cells. Nucleic Acids Res. 2004;32(12):3683-8. https://doi.org/10.1093/nar/gkh703.

Salman A, Kantor A, McClements ME, Marfany G, Trigueros S, MacLaren RE. Non-Viral Delivery of CRISPR/Cas Cargo to the Retina Using Nanoparticles: Current Possibilities, Challenges, and Limitations. Pharmaceutics. 2022;14(9):1842. https://doi.org/10.3390/pharmaceutics14091842.

Xue K, MacLaren RE. Antisense oligonucleotide therapeutics in clinical trials for the treatment of inherited retinal diseases. Expert Opin Investig Drugs. 2020;29(10):1163-1170. https://doi.org/10.1080/13543784.2020.1804853. Epub 2020 Sep 1.

CHAPTER 61

Global Treatment Approaches Part 1: Neuroprotection and Stem Cell Therapy

Gareth D. Mercer, Brian G. Ballios, and Peter J. Kertes

Neuroprotection 344
Stem Cell and Precursor Cell Therapies 347
References 350

G. D. Mercer · B. G. Ballios · P. J. Kertes (✉)
Department of Ophthalmology & Vision Sciences,
Temerty Faculty of Medicine, University of Toronto,
Toronto, ON, Canada
e-mail: peter.kertes@sunnybrook.ca

S. H. Tsang et al. (eds.), *Atlas of Inherited Retinal Diseases*, Advances in Experimental Medicine and Biology 1467, https://doi.org/10.1007/978-3-031-72230-1_61

Global treatment approaches, also called gene-agnostic therapies, are those that address pathophysiologic responses common to retinal degenerations, i.e., photoreceptor and RPE metabolic stress, dysfunction and, ultimately, cell death. Global treatment strategies may be classified as those that aim to maintain or replace dysfunctional photoreceptors and/or RPE (e.g., neuroprotective agents and stem cell therapies) or those that circumvent the dysfunctional photoreceptors entirely (e.g., bioelectronic retinal implants and optogenetics). The research and development costs of each gene-specific therapy are high relative to the number of individuals who can benefit from it. Gene agnostic treatments have the potential to benefit a significantly larger number of patients, including those for whom the genetic cause of their disease cannot be identified.

Neuroprotection

Retinal neuroprotection refers to treatment strategies aimed at slowing vision loss by preserving retinal neuronal structure and function. Oxidative stress and neuronal apoptosis are common endpoints leading to irreversible vision loss in a wide variety of conditions, including inherited and acquired retinal degenerations, diabetic retinopathy, and retinal detachment. Slowing these disease processes offers an attractive therapeutic strategy with the potential to benefit millions of patients worldwide. A variety of neuroprotective treatment have been evaluated in pre-clinical studies, ranging from traditional small molecules to gene therapy and stem cells. Although no neuroprotective treatments have yet been approved for clinical use, with numerous phase II and III trials ongoing, there is hope that this will soon be a part of the treatment armamentarium for retinal disease (Table 61.1).

In this chapter, we limit our discussion to neuroprotective agents in late phase clinical trials for patients with IRDs, namely N-acetyl cysteine (NAC) and Valproic acid for RP, and ALK-001 and Avacincaptad pegol for Stargardt macular dystrophy.

In RP, oxidative stress is felt to be one mechanism by which defects in rod photoreceptors eventually result in loss of cones. Rods account for 95% of retinal photoreceptors. As they degenerate, retinal oxygen consumption declines, and tissue oxygenation increases. In animal models of RP, rod loss is associated with progressive evidence of oxidative stress in cones, whereas administration of antioxidants appears to promote cone survival (Narayan et al. 2016). N-acetyl cysteine (NAC) is a molecule that directly reduces reactive oxygen species and is used by the body to synthesize glutathione, an important endogenous antioxidant. Its primary clinical use is as an emergency treatment for acetaminophen overdose. Oral administration of NAC to rd10$^{+/+}$ mice was associated with partial maintenance of cone cell survival and electroretinographic function (Lee et al. 2011). A phase I dose escalation trial of NAC in patients with clinically diagnosed RP demonstrated acceptable safety and tolerability up to 1800 mg three times daily (Campochiaro et al. 2020). In this uncontrolled trial, visual acuity improved significantly in all treatment groups, but macular sensitivity, measured by microperimetry, and ellipsoid zone width on OCT remained unchanged (Campochiaro et al. 2020). Based on these findings, a multicenter phase III clinical trial is planned for 2023. The study aims to enrol 438 RP patients to estimate the efficacy of NAC given orally at a dose of 1800 mg twice daily on OCT ellipsoid zone integrity (a marker of cone photoreceptor preservation) over 4 years (NCT05537220).

Valproic acid (VPA) is short chain fatty acid used to treat epilepsy, bipolar disorder, headache, and neuropathic pain. Its primary mechanism of action in these conditions is to enhance γ-aminobutyric acid (GABA) signalling in the central nervous system. However, it has been found to have numerous other potentially neuroprotective effects, including inhibition of histone deacetylase, upregulation of growth factor gene expression, and complement inhibition (Kimura et al. 2015; Birch et al. 2018). In a mouse model of ganglion cell degeneration, intravitreal VPA was found to increase Müller cell production of brain derived neurotrophic factor and nerve growth factor (Kimura et al. 2015). Based on these findings, it has been proposed that VPA could have neuroprotective effects in patients with retinitis pigmentosa. Small, uncontrolled clinical studies have shown mixed results (Chen et al. 2019). The best available evidence is from a phase II trial involving 90 patients with autosomal dominant RP (adRP), which found that oral valproic acid (500–1000 mg) taken daily for 12 months was associated with greater loss of visual field than was seen in patients taking a placebo (Birch et al. 2018). At this time, valproic acid should not be recommended to patients with adRP.

In autosomal recessive Stargardt macular dystrophy (STGD1), the affected gene is *ABCA4*, which encodes a photoreceptor outer segment membrane-associated protein called rim protein. This protein functions to transport all-trans-retinaldehyde-phosphatidylethanolamine (retinaldehyde-PE), a product of the retinoid cycle, out of photoreceptor outer segment discs. When the transporter protein is defective, retinaldehyde-PE may react to form vitamin A dimers (including A2E and ATR-dimer), which accumulate in RPE

Table 61.1 Stem cell and neuroprotection treatments for inherited retinal disease currently in clinical trials

Intervention type	Disease	Trial identifier	Phase	Status	Sponsor	Intervention description
Stem cell therapy	Dry AMD, DR, RVO, RP, hereditary macular degeneration	NCT04925687 and NCT01736059	1	Recruiting	University of California Davis, USA	Intravitreal injection of autologous bone marrow CD34+ stem cells
Stem cell therapy	RP	NCT04284293	1	Recruiting	Cedars-Sinai Medical Center, USA	Subretinal administration of human fetal cortical-derived neural progenitor cells (CNS10-NPC)
Stem cell therapy	RP	NCT03963154	1/2	Active, not recruiting	Centre d'Etude des Cellules Souches, France	Subretinal implantation of monolayer of human embryonic stem cell-derived RPE
Stem cell therapy	RP	NCT02464436	1/2	Complete, results pending	ReNeuron	Subretinal injection of human retinal progenitor cells
Stem cell therapy	RP	NCT0460489	2	Complete, results pending	jCtye, Inc.	Intravitreal injection of human retinal progenitor cells (jCell)
Neuroprotection	RP	NCT04864496	2	Active, not recruiting	Shahid Beheshti University, Iran	Oral N-acetylcysteine, 1200 mg twice daily
Neuroprotection	RP	NCT05537220	3	Recruiting	Johns Hopkins University, USA	Oral N-acetylcysteine, 1800 mg twice daily
Neuroprotection	RP	NCT05748873	1/2	Not yet recruiting	SparingVision	Subretinal SPVN06, a recombinant AAV vector carrying DNA for two isoforms of NXNL gene, to produce RdCVF, a neurotrophic factor, and rod-derived cone viability factor RdCVFL, an antioxidant protecting cones from oxidative stress
Neuroprotection	Stargardt's	NCT02402660 and NCT04239625	2	Recruiting	Alkeus Pharmaceuticals	Orally administered ALK-001 (C20-D3-retinyl acetate), a chemically modified vitamin A
Neuroprotection	Stargardt's	NCT03364153	2	Recruiting	IVERIC Bio	Intravitreal injection of Avacincaptad pegol (Zimura) a completent C5 inhibitor
Modifier gene therapy	Stargardt disease	NCT05956626	1/2	Recruiting	Ocugen	Subretinal injection of OCU410ST, an AAV5 vector expressing the human RORA gene, which plays a role in lipid metabolism.
Modifier gene therapy	RP and LCA	NCT05203939	1/2	Recruiting	Ocugen	Subretinal injection of OCU400, a gene therapy based on the nuclear hormone receptor gene, NR2E3

after photoreceptor outer segment phagocytosis, contributing to the Lipofuscin flecks characteristic of STGD1 (Cideciyan et al. 2004; Weng et al. 1999; Fig. 61.1). Vitamin A dimers have been found to potentially sensitize RPE cells to light-induced apoptosis (Charbel 2015), and to cause inflammation through complement pathway activation (Lenis et al.

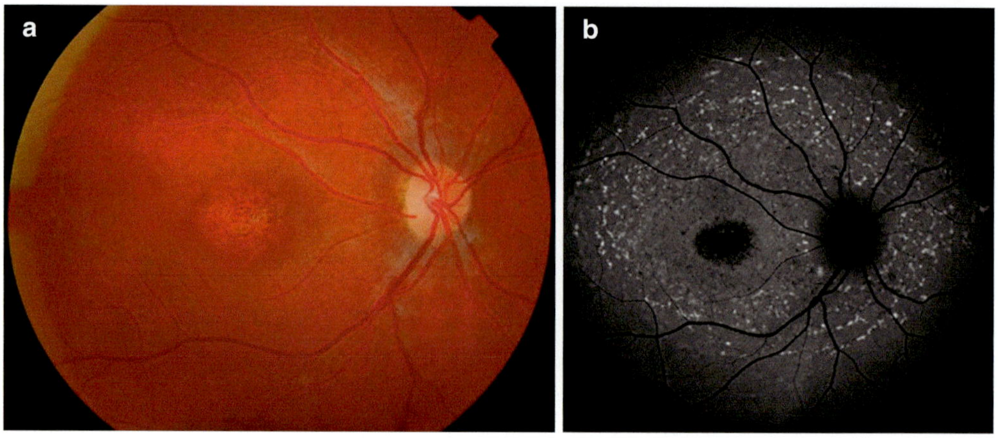

Fig. 61.1 (**a**) Color fundus photograph of a patient with Stargardt macular dystrophy showing typical yellow–white retinal flecks with macular atrophy. (**b**) Corresponding fundus autofluorescence image showing flecks of both increased and decreased autofluorescence and reduced central macular autofluorescence surrounded by an increased signal. (Reprinted from Tanna and Strauss 2017, under Creative Commons Attribution (CC BY) 4.0 licence)

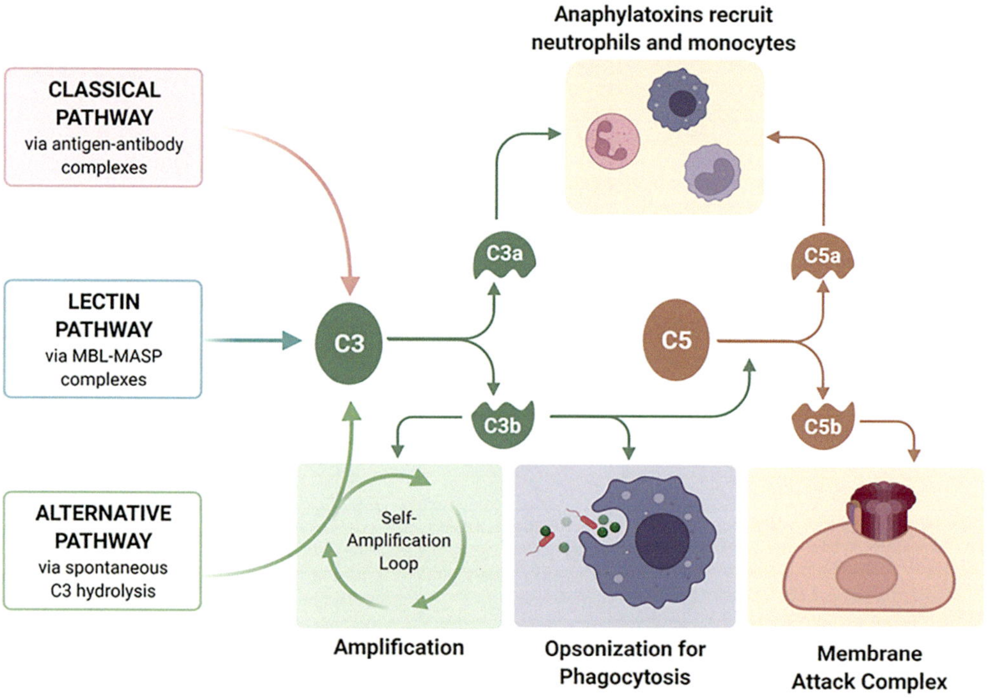

Fig. 61.2 Overview of the complement case and its role in innate immune responses. *MBL* mannose binding lectin, *MASP* MBL-associated serine protease. (Reprinted from wikimedia commons <https://commons.wikimedia.org/wiki/Category:Complement_system#/media/File:Complement_Overview.png> under Creative Commons Attribution (CC BY) 4.0 licence)

2017). Complement is an important component of the innate immune system, having pro-inflammatory and apoptotic effects. There are three pathways to complement activation, known as the classical, alternative and lectin pathways, which converge with activation of complement factor 3 (C3). C3, in turn, activates complement factor 5 (C5), which contributes to the inflammasome and membrane-attack complex, both of which can cause host cell death (Walport 2001; Fig. 61.2).

The alternative complement pathway has been implicated in the pathogenesis of geographic atrophy (GA) in age-related macular degeneration and recent phase III clinical trials of two different complement inhibitors demonstrated modest reductions in GA progression. The first of these molecules, Pegcetacoplan (trade name Syfovre), is a polypeptide inhibitor of C3 and was approved by the US FDA in February 2023 for use in patients with GA (Liao et al. 2020). The second molecule, Avacincaptap pegol (trade name Izervay), is a single-stranded oligonucleotide inhibitor of C5 (Jaffe et al. 2021).

In STGD1, findings from preclinical studies suggest that complement inhibition may reduce disease progression. For example, in an $ABCA4^{-/-}$ mouse model, over-expression of complement receptor 1-like protein, a negative regulator of complement, was associated with reduced accumulation of vitamin A dimers and improved photoreceptor survival (Lenis et al. 2017). A phase II clinical trial is currently underway to assess the safety and efficacy of intravitreal injections of Avacincaptap pegol in adult patients with Stargardt macular dystrophy (NCT03364153). Results are expected in late 2024.

Another potential neuroprotective strategy in STGD1 targets the upstream process of vitamin A dimerization. ALK-001 is a chemically modified vitamin A molecule, in which hydrogen atoms at position C20 are substituted with deuterium. Dietary replacement of vitamin A with ALK-001 in $ABCA4^{-/-}$ mice was found to decrease the levels of vitamin A dimers and lipofuscin granules to levels observed in wild type mice without any adverse effects on retinal function, as measured by ERG (Charbel et al. 2015). In a phase II clinical trial involving 50 pediatric and adult patients with STGD1 (NCT02402660), ALK-001 taken orally once daily for 2 years was found to significantly decrease the rate of expansion of atrophic lesions compared to placebo (by 21–28%) and was well tolerated without any severe treatment-related adverse events (Scholl 2022).

Stem Cell and Precursor Cell Therapies

Stem cells are progenitor cells capable of indefinite replication and differentiation into various mature cell types. They may be pluripotent, i.e., capable of giving rise to mature cell types in any of the three germ layers, or multipotent, meaning capable of giving rise to a limited number of related cell types. In IRDs, stem cells offer the potential to replenish dysfunctional and damaged photoreceptors and retinal pigment epithelium with healthy, functional cells (Nair and Thomas 2022). In addition, or alternatively, they release trophic factors, thereby rescuing and preserving damaged photoreceptors (Nair and Thomas 2022).

Two techniques have been studied for retinal stem cell transplantation: cell transplantation and sheet/patch transplantation (Fig. 61.3; Maeda et al. 2022; Gagliardi et al. 2019). In cell transplantation, purified cell suspensions are injected into the subretinal or intravitreal space. The injected cells cover a wide area but, especially in the case of photoreceptors, they do not form a structured layer like that of morphologically mature retina. In sheet transplantation, cells to be transplanted are grown in a structured layer on a synthetic scaffold made of biodegradable or non-biodegradable polymers and the entire complex is placed into the subretinal space. In sheet transplantation, the area treated is more limited, but the structured arrangement may allow cells to survive longer (Maeda et al. 2022).

There are no approved stem cell therapies for inherited retinal diseases, but the cell types with the most promise for retinal repair are stem-cell derived RPE, retinal progenitor cells, and stem-cell derived photoreceptors/photoreceptor precursors. Mesenchymal stem cells are being studied for their neurotrophic properties.

Human embryonic stem cell (hESC)-derived RPE cells have been studied in phase I/II clinical trials for Stargardt macular dystrophy and phase I/II trials are ongoing in patients with RP (Table 61.1). Two phase I/II clinical trials (total 16 patients), conducted in the USA and China, investigated subretinal transplantation of hESC derived RPE in patients with Stargardt macular dystrophy (Schwartz et al. 2015; Li et al. 2021) In both studies, hESCs were differentiated spontaneously into RPE through repeated passaging, then expanded and characterized morphologically using cell surface marker staining. Transplantation involved standard three-port pars plana vitrectomy with PVD-induction, followed by subretinal injection of RPE cell suspension using a 38–41 g cannula. All participants received peri- and post-operative oral immunosuppression medication (mycophenolate mofetil and tacrolimus) for 3 months. Overall, subretinal hESC-derived RPE transplantation therapy was found to be safe and well tolerated up to 5 years after surgery. Notably, no episodes of tumorigenesis or rejection were observed. Adverse events were deemed attributable to vitrectomy surgery and immunosuppression, including cataract, bacterial endophthalmitis, and non-infectious intraocular inflammation. Interestingly, most subjects in the US study, and those with more advanced disease in the Chinese study, developed areas of increased subretinal pigmentation, typically bordering areas of macular atrophy, assumed to reflect integration of transplanted RPE (Fig. 61.4). In the US study, visual acuity improved by an average of 8 letters in treated versus untreated eyes at 12 months, and improvement was reported in vision-related quality of life (Schwartz et al. 2015). Whereas, in the Chinese study, visual acuity was found to remain unchanged comparing treated to untreated eyes in all but one patient who had significant visual deterioration in the treated eye (Li et al. 2021). The latter study also found that some patients developed areas of subretinal hyper-reflective foci on OCT imaging with accompanying multifocal ERG improvements and that this treatment response could potentially be predicted using features on pre-operative autofluorescence imaging (Li et al. 2021). Hopefully, phase III trials will clarify the conflicting functional outcomes observed in these two early phase trials.

Retinal progenitor cell (RPC) transplantation has been studied as a treatment for RP in two phase II trials. One treatment strategy involves intravitreal injection of human allogenic RPCs (JCyte). In a phase IIb study involving 84 participants followed for 1 year, this strategy appeared to show a significant improvement in BCVA at the higher dose

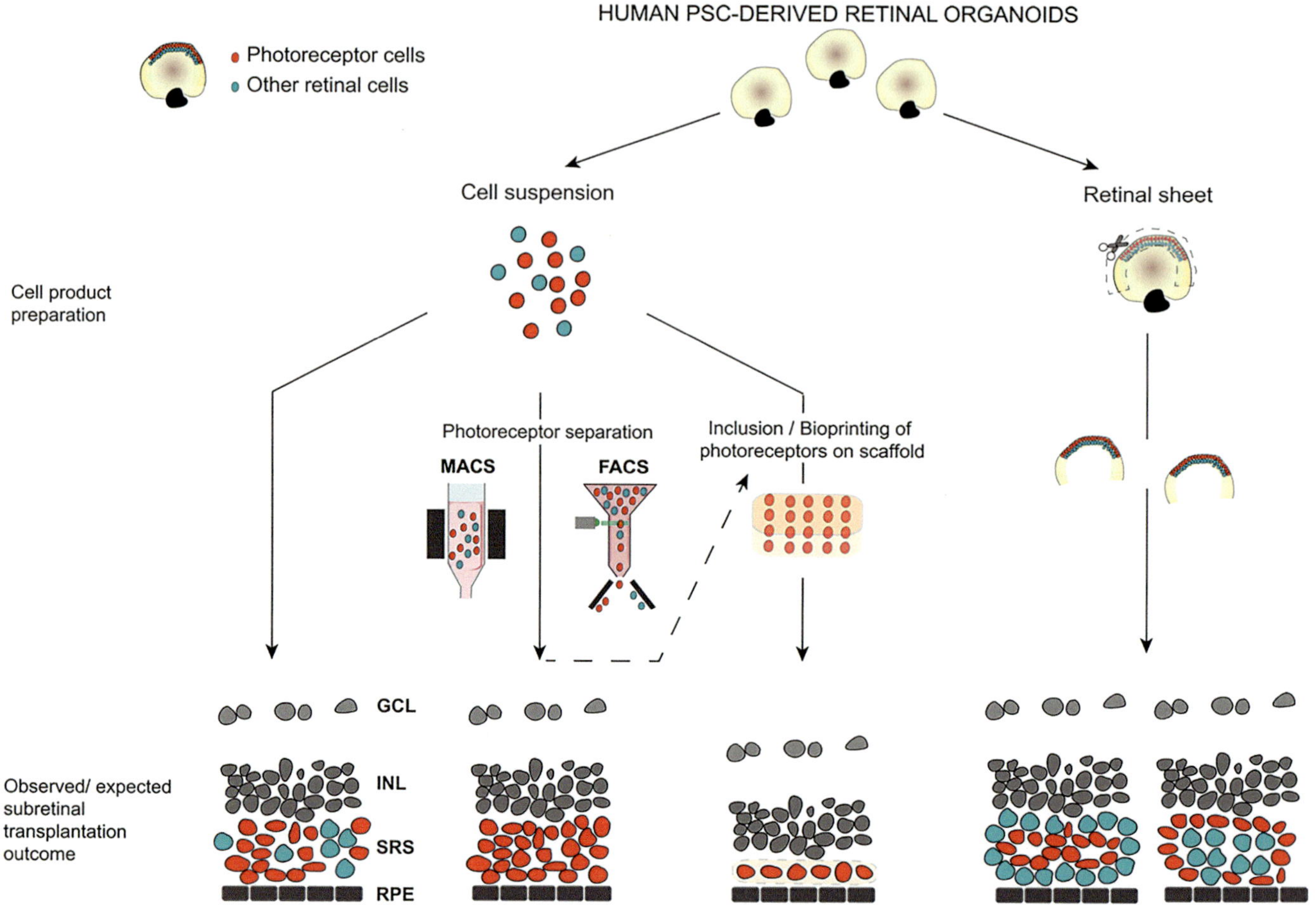

Fig. 61.3 Schematic illustrating the major strategies adopted so far in transplantation studies using human pluripotent stem cell (PSC) derived retinal organoids as source of retinal cells. *FACS* fluorescence-activated cell sorting, *GCL* ganglion cell layer, *INL* inner nuclear layer, *MACS* magnetic-activated cell sorting, *ONL* outer nuclear layer, *RPE* retinal pigment epithelium, *SRS* subretinal space. (Reprinted from Gagliardi et al. 2019, with permission from Elsevier)

tested with treatment being generally well tolerated (Liao et al. 2021). The beneficial effect of this treatment is proposed to be through the release of neutrotophic factors by transplanted RPCs. An alternative approach makes use of subretinal transplantation of human foetal-derived RPCs, offering the potential for trophic support and photoreceptor replacement (ReNeuron 2022). Ocular tissues of 12- to 16-week-old aborted fetuses are obtained, and retinal precursor cells are isolated and expanded using tissue culture, before being delivered via subretinal injection as described for hESC-derived RPE. In a phase IIa trial involving ten participants with clinically diagnosed rod-cone dystrophy, no treatment-related severe adverse events were observed. Interestingly, transplanted cells could be observed as an area of subretinal thickening on OCT up to 1-month post-transplantation but were no longer discernible at 1–2 years. Compared to untreated eyes, visual acuity improved by a mean of 9.9 ETDRS letters at 12 months. However, in an extension study involving a higher treatment dose, visual acuity gains were not maintained up to 24 months (ReNeuron 2022). The surgical procedure involving the higher dose was also associated with more complications. The sponsor has suspended further development of this product (ReNeuron 2022). Additional human studies are necessary to clarify the mechanism of action and durability of grafted cells, including in patients with earlier stage retinal degeneration.

For conditions with extensive photoreceptor loss, as is typical of late-stage IRDs, an ideal therapy would replenish the retina with healthy, functional photoreceptors. To this end, stem-cell derived photoreceptors and photoreceptor precursors offer promise but have yet to enter human clinical trials (Nair and Thomas 2022). Pioneering research has established the successive cell-signalling events responsible for pluripotent cell differentiation into photoreceptors (Swaroop et al. 2010; Brzezinski 2015). Exploiting this knowledge, various cell culture protocols have been developed to drive photoreceptor differentiation from embryonal and induced pluripotent stem cells. Cutting-edge strategies

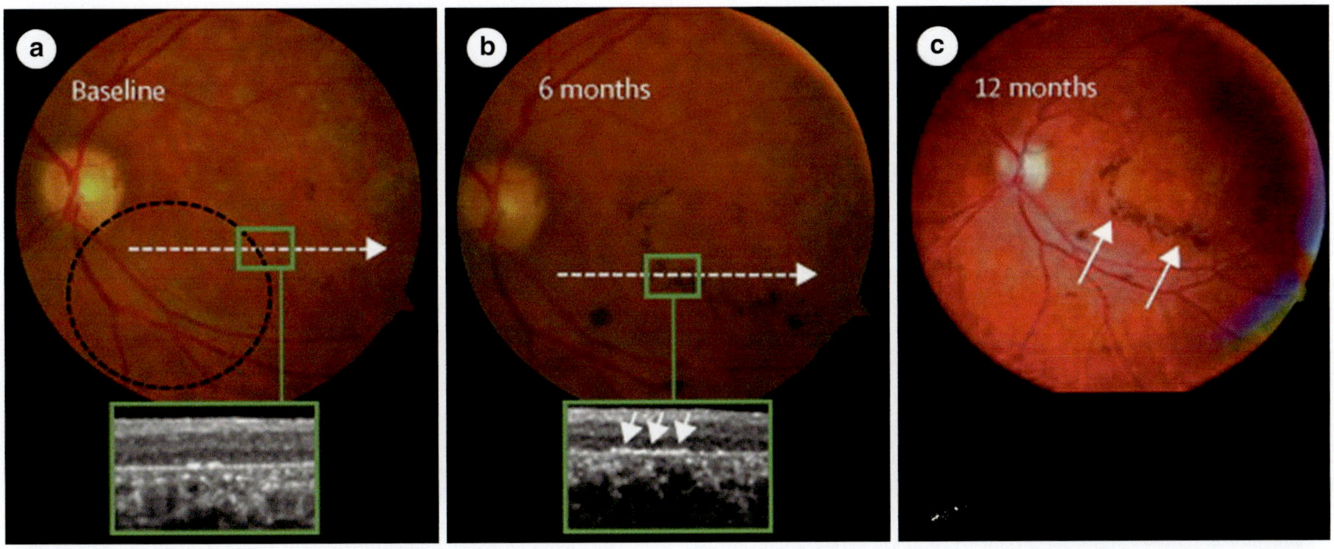

Fig. 61.4 Fundus images of eyes with pigmentation after transplantation with hESC-retinal pigment epithelium (**a–c**) Color fundus photographs and SD-OCT images at baseline of an eye from a patient with Stargardt macular dystrophy (dotted circle shows an outline of the transplanted area), 6 months, and 1 year. Note the absence of pigment in the preoperative photograph (**a**). Patches of pigmented cells are evident around the border of baseline atrophy in retinal pigment epithelium (**b**) that become more prominent at 1 year (**c**, arrows). SD-OCT images at baseline (**a**) and 6 months (**b**) show increasing pigmentation is at the level of the retinal pigment epithelium, normal monolayer retinal pigment epithelium engraftment, and survival at 6 months (**b**, arrows) adjacent to a region of bare Bruch's membrane devoid of native retinal pigment epithelium. (Reprinted from The Lancet, Schwartz et al. 2015, with permission from Elsevier)

involve isolation of self-forming 3-dimensional neuroepithelial cell arrangements, known as retinal organoids, which demonstrate a differentiation state similar to the optic vesicle (Zhong et al. 2014; Meyer et al. 2011). Cultured retinal organoids afford a source of large quantities of transplantable photoreceptors.

In animal models of IRDs, researchers have demonstrated the ability to transplant embryonic stem-cell derived photoreceptors into the subretinal space, with partial persistence, outer segment maturation, synapse formation with inner retinal cells and an absence of teratoma formation (Ribeiro et al. 2021; Waldron et al. 2018; Iraha et al. 2018; Shirai et al. 2016). Both dissociated photoreceptor cell suspensions and cell sheets have been successfully transplanted, with studies having identified factors potentially promoting successful integration with host retina, including persistence of host photoreceptor outer segments and the absence of inner retinal elements in the transplanted material (Nair and Thomas 2022). Unfortunately, promising early animal studies showing restoration of retinal light sensitivity, which was interpreted to be from functional integration of transplanted photoreceptors, have since been shown to be the result of cytoplasmic material transfer between graft and host cells (Waldron et al. 2018; Nickerson et al. 2018; Santos-Ferreira et al. 2016). This may, however, represent a new mechanism to enhance cell therapies in vivo. Considerable challenges remain before stem cell-derived photoreceptor transplantation enters human trials, including obtaining the precise arrangement of graft photoreceptors necessary for functional integration with both host RPE and inner retinal neurons.

Mesenchymal stem cells, derived from bone marrow and umbilical cord, are currently being studied in Phase I-III clinical trials for their potential to provide neurotrophic support in a variety of inherited and acquired retinal disorders (Table 3). One phase I study has been reported examining intravitreal transplantation of autologous mesenchymal stem cells in 14 participants with advanced retinitis pigmentosa (Tuekprakhon et al. 2021). Autologous bone marrow was aspirated from the posterior iliac crest and stem cells were enriched with three cell culture passages, followed by cell surface marker expression profiling and confirmation of trilineage differentiation ability. Early adverse events included self-limited intraocular inflammation and intra-ocular pressure elevation. Three years following treatment, one patient developed vitreous hemorrhage and focal tractional retinal detachment caused by osseous metaplasia involving the ciliary body, requiring vitrectomy surgery. On average, visual acuity improved transiently in treated eyes during the first year after treatment, but the change was not significant when compared to non-treated fellow eyes. Given that neurotrophic photoreceptor rescue is expected to be more effective in patients with early-stage disease, additional clinical trials are warranted.

References

Birch DG, Bernstein PS, Iannacone A, Pennesi ME, Lam BL, Heckenlively J, Csaky K, Hartnett ME, Winthrop KL, Jayasundera T, Hughbanks-Wheaton DK, Warner J, Yang P, Fish GE, Teske MP, Sklaver NL, Erker L, Chegarnov E, Smith T, Wahle A, VanVeldhuisen PC, McCormack J, Lindblad R, Bramer S, Rose S, Zilliox P, Francis PJ, Weleber RG. Effect of Oral Valproic Acid vs Placebo for Vision Loss in Patients With Autosomal Dominant Retinitis Pigmentosa: A Randomized Phase 2 Multicenter Placebo-Controlled Clinical Trial. JAMA Ophthalmol. 2018;136(8):849-856. https://doi.org/10.1001/jamaophthalmol.2018.1171. Erratum in: JAMA Ophthalmol. 2018 Dec 1;136(12):1432.

Brzezinski JA, Reh TA. Photoreceptor cell fate specification in vertebrates. Development. 2015;142(19):3263-3273. https://doi.org/10.1242/dev.127043.

Campochiaro PA, Iftikhar M, Hafiz G, Akhlaq A, Tsai G, Wehling D, Lu L, Wall GM, Singh MS, Kong X. Oral N-acetylcysteine improves cone function in retinitis pigmentosa patients in phase I trial. J Clin Invest. 2020;130(3):1527-1541. https://doi.org/10.1172/JCI132990.

Charbel IP, Barnard AR, Herrmann P, Washington I, MacLaren RE. Rescue of the Stargardt phenotype in Abca4 knockout mice through inhibition of vitamin A dimerization. Proc Natl Acad Sci U S A. 2015;112(27):8415-20. https://doi.org/10.1073/pnas.1506960112.

Chen WJ, Ma L, Li MS, Ma X. Valproic acid's effects on visual acuity in retinitis pigmentosa: a systemic review and Meta-analysis. Int J Ophthalmol. 2019;12(1):129-134. https://doi.org/10.18240/ijo.2019.01.20.

Cideciyan AV, Aleman TS, Swider M, Schwartz SB, Steinberg JD, Brucker AJ, Maguire AM, Bennett J, Stone EM, Jacobson SG. Mutations in ABCA4 result in accumulation of lipofuscin before slowing of the retinoid cycle: a reappraisal of the human disease sequence. Hum Mol Genet. 2004;13(5):525-34. https://doi.org/10.1093/hmg/ddh048. Epub 2004 Jan 6.

Gagliardi G, Ben M'Barek K, Goureau O. Photoreceptor cell replacement in macular degeneration and retinitis pigmentosa: A pluripotent stem cell-based approach. Prog Retin Eye Res. 2019;71:1-25. https://doi.org/10.1016/j.preteyeres.2019.03.001. Epub 2019 Mar 16.

Hendrik P Scholl, Gabrielle DeBartolomeo, Ilyas Washington, Leonide Saad; ALK-001 (C20-D3-Vitamin A) slows the growth of atrophic lesions in ABCA4-related Stargardt Disease: Results of a Phase 2 placebo-controlled clinical trial (TEASE study). Invest. Ophthalmol. Vis. Sci. 2022;63(7):38.

Iraha S, Tu HY, Yamasaki S, Kagawa T, Goto M, Takahashi R, Watanabe T, Sugita S, Yonemura S, Sunagawa GA, Matsuyama T, Fujii M, Kuwahara A, Kishino A, Koide N, Eiraku M, Tanihara H, Takahashi M, Mandai M. Establishment of Immunodeficient Retinal Degeneration Model Mice and Functional Maturation of Human ESC-Derived Retinal Sheets after Transplantation. Stem Cell Reports. 2018;10(3):1059-1074. https://doi.org/10.1016/j.stemcr.2018.01.032.

Jaffe GJ, Westby K, Csaky KG, Monés J, Pearlman JA, Patel SS, Joondeph BC, Randolph J, Masonson H, Rezaei KA. C5 Inhibitor Avacincaptad Pegol for Geographic Atrophy Due to Age-Related Macular Degeneration: A Randomized Pivotal Phase 2/3 Trial. Ophthalmology. 2021;128(4):576-586. https://doi.org/10.1016/j.ophtha.2020.08.027.

Kimura A, Namekata K, Guo X, Noro T, Harada C, Harada T. Valproic acid prevents NMDA-induced retinal ganglion cell death via stimulation of neuronal TrkB receptor signaling. Am J Pathol. 2015;185(3):756-64. https://doi.org/10.1016/j.ajpath.2014.11.005. Epub 2014 Dec 24.

Lee SY, Usui S, Zafar AB, Oveson BC, Jo YJ, Lu L, Masoudi S, Campochiaro PA. N-Acetylcysteine promotes long-term survival of cones in a model of retinitis pigmentosa. J Cell Physiol. 2011;226(7):1843-9. https://doi.org/10.1002/jcp.22508.

Lenis TL, Sarfare S, Jiang Z, Lloyd MB, Bok D, Radu RA. Complement modulation in the retinal pigment epithelium rescues photoreceptor degeneration in a mouse model of Stargardt disease. Proc Natl Acad Sci U S A. 2017;114(15):3987-3992. https://doi.org/10.1073/pnas.1620299114.

Li S, Liu Y, Wang L, Wang F, Zhao T, Li Q, et al. A phase I clinical trial of human embryonic stem cell-derived retinal pigment epithelial cells for early-stage Stargardt macular degeneration: 5-years' follow-up. Cell Prolif. 2021;54(9):e13100.

Liao D, Boyer DS, Kaiser P, Kuppermann BD, Heier J, Mehta M, Joseph A, Kammer R, Mills B, Yang J, Klassen H; Intravitreal Injection of Allogeneic Human Retinal Progenitor Cells (hRPC) for Treatment of Retinitis Pigmentosa: A Prospective Randomized Controlled Phase 2b Trial. Invest. Ophthalmol. Vis. Sci. 2021;62(8):3240.

Liao DS, Grossi FV, El Mehdi D, Gerber MR, Brown DM, Heier JS, Wykoff CC, Singerman LJ, Abraham P, Grassmann F, Nuernberg P, Weber BHF, Deschatelets P, Kim RY, Chung CY, Ribeiro RM, Hamdani M, Rosenfeld PJ, Boyer DS, Slakter JS, Francois CG. Complement C3 Inhibitor Pegcetacoplan for Geographic Atrophy Secondary to Age-Related Macular Degeneration: A Randomized Phase 2 Trial. Ophthalmology. 2020;127(2):186-195. https://doi.org/10.1016/j.ophtha.2019.07.011.

Maeda T, Mandai M, Sugita S, Kime C, Takahashi M. Strategies of pluripotent stem cell-based therapy for retinal degeneration: update and challenges. Trends Mol Med. 2022;28(5):388–404.

Meyer JS, Howden SE, Wallace KA, Verhoeven AD, Wright LS, Capowski EE, Pinilla I, Martin JM, Tian S, Stewart R, Pattnaik B, Thomson JA, Gamm DM. Optic vesicle-like structures derived from human pluripotent stem cells facilitate a customized approach to retinal disease treatment. Stem Cells. 2011;29(8):1206-18. https://doi.org/10.1002/stem.674.

Nair D, Thomas B. Stem Cell-based Treatment Strategies for Degenerative Diseases of the Retina. Curr Stem Cell Res Ther. 2022;17(3):214–25.

Narayan DS, Wood JP, Chidlow G, Casson RJ. A review of the mechanisms of cone degeneration in retinitis pigmentosa. Acta Ophthalmol. 2016;94(8):748-754. https://doi.org/10.1111/aos.13141. Epub 2016 Jun 27.

Nickerson PEB, Ortin-Martinez A, Wallace VA. Material Exchange in Photoreceptor Transplantation: Updating Our Understanding of Donor/Host Communication and the Future of Cell Engraftment Science. Front Neural Circuits. 2018;12:17. https://doi.org/10.3389/fncir.2018.00017.

Ribeiro J, Procyk CA, West EL, O'Hara-Wright M, Martins MF, Khorasani MM, Hare A, Basche M, Fernando M, Goh D, Jumbo N, Rizzi M, Powell K, Tariq M, Michaelides M, Bainbridge JWB, Smith AJ, Pearson RA, Gonzalez-Cordero A, Ali RR. Restoration of visual function in advanced disease after transplantation of purified human pluripotent stem cell-derived cone photoreceptors. Cell Rep. 2021;35(3):109022. https://doi.org/10.1016/j.celrep.2021.109022.

ReNeuron Decides to Stop Retinitis Pigmentosa Program; Focus on Exosome Drug Delivery Technology. ReNeuron, Bridgend, Wales, January 19, 2022. Webpage retrieved June 22, 2023. <https://eyewire.news/news/reneuron-decides-to-stop-retinitis-pigmentosa-program-focus-on-exosome-drug-delivery-technology>

Santos-Ferreira T, Llonch S, Borsch O, Postel K, Haas J, Ader M. Retinal transplantation of photoreceptors results in donor-host cytoplasmic exchange. Nat Commun. 2016;7:13028. https://doi.org/10.1038/ncomms13028.

Schwartz S, Regillo C, Lam B, Eliott D, Rosenfeld P, Gregori N, et al. Human embryonic stem cell-derived retinal pigment epithelium in patients with age-related macular degeneration and Stargardt's macular dystrophy: follow-up of two open-label phase 1/2 studies. Lancet. 2015;385(9967):509–16.

Shirai H, Mandai M, Matsushita K, Kuwahara A, Yonemura S, Nakano T, Assawachananont J, Kimura T, Saito K, Terasaki H, Eiraku M, Sasai Y, Takahashi M. Transplantation of human embryonic stem cell-derived retinal tissue in two primate models of retinal degeneration. Proc Natl Acad Sci U S A. 2016;113(1):E81-90. https://doi.org/10.1073/pnas.1512590113. Epub 2015 Dec 22.

Swaroop A, Kim D, Forrest D. Transcriptional regulation of photoreceptor development and homeostasis in the mammalian retina. Nat Rev Neurosci. 2010;11(8):563-76. https://doi.org/10.1038/nrn2880.

Tanna P, Strauss RW, Fujinami K, et al Stargardt disease: clinical features, molecular genetics, animal models and therapeutic options British Journal of Ophthalmology 2017;101:25-30.

Tuekprakhon A, Sangkitporn S, Trinavarat A, Pawestri A, Vamvanij V, Ruangchainikom M, et al. Intravitreal autologous mesenchymal stem cell transplantation: a non-randomized phase I clinical trial in patients with retinitis pigmentosa. Stem Cell Res Ther. 2021;12(1):52.

Waldron PV, Di Marco F, Kruczek K, Ribeiro J, Graca AB, Hippert C, Aghaizu ND, Kalargyrou AA, Barber AC, Grimaldi G, Duran Y, Blackford SJI, Kloc M, Goh D, Zabala Aldunate E, Sampson RD, Bainbridge JWB, Smith AJ, Gonzalez-Cordero A, Sowden JC, Ali RR, Pearson RA. Transplanted Donor- or Stem Cell-Derived Cone Photoreceptors Can Both Integrate and Undergo Material Transfer in an Environment-Dependent Manner. Stem Cell Reports. 2018;10(2):406-421. https://doi.org/10.1016/j.stemcr.2017.12.008. Epub 2018 Jan 4.

WalportMJ.Complement.Firstoftwoparts.NEnglJMed.2001;344(14):1058-66. https://doi.org/10.1056/NEJM200104053441406.

Weng J, Mata NL, Azarian SM, Tzekov RT, Birch DG, Travis GH. Insights into the function of Rim protein in photoreceptors and etiology of Stargardt's disease from the phenotype in abcr knockout mice. Cell. 1999;98(1):13-23. https://doi.org/10.1016/S0092-8674(00)80602-9.

Zhong X, Gutierrez C, Xue T, Hampton C, Vergara MN, Cao LH, Peters A, Park TS, Zambidis ET, Meyer JS, Gamm DM, Yau KW, Canto-Soler MV. Generation of three-dimensional retinal tissue with functional photoreceptors from human iPSCs. Nat Commun. 2014;5:4047. https://doi.org/10.1038/ncomms5047.

CHAPTER 62

Global Treatment Approaches Part 2: Optogenetics and Retinal Implants

Gareth D. Mercer, Brian G. Ballios, and Peter J. Kertes

Optogenetics 354
Bioelectronic Retinal Implants 355
References 356

G. D. Mercer · B. G. Ballios · P. J. Kertes (✉)
Department of Ophthalmology & Vision Sciences, Temerty Faculty of Medicine, University of Toronto, Toronto, ON, Canada
e-mail: peter.kertes@sunnybrook.ca

S. H. Tsang et al. (eds.), *Atlas of Inherited Retinal Diseases*, Advances in Experimental Medicine and Biology 1467, https://doi.org/10.1007/978-3-031-72230-1_62

Optogenetics and bioelectronic retinal implants are two gene agnostic treatment strategies that rely on the fact that cells of the inner retina (bipolar and retinal ganglion cells) tend to remain functional until late in IRD pathogenesis and can, therefore, be harnessed to restore vision.

Optogenetics

Optogenetics involves genetically modifying cells to become light sensitive through ectopic expression of microbial opsins (i.e., light-gated ion channels). As a treatment for IRDs, the strategy involves conferring light-sensitivity to bipolar cells or retinal ganglion cells, bypassing dysfunctional photoreceptors (Tomita and Sugano 2021).

Various opsin molecules have been studied for their potential to restore vision in animal models of IRDs (Lu and Pan 2021). However, the only one to have thus far reached the stage of human clinical trials is the Chlamydomonas-derived channelrhodopsin 2 (ChR2). ChR2 induces cell depolarization when exposed to blue wavelength light. Initial proof-of-concept studies showed that using an AAV2 vector to transfect retinal ganglion cells with the ChR2 gene restored visual evoked potentials in the rd1 mouse model of retinal degeneration (Bi et al. 2006). Subsequent studies have demonstrated similar vision restoration by inducing ChR2 expression in ON bipolar cells (Lagali et al. 2008).

There are currently three active phase I/II clinical trials examining the safety of gene therapy using recombinant ChR2 in patients with RP, and a single phase IIa trial in patients with autosomal recessive Stargardt disease (Table 62.1). Encouraging preliminary results have been presented for two of these trials. PIONEER is a phase I/IIa dose escalation trial of GS030 gene therapy in patients with nonsyndromic RP (Sahel et al. 2022). GS030 is an intravitreally delivered recombinant AAV2 vector designed to induce expression of channelrhodopsin ChrimsonR-tdTomato in retinal ganglion cells. Gene therapy is paired with the use of light-stimulating goggles to activate the retinal ganglion cells. The treatment was found to be safe up to 2.5 years after administration. Performance on a visual object localization task improved in two out of nine subjects. RESTORE is a phase Ib trial of MCO-010, another recombinant AAV2-mediated channelrhodopsin gene therapy for patients with advanced RP, in this case targeting bipolar cells. The investigators recently presented their 12-month study data, which found no serious adverse events and encouraging efficacy measures (Nanoscope Therapeutics 2023). Twice as many subjects receiving MCO-010 demonstrated a ≥2 luminance level improvement on a vision guided mobility test (MLYMT) or a near field object recognition test (MLSDT) compared to sham-treated subjects (89vs. 44%).

Table 62.1 Optogenetic treatments and retinal prostheses for inherited retinal disease currently in clinical trials

Intervention type	Disease	Trial identifier	Phase	Status	Sponsor	Intervention description
Optogenetics	Non-syndromic RP	NCT03326336	1/2a	Recruiting	GenSight Biologics	GS030-Drug Product (GS030-DP)—rAAV2.7 m8 expressing optimized channelrhodopsin ChrimsonR-tdTomato gene under the control of the ubiquitous CAG promoter; GS030-Medical Device (GS030-MD)—Visual Interface Stimulating Glasses to amplify the external visual stimulus
Optogenetics	RP	NCT02556736	1/2a	Active, not recruiting	AbbVie	Single intravitreal injection of RST-001 (a recombinant chanelrhodopsin gene therapy)
Optogenetics	RP	NCT04945772	2b	Active, not recruiting	Nanoscope Therapeutics	Single intravitreal injection of vMCO-010, an AAV2-based vector carrying multi-characteristics opsin (MCO) gene expression cassette
Optogenetics	STGD	NCT05417126	2a	Active, not recruiting	Nanoscope Therapeutics	Single intravitreal injection of vMCO-010, an AAV2-based vector carrying multi-characteristics opsin (MCO) gene expression cassette
Retinal prothesis	RP, other retinal degenerations	NCT04295304	1	Terminated (lack of funds)	Nano Retina	NR600 epiretinal implant

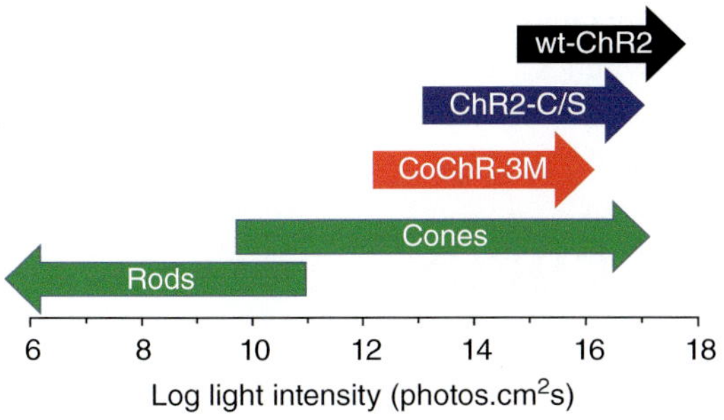

Fig. 62.1 Comparison of the light sensitivities of rod and cone photoreceptors with wild-type channelrhodopsin 2 (ChR2) and variants (ChR2-C/S, CoChR-3M). (Reprinted from Lu and Pan 2021, with permission from Springer Nature)

A significant advantage of optogenetics is the potential to restore intrinsic visual processing features of the retina, such as ON and OFF pathways, color-specificity, and center-surround antagonistic receptive fields (Lu and Pan 2021). However, the naturally occurring ChR2 receptor has a sensitivity 4–5 log units lower than that of cone photoreceptors (Fig. 62.1), and peak absorption in the blue wavelength. To overcome these limitations, some current clinical trials rely on the use of a prosthetic device to amplify incoming light to levels detectable by ChR2, but with an associated risk of retinal photochemical injury (Sahel et al. 2022; Tomita and Sugano 2021). Much research in optogenetics has thus focused on identifying ChRs with greater sensitivity and different wavelength specificities. The resulting receptors are soon expected to enter human trials (Tomita and Sugano 2021). More distant goals are to have receptors with the ability to hyperpolarize rather than depolarize target cells, which, when coupled with differential gene expression and protein targeting, may enable recapitulation of center-surround receptor organization (Lu and Pan 2021).

Bioelectronic Retinal Implants

The concept underlying bioelectronic retinal implants is to simulate light perception by applying electric currents to the retina to induce retinal ganglion cell depolarization (Yue et al. 2016). Electric currents are applied using an array of extracellular electrodes placed in proximity to the retina. The prothesis set-up requires an image acquisition device to capture information about the patient's visual field, which is then translated into electrical impulses and transmitted to the electrode array, often with intermediate image processing. Power and data are sent between the extra-ocular and intraocular portions of the device using radiofrequency telemetry,

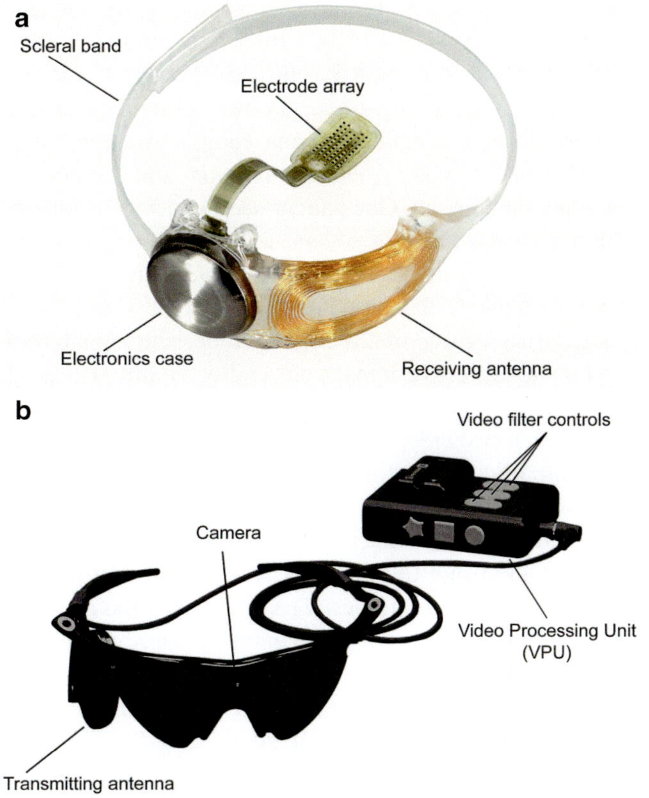

Fig. 62.2 Photographs of the Argus II Retinal Prosthesis System (Second Sight Medical Products, Inc., Sylmar, CA): (**a**) the implanted components of the system and (**b**) the external (body-worn) components of the system. (Reprinted from da Cruz 2016, with permission from the American Academy of Ophthalmology)

optical link, or transcleral cable. Products vary mainly in terms of the image acquisition device used, and the location and complexity of the implanted electrode array. Implants have been developed for the epiretinal, subretinal, and suprachoroidal spaces.

Three retinal implants have been approved for clinical use:

- Argus II and its predecessor Argus I (Second Sight, USA) were the first retinal protheses approved (European Union CE, 2011, and FDA, 2013) and the most widely used. The design (Fig. 62.2) comprises an epiretinal electrode array and a miniature camera mounted onto a pair of glasses. Information from the camera is sent to an external video processing unit, generating a pixelated image (Yue et al. 2016). The resulting pixel data are transmitted wirelessly to an extraocular receiver and power supply mounted on an episcleral band. Finally, electrical pulses, with amplitude corresponding to the intensity of each pixel, are transmitted to the intraocular electrode array via a transscleral cable. The epiretinal electrode array, consisting of a total of 60 electrodes, is fixated on the surface of the macula using a tack. The array subtends a visual angle of approximately 11°× 19°. In a three-year study involving 30 patients with end-stage retinal dystrophy, 29 patients retained a functioning device. In total, 11 patients experienced adverse events, including conjunctival erosion or dehiscence, hypotony, endophthalmitis, and the need to re-tack the implant. One patient had the device explanted during management of adverse events. In terms of visual function, 89% of patients had improved target localization with the device turned on versus off, 33% had a grating visual acuity of 2.5LogMAR or better, and 65% reported a positive impact on quality of life (Ho et al. 2015). After having been implanted into more than 350 patients worldwide, Second Sight announced that it was ending production of the Argus II device in 2019 due to financial difficulty.

- IRIS II (Pixium, France) is another epiretinal implant, similar in design to the Argus II, except that it uses an array with 150 electrodes. The company producing the device was granted European Union CE approval in 2016, but halted commercial production in 2018, after finding many implants stopped functioning within the first year after implantation (Ayton et al. 2020).
- Alpha AMS (Retina Implant, Germany) differs from the Argus and IRIS devices in two significant ways. First, it is implanted in the subretinal space (between the photoreceptors and RPE). Second, instead of using a camera for image detection, the implant is composed of an array of 1600 photodiodes, which transduce light entering the eye into electrical impulses. A transcleral cable carries power from an extraocular power pack to operate the subretinal chip (Ayton et al. 2020). The theoretical visual acuity afforded by the electrode design is 20/250, which was similar to the grating visual acuity found in a study of 29 patients who received the device (Stingl et al. 2013). However, a number of patients experienced device failure in the first 12 months after implantation. Although the device was approved for clinical use in Europe in 2016, the manufacturer ceased operations in 2019 (Ayton et al. 2020).

Globally, since their introduction, retinal implants have allowed over 500 patients with severe bilateral blindness to achieve some level of vision, yet this technology faces significant challenges, as evidenced by the fact that none of the devices approved for clinical use are still commercially available. These devices require invasive surgery, and their spatial resolution is limited by the physical constraints of electrode arrays. While clinical trials of new devices are underway (Table 62.1), it is foreseeable that this technology may be supplanted by advances in optogenetics and chemical photoswitches, which use light stimulation directly to induce retinal neuronal activation.

References

Ayton LN, Barnes N, Dagnelie G, Fujikado T, Goetz G, Hornig R, Jones BW, Muqit MMK, Rathbun DL, Stingl K, Weiland JD, Petoe MA. An update on retinal prostheses. Clin Neurophysiol. 2020;131(6):1383-1398. https://doi.org/10.1016/j.clinph.2019.11.029.

Bi A, Cui J, Ma Y, Olshevskaya E, Pu M, Dizhoor A, et al. Ectopic expression of a microbial-type rhodopsin restores visual responses in mice with photoreceptor degeneration. Neuron. 2006;50:23–33. https://doi.org/10.1016/j.neuron.2006.02.026

da Cruz L, Dorn JD, Humayun MS, Dagnelie G, Handa J, Barale PO, Sahel JA, Stanga PE, Hafezi F, Safran AB, Salzmann J, Santos A, Birch D, Spencer R, Cideciyan AV, de Juan E, Duncan JL, Eliott D, Fawzi A, Olmos de Koo LC, Ho AC, Brown G, Haller J, Regillo C, Del Priore LV, Arditi A, Greenberg RJ; Argus II Study Group. Five-Year Safety and Performance Results from the Argus II Retinal Prosthesis System Clinical Trial. Ophthalmology. 2016;123(10):2248-54. https://doi.org/10.1016/j.ophtha.2016.06.049.

Ho AC, Humayun MS, Dorn JD, da Cruz L, Dagnelie G, Handa J, Barale PO, Sahel JA, Stanga PE, Hafezi F, Safran AB, Salzmann J, Santos A, Birch D, Spencer R, Cideciyan AV, de Juan E, Duncan JL, Eliott D, Fawzi A, Olmos de Koo LC, Brown GC, Haller JA, Regillo CD, Del Priore LV, Arditi A, Geruschat DR, Greenberg RJ; Argus II Study Group. Long-Term Results from an Epiretinal Prosthesis to Restore Sight to the Blind. Ophthalmology. 2015;122(8):1547-54. https://doi.org/10.1016/j.ophtha.2015.04.032.

Lagali P, Balya D, Awatramani G, Münch T, Kim D, Busskamp V, et al. Light-activated channels targeted to ON bipolar cells restore visual function in retinal degeneration. Nat Neurosci. 2008;11(6):667–75.

Lu Q, Pan Z. Optogenetic Strategies for Vision Restoration. Adv Exp Med Biol. 2021;1293:545–55.

Nanoscope Therapeutics Announces Presentation of Key Results from Phase 2b RESTORE Trial of MCO-010 for the Treatment of Retinitis Pigmentosa at the ARVO Annual Meeting. Nanoscope therapeutics, Dallas TX, April 27, 2023. Webpage retrieved May 28, 2023. <https://nanostherapeutics.com/2023/04/27/nanoscope-therapeutics-announces-presentation-of-key-results-from-phase-2b-restore-trial-of-mco-010-for-the-treatment-of-retinitis-pigmentosa-at-the-arvo-annual-meeting/>

Sahel JS, Audo IS, Boulanger-Scemama E, Pagot C, Arleo A, Martel JN, Esposti SD, Delaux A, de Saint Aubert JP, de Montleau C, Gutman E, Duebel J, Picaud S, Dalkara D, Taiel M, Roska B. Optogenetics in the clinic: safety and efficacy updates on the phase 1/2 clinical trial PIONEER. Invest. Ophthalmol. Vis. Sci. 2022;63(7):1106.

Stingl K, Bartz-Schmidt K-U, Gekeler F, Kusnyerik A, Sachs H, Zrenner E. Functional outcome in subretinal electronic implants depends on foveal eccentricity. Invest Ophthalmol Vis Sci. 2013;54:7658–7665. https://doi.org/10.1167/iovs.13-12835.

Tomita H, Sugano E. Optogenetics-Mediated Gene Therapy for Retinal Diseases. Adv Exp Med Biol. 2021;1293:535–43.

Yue L, Weiland JD, Roska B, Humayun MS. Retinal stimulation strategies to restore vision: Fundamentals and systems. Prog Retin Eye Res. 2016;53:21-47. https://doi.org/10.1016/j.preteyeres.2016.05.002

Index

A
A3243G mtDNA point mutation, 180
ABCA4 gene, 65–66, 140, 149, 158
Abnormal fluorescence, 10
Abnormal gait, 186
Achromatopsia, 128
clinical trials, 133
complete achromatopsia, 128
electrophysiology, 131
fundus, 128
incomplete achromatopsia, 128
molecular genetics, 132–133
stages, 128–132
symptoms, 128
Acquired syphilis, 246
Active electrode, 22
Acute exudative polymorphous vitelliform maculopathy (AEPVM), 250
Acute idiopathic maculopathy, 270
Acute macular neuroretinopathy, 270
Acute posterior multifocal placoid pigment epitheliopathy, 264
Acute syphilitic posterior placoid chorioretinopathy (ASPPC), 246, 247
Acute zonal occult outer retinopathy (AZOOR), 19, 28, 264
ADAM9 (CORD9) gene, 66
Adeno-associated viral (AAV), 326, 327
AD vitreoretinochoroidopathy (ADVIRC), 294
Adult-onset foveomacular vitelliform dystrophy (AFMD), 90, 97
Adult vitelliform dystrophy, 19
Age-related macular degeneration (AMD), 19, 112
Alendronate, 262
Allelic heterogeneity, 33
Alport syndrome, 212
COL4A3, 212
COL4A4, 212
COL4A5, 212
nephropathy, 212
neurosensory deafness, 212
ocular features, 212
Alström syndrome, 194
Aneuploidy, 31
Angiomyolipomas, 230
Anticipation, 33
Antisense oligonucleotide (ASO), 339, 340
Anti–vascular endothelial growth factor (anti-VEGF), 227
Astigmatism, 196
Astrocytic hamartoma, 230, 231
Astrocytoma, 230
ATP-binding cassette (ABC) transporter protein, 158
Autofluorescence, 10
Autoimmune retinopathies (AIRs), 250
Autosomal dominant (AD), 80, 290, 320
cone dystrophy, 65
drusen, 100
inheritance, 33
Autosomal dominant vitreoretinochoroidopathy (ADVIRC), 90
Autosomal recessive (AR), 80
Autosomal recessive bestrophinopathy (ARB), 90, 170
Autosomal recessive inheritance, 33
Autosomal recessive retinitis pigmentosa (AR-RP), 136, 139

B
Bardet-Biedl syndrome (BBS), 186
advanced case of, 186
criteria for diagnosis, 187
early case of, 187
management, 188
molecular genetics, 187–188
Batten disease, 25
BBS-related genes (*BBS1–BBS19*), 187
BCG vaccine, 262
Best corrected visual acuity (BCVA), 144
Best vitelliform macular dystrophy (BVMD), 88
BEST1 gene, 91

S. H. Tsang et al. (eds.), *Atlas of Inherited Retinal Diseases*, Advances in Experimental Medicine and Biology 1467, https://doi.org/10.1007/978-3-031-72230-1

Bestrophin-1 associated disease, 19
Bevacizumab, 262
Biallelic null *ABCA4*, 159
Bietti crystalline dystrophy (BCD), 208
Bilateral diffuse uveal melanocytic proliferation (BDUMP), 250
Bilateral doyne, 103
Biochemical genetic tests, 34
Bioelectronic retinal implants, 355, 356
Bisretinoid compounds accumulation, 18
Blocked fluorescence, 10
Blood-retina barriers (BRB), 8
Blue cone monochromatism (BCM), 74
Brimonidine, 262
Bruch's membrane, 4
Bull's eye maculopathy (BEM), 63, 65, 140, 152
Butterfly-type pattern dystrophy, 94

C

C1QTNF5 mutation (*CTRP5*), 124
Cancer-associated retinopathy (CAR), 250, 264, 292
Canthaxanthin, 261
Cardiac rhabdomyoma, 230
Cataracts, 206
Central dogma, 32
Central dystrophies, 293
Central retina, 5
Central serous chorioretinopathy (CSCR), 278
  choroid, 278
  classification, 283
  clinical symptoms, 278
  color fundus photographs, 280
  complications, 278
  multimodal imaging, 279
  pathophysiology, 278
  retinal pigment epithelium dysfunction, 278
  treatment, 278
CEP290/*ICQB1* mutations, 144, 148
Ceramide kinase-like (*CERKL*) gene, 140, 141
Chédiak–Higashi syndrome, 60
Chlamydomonas-derived channelrhodopsin 2 (ChR2), 354
Chloroquine, 254
Choriocapillaris (CC), 198, 200
Chorioretinal atrophy, 282
Choroid, 4
Choroidal neovascular membranes (CNVMs), 116
Choroideremia (CHM), 44–46, 48, 49
  clinical trials, 49
  female carriers, 48–49
  imaging and tests, 45–48
  molecular genetics, 49
Chromosomal genetic tests, 34
Chromosome, 30
Cidofovir, 262
Ciliopathies, 182
Cisplatin, 256
Clinical heterogeneity, 33
Clofazimine, 256
CNGA3, 132
CNGB3, 132
CNS-related ataxia, 186
Codominant inheritance, 33
Combined pituitary hormone deficiency type 6 (CPHD6), 220
Complete achromatopsia, 128
Cone-rod dystrophy (CORD), 63, 148
Confluent drusen, 104
Congenital heart disease, 243
Congenital stationary night blindness (CSNB), 70, 290
  fundus albipunctatus, 71
  imaging and tests, 71
  molecular genetics, 71
  Oguchi disease, 71, 72
  types, 70–71
Congenital syphilis, 246
Corneal stromal opacities, 196
Cortical dysplasias, 230
Corticosteroid, 256
CRB1 mutation, 147
CRISPR activation (CRISPRa), 321
CRX gene, 66–67, 146
Cutaneous melanoma-associated retinopathy, 250
Cystoid macular edema (CME), 292, 300
Cytoplasm, 30
Cytotoxic T lymphocyte (CTL), 328

D

DCDC4B gene, 109
De novo (new) mutation, 31
Deferoxamine, 256
Deoxyribonucleic acid (DNA), 30, 306
  analysis, 314
  repair, 31
  sequence, 30, 32, 316
  structure, 30
Dideoxyinosine (DDI), 256
Diffuse unilateral subacute neuroretinitis (DUSN), 274
Digenic inheritance, 33
Doyne honeycomb retinal dystrophy, 100
Duchenne muscular dystrophy, 25
Dye-transit phases, 8–10

E

Early-childhood onset retinal dystrophy (ECORD), 328
EFEMP1 gene, 105
Electrodes, 28
Electronegative ffERG, 23–24
Electronegative wave pattern, 70
Electrooculography/electrooculogram (EOG), 28, 90, 170, 294
Electrophysiology, 74, 100, 131
Electroretinography/electroretinograms (ERG), 22, 38, 39, 44, 53, 63, 70, 90, 130, 144, 155, 166, 170, 186, 209, 221
ELOVL4 protein, 158
Encephalomyopathy, 178
Endoplasmic reticulum, 30
Enhanced S-cone syndrome (ESCS), 164, 166
Epigenome, 32
Epigenomic modification, 32
Equine infectious anemia virus (EIAV), 327
Ergot alkaloids, 260
European Bioinformatics Institute (EBI), 315
Eyes shut homolog (*EYS*) mutations, 139

F

Facial hypotonia, 186
Familial disease, 33
Familial dominant drusen, 100
Fatty aldehyde dehydrogenase (FALDH), 196
Female carrier, 39
Film-based images, 8

Fluorescein angiography (FA), 9, 45, 254, 264
  abnormal fluorescence, 10
  contraindications, 11
  dye-transit phases, 8–10
  FAF imaging, 8
  film-based images, 8
  image acquisition, 8
  side effects, 11
Fluorescein molecules, 8
Fovea, 5
Foveoschisis, 282
Frameshift mutation, 32
Full-field electroretinography (ERG), 22–24, 28, 108, 131
Fundus albipunctatus, 71
Fundus autofluorescence (FAF), 8, 19, 45, 47, 53, 71, 74, 90, 94, 128, 152, 166, 170, 178, 186, 190, 202, 209, 264, 265
  clinical application, 18–19
  interpretation, 18–19
  OPTOS autofluorescence, 19
  short-wavelength autofluorescence, 18
Fundus pulverulentus, 97

G

Ganglion cell layer (GCL), 4
Ganzfeld bowl, 22, 28
Gemcitabine, 260
Gene editing, 339
Gene expression, 32
Gene mutation, 31
Gene regulation, 32
Gene therapy, 34, 41, 56, 324–326
Genetics, 30–31
  code, 307
  counselors, 308
  disease, 33
  disorders, 31–32
  testing, 33, 307, 309–311, 316
  variants, 314
Genome, 30, 306
Genome-wide association study (GWAS), 34
Genomic variation, 306, 307
Genotype, 30, 144
Golgi apparatus, 30
GUCA1A gene, 66
GUCY2D gene, 65, 66, 144
Gyrate atrophy, 198

H

Hematoxylin/eosin, 4
Hereditary disease, 33
Hereditary/germline mutations, 31
Hermansky-Pudlak syndrome, 60
Homology-directed repair (HDR), 321, 339
HUGO Gene Nomenclature Committee (HGNC), 32
Human embryonic stem cell (hESC), 347
Human Genome Project, 30
Human Genome Variation Society (HGVS), 314, 315
Hyperautofluorescent ring, 182, 183
Hyperfluorescence, 10
Hypofluorescence, 10

I

Incomplete achromatopsia, 128
Indocyanine green angiography (ICGA), 11, 283
Inheritance, 33
Inherited retinal dystrophies (IRDs), 14, 308, 320, 321
Inner nuclear layer (INL), 4
Inner plexiform layer (IPL), 4
Interferon alfa-2a, 260
Internal limiting membrane (ILM), 4
Intracameral vancomycin, 260
Intraocular gentamicin, 260
Intravenous fluorescein angiography (IVFA), 8
Intuitional review board (IRB), 307–308
Inverted terminal repeats (ITRs), 326
Irdshot chorioretinopathy, 266

J

Juvenile X-linked retinoschisis, 23

K

Karyotype, 30, 311
KCNV2 gene, 66
Kearns-Sayre syndrome, 174

L

Lactic acidosis and stroke-like episodes (MELAS), 178
Laser power, 19
Late-onset retinal degeneration (LORD), 124
LCA1 gene, 144
LCA2 gene, 144
LCA3 gene, 144–146
LCA4 gene, 146
LCA5 gene, 146
LCA6 gene, 146
LCA7 gene, 146
LCA8 gene, 146–147
LCA9 gene, 147
LCA10 gene, 147
LCA11 gene, 147
LCA12 gene, 147–148
LCA14 gene, 148
LCA15 gene, 148, 149
LCA57 gene, 149
Leber congenital amaurosis (LCA), 144, 145, 194, 308
Locus heterogeneity, 33
Locus Reference Genomic (LRG), 315
Long-wave sensitive (LWS) type 1 opsin, 75
Lymphangioleiomyomatosis, 230
Lyonization, 48

M

Macula, anatomical landmarks of, 5
Macula lutea, 5
Macular dystrophy, 19
Macular integrity assessment (MAIA), 334
Malattia Leventinese, 100
Maternally inherited diabetes and deafness (MIDD), 178
Medical record number (MRN), 314
Medium-wave sensitive (MWS) type 1 opsin, 75
Melanoma-associated retinopathy, 23
Mendelian disorder, 33
Mesenchymal stem cells, 349
Messenger ribonucleic acid (mRNA), 306
Methanol, 260
Methoxyflurane, 261
Microphthalmia type 5 (MCOPS5), 220

Missense mutation, 31
Mitochondria, 30
Mitochondrial diseases, 174
Mitochondrial DNA (mtDNA), 31, 175
Mitochondrial/maternal inheritance, 33
Miyake disease, 108
Molecular biology, 32–33
Molecular genetics, 34, 40–41
Monosomy, 31
Multifactorial disorders, 32
Multifocal electroretinography (mfERG), 24–25
Multi Luminance Mobility Test (MLMT), 331, 332
Multimodal imaging, 18, 132
Multiple evanescent white dot syndrome, 264
Myopia, 74, 196
Myopic traction maculopathy (MTM), 282

N
N-acetyl cysteine (NAC), 344
Near-infrared auto-fluorescence (NIR-FAF) images, 18
Negative genetic test, 34
Nerve fiber layer (NFL), 4
Neural cells, 4
Neurofibromatosis, 236
  molecular genetics, 237
  type 1, 236
  type 2, 236–237
Neuroprotection, 344
Night blindness, 112
Non-homologous end-joining (NHEJ), 321
Noninfectious uveitis syndromes, 264
Nonsense mutation, 31
Normal macula, 18
North Carolina Macular Dystrophy (NCMD), 116, 117
NR2E3 gene, 165, 166
Nucleotide, 30
Nucleus, 30
Nyctalopia, 206

O
Occult macular dystrophy (OMD), 108
Ocular albinism, 58–60
Ocular syphilis, 247
Oculocutaneous albinism (OCA), 60
Oculodigital reflex, 144
Oguchi disease, 71, 72
Open reading frame (ORF), 34
Optical coherence tomography (OCT), 14, 45, 47, 71, 74, 90, 108, 155, 170, 190, 208, 209, 226, 242, 264, 265, 283, 331
Optic neuritis, 274
Optogenetics, 354, 355
Optos, 18
Oral acetazolamide, 55
Oral contraceptives, 260
Ornithine aminotransferase, 200
OTX2 syndromes, 220, 221
Outer limiting membrane (OLM), 4
Outer nuclear layer (ONL), 4, 251
Outer plexiform layer (OPL), 4

P
Pamidronate, 262
Parafovea, 5
Paraneoplastic autoimmune retinopathy (PAIR), 23
Paraneoplastic vitelliform maculopathy (PVM), 250
Pathologic myopia, 282
Pattern dystrophy, 19, 94
Pattern electroretinography (pERG), 25–26
PDE6 gene, 138
Pendular nystagmus, 74
Perifovea, 5
Peripapillary atrophy, 282
Peripheral retinoschisis, 52
Persistent placoid maculopathy (PPM), 269, 271
Phakomatoses, 226
Phenothiazines, 254
Phenotype, 30, 144
Photophobia, 74
Photopic On-Off ERG, 25–26
Photoreceptor dystrophy, 23
Photoreceptor layer (PL), 4
Pigmentary deposits, 39
Pigmentary retinopathy, 247
Pigmented paravenous chorioretinal atrophy (PPCRA), 120
Pigment epithelial detachment (PED), 10
PITPNM3 gene, 67
Plasmapheresis, 206
Platelet-derived growth factor (PDGF), 227
Polycystic ovary syndrome (PCOS), 194
Polymorphism, 307
Pooling, 10
Positive genetic test, 34
Posterior pole, 5
Posterior staphyloma, 282
Precision medicine, 34
Progressive panretinal degeneration, 206
Prostaglandin analogues (topical), 262
Proteins, 32
Protospacer adjacent motif (PAM), 338
PRPF3 gene, 82
PRPF8 gene, 82
PRPH2 gene, 66, 97–98
Pseudoxanthoma elasticum (PXE), 202
  autosomal recessive multisystem disorder, 202
  cardiovascular changes, 202
  fragmentation, 202
  molecular genetics, 203
  progressive calcification, 202
  skin changes, 202
Punctate inner choroidopathy, 266
Punctate keratitis, 196

Q
Quantitative autofluorescence (qAF), 19
Quinine sulfate, 260
Quinine toxicity, 25

R
R218H mutation, 28
Rab Escort Protein 1 (REP1), 334
Ranibizumab, 262
RDH12 mutation, 148, 149
Reference electrode, 22
Refsum disease, 206
Retina, 44
Retinal degeneration slow (RDS), 291
Retinal dystrophies, 301, 302
  classification, 288
  examination, 289

gene therapy, 300, 301
stationary cone dysfunction syndromes, 291
stationary rod dystrophies, 290
treatments, 299, 300
Retinal hamartoma, 230
Retinal pigment epithelium (RPE), 4, 8, 18, 44, 71, 94, 100, 112, 124, 128, 198, 200, 202, 208, 278
Retinal progenitor cell (RPC), 347
Retinitis pigmentosa (RP), 19, 44, 138, 190, 291
Retinitis Pigmentosa GTPase Regulator (RPGR), 40, 191
Retinitis pigmentosa 1-like 1 (RP1L1) mutation, 109
RHO gene, 81–82
Ribosomes, 30
Rifabutin (oral), 262
Riggs type, 70
RIMS1 gene, 67
Rod-cone dystrophy, 141, 144
RP1 gene, 82–85
RP2 gene, 41
RP38 gene, 149
RPE65 gene, 140
RPGR gene, 40, 67
RPGRIP1 gene, 66
Rubella retinopathy, 242, 243
complications, 242
ocular manifestation, 242
optical coherence tomography, 242

S

Scanning laser ophthalmoscopy (SLO), 18, 289
Schubert-Bornschein type, 70
SEMA4A gene, 66
Senior-Løken syndrome, 190, 191
Serpiginous chorioretinitis, 274
Serpiginous choroiditis, 267
Severe cone-rod dystrophy, 144
Short-wavelength (SW) light, 18
Short-wavelength fundus autofluorescence (SW-AF), 18
Single-gene disorder, 33
Single gene sequencing, 311
Single nucleotide polymorphisms (SNPs), 34
Sjogren's reticular-type pattern dystrophy, 97
Sjögren-Larsson syndrome, 196
SLSN5 gene, 149
Sorsby fundus dystrophy (SFD), 112
Sorsby macular dystrophy, 113
SPATA7 mutation, 145
Spectral domain optical coherence tomography (SD-OCT), 14, 128, 251
Spirochete, 246
Staining, 10
Stargardt disease (STGD), 19, 152–161, 293
fundus appearance, 152–155
groupings, 155
stages, 152–155
types, 155
Stationary cone dysfunction syndromes, 291
Stationary rod dystrophies, 290
Stem cells, 345, 347
STGD1, 344, 346, 347
Stickler syndrome, 216
Subependymal giant cell, 230
Subependymal nodules, 230
Sulfonamides (oral), 262
Swept-source OCT, 14
Syphilis, 246
acquired, 246
congenital, 246
ocular manifestations, 246, 247

T

Talc, 260
Tamoxifen, 260
Tapetal reflex, 40
TIMP3 gene, 113, 114
Transcription, 32
Transfer RNA (tRNA), 32
Translation, 32
Translational read through inducing drugs (TRADs), 338
Treponema pallidum, 246
Trisomy, 31
Tuberous sclerosis, 230, 233
TULP1 gene, 138
Tyrosinase, 60

U

Uninformative/indeterminate test, 34
Uniparental disomy (UPD), 311
Usher syndrome, 182
Usher syndrome 1 (USH1), 182, 183
Usher syndrome 2 (USH2), 182, 183
Usher syndrome 3 (USH3), 182, 183

V

Valproic acid (VPA), 344
Variable expressivity, 33
Variants of unknown significance (VOUS), 31
Vascular filling defects, 10
Visual field, 45
Vitreoretinal surgery, 55
Von Hippel-Lindau disease, 226
general features, 226, 227
molecular genetics, 227
Voretigene neparvovec-rzyl, 331

W

Whole exome sequencing (WES), 34, 311
Whole genome sequencing, 34
Window defect, 10

X

X-linked cone dystrophy, 63
X-linked cone-rod dystrophy (CORD), 62
X-linked CSNB, 71
X-linked juvenile retinoschisis (XLRS), 52, 55
X-linked ocular albinism (XLOA)
female carriers, 58–60
non-progressive disorder, 58
OCA, 60
ocular albinism, 60
oculocutaneous albinism, 60
slit-lamp examination, 58
X-linked recessive (XLR), 33, 290
X-linked retinitis pigmentosa (XLRP)
female carrier, 39
gene therapy trials, 41
molecular genetics, 40–41
X-linked retinoschisis, 25

GPSR Compliance

The European Union's (EU) General Product Safety Regulation (GPSR) is a set of rules that requires consumer products to be safe and our obligations to ensure this.

If you have any concerns about our products, you can contact us on ProductSafety@springernature.com

In case Publisher is established outside the EU, the EU authorized representative is:

Springer Nature Customer Service Center GmbH
Europaplatz 3
69115 Heidelberg, Germany

Batch number: 09625898

Printed by Printforce, the Netherlands